DICTIONARY
OF BIOCHEMISTRY

DICTIONARY OF BIOCHEMISTRY

J. STENESH
Professor of Chemistry
Western Michigan University

A WILEY-INTERSCIENCE PUBLICATION

JOHN WILEY & SONS, New York • London • Sydney • Toronto

Library of Congress Cataloging in Publication Data:

Stenesh, J   1927–
    Dictionary of biochemistry.

    "A Wiley-Interscience publication."
    1. Biological chemistry—Dictionaries.   I. Title.

QP512.S73   574.1′92′03   75-23037
ISBN 0-471-82105-5

Printed in the United States of America

10 9 8 7 6 5 4

PREFACE

This dictionary was written to provide scientists and students in the life sciences with a reference work on the terminology of biochemistry. The dictionary contains approximately 12,000 entries drawn from over 200 textbooks and reference books of various kinds and from the research literature of biochemical journals; all of the source material consulted has been published since 1962. The recommendations of the Commission on Biochemical Nomenclature of the International Union of Pure and Applied Chemistry and the International Union of Biochemistry were among the sources used in the compilation of this dictionary.

The selection of entries was influenced by characteristics of the terminology of biochemistry. One of these characteristics is the extensive use of terms from other sciences, since biochemistry, by its very nature, draws heavily on allied sciences. For this reason, many terms from these sciences are included in the dictionary. The sciences emphasized most in this connection are chemistry, immunology, genetics, virology, biophysics, and bacteriology. A second characteristic is the extensive use of abbreviations, both standard and nonstandard. Many of these are included to aid the reader of biochemical literature and to account for the likelihood that some of the nonstandard abbreviations will become standard ones in the future. A third characteristic is the extensive use of synonymous expressions, frequently differing from each other only by minor variations. Since the synonymous nature of one expression to another may not always be apparent to the reader, principal synonymous expressions are included and cross-referenced.

An effort was made to include terms recently introduced into the biochemical literature and to exclude obsolete ones, except for a few of historical interest. While it was considered essential to include the names of specific compounds and other substances, no attempt was made to be exhaustive in this respect. Terms are generally defined in a concise manner except for those where a somewhat more expanded definition was considered to be useful; comprehensive, encyclopedic treatment was, however, purposely avoided.

I would like to acknowledge Drs. Irwin H. and Leigh D. Segel for their critical review of the manuscript, and Dr. Mary Conway and Dr. Robert Badger, my editors at Wiley-Interscience, for their cooperation and helpful suggestions. My thanks go to Mrs. Janet Springer and Mrs. Jamie Jeremy, secretaries of the Chemistry Department of Western Michigan University, for their typing of the manuscript. I am especially grateful to my wife, Mabel, for her advice pertaining to the technical aspects of lexicography and for her critical evaluation of style. Above all, I am indebted to her and to my son, Ilan, for helping to make this book a reality by willingly accepting the sacrifices which the work demanded.

J. STENESH

Kalamazoo, Michigan
April 1975

EXPLANATORY NOTES

Arrangement of Entries. The entries are arranged in alphabetical order, letter by letter; thus "acidimetry" precedes "acid number," and "waterfall sequence" precedes "water hydrate model." Identical alphabetical listings are entered so that lowercase letters precede capital ones and subscripts precede superscripts.

Chemical prefixes, in either abbreviated or unabbreviated form, are disregarded in alphabetizing when they are used in the ordinary sense of denoting structure of organic compounds. These include ortho-, meta-, para-, alpha-, beta-, gamma-, delta-, cis-, trans-, N-, O-, and S-. Such prefixes are, however, included in alphabetizing when they form integral parts of entries and are used in ways other than for the indication of structure of organic compounds, as in "alpha helix," "beta configuration," and "N-terminal." The prefixes mono-, di-, tri-, tetra-, and poly-, which form integral parts of entries, are included in alphabetizing, as in "monoglyceride" and "tetrahydrofolic acid."

All numbers are disregarded in alphabetizing; these include numbers denoting chemical structure, as in "glucose-6-phosphate dehydrogenase" and "5-HT," and numbers used for other purposes, as in "factor IV" and "S-100 fraction."

The letters D and L, denoting configuration, are omitted from names of terms as entered and are usually omitted from the definitions themselves.

Form of Entries. All entries are direct entries so that, for example, "first law of cancer biochemistry" is entered as such and not as "cancer biochemistry, first law of." The entries are generally in the singular, with the plural indicated only when considered necessary. When several parts of speech of a term are in use, the term is generally entered in the noun form, and other parts of speech are entered only to the extent deemed useful. The different meanings of a term are numbered, chemical formulas are generally omitted, and the spelling is American.

Cross References. Four types of cross-references are used in this dictionary; they are indicated by the use of *see, aka, see also,* all in italics, and by the use of words in capital letters. The word *see* is used either in a directive sense, as in "coat—*see* spore coat; viral coat" and "hereditary code—*see* genetic code," or to indicate that the term is defined within the definition of another, separately entered term, as in "E_0'—*see* standard electrode potential" and "MIH—*see* melanocyte-stimulating hormone regulatory hormone." The abbreviation *aka* (also known as) is used at the end of a definition to indicate expressions that are synonymous to the entry; principal synonymous expressions are entered separately in the text. The phrase *see also* is used at the end of a definition where it is considered useful to point out to the reader comparable, contrasting, or other kinds of related entries. Capital letters are used to indicate an expression that is synonymous to the entry and that is defined in its alphabetical place in the book. Thus the definition of the entry "amphiphilic" by the word "AMPHIPATHIC," and the definition of the entry "pentose oxidation cycle" by the term "HEXOSE MONOPHOSPHATE SHUNT" indicate that the capitalized terms are expressions that are synonymous to the entries and that are themselves defined in their appropriate alphabetical places in the text.

Abbreviations and Symbols. The following standard abbreviations and symbols are used in this dictionary:

A	ampere
Å	angstrom unit
abbr	abbreviation
adj	adjective
adv	adverb
aka	also known as
atm	atmosphere
°C	degree centigrade
cc	cubic centimeter
cm	centimeter
cps	cycles per second
deg	degree
dm	decimeter
e.g.	for example
esu	electrostatic unit
g	gram
Icd	international candle
i.e.	that is
J	joule
kg	kilogram
l	liter
lb	pound
lm	lumen
m	meter
mg	milligram
min	minute
ml	milliliter
mm	millimeter
n	noun
nm	nanometer

pl	plural
sec	second
sym	symbol
v	verb
var sp	variant spelling
%	percent
μ	micro
Ω	ohm

Various letters of the Greek alphabet are also used in this dictionary. For completeness, the entire Greek alphabet is listed below:

Capital	Lowercase	Name
A	α	alpha
B	β	beta
Γ	γ	gamma
Δ	δ, ∂	delta
E	ϵ	epsilon
Z	ζ	zeta
H	η	eta
Θ	θ, ϑ	theta
I	ι	iota
K	κ	kappa
Λ	λ	lambda
M	μ	mu
N	ν	nu
Ξ	ξ	xi
O	o	omicron
Π	π	pi
P	ρ	rho
Σ	σ, ς	sigma
T	τ	tau
Υ	υ	upsilon
Φ	ϕ	phi
X	χ	chi
Ψ	ψ	psi
Ω	ω	omega

A

a. Atto.

A. 1. Adenine. 2. Adenosine. 3. Absorbance. 4. Angstrom unit. 5. Mass number. 6. Alanine. 7. Helmholtz free energy. 8. Ampere.

Å. Angstrom unit.

AA. Amino acid.

AA-AMP. Aminoacyl adenylate.

AAN. Amino acid nitrogen.

AA-tRNA. Aminoacyl transfer RNA.

AA-tRNA^{AA}. Aminoacyl transfer RNA; the prefix AA denotes the aminoacyl group attached to the transfer RNA molecule, while the superscript AA denotes the amino acid for which the transfer RNA is specific.

Ab. Antibody.

A band. A transverse dark band that is seen in electron microscope preparations of myofibrils from striated muscle and that consists of thick and thin filaments.

Abbe refractometer. A refractometer for the direct measurement of the refractive index of a solution.

ABC. Antigen binding capacity.

a × b × c code. An early version of the genetic code according to which there exist, respectively, *a, b,* and *c* distinguishable and nonequivalent bases for each of the three positions of the codon, so that the product $a \times b \times c$ is equal to the number of categories into which the triplet codons are divided. The original $a \times b \times c$ code was thought to be a $4 \times 3 \times 2$ code.

aberration. *See* chromosomal aberration.

abetalipoproteinemia. A genetically inherited metabolic defect in man that is characterized by either a decrease in, or a complete absence of, low-density lipoproteins.

abiogenesis. 1. The formation of a substance other than by a living organism. 2. The doctrine that living organisms can come from nonliving matter; spontaneous generation.

abiogenetic. Of, or pertaining to, abiogenesis.

abiogenic. Of, or pertaining to, abiogenesis (1).

abiological. Of, or pertaining to, nonliving matter.

abiosis. The absence of life.

abiotic. Of, or pertaining to, abiosis.

abnormal hemoglobin. A hemoglobin that differs from normal hemoglobin in its amino acid sequence.

ABO blood group system. A human blood group system in which there are two antigens, denoted A and B, that give rise to four serum groups, denoted A, B, AB, and O. The antigens are mucopeptides and contain a mucopolysaccharide that is identical in both antigens except for its nonreducing end. The serum groups A, B, AB, and O are characterized, respectively, by having red blood cells that carry A antigens, B antigens, both A and B antigens, and neither A nor B antigens.

abortive infection. A viral infection that either does not lead to the formation of viral particles or leads to the formation of noninfectious viral particles.

abortive transduction. Bacterial transduction in which the DNA from the donor cell is introduced into the recipient cell, but fails to become integrated into the chromosome of the recipient bacterium.

abrin. A plant protein in the seeds of *Abrus precatorius* that is toxic to animals and man, inhibits protein synthesis, and has antitumor activity.

abscissa. The horizontal axis, or x-axis, in a plane rectangular coordinate system.

absolute configuration. The actual spatial arrangement of the atoms about the asymmetric carbon atoms in a molecule.

absolute counting. The counting of radiation that includes every disintegration that occurs in the sample; such counts are expressed as disintegrations per minute.

absolute deviation. The numerical difference between an experimental value and either the true, or the best, value of the quantity being measured.

absolute plating efficiency. The percentage of cells that give rise to colonies when a given number of cells is plated on a nutrient medium.

absolute reaction rates. *See* theory of absolute reaction rates.

absolute specificity. The extreme selectivity of an enzyme that allows it to catalyze only the reaction with a single substrate in the case of a monomolecular reaction, or the reaction with a single pair of substrates in the case of a bimolecular reaction. *Aka* absolute group specificity.

absolute temperature scale. A temperature scale on which the zero point is the absolute zero, and the degrees, denoted $°T$ or $°K$, match those of the centigrade scale. *Aka* Kelvin temperature scale.

absolute zero. The zero point on the absolute temperature scale; $-273.2°C$.

absorb. To engage in the process of absorption.

absorbance. A measure of the light absorbed by a solution that is equal to log I_0/I, where I_0 is the intensity of the incident light, and I is the intensity of the transmitted light. *Sym* A. *Aka* optical density.

absorbance index. ABSORPTIVITY.

absorbance unit. The amount of absorbing material contained in 1 ml of a solution that has an absorbance of 1.0 when measured with an optical path length of 1.0 cm.

absorbancy. Variant spelling of absorbance.

absorbate. The substance that is absorbed by another substance.

absorbed antiserum. An antiserum from which antibodies have been removed by the addition of soluble antigens.

absorbed dose. *See* radiation absorbed dose.

absorbent. 1. *n* The substance that absorbs another substance. 2. *adj* Having the capacity to absorb.

absorber. A material used to absorb radioactive radiation.

absorptiometer. 1. An instrument for measuring the amount of gas absorbed by a liquid. 2. A device for measuring the thickness of a layer of liquid between parallel glass plates. 3. COLORIMETER.

absorption. 1. The uptake of one substance by another substance. 2. The passage of materials across a biological membrane. 3. The process by which all or part of the energy of incident radiation (includes heat, electromagnetic, and radioactive radiation) is transferred to the matter through which it passes. 4. The removal of antibodies from a mixture by the addition of soluble antigens, or the removal of soluble antigens from a mixture by the addition of antibodies.

absorption band. A portion of the electromagnetic spectrum in which a molecule absorbs radiant energy.

absorption cell. CUVETTE.

absorption coefficient. 1. ABSORPTIVITY. 2. BUNSEN ABSORPTION COEFFICIENT. 3. The rate of change in the intensity of a beam of radiation as it passes through matter.

absorption cross-section. The product of the probability that a photon passing through a molecule will be absorbed by that molecule and the average cross-sectional area of the molecule; the absorption cross-section s is related to the molar absorptivity ϵ by $s = 3.8 \times 10^{-21} \epsilon$.

absorption optical system. An optical system that focuses ultraviolet light passing through a solution in such a fashion that a photograph is obtained in which the darkening of the photographic film depends on the amount of light transmitted by the solution. A boundary in the solution appears as a transition between a lighter and a darker region, and measurements are made on the film by means of a densitometer tracing. The optical system is used in the analytical ultracentrifuge.

absorption ratio. The ratio of the concentration of a compound in solution to its absorptivity.

absorption spectrum. A plot of the absorption of electromagnetic radiation by a molecule as a function of either the frequency or the wavelength of the radiation.

absorptive lipemia. The transient lipemia that follows the ingestion of fat.

absorptivity. The proportionality constant ϵ in Beer's law, $A = \epsilon lc$, where A is the absorbance, l is the length of the light path, and c is the concentration.

abstraction. The removal of either an atom or an electron from a compound.

Ac. Acetyl group.

acanthocytosis. 1. A condition characterized by blood that contains spherical erythrocytes which have numerous projecting spines. 2. ABETALIPOPROTEINEMIA.

acatalasia. A genetically inherited metabolic defect in man that is due to a deficiency of the enzyme catalase.

acceleration. A stage in carcinogenesis in which, according to the Busch theory, an accelerator protein is synthesized which functions in accelerating the production of cancer RNA from cancer DNA.

accelerator. An instrument for imparting high kinetic energy to subatomic particles by means of electric and magnetic fields.

accelerator globulin. PROACCELERIN.

accelerator protein. *See* acceleration.

accelerin. The activated form of proaccelerin that converts prothrombin to thrombin during blood clotting.

acceptor control. The dependence of the respiratory rate of mitochondria on the ADP concentration. *See also* loose coupling; tight coupling.

acceptor-control ratio. The rate of respiration, in terms of oxygen uptake per unit time, in the presence of ADP, divided by the rate in the absence of ADP; measured either in the intact cell or in isolated mitochondria.

acceptor RNA. TRANSFER RNA.

acceptor site. AMINOACYL SITE.

accessory pigment. A photosynthetic pigment, such as a carotenoid or a phycobilin, that functions in conjunction with a primary photosynthetic pigment.

AcCoA. Acetyl coenzyme A.

accumulation theory. A theory of aging according to which aging is due to the accumulation of either a deleterious or a toxic substance.

accumulator organism. An organism capable of

absorbing and retaining large amounts of specific chemical elements.

accuracy. The nearness of an experimental value to either the true, or the best, value of the quantity being measured.

ACD solution. Acid–citrate–dextrose solution.

acellular. Not composed of cells.

ACES. N-(2-Acetamido)-2-aminoethanesulfonic acid; used for the preparation of buffers in the pH range of 6.0 to 7.5.

acetal. A compound derived from an aldehyde and two alcohol molecules by splitting out a molecule of water.

acetate hypothesis. The hypothesis that a multitude of complex substances may be formed naturally as a result of modifications of the linear chains formed by repeated head-to-tail condensation of acetic acid residues; typical modifications are cyclization, oxidation, and alkylation.

acetate-replacing factor. LIPOIC ACID.

acetate thiokinase. A fatty acid thiokinase that catalyzes the activation of fatty acids having two or three carbon atoms to fatty acyl coenzyme A.

acetification. The spoilage of beverages, such as wine and beer, due to the aerobic oxidation of ethyl alcohol to acetic acid by microorganisms.

acetoacetic acid. A ketoacid that can be formed from acetyl coenzyme A and that is one of the ketone bodies.

acetogenin. A compound formed by the head-to-tail condensation of either acetic acid residues or derivatives of acetic acid residues.

acetoin. A compound that can be formed by air oxidation of butylene glycol in the course of butylene glycol fermentation.

acetoin fermentation. BUTYLENE GLYCOL FERMENTATION.

acetone. A ketone that can be formed from acetyl coenzyme A and that is one of the ketone bodies.

acetone body. KETONE BODY.

acetonemia. 1. The presence of excessive amounts of acetone in the blood. 2. The presence of excessive amounts of ketone bodies in the blood.

acetone powder. A preparation of one or more proteins that is produced by removal of acetone by vacuum filtration from an acetone extract of a tissue; used in the course of isolating and purifying an enzyme or other protein.

acetonuria. 1. The presence of excessive amounts of acetone in the urine. 2. The presence of excessive amounts of ketone bodies in the urine.

acetylation. An acylation reaction in which an acetyl radical CH_3CO— is introduced into an organic compound.

acetylcholine. The acetylated form of choline; the hydrolysis of acetylcholine to choline and acetic acid is catalyzed by acetylcholinesterase and is a key reaction in the transmission of the nerve impulse. *Abbr* ACh.

acetylcholinesterase. The enzyme that catalyzes the hydrolysis of acetylcholine to choline and acetic acid during the transmission of a nerve impulse. *Abbr* AChE. *Aka* true cholinesterase; choline esterase I; specific cholinesterase. *See also* cholinesterase.

acetyl CoA. Acetyl coenzyme A.

acetyl coenzyme A. The acetylated form of coenzyme A; a key intermediate in the citric acid cycle, in fatty acid oxidation, in fatty acid synthesis, and in other metabolic reactions. Variously abbreviated as acetyl-S-CoA, acetyl-CoA, CoASAc, AcSCoA, and AcCoA.

acetyl coenzyme A carboxylase. The enzyme that catalyzes the synthesis of malonyl coenzyme A from acetyl coenzyme A and carbon dioxide, and that contains biotin as a prosthetic group.

acetylene. 1. The hydrocarbon $CH \equiv CH$. 2. ALKYNE.

acetyl group. The acyl group of acetic acid; the radical CH_3CO—.

N-acetylmuramic acid. A compound derived from acetic acid, glucosamine, and lactic acid that is a major building block of bacterial cell walls.

N-acetylneuraminic acid. A compound derived from acetic acid, mannosamine, and pyruvic acid that is a major building block of animal cell coats. *Abbr* NANA; NAcneu.

acetyl number. A measure of the number of hydroxyl groups in a fat; equal to the number of milligrams of potassium hydroxide required to neutralize the acetic acid in 1 gram of acetylated fat. *Aka* acetyl value.

acetylornithine cycle. A cyclic set of reactions in bacteria and plants that constitutes a major pathway for the synthesis of ornithine from glutamic acid and N-acetylornithine.

acetyl-S-CoA. Acetyl coenzyme A.

N-acetyl serine. The acetylated form of serine that is believed to function in the initiation of translation in mammalian systems, much as N-formylmethionine functions in the initiation of translation in bacterial systems.

acetyltransferase. An enzyme that catalyzes the transfer of an acetyl group from acetyl coenzyme A to another compound.

AcG. Accelerator globulin.

Ac globulin. Accelerator globulin.

ACh. Acetylcholine.

A chain. 1. The shorter of the two polypeptide chains of insulin, containing 21 amino acids and one intrachain disulfide bond. 2. The heavy chain (H chain) of the immunoglobulins.

AChE. Acetylcholinesterase.

achromic. Devoid of color.

achromic point. A stage in the hydrolysis of starch at which the addition of iodine fails to produce a blue color.

achromotrichia factor. *p*-AMINOBENZOIC ACID.

achromycin. *See* tetracycline.

acid. *See* Bronsted acid; Lewis acid.

acidaminuria. AMINOACIDURIA.

acid anhydride. A compound containing two acyl groups bound to an oxygen atom. The compound is referred to as either a simple or a mixed anhydride depending on whether the two acyl groups are identical or different. In biochemistry, both simple and mixed anhydrides frequently contain the phosphoryl group.

acid-base balance. The reactions and factors involved in maintaining a constant internal environment in the body with respect to the buffer systems and the pH of the various fluid compartments.

acid-base titration. A titration in which either acid or base is added to a solution, and the titration is followed by means of pH measurements or by means of indicators.

acid–citrate–dextrose solution. An aqueous solution of citric acid, sodium citrate, and dextrose, that is used as an anticoagulant in the collection and storage of blood.

acidemia. A condition characterized by an increase in the hydrogen-ion concentration of the blood.

acid-fast. Descriptive of the lipid-rich cell walls of some bacteria that resist decolorization by mineral acids after having been stained with basic aniline dyes.

acid hematin. A hematin formed from hemoglobin by treatment with acid below pH 3.

acidic. 1. Of, or pertaining to, an acid. 2. Of, or pertaining to, a solution having a pH less than 7.0.

acidic amino acid. An amino acid that has one amino and two carboxyl groups.

acidic dye. An anionic dye that binds to, and stains, positively charged macromolecules. *Aka* acidic stain.

acidification of urine. The process whereby the glomerular filtrate of the kidney that has an approximate pH of 7.4 is converted to urine that has a lower pH and may have a pH as low as 4.8.

acidimetry. 1. The chemical analysis of solutions by means of titrations, the end points of which are recognized by a change in the hydrogen-ion concentration. 2. A determination of the amount of an acid by titration against a standard alkaline solution.

acid number. The number of milligrams of potassium hydroxide required to neutralize the free fatty acids in 1 gram of fat. *Aka* acid value.

acidolysis. Hydrolysis by means of an acid.

acidophil. A cell that stains with an acidic dye.

acidosis. A deviation from the normal acid-base balance in the body that is due to a disturbance which, by itself and in the absence of compensatory mechanisms, would tend to lower the pH of the blood. The actual change in pH depends on whether and to what extent the disturbance is compensated for. The disturbances and the compensatory mechanisms are considered primarily with respect to their effect on the bicarbonate/carbonic acid ratio of blood plasma. *See also* metabolic acidosis; primary acidosis; etc.

acidotic. Of, or pertaining to, acidosis.

acid pH. A pH value below 7.0.

acid phosphatase. A phosphatase, the optimum pH of which is below 7.0.

aciduria. A condition characterized by the excretion of an excessively acidic urine.

***cis*-aconitic acid.** A tricarboxylic acid formed from citric acid in one of the reactions of the citric acid cycle.

ACP. Acyl carrier protein.

a-c polarography. Alternating-current polarography; a polarographic method in which a small alternating potential is superimposed on the normal, direct-current applied potential, and the a-c component of the resulting current is measured.

acquired antibody. An antibody produced by an immune reaction as distinct from one occurring naturally.

acquired hemolytic anemia. An autoimmune disease in which an individual forms antibodies to his own red blood cells.

acquired immunity. The immunity established in an animal organism during its lifetime.

acquired tolerance. The immunological tolerance produced in an animal organism by the injection of antigen into it; acquired tolerance persists only as long as the antigen remains in the organism.

acridine dye. A planar heterocyclic molecule used to stain DNA and RNA. Acridine dyes are basic dyes that become intercalated into the nucleic acid molecule; they are mutagenic, since their intercalation produces insertions or deletions.

acridine orange. An acridine dye that functions both as a fluorochrome for staining nucleic acids and as a mutagen, producing insertions or deletions.

acriflavin. An acridine dye.

acrolein test. A qualitative test for glycerol, based on the dehydration and oxidation of glycerol to acrolein by heating with potassium bisulfate.

acromegaly. A condition characterized by overgrowth of skeletal structures due to the excessive production of growth hormone.

acrylamide. *See* polyacrylamide gel.

AcSCoA. Acetyl coenzyme A.

ACTH. Adrenocorticotropic hormone.

actidione. CYCLOHEXIMIDE.

actin. A major protein component of the myofilaments of striated muscle and the principal constituent of the thin filaments.

actin filament. A thin filament of striated muscle that consists largely of actin and that is linked to thick filaments by means of cross-bridges which protrude from them; a myofilament.

actinin. A minor protein component of striated muscle, believed to be part of the thin filaments and to be concentrated in both the Z line and the I band. Two components, denoted α and β actinin, have been identified.

actinometer. A device for the determination of absorbed light by means of a photochemical reaction of known quantum yield.

actinometry. A method of chemical analysis by means of an actinometer.

Actinomycin D. An antibiotic, produced by *Streptomyces chrysomallus,* that inhibits the transcription of DNA to RNA by binding to DNA and that also has immunosuppressive activity. *Aka* actinomycin C1.

action potential. The membrane potential of a stimulated membrane, produced by the ion flux across the membrane when its permeability is changed upon stimulation.

action spectrum. A plot of a quantitative biological or chemical response as a function of the wavelength of the radiation producing the response; the death of bacteria, the occurrence of mutations, the occurrence of fluorescence, and photosynthetic efficiency are examples of responses.

activated alumina. Alumina that has been thoroughly dried.

activated complex. *See* theory of absolute reaction rates.

activated form. *See* active form.

activating enzyme. 1. FATTY ACID THIO-KINASE. 2. AMINOACYL-tRNA SYNTHETASE.

activation. 1. The conversion of a compound to a more reactive form; the change of an amino acid to aminoacyl transfer RNA, the change of a fatty acid to fatty acyl coenzyme A, and the change of an inactive enzyme precursor to the active enzyme are some examples. 2. The increase in the extent, and/or the rate, of an enzymatic reaction. 3. The drying of chromatographic supports. 4. The first stage in the conversion of a spore to a vegetative cell; this stage can frequently be produced by heat or

aging and is believed to involve damage to an outer layer of the spore.

activation analysis. A method for the qualitative and quantitative analysis of the chemical elements in a sample; based on identification and determination of the radionuclides formed when the sample is bombarded with neutrons or other particles.

activation energy. The difference in energy between that of the activated complex and that of the reactants; the energy that must be supplied to the reactants before they can undergo transformation to products.

activation stage. That part of the blood clotting process that consists of the formation of active thrombin.

activator. A metal ion that serves as a cofactor for an enzyme.

activator constant. The equilibrium constant for the reaction $EA \rightleftarrows E + A$, where E is an enzyme and A is an activator.

active acetaldehyde. An acetaldehyde molecule attached to thiamine pyrophosphate; α-hydroxyethyl thiamine pyrophosphate.

active acetate. ACETYL COENZYME A.

active acetyl. 1. ACETYL COENZYME A. 2. Acetyl lipoic acid.

active acyl. 1. An acyl coenzyme A. 2. An acyl lipoic acid.

active adenosyl. ADENOSINE-5´-TRIPHOS-PHATE.

active adenylate. ADENOSINE-5´-TRIPHOS-PHATE.

active aldehyde. An aldehyde molecule attached to thiamine pyrophosphate; α-hydroxyalkyl thiamine pyrophosphate.

active aldehyde theory. The theory according to which the nonenzymatic browning of foods is due to reactions involving very active aldehydes that are formed by the dehydration of sugars.

active amino acid. 1. AMINOACYL ADENYLATE. 2. AMINOACYL-tRNA. 3. A Schiff base of an amino acid as that formed in transamination.

active anaphylaxis. The anaphylactic reaction produced in an animal organism as a result of the injection of antigen.

active carbohydrate. 1. A UDP-sugar. 2. A GDP-sugar.

active carbon dioxide. CARBOXYBIOTIN.

active center. ACTIVE SITE.

active concentration. ACTIVE TRANSPORT.

active form. 1. That derivative of a metabolite that can serve as a high energy compound and/or as a compound that initiates a reaction or a series of reactions. 2. That form of a macromolecule that possesses biological activity.

active formaldehyde. ACTIVE FORMYL.

active formate. 1. ACTIVE FORMYL. 2. ACTIVE FORMIMINO.

active formimino. A formimino group $NH{=}CH{-}$ attached to tetrahydrofolic acid.

active formyl. A formyl group $O{=}CH{-}$ attached to tetrahydrofolic acid.

active fructose. FRUCTOSE-1,6-DIPHOSPHATE.

active glycolaldehyde. A glycolaldehyde group $CH_2OH{-}CO{-}$ attached to thiamine pyrophosphate; α,β-dihydroxyethyl thiamine pyrophosphate.

active hydroxyethyl. ACTIVE ACETALDEHYDE.

active hydroxymethyl. 5,10-Methylene tetrahydrofolic acid.

active immunity. The immunity acquired by an animal organism as a result of the injection of antigens into it.

active iodine. That form of iodine, possibly an iodinium ion I^+, which reacts with tyrosine to form iodotyrosines in the thyroid gland.

active mediated transport. An active transport that requires one or more transport agents.

active methionine. S-ADENOSYLMETHIONINE.

active methyl. 1. 5-Methyl-tetrahydrofolic acid. 2. S-ADENOSYLMETHIONINE.

active patch. ANTIBODY COMBINING SITE.

active phosphate. 1. ADENOSINE-5′-TRIPHOSPHATE. 2. GUANOSINE-5′-TRIPHOSPHATE.

active phospholipid. A cytidine-5′-diphosphate derivative of either a phospholipid or a component of phospholipids.

active pyrophosphate. ADENOSINE-5′-TRIPHOSPHATE.

active pyruvate. A pyruvic acid molecule attached to thiamine pyrophosphate.

active site. 1. That portion of the enzyme molecule that interacts with, and binds, the substrate, thereby forming an enzyme-substrate complex. 2. That portion of the antibody molecule that interacts with, and binds, the antigen, thereby forming an antigen-antibody complex.

active site-directed irreversible inhibitor. An artificially designed inhibitor for the irreversible inhibition of a given enzyme. The inhibitor is a trifunctional molecule that contains (a) a functional group that can bind to the active site of the enzyme, (b) a nonpolar fragment that can attach to a nonpolar region just outside the active site, and (c) a group, such as a sulfonyl chloride, that can alkylate a functional group of the enzyme just outside the nonpolar region. The first functional group serves to direct the inhibitor to the active site of the enzyme; the nonpolar fragment serves to align the inhibitor so that the alkylating group is brought into contact with a susceptible group on the enzyme; and the third functional group then leads to an alkylation reaction that results in the irreversible inhibition of the enzyme. *See also* affinity labeling.

active sulfate. 1. The compound 3′-phosphoadenosine-5′-phosphosulfate that serves as a sulfating agent in the esterification of sulfate with alcoholic and phenolic hydroxyl groups. 2. The compound adenosine-5′-phosphosulfate that serves as an intermediate in the synthesis of 3′-phosphoadenosine-5′-phosphosulfate and that can be reduced directly to sulfite in *Desulfovibrio desulfuricans*.

active translocation. ACTIVE TRANSPORT.

active transport. The movement of a solute across a biological membrane such that the movement is directed upward in a concentration gradient (i.e., against the gradient) and requires the expenditure of energy.

activity. 1. A measure of the effective concentration of an enzyme, drug, hormone, or other substance, and by extension, the substance the effectiveness of which is being measured. 2. The product of the molar concentration of an ionic solute and its activity coefficient; activities must be used in place of molar concentrations for nonideal solutions.

activity coefficient. The ratio of the activity of an ion to its molar concentration; the logarithm of the activity coefficient is equal to $-0.5Z^2\ \sqrt{\Gamma/2}$, where Z is the charge of the ion and $\Gamma/2$ is the ionic strength. *See also* mean activity coefficient.

actomyosin. The complex formed between myosin and actin, either as extracted from muscle or as prepared from the purified components.

acute porphyria. A porphyria that is of short duration and that is characterized by the excretion of excessive amounts of uroporphyrin III, coproporphyrin III, and porphobilinogen.

acute serum. A serum obtained soon after the onset of a disease.

acute test. A toxicity test that is performed on laboratory animals and that requires only a single dose of a chemical, administered in a single application.

acyclic. ALIPHATIC.

acylation. The introduction of an acyl radical $RCO{-}$ into an organic compound.

acyl-carrier protein. A low-molecular weight protein that constitutes part of the fatty acid synthetase complex and that serves as a carrier of acyl groups during fatty acid biosynthesis. The acyl group is joined as a thioester to the sulfhydryl group of 4′-phosphopantetheine which, in turn, is esterified to a serine residue of the acyl carrier protein. *Abbr* ACP.

acyl enzyme. A postulated intermediate in the reaction catalyzed by the enzyme glyceraldehyde-3-phosphate dehydrogenase in which both NAD^+ and the acyl form of glyceraldehyde-3-phosphate are bound to the enzyme.

acylglycerol. A glyceride; mono-, di-, and triacylglycerols are referred to, respectively, as mono-, di-, and triglycerides.

acyl group. The radical $RCO-$ that is derived from an organic acid by removal of the OH from the carboxyl group.

acyl-S-CoA. Acyl coenzyme A.

acyltransferase. An enzyme that catalyzes the transfer of an acyl group from acyl coenzyme A to another compound.

ADA. N-(2-Acetamido)-iminodiacetic acid; used for the preparation of buffers in the pH range of 5.8 to 7.4.

Adair equation. An equation used for calculating the average number of bound ligand molecules per molecule of total protein from binding data; the protein may have n binding sites with, or without, interaction between the sites.

Adamkiewicz reaction. The production of a violet color upon treatment of a solution containing protein with acetic acid and sulfuric acid.

Adam's catalyst. Platinum oxide, a catalyst for hydrogenation reactions.

adapter hypothesis. The hypothesis, suggested by Crick in 1958, that an amino acid is joined to a specific adapter molecule during protein synthesis. The adapter serves to carry the amino acid to the ribosome and becomes bound to the codon of the amino acid in the messenger RNA which is attached to the ribosome. In this fashion the adapter, now known to be transfer RNA, assures the insertion of the amino acid into its proper place in the growing polypeptide chain.

adapter RNA. TRANSFER RNA.

adaptive enzyme. INDUCIBLE ENZYME.

adaptor RNA. Variant spelling of adapter RNA.

Addison's disease. The pathological condition resulting from adrenal insufficiency.

addition polymer. A polymer formed by the addition of monomers to the growing chain through the breaking of double bonds in the monomers.

addition reaction. A chemical reaction in which there is an increase in the number of groups attached to carbon atoms so that the molecule becomes more saturated.

adduct. The product formed by the chemical addition of one substance to another.

adductor muscle. CATCH MUSCLE.

adenine. The purine 6-aminopurine that occurs in both RNA and DNA. *Abbr* A.

adenine nucleotide barrier. ATRACTYLOSIDE BARRIER.

adenohypophyseal. Of, or pertaining to, the anterior lobe of the pituitary gland.

adenohypophysis. The anterior lobe of the pituitary gland which produces the adrenocorticotropic, gonadotropic, lipotropic, somatotropic, and thyrotropic hormones.

adenoma. A tumor of epithelial tissue that is generally benign and in which the cells form glands or glandlike structures.

adenosine. The ribonucleoside of adenine. Adenosine mono-, di-, and triphosphate are abbreviated, respectively, as AMP, ADP, and ATP. The abbreviations refer to the 5′-nucleoside phosphates unless otherwise indicated. *Abbr* Ado; A.

adenosine-3′,5′-cyclic monophosphate. A cyclic nucleotide, commonly called cyclic AMP, that is formed from ATP in a reaction catalyzed by the enzyme adenyl cyclase. Cyclic AMP functions as a second messenger and mediates the effect of a large number of hormones. The hormones interact with the adenyl cyclase system in the cell membrane, and the intracellular cyclic AMP then interacts with specific enzymes or other intracellular components. *Abbr* cAMP. *Aka* cyclic adenylic acid.

adenosine diphosphate. The high-energy compound, adenosine-5′-diphosphate, that can undergo hydrolysis to adenosine-5′-monophosphate and inorganic phosphate. *Abbr* ADP.

adenosine monophosphate. The nucleotide, adenosine-5′-monophosphate, that can be formed by hydrolysis of either of the high-energy compounds, ATP or ADP. *Abbr* AMP.

adenosine-5′-phosphosulfate. *See* active sulfate (2).

adenosine triphosphatase. An enzyme, or a group of enzymes, that catalyzes the hydrolysis of ATP to either ADP and inorganic phosphate or to AMP and pyrophosphate. The enzyme is widely distributed in biological membranes and is implicated in the active transport of sodium and potassium ions. *Abbr* ATPase. *See also* Na,K-ATPase.

adenosine triphosphate. The high-energy compound, adenosine-5′-triphosphate, that functions in many biochemical systems. It can be hydrolyzed to either adenosine-5′-monophosphate or adenosine-5′-diphosphate; the hydrolysis reaction is accompanied by the release of a large amount of free energy which is used to drive a variety of metabolic reactions. *Abbr* ATP.

S-Adenosylmethionine. A high-energy compound that is derived from ATP and methionine and that functions as a biological methylating agent. *Abbr* SAM.

adenovirus. A naked, icosahedral virus that

contains double-stranded DNA. Adenoviruses infect mammals, often leading to respiratory infections; some are oncogenic.

adenovirus-associated virus. A small, naked, icosahedral virus that contains single-stranded DNA and that is found in association with adenoviruses.

adenylate. A compound consisting of adenylic acid that is esterified through its phosphate group to another molecule.

adenylate control hypothesis. The hypothesis that cellular metabolism is regulated by the relative amounts of AMP, ADP, and ATP in the cell. *See also* energy charge.

adenylate cyclase. *See* adenyl cyclase.

adenylate kinase. MYOKINASE.

adenylate pool. The total intracellular concentration of AMP, ADP, and ATP.

adenyl cyclase. The enzyme that catalyzes the formation of cyclic AMP from ATP by the splitting out of pyrophosphate.

adenylic acid. The ribonucleotide of adenine.

ADH. 1. ALCOHOL DEHYDROGENASE. 2. ANTIDIURETIC HORMONE.

adiabatic process. A process conducted without either a gain or a loss of heat; a process conducted in an isolated system.

adiabatic system. A thermodynamic system that is thermally insulated from its surroundings.

adipocyte. A fat cell; a cell of adipose tissue.

adipokinetic hormone. LIPOTROPIN.

adipose tissue. Lipid tissue; fat deposits in an organism.

adiposis. A condition characterized by excessive accumulation of fat in the body; the accumulation may be local or general.

adiposity. OBESITY.

adjuvant. A substance that increases the immune response of an animal to an antigen when injected together with the antigen.

adjuvanticity. The capacity of a substance to function as an adjuvant.

ad libitum. Referring to the feeding of experimental animals where the animals are allowed to eat without any imposed restrictions. *Abbr* ad lib.

admix. To mix one substance with another.

admixture. 1. A mixture. 2. The act of mixing.

Ado. Adenosine.

adoptive immunity. The immunity acquired by an animal organism when it is injected with lymphocytes from another organism.

adoptive tolerance. The immunological tolerance acquired by an animal organism when it is injected with lymphocytes from another organism.

ADP. 1. ADENOSINE DIPHOSPHATE. 2. Adenosine-5′-diphosphate.

ADPG. ADP-glucose.

ADP-glucose. A nucleoside diphosphate sugar

that is the donor of a glucose residue in the biosynthesis of starch in plants and in the biosynthesis of $\alpha(1 \rightarrow 4)$ glucans in bacteria. *Abbr* ADPG.

ADR. Adrenaline.

adrenal cortex. That part of the adrenal gland, derived from mesodermal tissue, which secretes the adrenal cortical hormones.

adrenal cortical hormone. A steroid hormone secreted by the adrenal cortex. Major adrenal cortical hormones are the glucocorticoids, cortisol and corticosterone, and the mineralocorticoid, aldosterone; minor adrenal cortical hormones are the sex hormones.

adrenal cortical steroid. A steroid produced by the adrenal cortex. Many of these steroids are hormones, such as the glucocorticoids, mineralocorticoids, and sex hormones; some, such as cholesterol, are not hormones.

adrenalectomy. The surgical removal of an adrenal gland.

adrenal gland. The endocrine gland located near the kidney and composed of two parts, a medulla that secretes epinephrine and nor-epinephrine, and a cortex that secretes the adrenal cortical hormones.

adrenaline. EPINEPHRINE.

adrenaline tolerance test. A test used in the diagnosis of glycogen storage disease type I; the test is based on measuring the level of blood glucose as a function of time following the injection of an individual with adrenaline.

adrenalism. A condition resulting from insufficient function of the adrenal glands.

adrenal medulla. That part of the adrenal gland, derived from ectodermal tissue, which secretes the hormones epinephrine and norepinephrine.

adrenal virilism. The appearance of male secondary sexual characteristics in a female as a result of excessive secretion of androgens by the adrenal cortex.

adrenergic. Of, or pertaining to, nerve fibers that release norepinephrine at the nerve endings.

adrenocortical steroid. CORTICOSTEROID.

adrenocorticotrophin. Variant spelling of adrenocorticotropin.

adrenocorticotropic hormone. A polypeptide hormone of 39 amino acids that stimulates the synthesis and secretion of adrenal cortical hormones by the adrenal cortex. The adrenocorticotropic hormone is secreted by the anterior lobe of the pituitary gland. *Var sp* adrenocorticotrophic hormone. *Abbr* ACTH.

adrenocorticotropin. ADRENOCORTICOTROPIC HORMONE.

adrenodoxin. A nonheme iron protein that functions as an electron carrier in microsomal, non-phosphorylating electron transport systems.

adrenosterone. An androgen produced by the adrenal gland.

adsorb. To attract and hold a substance to the surface of another substance.

adsorbate. A substance that is adsorbed to the surface of another substance from either a solution or a gas phase.

adsorbed antiserum. An antiserum from which antibodies have been removed by the addition of particulate antigens.

adsorbent. 1. *n* A substance that adsorbs another substance from either a solution or a gas phase. 2. *adj* Having the capacity to adsorb.

adsorption. 1. The adhesion of molecules to surfaces of solids. 2. The removal of antibodies from a mixture by the addition of particulate antigens, or the removal of particulate antigens from a mixture by the addition of antibodies. 3. The attachment of phage particles to a bacterial cell.

adsorption chromatography. A chromatographic technique in which molecules are separated on the basis of their adsorption properties. The stationary phase is a solid adsorbent, generally in the form of a column; the mobile phase is either an aqueous or an organic solution. The rate of movement of the molecules through the column depends on the degree of their adsorption to the solid adsorbent.

adsorption coefficient. A constant, under defined conditions, that relates the elution of a substance from a chromatographic column to the weight of adsorbent.

adult hemoglobin. The major form of hemoglobin in normal adults that is designated HbA; a minor form is designated HbA_2.

AE-cellulose. Aminoethyl-cellulose, an anion exchanger.

aerial mycelium. That portion of a fungal mycelium that projects above the surface of the medium and frequently bears either reproductive cells or spores.

aerobe. *See* facultative aerobe; obligate aerobe.

aerobic. 1. In the presence of oxygen; in an environment or an atmosphere containing oxygen. 2. Requiring the presence of molecular oxygen for growth. 3. Capable of using molecular oxygen for growth. *See also* oxybiontic.

aerobic glycolysis. The group of cellular reactions whereby glucose is converted to pyruvic acid.

aerobic respiration. RESPIRATION (3).

aerobiosis. Life under aerobic conditions.

aerobiotic. Of, or pertaining to, aerobiosis.

aerogel. A rigid gel that has maintained its original structure despite the loss of solvent.

aerosporin. POLYMYXIN.

aerotaxis. A taxis in which the stimulus is air; used particularly for the taxis of bacteria in response to oxygen.

afferent inhibition. The prevention of transplantation immunity through the binding of antibodies from the recipient animal to antigens in the transplant; as a result, the transplant antigens are unable to reach and/or to stimulate the antibody-forming cells in the recipient animal.

affinity. 1. The capacity of an enzyme to bind substrate; generally measured by the affinity constant. 2. The capacity of an antibody to bind either antigens or haptens; frequently measured by the average intrinsic association constant for the binding reaction.

affinity chromatography. A column chromatographic technique in which desired molecules are separated from a mixture of molecules by using a modified chromatographic support that is either biochemically or immunochemically specific for the molecules of interest. Biochemically or immunochemically reactive molecules are linked covalently to a support without destroying the activity and the specificity of the molecules. These covalently linked molecules will then bind specifically the molecules of interest when a mixture of molecules is passed through the column. Two examples are the use of DNA-cellulose for the isolation of DNA-dependent DNA polymerase, and the use of agarose-antibody preparations for the isolation of antigens.

affinity constant. The reciprocal of the dissociation constant for the complex *PL* in the reversible system $P + L \rightleftarrows PL$, where *P* is usually a protein and *L* is a ligand such as a substrate, an inhibitor, or an activator. *See* also association constant.

affinity elution. A chromatographic technique in which compounds are adsorbed nonspecifically to a column and the compound of interest is then eluted specifically through its binding to a ligand in the eluting solvent.

affinity labeling. A method for the specific labeling of the active site of an enzyme, antibody, or other protein. A reagent *A-X* that can bind specifically, reversibly, and noncovalently to the active site through its *A* group is first allowed to bind to the active site. The reagent is then linked covalently through its chemically reactive group *X* to an amino acid at or close to the active site. *See also* active site-directed irreversible inhibitor.

affinity ratio. The ratio of the substrate constant for one reaction to the substrate constant for a second reaction that is catalyzed by the same enzyme but involves a different substrate.

afibrinogenemia. A genetically inherited metabolic defect in man that is characterized either by the complete absence of fibrinogen or by the presence of a defective fibrinogen.

aflatoxin. A toxic and carcinogenic compound produced by fungi.

AFP. Alpha-fetoprotein.

Ag. 1. Antigen. 2. Silver.

agammaglobulinemia. A genetically inherited metabolic defect in man that is characterized by the complete absence of immunoglobulins. *See also* hypogammaglobulinemia.

agar. An acidic polysaccharide extracted from certain seaweeds; used as a solidifying agent of culture media in microbiology and as a support medium for zone electrophoresis.

agar diffusion method. A method for determining the sensitivity of a microorganism to an antimicrobial drug; based on measuring the zone of growth inhibition when the drug is placed in a cylinder, a hole, or a filter paper disk on a petri plate that has been seeded with the microorganism.

agar gel electrophoresis. Zone electrophoresis in which the supporting medium consists of a gel prepared from agar.

agarose. A sulfate-free, neutral fraction of agar used in gel filtration.

agar plate count. A plate count in which the solid nutrient medium contains agar.

age. The length of time that a preparation of cells or a subcellular fraction has been stored.

age pigment. An insoluble granule that accumulates in certain tissues upon aging.

agglutinating antibody. AGGLUTININ.

agglutination. The clumping of bacterial and other cells that is brought about by an antigen-antibody reaction between the particulate antigens on the cell surface and added antibodies.

agglutinin. An antibody that can bind to particulate antigens on the surface of cells to produce an agglutination reaction.

agglutinogen. A surface antigen of bacterial and other cells that can induce the formation of agglutinins and can bind to them to produce an agglutination reaction.

aggregate anaphylaxis. An anaphylactic shock that is produced by a single injection of antigen.

aggressin. A substance that is produced by a microorganism and that, though not necessarily toxic by itself, promotes the invasiveness of the microorganism in the host; the enzymes hyaluronidase and collagenase are two examples.

aglucone. The noncarbohydrate portion of a glucoside.

aglycone. The noncarbohydrate portion of a glycoside.

agonist. A molecule, such as a drug, an enzyme activator, or a hormone, that enhances the activity of another molecule.

A/G ratio. Albumin/globulin ratio.

AHF. Antihemophilic factor.

AHG. 1. Antihemophilic globulin. 2. Antihuman globulin.

AICAR. 5-Aminoimidazole-4-carboxamide ribonucleotide; an intermediate in the biosynthesis of purines.

AICF. Autoimmune complement fixation.

air dose. The dose of radiation delivered to a specified point in air.

air peak. The gas chromatographic peak that is produced when a small amount of air is injected with the sample into the chromatographic column.

Akabori hypothesis. The hypothesis that the origin of proteins is based on the polymerization of nonamino acid building blocks to form polyglycine and on the subsequent replacement of the alpha hydrogens in polyglycine by various *R* groups in secondary reactions.

Akabori reaction. The formation of an alkamine by the reaction of an aldehyde with the amino group of an amino acid.

Al. Aluminum.

Ala. 1. Alanine. 2. Alanyl.

alanine. An aliphatic nonpolar amino acid; α-alanine occurs in proteins and β-alanine occurs in the peptides anserine and carnosine. *Abbr* Ala; A.

alarm reaction. GENERAL ADAPTATION SYNDROME.

albinism. A genetically inherited metabolic defect in man that is characterized by the lack of skin pigmentation and that is due to a deficiency of the enzyme tyrosinase.

albino. A person or an animal that is deficient in skin pigmentation.

albumin. A water-soluble, globular, and simple protein that is not precipitated by ammonium sulfate at 50% saturation.

albumin/globulin ratio. The ratio of the concentration of serum albumin to that of serum globulin. *Abbr* A/G ratio.

albuminimeter. An apparatus for determining protein in biological fluids on the basis of the volume of the precipitated protein.

albuminuria. The presence of excessive amounts of protein, mainly albumin, in the urine.

Albustix test. A rapid, semiquantitative test for protein in urine by means of paper strips impregnated with buffer and indicator. *See also* protein error.

alcohol. 1. An alkyl compound containing a hydroxyl group. The alcohol is designated as a primary, a secondary, or a tertiary alcohol depending on whether the hydroxyl group is attached to a carbon atom that is linked to one, two, or three other carbon atoms. 2. Ethyl alcohol; ethanol.

alcohol dehydrogenase. A pyridine-linked dehydrogenase that catalyzes the reduction of acetaldehyde to ethanol.

alcoholic fermentation. The group of reactions, characteristic of yeast, whereby glucose is fermented to ethyl alcohol.

alcoholic hydroxyl group. A hydroxyl group attached to an aliphatic carbon chain.

alcoholic steroid. STEROL.

alcoholysis. The cleavage of a covalent bond of an acid derivative by reaction with an alcohol *ROH* so that one of the products combines with the *H* of the alcohol and the other product combines with the *OR* group of the alcohol.

aldaric acid. A dicarboxylic sugar acid of an aldose in which both the aldehyde group and the primary alcohol group have been oxidized to carboxyl groups.

aldehyde. An organic compound that contains an aldehyde group.

aldehyde group. The carbonyl group attached to one carbon and one hydrogen atom; the grouping $-CHO$.

aldehyde indicator. SCHIFF'S REAGENT.

aldimine. An organic compound of the general formula $R-CH=NH$.

alditol. GLYCITOL.

aldo-. 1. Combining form meaning aldose. 2. Combining form meaning aldehyde.

aldofuranose. An aldose in furanose form.

aldolase. 1. An aldehyde-lyase. 2. The enzyme of glycolysis that catalyzes the interconversion of fructose-1,6-diphosphate to dihydroxyacetone phosphate and glyceraldehyde-3-phosphate.

aldol condensation. An addition reaction of two ketones, or two aldehydes, or an aldehyde and a ketone.

aldonic acid. A monocarboxylic sugar acid of an aldose in which the aldehyde group has been oxidized to a carboxyl group.

aldopyranose. An aldose in pyranose form.

aldose. A monosaccharide, or its derivative, that has an aldehyde group.

aldosterone. The major mineralocorticoid in man.

aldosteronism. A pathological condition characterized by the excessive production and secretion of aldosterone.

alexin. COMPLEMENT.

ALG. Antilymphocyte globulin.

alga (*pl* algae). A chlorophyll-containing, photosynthetic protist; algae are unicellular or multicellular, are generally aquatic, and are either eucaryotic or procaryotic.

algal. Of, or pertaining to, algae.

alginic acid. An algal polysaccharide of mannuronic acid.

algorithm. 1. A computational method or a set of rules for obtaining the solution of all problems of a specified type in a finite number of operations; a fixed sequence of formulas and/or algebraic and/or logical steps for calculations

of a given problem. 2. A defined process consisting of a number of fixed step-by-step procedures for accomplishing a given result in a finite number of steps. *See also* heuristic process; stochastic process.

alicyclic. Designating a compound derived from a saturated cyclic hydrocarbon.

ali-esterase. CARBOXYLESTERASE.

alimentary. 1. Of, or pertaining to, food or nutrition. 2. Nutritious.

alimentary canal. DIGESTIVE TRACT.

alimentary glycosuria. The temporary increase in the level of glucose in the urine that follows a meal rich in carbohydrates.

aliphatic. Of, or pertaining to, an organic compound that has an open chain structure. *Aka* acyclic.

aliquot. 1. A part of a whole that divides the whole without a remainder; thus 4 ml, but not 7 ml, is an aliquot of 12 ml. 2. Any part or fraction of a whole.

alkalemia. A condition characterized by a decrease in the hydrogen-ion concentration of the blood.

alkali. A base, specifically one of an alkali metal.

alkali disease. One of a number of animal poisonings of either plant or mineral origin.

alkali metal. An element of group *IA* in the periodic table that consists of the elements lithium *Li*, sodium *Na*, potassium *K*, rubidium *Rb*, cesium *Cs*, and francium *Fr*.

alkalimetry. 1. The chemical analysis of solutions by means of titrations, the end points of which are recognized by a change in the hydrogen-ion concentration. 2. A determination of the amount of a base by titration against a standard acid solution.

alkaline. BASIC.

alkaline earth. An element of group *IIA* in the periodic table that consists of the elements beryllium *Be*, magnesium *Mg*, calcium *Ca*, strontium *Sr*, barium *Ba*, and radium *Ra*.

alkaline hematin. A hematin formed from hemoglobin by treatment with alkali above pH 11.

alkaline pH. A pH value above 7.0.

alkaline phosphatase. A phosphatase, the optimum pH of which is above 7.0.

alkaline reserve. The plasma bicarbonate concentration that is determined either from the carbon dioxide combining power of plasma or from the direct titration of plasma. *Aka* alkali reserve.

alkaline rigor. The increase in pH upon death that occurs in some species of fish where death was preceded by struggling.

alkaline tide. The increase in the pH of the blood and of the urine that occurs shortly after a meal; thought to be due to the withdrawal of chlorides from the blood for the formation of hydrochloric acid in the stomach.

alkaloid. A basic, nitrogenous organic compound of plant origin; alkaloids are generally heterocyclic compounds of complex structure and almost invariably have intense pharmacological activity.

alkalosis. A deviation from the normal acid-base balance in the body that is due to a disturbance which, by itself and in the absence of compensatory factors, would tend to raise the pH of the blood. The actual change in pH depends on whether and to what extent the disturbance is compensated for. The disturbances and the compensatory mechanisms are considered primarily with respect to their effect on the bicarbonate/carbonic acid ratio of blood plasma. *See also* metabolic alkalosis; primary alkalosis; etc.

alkalotic. Of, or pertaining to, alkalosis.

alkane. A saturated aliphatic hydrocarbon.

alkaptonuria. A genetically inherited metabolic defect in man that is characterized by the urinary excretion of black melanin pigments formed from homogentisic acid; the defect is due to a deficiency of the enzyme homogentisic acid oxidase which functions in the metabolism of phenylalanine and tyrosine.

alkene. An unsaturated aliphatic hydrocarbon that contains one or more double bonds.

alkenyl group. The radical derived from an alkene, or from a derivative of an alkene, by removal of a hydrogen atom.

alkylating agent. One of a group of compounds, including the nitrogen and sulfur mustards, that alkylates specific sites of biologically important molecules such as DNA and protein. Alkylating agents are frequently carcinogenic, mutagenic, and immunosuppressive; they are classified as mono-, bi-, and polyfunctional depending on the number of reactive groups per molecule of alkylating agent.

alkylation. The introduction of an alkyl group into an organic compound.

alkyl group. The radical derived from an alkane, or from a derivative of an alkane, by the removal of a hydrogen atom.

alkyne. An unsaturated aliphatic hydrocarbon that contains one or more triple bonds.

alkynyl group. The radical derived from an alkyne, or from a derivative of an alkyne, by the removal of a hydrogen atom.

allantoic acid. The carboxylic acid that is the end product of purine catabolism in some teleost fishes.

allantoin. The heterocyclic compound that is the end product of purine catabolism in mammals other than primates, and in some reptiles.

allatum hormone. An insect hormone that affects differentiation after molting and that is required for vitellogenesis in the adult female.

allele. A specific form of a gene; one of several possible mutational forms of a gene.

allelic. Of, or pertaining to, an allele.

allelic allotype. ALLOTYPE.

allelic complementation. INTRAGENIC COMPLEMENTATION.

allelism test. COMPLEMENTATION TEST.

allelomorph. ALLELE.

Allen correction. A method of correcting absorbance measurements for the absorbance due to interfering substances. The absorbance is measured at the peak wavelength and at two other wavelengths, generally equidistant from the peak. A baseline is drawn by connecting the measurements on either side of the peak, and the absorbance at the peak is corrected by subtracting the baseline value at the peak. The correction assumes that the absorbance change is linear between the three points.

Allen's test. A modification of Fehling's test for glucose in urine; the urine is added to boiling Fehling's solution and turbidity develops as the solution is cooled.

allergen. An antigen that produces an allergic response.

allergic contact dermatitis. An inflammation of the skin that is due to an allergic response brought about by exposure of the skin to a chemical sensitizer.

allergic response. The formation and the reactions of antibodies that occur when a sensitized animal is exposed to an allergen.

allergy. HYPERSENSITIVITY.

allo-. 1. Combining form meaning other or dissimilar. 2. Combining form referring to an isomeric form such as an enantiomer of a compound that has more than one pair of enantiomers, or the more stable form of two geometrical isomers. 3. Combining form referring to a dissimilar genome.

allogeneic. Referring to genetically dissimilar individuals of the same species.

allogeneic inhibition. The destruction of cells that is apparently nonimmunological and that is brought about by contact with genetically different cells or with extracts from such cells.

allogenic. Variant spelling of allogeneic.

allograft. A transplant from one individual to a genetically dissimilar individual of the same species.

allograft reaction. The immune reaction whereby an allograft is rejected.

allomerism. The variation in the chemical composition of substances that have the same crystalline form.

allomerization. The oxidation of chlorophyll by air in the presence of alkali.

allomorphism. The variation in the crystalline form of substances that have the same chemical composition.

allophycocyanin. A red accessory pigment of algal chloroplasts that consists of a protein conjugated to a phycobilin.

alloplex interaction. The interaction that takes place when a disordered protein molecule undergoes refolding upon contact with another protein molecule.

all-or-none. Descriptive of a reaction or a response that occurs either to its fullest extent or does not occur at all. The highly cooperative, thermal denaturation of DNA and the dose-response of an animal to a drug are two examples.

all-or-none model. SYMMETRY MODEL.

allosteric. Pertaining to two or more topologically distinct sites on the same protein molecule.

allosteric activation. The activation of a regulatory enzyme by a positive effector.

allosteric effector. *See* effector.

allosteric enzyme. 1.REGULATORY ENZYME. 2. An enzyme that is an allosteric protein.

allosteric inhibition. The inhibition of a regulatory enzyme by a negative effector.

allosteric interaction. The interaction of either a regulatory enzyme or an allosteric protein with an effector.

allosteric protein. A protein that has two or more topologically distinct binding sites such that the binding of ligands (effectors) to these sites alters the properties of the protein.

allosteric site. REGULATORY SITE.

allosteric transition. The conformational change of a regulatory enzyme or of an allosteric protein as a result of its interaction with an effector.

allotopic. Of, or pertaining to, allotopy.

allotopy. The phenomenon of a substance, such as an enzyme, possessing different properties when it exists in a particulate or in a soluble form.

allotropy. The phenomenon of an element existing in different forms in the same phase; the different crystal forms of phosphorus and the molecular forms of oxygen and ozone are two examples.

allotype. One of a group of structurally and functionally similar proteins of the same species, such as the immunoglobulins, that have different antigenic properties; such proteins are under the control of one genetic locus but are produced by different alleles of the same gene. *Aka* allelic allotype.

allotypy. The occurrence of allotypes.

alloxan diabetes. An experimentally produced diabetes in which the level of insulin in an animal is lowered through preferential destruction of the insulin-producing cells of the pancreas by the administration of the pyrimidine drug alloxan.

allozyme. One of a group of enzymes that are produced by alleles of the same gene.

all-trans retinal. The isomeric form of retinal that is produced by light from the 11-*cis* isomer.

allysine. A derivative of lysine in which the ϵ-amino group has been converted to an aldehyde group; allysine undergoes an aldol condensation with hydroxyallysine during the cross-linking of collagen chains.

alpha. 1. Denoting the first carbon atom next to the carbon atom that carries the principal functional group of the molecule. 2. Denoting a specific configuration of the substituents at the anomeric carbon in ring structures of carbohydrates. 3. Denoting observed rotation (α) and specific rotation ($[\alpha]$) in optical rotation. *Sym* α.

alpha amylase. The enzyme that catalyzes the random hydrolysis of α (1 $\rightarrow$ 4) glycosidic bonds in starch.

alpha amylose. AMYLOSE.

alpha chain. 1. The heavy chain of the IgA immunoglobulins. 2. One of the two types of polypeptide chains present in adult hemoglobin.

alpha decay. The radioactive disintegration of an atomic nucleus that results in emission of an alpha particle.

alpha-fetoprotein. A protein in blood plasma that, when present at higher than normal concentrations, has been linked to the occurrence of certain cancers. *Abbr* AFP.

alpha helix. A coil- or spring-like configuration of protein molecules that occurs particularly in globular proteins. In this configuration, the polypeptide chain is held together by means of intrachain hydrogen bonds between the $>CO$ and $>NH$ groups of peptide bonds in such a fashion that there are 3.6 amino acid residues per turn of the helix, that the rise per residue is 1.5 Å, and that the pitch of the helix is 5.4 Å; each $>CO$ group is hydrogen-bonded to the $>NH$ group of the third residue behind it in the chain. The helix may be left- or right-handed depending on whether it is twisted in the manner of a left- or a right-handed screw. The right-handed alpha helix is the configuration most commonly encountered in proteins.

alpha keratin. The helical form of keratin in which the polypeptide chains are in the alpha helical configuration.

alpha lactalbumin. The *B* protein of the enzyme lactose synthetase.

alpha lipoprotein. HIGH-DENSITY LIPO-PROTEIN.

alpha method. DEAN AND WEBB METHOD.

alpha oxidation. A minor pathway for the oxidation of fatty acids in germinating plant seeds.

alpha particle. 1. A subatomic particle consisting of two protons and two neutrons; the alpha particle is identical to the nucleus of the helium atom and is frequently emitted by radioactive isotopes. 2. A cluster of glycogen granules in the liver; the granules are referred to as beta particles.

alpha plateau. The low-potential portion of the characteristic curve of a proportional radiation detector at which the count rate is almost independent of the applied voltage, and at which the potential is of sufficient magnitude to detect alpha particles.

alpha radiation. A radiation consisting of alpha particles.

alpha ray. A beam of alpha particles.

alpha threshold. The lowest potential at which alpha particles can be detected with a proportional radiation detector.

ALS. Antilymphocyte serum.

alum. A double sulfate salt of aluminum and either a monovalent metal or an ammonium ion.

alumina. Aluminum oxide; an adsorbent used in column chromatography.

alumina gel. A gel prepared from ammonium sulfate and aluminum sulfate and used in the purification of proteins by adsorption chromatography.

aluminum adjuvant. An aluminum compound, such as aluminum hydroxide, aluminum phosphate, or alum, that functions as an adjuvant in alum precipitation.

alum precipitated toxoid. A toxoid precipitated with an aluminum adjuvant. *Abbr* APT.

alum precipitation. An immunochemical technique in which soluble antigens are mixed with aluminum adjuvants to form a precipitate. When injected into an animal, the precipitate forms a depot from which the antigen is slowly released.

alveolar. Of, or pertaining to, alveoli.

alveolus (*pl* alveoli). One of a large number of air cells in the lung through which the gas exchange of respiration takes place.

Amadori rearrangement. The isomerization of *N*-substituted aldosylamines into *N*-substituted 1-amino-1-deoxy-2-ketoses; occurs in the Maillard reaction, in the reaction of carbohydrates with phenylhydrazine, and in the biosynthesis of pteridines.

amaurotic familial idiocy. TAY-SACHS DISEASE.

amber codon. The codon UAG, one of the three termination codons.

Amberlite. Trademark for a group of ion-exchange resins.

amber mutant. A conditional lethal mutant that contains an amber codon in a gene with a vital function.

amber mutation. A mutation in which a codon is mutated to the amber codon, thereby causing the premature termination of the synthesis of a polypeptide chain.

amber suppression. The suppression of an amber codon.

ambient conditions. The conditions, such as temperature and pressure, of the surrounding environment.

ambiguity. The occurrence of mistakes in protein synthesis, particularly in in vitro systems, such as the incorporation of one amino acid in response to a codon for a different amino acid.

ambiguous codon. A codon that can lead to the incorporation of more than one amino acid.

ambivalent codon. A codon that is expressed in some mutants as a result of suppression but that is not expressed in other mutants; a nonsense codon; a termination codon.

ambivalent mutation. NONSENSE MUTATION.

amboceptor. A term introduced by Ehrlich to describe hemolysin, an antibody that possesses two different binding sites, one for the antigen and one for complement.

amelogenin. A protein in dental enamel.

amethopterin. A folic acid analogue that inhibits the enzyme dihydrofolate reductase and that is used in the treatment of leukemia.

amidation. The introduction of an amide group into an organic compound.

amide group. The radical $-CONH_2$, derived from an acid by replacement of the OH of the carboxyl group with an amino group.

amidinotransferase. The enzyme that catalyzes the transamidination reaction in which a guanido group is transferred from arginine to glycine.

amination. The introduction of an amino group into an organic compound.

amine. A basic organic compound derived from ammonia by substitution of one or more organic radicals for the hydrogens. The amine is designated as a primary, a secondary, or a tertiary amine depending on whether one, two, or three organic radicals have been substituted for the hydrogen atoms in ammonia.

amino acid. An organic compound that contains both a basic amino group and an acidic carboxyl group. The alpha amino acids, in which the amino group is attached to the alpha carbon, are the building blocks of peptides and proteins. The amino acids are commonly classified either as (a) neutral, basic, or acidic, or as (b) nonpolar, polar and uncharged, or polar and charged; the presence or absence of a charge on the amino acid refers to that at pH 7.0.

amino acid accepting RNA. TRANSFER RNA.

amino acid activating enzyme. AMINOACYL-tRNA SYNTHETASE.

amino acid activation. A set of two reactions, catalyzed by an aminoacyl-tRNA synthetase, whereby an amino acid becomes covalently linked first to AMP and then to a specific transfer RNA molecule.

amino acid analysis. The analytical determination of both the relative amounts and the types of the amino acids in a peptide or in a protein.

amino acid analyzer. An instrument for the automated amino acid analysis of peptide and protein hydrolysates. The amino acids are separated by ion-exchange chromatography and are quantitatively determined by colorimetry.

amino acid arm. The base-paired segment in the clover leaf model of transfer RNA to which the amino acid is covalently linked; the segment contains both the 5′- and the 3′-ends of the transfer RNA and the amino acid is attached to the 3′—OH at the 3′-end.

amino acid composition. The makeup of a peptide or a protein in terms of both the relative amounts and the types of its constituent amino acids.

amino acid incorporation. The in vivo or in vitro reactions whereby amino acids become constituents of proteins as a result of protein synthesis.

amino acid nitrogen. The nitrogen of the amino acids in serum. *Abbr* AAN.

amino acid oxidase. An enzyme that catalyzes the oxidative deamination of amino acids. An L-amino acid oxidase is specific for L-amino acids and is a flavoprotein having FMN as a prosthetic group; a D-amino acid oxidase is specific for D-amino acids and is a flavoprotein having FAD as a prosthetic group.

Amino acid replacement. The substitution of one amino acid for another at a position in a polypeptide chain as a result of a mutation in the corresponding codon. *See also* conservative substitution; radical substitution.

amino acid residue. That portion of an amino acid that is present in a peptide or a polypeptide; the amino acid minus the atoms that are removed from it in the process of linking it to other amino acids by means of peptide bonds. Depending on its position in the peptide or in the polypeptide chain, the amino acid loses a hydrogen atom, a hydroxyl group, or a molecule of water as it becomes linked to the other amino acids.

amino acid sequence. The linear order of the amino acids as they occur in a peptide or in a protein; the amino acid sequence is conven-tionally written with the N-terminal amino acid on the left and with the C-terminal amino acid on the right.

amino acid sequencer. *See* sequenator.

amino acid side chain. The atoms of the amino acid molecule exclusive of the alpha carbon atom and its hydrogen atom, the alpha amino group, and the carboxyl group.

amino acid starvation. *See* starvation (2).

amino acid:tRNA ligase. AMINOACYL-tRNA SYNTHETASE.

aminoaciduria. The presence of excessive amounts of amino acids in the urine.

aminoacyl-. Combining form denoting an amino acid that is esterified through its carboxyl group to another molecule.

aminoacyl adenylate. An amino acid that has been esterified through its carboxyl group to the phosphate group of AMP; an intermediate in the activation of an amino acid to the aminoacyl-tRNA. *Abbr* AA-AMP.

aminoacyl site. The site on the ribosome at which the incoming aminoacyl-tRNA is bound during protein synthesis.

aminoacyl-tRNA. An amino acid that has been esterified through its carboxyl group to the 3′-hydroxyl group of the terminal adenosine at the 3′-end of a transfer RNA molecule; aminoacyl-tRNA is the form in which an amino acid is transported to the ribosomes for protein synthesis. *Abbr* AA-tRNA; AA-tRNA^{AA}. *Aka* aminoacylated-tRNA.

aminoacyl-tRNA site. AMINOACYL SITE.

aminoacyl-tRNA synthetase. The enzyme that catalyzes the coupled reactions of amino acid activation whereby an amino acid is first attached to AMP to form an aminoacyl adenylate, and is then attached to a transfer RNA molecule to form an aminoacyl-tRNA molecule.

aminoadipic pathway. A biosynthetic pathway of lysine that proceeds by way of α-aminoadipic acid and occurs in fungi.

p-**aminobenzoic acid.** A component of folic acid that is generally classified with the B vitamins, since it is a growth factor for certain microorganisms. *Abbr* PABA; PAB.

γ-**aminobutyrate bypass.** A reaction sequence for the conversion of α-ketoglutaric acid to succinic acid that differs from the normal sequence in the citric acid cycle and occurs in brain tissue.

γ-**aminobutyric acid.** A fatty acid derivative that functions in the metabolism of brain.

aminoglycoside antibiotics. A group of antibiotics that includes streptomycin, kanamycin, and neomycin.

amino group. The radical—NH_2.

p-**aminohippuric acid.** A compound used for renal clearance tests. *Abbr* PAH.

δ-**aminolevulinic acid.** A key intermediate in the

biosynthesis of porphyrins in which two molecules of δ-aminolevulinic acid condense to form the pyrrole porphobilinogen. *Abbr* DALA.

aminopeptidase. An exopeptidase that catalyzes the sequential hydrolysis of amino acids in a polypeptide chain from the N-terminal.

aminopterin. A folic acid analogue that inhibits the enzyme dihydrofolate reductase and that is used in the treatment of leukemia.

2-aminopurine. A purine analogue that is incorporated into nucleic acids and thereby produces transitions. *Abbr* AP.

p-aminosalicylic acid. An analogue of *p*-aminobenzoic acid that is used in the treatment of tuberculosis. *Abbr* PAS.

amino sugar. A monosaccharide in which one or more hydroxyl groups have been replaced by an amino group.

amino terminal. N-TERMINAL.

aminotransferase. TRANSAMINASE.

ammonia. A colorless gas that is the major form in which nitrogen is utilizable by living cells. Ammonia is the first compound formed in biological nitrogen fixation and is also the end product of purine catabolism in some marine invertebrates and in crustaceans.

ammonia fixation. A group of three reactions, one or more of which occur in every organism, whereby ammonia is converted to glutamic acid, glutamine, or carbamyl phosphate.

ammonification. The formation of ammonia by the degradation of organic compounds.

ammonium sulfate fractionation. A fractional precipitation by means of ammonium sulfate that is used in the purification of enzymes and other proteins.

ammonolysis. The cleavage of a covalent bond of an acid derivative by reaction with ammonia so that one of the products combines with the hydrogen atom and the other combines with the amino group of ammonia.

ammonotelic organism. An organism, such as a teleost fish, which excretes the nitrogen from amino acid catabolism primarily in the form of ammonia.

amnion cell. A cell of the epithelial membrane that forms the fluid-filled sac in which the embryo of mammals and higher vertebrates develops.

amniotic fluid. The fluid that fills the membranous sac enclosing the fetus.

amorph. A mutant allele that has little or no effect on the expression of a trait compared to the effect that the wild-type allele has.

amorphous. 1. Noncrystalline; devoid of a regular shape and a molecular lattice structure. 2. Lacking a definite form or organization; descriptive of nonhelical regions in macromolecules.

AMP. 1. Adenosine monophosphate (adenylic acid). 2. Adenosine-5′-monophosphate (5′-adenylic acid).

ampere. A unit of electrical current intensity; equal to the constant current that, when passed through a standard aqueous solution of silver nitrate, deposits silver at the rate of 0.001118 g/sec. *Sym* A.

amperometric titration. A titration in which either the titrant or the substance being titrated is electroactive and the limiting current is plotted as a function of added titrant.

amphetamine. The drug, 1-phenyl-2-aminopropane, that stimulates the central nervous system and inhibits sleep.

amphibaric. Descriptive of a pharmacologically active substance that can either lower or raise the blood pressure depending on its dose or concentration.

amphibolic metabolic pathway. CENTRAL METABOLIC PATHWAY.

amphipathic. Descriptive of a molecule that has both pronounced polar and pronounced nonpolar groups.

amphiphilic. AMPHIPATHIC.

amphiprotic. Descriptive of a compound that can either gain or lose protons; synonymous with amphoteric if acids are defined as proton donors and bases as proton acceptors.

amphiprotic solvent. A nonaqueous solvent that can act either as a proton donor or as a proton acceptor with respect to the solute.

ampholyte. An amphoteric electrolyte.

amphoteric. Descriptive of a compound that has at least one group that can act as an acid and one group that can act as a base; a compound that can act either as a proton donor or as a proton acceptor. Synonymous with amphiprotic if acids are defined as proton donors and bases as proton acceptors.

amplification. *See* cascade mechanism; gas amplification; gene amplification.

amplitude. The maximum displacement of an oscillation, a vibration, or a wave.

amu. Atomic mass unit.

amyelination. The failure to form myelin.

A myeloma protein. An abnormal immunoglobulin of the IgA type that is produced by individuals suffering from multiple myeloma.

amylase. An enzyme that catalyzes the hydrolysis of starch.

amylo-. Combining form meaning starch.

amyloclastic. AMYLOLYTIC.

amyloclastic method. A method of assaying for the enzyme amylase by determining the amount of unhydrolyzed starch that remains after incubation of the starch with the enzyme.

amylodextrin. SOLUBLE STARCH.

amyloid. A complex proteinaceous material, believed to be a glycoprotein, that gives a starch-like reaction with iodine and that is deposited in blood vessels and other tissues under certain pathological conditions.

amyloidosis. A pathological condition characterized by the formation of amyloid deposits.

amylolysis. The hydrolysis of starch.

amylolytic. Of, or pertaining to, amylolysis.

amylometric method. A method of assaying for the enzyme amylase by determining the amount of starch that is hydrolyzed during incubation of the starch with the enzyme.

amylopectin. The form of starch that is composed of branched chains of glucose units which are joined by means of $\alpha(1 \rightarrow 4)$ and α $(1 \rightarrow 6)$ glycosidic bonds.

amyloplast. A starch-storing plastid.

amylopsin. The α-amylase present in the pancreatic juice.

amylose. The form of starch that is composed of long, unbranched chains of glucose units which are joined by means of α $(1 \rightarrow 4)$ glycosidic bonds. *Aka* alpha amylose.

amylose synthetase. The enzyme that catalyzes the synthesis of amylose from ADP-glucose.

amytal. The barbiturate drug, 5-ethyl-5-isoamylbarbituric acid, that inhibits the electron transport system between the flavoproteins and coenzyme Q.

anabiosis. ANHYDROBIOSIS (2).

anabolic. Of, or pertaining to, anabolism.

anabolism. 1. The phase of intermediary metabolism that encompasses the biosynthetic and energy-requiring reactions whereby cell components are produced. 2. The cellular assimilation of macromolecules and complex substances from low-molecular weight precursors.

anacidity. 1. A lack of acidity, particularly the lack of gastric hydrochloric acid. 2. The pathological condition due to a lack of gastric hydrochloric acid.

anaerobe. *See* facultative anaerobe; obligate anaerobe.

anaerobic. 1. In the absence of oxygen; in an environment or an atmosphere devoid of oxygen. 2. Not requiring the presence of molecular oxygen for growth. 3. Not capable of using molecular oxygen for growth. *See also* anoxybiontic.

anaerobic-aerotolerant. MICROAEROPHILIC.

anaerobic fermentation. *See* fermentation (2).

anaerobic glycolysis. The group of cellular reactions whereby glucose is converted to lactic acid.

anaerobic respiration. The energy-yielding metabolic breakdown of organic compounds in an organism that proceeds in the absence of molecular oxygen and with the use of inorganic compounds as oxidizing agents. *See also* fermentation (2).

anaerobiosis. Life under anaerobic conditions.

anaerobiotic. Of, or pertaining to, anaerobiosis.

analbuminemia. A genetically inherited metabolic defect in man that is characterized by an impaired synthesis of serum albumin.

analgesia. The relief of pain without loss of consciousness.

analgesic. 1. *n* An agent that brings about analgesia. 2. *adj* Of, or pertaining to, analgesia.

analog computer. A computer that receives information in the form of continuous variables, such as temperature, pressure, and flow, and that processes the information by translating each variable into an analogous or a related mechanical or electrical variable, such as voltage.

analogous. Having a similar function and a similar, but not identical, structure.

analogous enzyme variants. Enzyme variants that differ significantly in their molecular structures and catalytic properties.

analogue. A compound that is structurally similar to another compound and that is used for such purposes as the determination of structural prerequisities of enzyme substrates, the competitive inhibition of specific enzymatic and other reactions, and the synthesis of altered macromolecules. *Var sp* analog.

analysis of covariance. A statistical analysis for determining the variability in the principal variable that is due to variability in some other factor.

analysis of variance. A statistical analysis for segregating the sources of variability in measurements, as in determining the extent to which the variability in sets of observations is due to differences between the sets and the extent to which it is due to random variations.

analytical biochemistry. A branch of biochemistry that deals with the qualitative and quantitative determination of substances in living systems.

analytical method. A method, such as ultracentrifugation, electrophoresis, or chromatography, that requires relatively small amounts of sample and that is used primarily for the identification and characterization of specific substances.

analytical ultracentrifuge. A high-speed centrifuge, equipped with one or more optical systems, that is used for measurements of sedimentation coefficients and molecular weights as well as for a variety of studies of macromolecules. The centrifuge is capable of generating speeds of approximately 60,000 rpm and centrifugal forces of approximately 500,000 $\times$ g. The optical systems used in conjunction with the analytical ultracentrifuge are a schlieren optical system, an absorption optical system, and an interferometric optical system.

analyzer. The nicol prism in a polarimeter that is used for determining the rotation of the plane-polarized light. *See also* polarizer.

anamnestic response. SECONDARY IMMUNE RESPONSE.

anaphase. The third stage in mitosis during which the chromosomes move to opposite poles.

anaphoresis. 1. The movement of charged particles toward the anode. 2. ELECTROPHORESIS.

anaphylactic response. The immune reactions of anaphylaxis.

anaphylactic shock. A severe and generalized form of anaphylaxis that is characterized by violent cardiac and respiratory symptoms and that may be produced by the injection of a substance to which an individual is either allergic or sensitized.

anaphylactoid reaction. A condition that resembles an anaphylactic shock but that is not caused by an immunological reaction.

anaphylatoxin. A pharmacologically active substance, apparently a polypeptide fragment of complement, that can cause the release of histamine from mast cells in anaphylaxis.

anaphylaxis. A hypersensitivity in which the first administration of an antigen to an animal is harmless, but the second administration leads to an intense secondary immune response accompanied by pathological reactions. *See also* active anaphylaxis; passive anaphylaxis; reverse passive anaphylaxis.

anaplasia. The loss by a cell of its characteristic structure accompanied by its reversion to a more primitive, embryonic type.

anaplastic. Of, or pertaining to, anaplasia.

anaplerotic reaction. A reaction whereby a metabolic intermediate is replenished; this is generally achieved through the insertion of either a one-carbon fragment, in the form of carbon dioxide, or a two-carbon fragment, in the form of acetyl coenzyme A, into the appropriate metabolic reaction.

anatoxin. TOXOID.

anchimeric assistance. The enhanced reactivity of an ion, the charge of which is divided between two adjacent atoms. *Aka* anchimeric acceleration.

anchorage dependence. The difference between the extent of cellular transformation that is produced by polyoma virus with cells that are planted in agar and with cells that are suspended in a viscous medium.

Andersen's disease. GLYCOGEN STORAGE DISEASE TYPE IV.

androgen. 1. A 19-carbon steroid that is a male sex hormone or one of its metabolites. 2.

Any 19-carbon steroid. *See also* male sex hormone.

androstane. The parent ring system of the androgens.

androsterone. A major metabolite of testosterone.

anemia. A condition in which the number of red blood cells, the volume of red blood cells, or the hemoglobin content of the blood are below normal levels. *See also* hemolytic anemia; hypochromic anemia; pernicious anemia; sickle cell anemia.

anemic. Of, or pertaining to, anemia.

anergy. The total absence of an allergic response in an animal under conditions that would otherwise be expected to lead to such a response.

anesthetic drug. A drug that induces either a local, or a total, loss of sensation in the body.

aneuploid state. The chromosome state in which there is a loss or a gain of single chromosomes, and the chromosome number is not an exact multiple of the basic number in the genome. *Aka* aneuploidy.

aneurin. THIAMINE.

aneurysm. 1. A blood-containing tumor connected directly with the lumen of an artery. 2. A circumscribed dilatation of an artery.

ANF. Antinuclear factor.

angioma. A tumor consisting chiefly of blood or lymphatic vessels.

angiotensin I. The inactive decapeptide precursor of angiotensin II; it is cleaved off from angiotensinogen in a reaction catalyzed by the enzyme renin.

angiotensin II. The active octapeptide formed from angiotensin I by hydrolytic removal of two amino acids in a reaction catalyzed by the serum converting enzyme; a powerful hypertensive agent.

angiotensinogen. The α_2-globulin from which the decapeptide angiotensin I is cleaved off in a reaction catalyzed by the enzyme renin.

angiotonin. ANGIOTENSIN.

angle rotor. A centrifuge rotor in which the tubes containing solution are held at a fixed angle. Such rotors are used for the preparative fractionation of macromolecules and their efficiency is due to the fact that convection is superimposed upon sedimentation in the tube. *Aka* angle head.

angstrom unit. A unit of length equal to 10^{-8} cm and used in describing atomic and molecular dimensions. *Abbr* A.U.; Å; A. *Aka* angstrom.

angular methyl group. A methyl group attached to the perhydrocyclopentanophenanthrene ring system of steroids.

angular velocity. The velocity of rotation expressed in terms of the central angle, in radians, transversed per unit time.

anhaptoglobinemia. The lack of sufficient amounts of haptoglobins in the blood.

anhydride. *See* acid anhydride.

anhydrobiosis. 1. Life in the absence of water. 2. A state of suspended animation shown by some organisms in which they can sustain the removal of all, or almost all, of their cellular water and return to normal living conditions when resupplied with water.

anhydrous. Devoid of water.

animal cephalin. PHOSPHATIDYL SERINE.

animal charcoal. BONEBLACK.

animal hormone. *See* hormone.

animal protein factor. VITAMIN B_{12}.

animal saponin. A sulfur-containing steroid glycoside that has properties of a plant saponin but is isolated from a marine invertebrate.

animal starch. GLYCOGEN.

animal toxin. A toxin of animal origin, such as that in snake venom.

animal virus. A virus that infects animal cells and multiplies in them.

anion. A negatively charged ion.

anion exchanger. A positively charged ion-exchange resin that binds anions.

anionic detergent. A surface-active agent in which the surface-active part of the molecule carries a negative charge. *Aka* anionic surface-active agent.

anion respiration. The phenomenon that exposure of plant tissues to salt solutions frequently leads to an increase in respiration which appears to be proportional to the rate of anion absorption by the plant.

anisotropic. Of, or pertaining to, anisotropy.

anisotropic band. A BAND.

anisotropy. The variation in the physical properties of a substance as a function of the direction in which these properties are measured. *Aka* anisotropism.

annealing. 1. The renaturation of heat-denatured proteins or heat-denatured nucleic acids by slow cooling. 2. The formation of hybrid nucleic acid molecules, containing paired strands from different sources, by slow cooling of a mixture of denatured nucleic acids. 3. The tempering of glass in glass blowing by slow cooling.

annular. Ring-shaped.

anode. The positive electrode by which electrons leave the solution of an electrolyte and toward which the anions move in solution.

anodic. 1. Of, or pertaining to, the anode. 2. Descriptive of a component that moves toward the anode in electrophoresis.

anomalous dispersion. An optical rotatory dispersion that cannot be expressed by a simple, one-term Drude equation; such a dispersion is generally expressed as $[m'] = a_0\lambda_0^2/(\lambda^2 - \lambda_0^2) + b_0\lambda_0^4/(\lambda^2 - \lambda_0^2)^2$, where $[m']$ is the reduced mean residue rotation, λ is the wavelength, and a_0, b_0, and λ_0 are constants.

anomalous osmosis. The electroosmotic flow of water through a charged membrane that is caused by a potential gradient across the membrane. The anomalous osmosis is said to be positive when the water moves from a dilute to a concentrated solution and is said to be negative when the flow of water is in the opposite direction.

anomer. One of two isomeric carbohydrates that differ from each other only in the configuration about the anomeric carbon of the ring structure.

anomeric carbon. The carbon atom of the carbonyl group in a carbohydrate.

anomeric effect. The stereochemical effect in carbohydrate chemistry in which the interaction between the oxygen of the monosaccharide ring and the substituent ($-OR$; $-O-CO-R$; or halogen) at the anomeric carbon is such as to favor the maximum separation between the oxygen and the substituent; as a result, the axial substituent, or α-anomer, is favored over the equatorial substituent, or β-anomer.

anovar. Acronym for analysis of variance.

anoxia. HYPOXIA.

anoxybiontic. Not capable of using molecular oxygen for growth. *Aka* anoxybiotic. *See also* anaerobic (2,3).

anserine. A dipeptide of β-alanine and methyl histidine that occurs in vertebrate muscle.

antagonism. The phenomenon in which the action of one agent is counteracted by the action of another agent that is present at the same time.

antagonist. A molecule, such as a drug, enzyme inhibitor, or hormone, that diminishes or prevents the action of another molecule.

ante-iso fatty acid. A fatty acid that is branched at the carbon atom preceding the penultimate carbon atom at the hydrocarbon end of the molecule.

ante-penultimate carbon. The third carbon atom from the end of a chain.

anterior. 1. In front of, or in the front part of, a structure. 2. Before, in relation to time.

anthocyanidin. The aglycone of an anthocyanin.

anthocyanin. A water-soluble plant pigment that occurs largely in the form of a glycoside of an anthocyanidin. Anthocyanins are flavonoids and are responsible for most of the red, pink purple, and blue colors of higher plants.

anthrone reaction. A colorimetric reaction for carbohydrates, particularly hexoses, that is based on the production of a green color on treatment of the sample with anthrone.

anti. 1. Referring to a nucleoside conformation in which the base has been rotated around the sugar, using the $C-N$ glycosidic bond as a pivot, so that the sugar is in direct opposition to the base. This represents a sterically less hindered conformation than the syn conformation; in polynucleotides, it leads to the bulky portions of the bases being pointed away from the sugar-phosphate backbone of the chain. 2. Referring to a trans configuration for certain compounds containing double bonds, such as the oximes

which contain the group $\diagdown C{=}N{-}OH$. 3. Referring to the position occupied by two radicals of a stereoisomer in which the radicals are farther apart as opposed to the syn position in which they are closer together. *See also* syn.

antiacrodynia factor. VITAMIN B_6.

antiantibody. An antibody produced in response to an antigenic determinant of an antibody molecule.

antiauxin. A compound that functions as a competitive inhibitor of auxin.

antiberiberi factor. VITAMIN B_1.

antibiosis. The association of two organisms in which one produces a substance, such as an antibiotic, or a condition that is harmful to the other.

antibiotic. Originally defined as a compound produced by a microorganism that inhibits the reproduction, or causes the destruction, of other microorganisms. Now more generally defined as a compound produced by a microorganism or a plant, or a close chemical derivative of such a compound, that is toxic to microorganisms from a number of other species. *See also* under individual antibiotics and classes of antibiotics, such as streptomycin and macrolide antibiotic.

anti-black-tongue factor. NICOTINIC ACID.

antibody. A protein of the globulin type that is formed in an animal organism in response to the administration of an antigen and that is capable of combining specifically with that antigen. *Abbr* Ab. *See also* immunoglobulin.

antibody-binding fraction. Fab FRAGMENT.

antibody combining site. One of at least two sites on the antibody molecule to which a complementary portion of an antigen, the antigenic determinant, becomes bound in the course of an antigen-antibody interaction; the active site of an antibody.

antibody diversity. ANTIBODY HETEROGENEITY.

antibody-excess zone. A zone in the precipitin curve of the antigen-antibody reaction in which the amount of antibody precipitated increases with increasing amounts of antigen.

antibody fixation. The binding of antibodies to cell receptors in immediate-type hypersensitivity.

antibody formation. *See* theory of antibody formation.

antibody heterogeneity. The state of a given preparation of antibodies in which the antibodies differ with respect to size, structure, charge, or other properties.

antibody response. IMMUNE RESPONSE.

antibody specificity. *See* specificity (2).

antibody titer. The highest dilution of an antiserum that will produce detectable precipitation or agglutination when reacted with antigens.

antibody valence. The number of combining sites, of which there are at least two, per antibody molecule.

antibonding orbital. A molecular orbital in which there is a node of electron density between the bonding atomic nuclei, resulting in a weakening of the bond between the nuclei. Antibonding orbitals are generally of higher energy than sigma (σ) and pi (π) orbitals and are designated sigma star (σ^*) and pi star (π^*).

anticancer compound. A compound that arrests or reverses the growth of a malignant tumor.

anticarcinogenesis. The inhibition of the action of one carcinogen by the simultaneous administration of a second carcinogen.

anticholinesterase. An inhibitor of cholinesterase.

anticoagulant. A substance that prevents the clotting of blood; most anticoagulants function by binding calcium ions.

anticodon. A sequence of three nucleotides in transfer RNA that, in the process of protein synthesis, binds to a specific codon in messenger RNA by complementary base pairing.

anticodon arm. The base-paired segment in the clover leaf model of transfer RNA to which the loop, containing the anticodon, is attached.

anticompetitive inhibition. UNCOMPETITIVE INHIBITION.

anticomplementary. Referring to a treatment or an agent that either removes or inactivates a component of complement.

anticomplement fluorescent antibody technique. A fluorescent antibody technique in which an antigen-antibody complex is reacted with complement and the entire aggregate is then stained by means of fluorescent antibodies to complement.

antidermatosis vitamin. PANTOTHENIC ACID.

antidiuresis. A decrease in the excretion of urine.

antidiuretic. 1. *n* An agent that decreases the excretion of urine. 2. *adj* Of, or pertaining to, antidiuresis.

antidiuretic hormone. VASOPRESSIN.

antidotal agent. ANTIDOTE.

antidotal therapy. Therapy by means of antidotes.

antidote. An agent that limits or reverses the effect of a poison.

anti-egg-white-injury factor. BIOTIN.

anti-enzyme. An antibody to an enzyme.

antifatty-liver factor. LIPOCAIC.

antifoam. A chemical substance added to liquid cultures of microorganisms to minimize foam formation during growth.

antifolate. An antimetabolite of folic acid or of a derivative of folic acid. *Aka* antifolic acid agent.

antigen. A substance, frequently a protein, that can stimulate an animal organism to produce antibodies and that can combine specifically with the antibodies thus produced; called also complete antigen as distinct from a hapten. *Abbr* Ag.

antigen-antibody complex. The generally insoluble molecular aggregate that is formed by the specific interaction of antigens and antibodies.

antigen-antibody lattice. *See* lattice theory.

antigen-antibody reaction. PRECIPITIN REACTION.

antigen binding capacity. The total antibody concentration in an antiserum based on a determination of the amount of antigen bound by a given volume of the antiserum. *Abbr* ABC.

antigen binding site. ANTIBODY COMBINING SITE.

antigen-excess zone. A zone in the precipitin curve of the antigen-antibody reaction in which the amount of antibody precipitated decreases with increasing amounts of antigen.

antigenic competition. The decrease in the immune response to one antigen that is produced by the administration of a second antigen.

antigenic conversion. *See* conversion.

antigenic deletion. The cellular loss of antigenic determinants, or the masking of existing cellular antigenic determinants.

antigenic determinant. That portion of the antigen molecule that is responsible for the specificity of the antigen in an antigen-antibody reaction and that combines with the antibody combining site to which it is complementary.

antigenic gain. The cellular acquisition of new antigenic determinants, or the unmasking of existing cellular antigenic determinants.

antigenicity. The capacity of an antigen to stimulate the formation of specific antibodies.

antigenic sin. *See* doctrine of original antigenic sin.

antigen template theory. An instructive theory of antibody formation according to which antigens taken up by a cell serve as templates for the synthesis of antibodies by that cell. The

antigens are considered to bind to ribosomes or to messenger RNA, thereby modifying translation so that antibodies are formed, the combining sites of which are complementary to the antigenic determinants of the bound antigens.

antigen tolerance. IMMUNOLOGICAL TOLERANCE.

antigen valence. The number of antigenic determinants per antigen molecule; an antigen molecule may have one valence with respect to one antibody and have a different valence with respect to another antibody.

antiglobulin method. INDIRECT FLUORESCENT ANTIBODY TECHNIQUE.

antiglobulin test. COOMBS' TEST.

anti-gray-hair factor. *p*-AMINOBENZOIC ACID.

antihemophilic factor. A factor in the intrinsic system of blood clotting that is activated by the Christmas factor. *Abbr* AHF. *Aka* antihemophilic factor A; antihemophilic globulin.

antihemophilic factor B. CHRISTMAS FACTOR.

antihemophilic factor C. PLASMA THROMBOPLASTIN ANTECEDENT.

antihemophilic globulin. ANTIHEMOPHILIC FACTOR.

antihemorrhagic vitamin. VITAMIN K.

antihistamine. A drug that blocks the action of histamine and that is used in the treatment of immediate-type hypersensitivity.

antihormone. An antibody to a hormone.

anti-infective vitamin. VITAMIN A.

anti-insulin. A compound, such as a sex hormone or a corticosteroid, that decreases the activity of insulin.

antilipotropic. Descriptive of a substance that has the capacity of diverting methyl groups from the synthesis of choline.

antilogarithm. The antilogarithm of X is that number the logarithm of which is X. *Abbr* antilog.

antilymphocyte globulin. The globulin fraction of antilymphocyte serum. *Abbr* ALG.

antilymphocyte serum. A serum that contains antibodies to lymphocytes and that is used as an immunosuppressive agent. *Abbr* ALS.

antimalarial. 1. *n* A drug used to prevent or treat malaria. 2. *adj* Preventing or curing malaria.

antimer. ENANTIOMER.

antimetabolite. A compound that competitively inhibits a specific enzymatic or other reaction in metabolism because of its similarity in structure to the natural metabolite which participates in the reaction. *See also* competitive inhibitor.

antimicrobial spectrum. The types of

microorganisms against which an antimicrobial drug is effective. *See also* sensitivity spectrum.

antimorph. 1. ENANTIOMER. 2. A mutant gene that has an effect opposite that of its corresponding wild-type gene.

antimutagen. A substance that counteracts the action of a mutagen by decreasing the rate of induced, and occasionally of spontaneous, mutations.

antimycin A. An antibiotic, produced by *Streptomyces griseus*, that inhibits the electron transport system between cytochromes *b* and c_1.

antineuritic factor. VITAMIN B_1.

antinuclear factor. An antibody against a constituent of the cell nucleus. *Abbr* ANF. *Aka* antinuclear antibody.

antioxidant. A substance, generally an organic compound, that is more readily oxidized than a second substance and hence can retard or inhibit the autoxidation of the second substance when added to it.

antiparallel chains. 1. Two polypeptide chains running in opposite directions, with the one progressing from the C-terminal to the N-terminal, and the other progressing in the opposite direction. 2. ANTIPARALLEL STRANDS.

antiparallel spin. The spin of two particles in opposite directions.

antiparallel strands. Two polynucleotide strands running in opposite directions, with the one progressing from the 3'-terminal to the 5'-terminal, and the other progressing in the opposite direction.

antipellagra factor. NICOTINIC ACID.

antipernicious anemia factor. VITAMIN B_{12}.

antipode. OPTICAL ANTIPODE.

antipolarity. The decrease that may occur in the synthesis of an enzyme if the enzyme is specified by a gene which precedes another gene that has undergone a polar mutation.

antiport. The linked transport in opposite directions of two substances across a membrane. *See also* symport; uniport.

antirachitic vitamin. VITAMIN D.

antiscorbutic factor. VITAMIN C.

antisense strand. CRICK STRAND.

antisepsis. The destruction and the prevention of growth of microorganisms causing disease, decay, or putrefaction.

antiseptic. Of, or pertaining to, antisepsis.

antiserum. A serum that contains antibodies and that has been obtained from an animal organism subsequent to its immunization with an antigen.

antisigma factor. A protein that is produced during the infection of *Escherichia coli* with T4 phage and that prevents recognition of the promoter by the sigma factor of DNA-dependent RNA polymerase.

antisterility factor. VITAMIN E.

antitermination factor. A protein that prevents the termination of RNA synthesis by DNA-dependent RNA polymerase.

antithyroid agent. An agent that inhibits thyroid function by affecting the synthesis, release, or utilization of thyroxine.

antitoxin. An antibody to a toxin that is capable of neutralizing a toxin.

antitumor antibiotic. An antibiotic that arrests or reverses the growth of a malignant tumor.

antitumor antimetabolite. An antimetabolite that arrests or reverses the growth of a malignant tumor.

antiviral protein. A protein, induced by interferon, that binds to ribosomes and inhibits the translation of either viral RNA or of the messenger RNA which is derived from viral DNA. *Abbr* AVP.

antivitamin. A structural analogue of a vitamin; a vitamin antagonist that acts as a competitive inhibitor of a vitamin.

antixerophthalmic factor. VITAMIN A.

anucleate. Lacking a nucleus.

anucleolate. Lacking a nucleolus.

anucleolate mutation. A mutation that produces a cell lacking the nucleolus organizer.

anuresis. The failure or the inability to void urine that is formed but which is retained in the urinary bladder.

anuria. The lacking or the defective excretion of urine due to a failure in the function of the kidneys.

AP. Aminopurine.

APA. Apurinic acid.

aperiodic polymer. A polymer consisting of nonidentical repeating units.

apholate. An aziridine mutagen that is used for the sterilization of insects.

aplasia. 1. The defective development of an organ or a tissue. 2. The absence of an organ or a tissue from the body.

apnea. *See* drug-induced apnea.

apo-. Combining form denoting the protein portion of a conjugated protein.

apocrine. A gland that contributes a part of its cytoplasm to its secretion.

apoenzyme. The protein portion of a conjugated enzyme.

apoferritin. The protein component of ferritin.

apolar. NONPOLAR.

apoprotein. The protein component of a conjugated protein.

aporepressor. REPRESSOR (1).

apparent competitive inhibition. A competitive inhibition of enzyme activity in which the presence of the inhibitor on the enzyme affects the affinity of the enzyme for the substrate.

apparent equilibrium constant. 1. An equilibrium constant based on molar concentrations rather than on activities. 2. An equilibrium constant calculated for a fixed pH for a reaction in which protons are produced; $K_{eq}/[H^+]$. *Sym* K′.

apparent specific volume. The change in volume per gram of solute when a known weight of solute is added to a known volume of solvent.

approach to sedimentation equilibrium. ARCHIBALD METHOD.

A protein. 1. A minor protein of the capsid of M13 phage. 2. A protein subunit of the enzyme tryptophan synthetase. 3. MATURATION PROTEIN.

aprotic solvent. A nonaqueous solvent that acts neither as a proton acceptor nor as a proton donor with respect to the solute.

APS. Adenosine-5′-phosphosulfate.

APT. Alum precipitated toxoid.

aptitude. The physiological state of a lysogenic bacterium that enables it, upon induction, to produce infectious phage particles.

apurinic acid. A DNA molecule from which the purines have been removed by mild acid hydrolysis. *Abbr* APA.

apyrase. The enzyme that catalyzes the hydrolysis of ATP to AMP and two molecules of orthophosphate.

apyrimidinic acid. A DNA molecule from which the pyrimidines have been removed by treatment with hydrazine.

aq. AQUEOUS.

aquametry. The quantitative determination of water.

aquated ion. AQUO-ION.

aquation. The formation of aquo-ions.

aqueous. Of, or pertaining to, water. *Abbr* aq.

aqueous humor. The clear fluid that fills the anterior chamber of the eye.

aqueous phase separator centrifugation. An interface centrifugation in which particles are selectively transferred from a crude aqueous solution to a short isodensity column; used for the purification of viruses.

aqueous solution. A solution having water as its principal solvent.

aquo-ion. A complex ion containing one or more water molecules.

A.R. Analytical reagent; denotes a chemical reagent for which either the actual or the maximum permissible concentrations of impurities are known.

arabinan. A homopolysaccharide of arabinose.

arabinose. An aldose having five carbon atoms.

arbovirus. An enveloped virus containing singlestranded RNA. Arboviruses multiply in both vertebrates and arthropods, with the arthropods generally serving as vectors.

arc. The line of antigen-antibody precipitate obtained in immunodiffusion and in immunoelectrophoresis.

Archeozoic era. One of the two subdivisions of the Precambrian era; the earliest period of geologic time and an era that is devoid of fossil remains, that extended over a period of about 600 million years, and that ended about 1.6 billion years ago.

Archibald method. A centrifugal method for determining molecular weights and assessing size homogeneity of macromolecules that is generally performed in the analytical ultracentrifuge, using relatively low speeds of rotation. The method is based on applying sedimentation equilibrium criteria to both the meniscus and the bottom positions of the cell, and on measuring the curvature of the gradient curve at those positions.

Arg. 1. Arginine. 2. Arginyl.

argentation chromatography. A chromatographic technique based on the rapid formation of loose complexes between silver ions in an adsorbent and the π-electrons of double and triple bonds in the molecules being separated; the greater the extent of unsaturation in a molecule, the greater is the extent of complex formation, and hence the slower is the rate of chromatographic migration. The method is used particularly for the separation of lipids.

arginase. The enzyme that catalyzes the hydrolysis of arginine to urea and ornithine in the urea cycle.

arginine. An aliphatic, basic, and polar alpha amino acid that contains the guanido group. *Abbr* Arg; R.

arginine cycle. UREA CYCLE.

arginine vasopressin. A vasopressin molecule in which the eighth amino acid residue has been replaced by an arginine residue. *Abbr* AVP.

arginine vasotocin. A vasotocin molecule in which the eighth amino acid residue has been replaced by an arginine residue.

argininosuccinic acidemia. A genetically inherited metabolic defect in man that is due to a deficiency of the enzyme argininosuccinase.

argininosuccinuria. A genetically inherited metabolic defect in man that is associated with mental retardation and that is characterized by a high blood concentration and a large renal excretion of argininosuccinic acid.

argon detector. An ionization detector, employed in gas chromatography, in which argon is used to ionize the organic compounds being separated; useful for trace analysis of steroids, fatty acids, and related compounds of relatively high molecular weights.

arithmetic mean. MEAN.

arm. A base-paired, helical segment in the clover leaf model of transfer RNA. There are four or five such segments per molecule, and they are referred to as amino acid arm, anticodon arm, variable arm, TψC arm, and dihydro-U arm.

Aroclor. Trademark for a mixture of polychlorinated biphenyls.

aromatic. Of, or pertaining to, a carbocyclic organic compound that contains the benzene nucleus.

aromatic amino acid. An amino acid that contains the benzene ring.

Arrhenius activation energy. *See* Arrhenius equation.

Arrhenius equation. An equation relating the rate constant k of a reaction to the absolute temperature T; specifically, $ln\ k = ln\ A - E/RT$, where R is the gas constant, A is a constant, and E is the activation energy of the reaction.

Arrhenius plot. A plot of the logarithm of the rate constant of a reaction versus the reciprocal of the absolute temperature; used for determining the activation energy of the reaction. *See also* Arrhenius equation.

arsenolysis. The cleavage of a covalent bond of an acid derivative by reaction with arsenic acid H_3AsO_4 so that one of the products combines with the H and the other product combines with the H_2ASO_4 group of the arsenic acid.

artefact. Variant spelling of artifact.

arterial. Of, or pertaining to, arteries.

arteriosclerosis. The hardening of the arteries.

artery. A blood vessel that transports blood from the heart to the tissues.

arthropod-borne virus. ARBOVIRUS.

Arthus reaction. An allergic reaction characterized by skin inflammation in response to repeated subcutaneous injections of antigen; similar reactions can also be produced in tissues other than skin. The Arthus reaction may be active, passive, or reverse passive; *see* active, passive, and reverse passive anaphylaxis for a definition of these terms.

artifact. Any structure or substance that is not representative of the in vivo state of the specimen or of the makeup of the original sample but which is, instead, a result of the isolation procedure, the handling, or other factors.

artificial induction. The induction of a prophage by a change in the conditions of the bacterial culture such that the immunity substance is either inactivated or not synthesized.

artificial kidney. HEMODIALYZER.

artificial nitrogen fixation. A synthetic reaction, such as the Haber process, that converts atmospheric nitrogen to ammonia.

artificial pH gradient. A pH gradient formed by the layering of two or more buffers of different pH values; used originally for isoelectric focusing but since it changes with time upon the ap-

plication of an electric field, it is useful only for experiments of short duration.

aryl group. An organic radical derived from an aromatic compound by loss of a hydrogen atom.

ascending boundary. The electrophoretic boundary that moves upward in one of the arms of a Tiselius electrophoresis cell. *See also* Tiselius apparatus.

ascending chromatography. A chromatographic technique in which the mobile phase moves upward along the support.

Aschheim-Zondek test. A test for pregnancy based on the injection of urine, voided during the early stages of pregnancy, into mice.

ascites. The abnormal accumulation of serous fluid in the abdominal cavity.

ascitic. Of, or pertaining to, ascites.

Ascoli test. 1. RING TEST. 2. A precipitin test for anthrax antigens.

ascorbic acid. VITAMIN C.

-ase. Combining form denoting an enzyme.

asepsis. The prevention of access of microorganisms causing disease, decay, or putrefaction to the site of a potential infection.

aseptic. Of, or pertaining to, asepsis.

A-site. AMINOACYL SITE.

A-site-P-site model. The model of translation according to which a ribosome possesses two binding sites; the aminoacyl-, or A-, site binds the incoming aminoacyl-tRNA and the peptidyl-, or P-, site binds the peptidyl-tRNA subsequent to the addition of each new amino acid to the growing polypeptide chain.

Asn. 1. Asparagine. 2. Asparaginyl.

Asp. 1. Aspartic acid. 2. Aspartyl.

asparagine. An aliphatic, polar alpha amino acid that is the amide of aspartic acid. *Abbr* Asn; AspNH$_2$; N.

aspartate transcarbamylase. A regulatory enzyme that catalyzes the first step in the biosynthesis of the pyrimidines in which N-carbamyl aspartic acid is formed from carbamyl phosphate and aspartic acid. The enzyme consists of two classes of subunits referred to as catalytic and regulatory subunits.

aspartic acid. An aliphatic, acidic, and polar alpha amino acid. *Abbr* Asp; D.

aspartic semialdehyde. A derivative of aspartic acid in which only one of the two carboxyl groups has been converted to an aldehyde group.

aspirin. Acetylsalicylic acid, an analgesic.

AspNH$_2$. Asparagine.

asporogenous mutant. A mutant that is unable to form spores. *Aka* asporogenic mutant.

assay. A measurement of either the concentration or the activity of a substance. *See also* enzyme assay.

assimilation. The conversion of nutrients by an organism into intra- and extracellular compounds utilized by that organism.

assimilation time. The average time required for the reduction of one molecule of carbon dioxide by a molecule of chlorophyll when a plant is exposed to bright light.

assimilatory reduction. The process in plants whereby sulfate and nitrate, after reduction, are assimilated into cellular organic compounds. *See also* respiratory reduction.

association colloid. A surface-active agent that tends to aggregate and to form micelles in solution.

association constant. 1. The equilibrium constant for the formation of a more complex compound from simpler components, as the association of a proton and an anion to form an acid. 2. The equilibrium constant for the formation of a complex containing one or more macromolecules, as the binding of an inhibitor to an enzyme, or the binding of an antigen to an antibody. *See also* affinity constant.

Asx. The sum of aspartic acid and asparagine when the amide content is either unknown or unspecified.

asymmetric. 1. Lacking symmetry; unsymmetric. 2. Descriptive of a molecule that is totally lacking in symmetry as contrasted with a dissymmetric one. 3. Descriptive of a macromolecule, the shape of which differs significantly from that of a sphere.

asymmetric carbon. A carbon atom to which are attached four different substituents.

asymmetric center. An asymmetric carbon atom or one of several identical asymmetric carbon atoms in a molecule.

asymmetric synthesis. The synthesis of only one of two optical isomers; generally the case for enzymatic, but not for nonenzymatic, reactions.

asymmetric unit. The smallest part of a structure which, when operated on by symmetry elements, will reproduce the complete structure. The asymmetric unit is equal to, or smaller than, the unit cell.

asymmetry factor. SHAPE FACTOR.

asymmetry potential. 1. The potential across two permselective membranes (M_1 and M_2) which are separated by a concentrated polyelectrolyte solution (B), and which have another, but identical, electrolyte solution (A) on the other side of each membrane (i.e., A-M_1-B-M_2-A). 2. The potential of a membrane electrode, such as the glass electrode, that arises from slight imperfections in the membrane.

asymmetry ratio. The sum of the concentrations of adenine and thymine divided by the sum of the concentrations of cytosine and guanine for a given DNA; the concentrations are expressed in terms of mole percent.

asynchronous growth. The growth of cells that are randomly distributed with respect to their stage in cell division.

asynchronous muscle. A muscle that yields a number of contractions for every motor nerve impulse which it receives.

ATA. Aurintricarboxylic acid.

atactic polymer. A polymer in which the R groups of the monomer are randomly distributed on both sides of the plane that contains the main chain.

ATCase. Aspartate transcarbamylase.

A + T/G + C ratio. ASYMMETRY RATIO.

atherogenesis. The development of atherosclerosis; the formation of atheromas.

atherogenic. Having the capacity to initiate, or to increase, the process of atherogenesis.

atheroma. 1. A lipid-containing deposit in arteries undergoing hardening. 2. ATHEROSCLEROSIS.

atheromatous. Of, or pertaining to, atheromas.

atherosclerosis. The formation of lipid deposits in, and the hardening of the inner lining or intima, of an artery; a form of arteriosclerosis.

Atmungsferment. CYTOCHROME OXIDASE.

atom. The smallest part of an element that is capable of undergoing a chemical reaction and that is chemically indestructible and indivisible; a structural unit of matter that remains unchanged in chemical reactions except for the loss or gain of electrons.

atomic absorption spectrophotometry. A sensitive analytical method for the spectrophotometric determination of elements; based on a measurement of the radiation absorbed by unexcited, nonionized, and ground-state atoms which are produced when compounds are dissociated into atoms by means of a flame.

atomic mass. The mass of a neutral atom expressed in terms of atomic mass units.

atomic mass unit. One twelfth of the mass of the carbon isotope, $_6C^{12}$, and equal to 1.67×10^{-24} gram; prior to 1961 the atomic mass unit was defined as one sixteenth of the mass of the oxygen isotope, $_8O^{16}$. *Abbr* amu. *See also* atomic weight unit.

atomic number. The number of protons in the nucleus of an atom which is also equal to the number of orbital electrons surrounding the nucleus of the neutral atom. *Sym* Z.

atomic orbital. The orbital of an electron about the nucleus of an atom.

atomic radius. The distance between the nucleus and the outermost electron shell of an atom. *See also* Van der Waals radius.

atomic weight. The average weight for the neutral atoms of an element, existing as a mixture of isotopes identical to that found in nature; expressed in atomic mass units.

atomic weight unit. One twelfth of the mass of the carbon isotope, $_6C^{12}$, and equal to 1.67×10^{-24} gram; identical to the atomic mass unit. Prior to 1961 the atomic weight unit was defined as one sixteenth of the average mass of the oxygen isotopes weighted in the same ratio as they occur in nature. *Abbr* awu.

atomizer. A spraying device for breaking up a solution into fine droplets.

atom percent excess. A measure of the concentration of a stable isotope expressed in terms of the excess, in percent of atoms, of the isotope over its natural abundance.

atom smasher. ACCELERATOR.

atopic. Of, or pertaining to, atopy.

atopic reagin. REAGIN.

atopy. An immediate-type hypersensitivity, such as asthma or hay fever, that is due to the production of reagins and that tends to occur as an inherited tendency.

ATP. 1. Adenosine triphosphate. 2. Adenosine-5′-triphosphate.

ATP-ADP cycle. The sum of the reactions by which (a) ADP is converted to ATP by means of the energy derived from food in the course of catabolism, and (b) ATP is hydrolyzed to ADP with the release of energy which is used to drive the energy-requiring reactions of an organism.

ATPase. Adenosine triphosphatase.

ATP regenerating system. An enzymatic system for the synthesis of ATP from ADP that is used in cell-free amino acid incorporation experiments for replenishing the ATP that is used up in the course of amino acid activation. The system consists either of phosphocreatine and creatine kinase or of phosphoenolpyruvate and pyruvate kinase.

atractyloside. A toxic glycoside that inhibits the ADP-ATP carrier system of the inner mitochondrial membrane.

atractyloside barrier. The block to adenine nucleotide transport across the inner mitochondrial membrane that is produced by atractyloside.

atrophy. The reduction in size of a tissue or an organ as a result of nutritional deficiencies and/or decreased functional activity.

attenuate. 1. To decrease the virulence of a bacterium or a virus. 2. To decrease the intensity of a beam of radiation by passage of the beam through matter.

atto-. Combining form meaning 10^{-18} and used with metric units of measurement. *Sym* a.

atypical insulin. INSULIN-LIKE ACTIVITY.

A.U. Angstrom unit.

Auger effect. The process whereby an orbital electron passes from an excited to a lower energy level and thereby produces an x-ray

which collides with, and ejects, an orbital electron (Auger electron); subsequent to the ejection of the electron, the x-ray escapes from the sphere of influence of the atom.

Auger electron. *See* Auger effect.

aureomycin. *See* tetracycline.

AU-rich DNA. DNA-LIKE RNA.

aurintricarboxylic acid. A dye that inhibits protein synthesis by blocking the attachment of messenger RNA to ribosomes. *Abbr* ATA.

Australia antigen. An antigen in the serum of individuals suffering from viral hepatitis; the antigen shows up as a virus-like particle in electron micrographs and was originally detected in the serum of an Australian aborigine.

autacoid. Any internal secretion of the body that acts on a restricted area within the body; hormones (excitatory autacoids), chalones (inhibitory or restraining autacoids), and kinins are some examples.

autarky. Self-sufficiency as opposed to parasitic existence.

autoallergic disease. AUTOIMMUNE DISEASE.

autoallergy. Allergy to autoantigens.

autoanalyzer. An instrument for the automated analysis of elements and compounds. The instrument consists of automated devices that replace such manual steps as the pipetting of reagents, the preparation of protein-free filtrates, and the heating of solutions for the development of a color.

autoantibody. An antibody that is formed in an individual in response to antigens of the same individual.

autoantigen. An antigen that is a normal constituent of an individual and that has the capacity of producing an immune response against itself in the same individual.

autocatalysis. The phenomenon in which the product of a reaction serves as a catalyst for the reaction that forms the product so that the velocity of the reaction keeps increasing with time.

autocatalytic. Of, or pertaining to, autocatalysis.

autocatalytic induction. A self-accelerating enzyme induction as that occurring when (a) an inducer induces both an enzyme and a transport system that actively transports the inducer, or (b) an enzyme is induced by the product of the reaction that it catalyzes.

autochthonous. Originating or formed in the place where found; said of a disease or a tumor that originated in that part of the body in which it is found.

autoclave. An instrument for sterilizing materials in an airtight chamber by the use of steam at high pressure.

autocoid. Variant spelling of autacoid.

autocoupling hapten. A reactive low-molecular weight compound, such as a diazonium salt or an acid anhydride, that, when injected into an animal, will combine with tissue antigens to form hapten-antigen compounds which then lead to the formation of antibodies.

autocytolysis. AUTOLYSIS.

autofluorescence. The fluorescence of tissues that is due to molecules naturally present in the tissues and that is unrelated to the treatment of the tissues with fluorochromes. *See also* intrinsic fluorescence.

autogenous. AUTOLOGOUS (1).

autogenous vaccine. A vaccine made from the microorganisms that have infected an individual which is then used to reinoculate the same individual.

autograft. A transplant from one site to another in the same individual.

autoimmune complement fixation. A complement fixation test in which both the antigens and the antibodies are derived from the same individual. *Abbr* AICF.

autoimmune disease. A disease produced by an autoimmune response.

autoimmune response. 1. The formation of antibodies to autoantigens. 2. The allergic response produced by the injection of autoantibodies.

autoimmunity. The immune reactions in an individual in which both the antigens and the antibodies are derived from that individual.

autoinduction. The induction of a drug-metabolizing enzyme by the chronic administration of that drug.

autointerference. The decrease in viral multiplication in animal cells that are infected with a large dose of virus particles as compared to the extent of viral multiplication in cells that are infected with a small dose. The effect is due to the appreciable concentration of interferon that is produced when the cells are infected with the larger viral dose.

autointoxication theory. The theory that psychoses and other mental diseases are caused by the endogenous production of toxins.

autologous. 1. Derived from the same organism. 2. Designating a transplant from one site to another in the same individual.

autolysate. The suspension of broken cells obtained upon autolysis. *Var sp* autolyzate.

autolysis. The self-destruction of a cell as a result of the action of its own hydrolytic enzymes.

autolytic. Of, or pertaining to, autolysis.

automatic pipet. A pipet that is connected to a reservoir of liquid and that is automatically refilled after delivery of a given volume of liquid; used for the repeated delivery of a fixed volume of liquid.

automutagen. A compound that is produced by the metabolic reactions of an organism and that is mutagenic for the same organism.

autonomous. Existing and functioning independently; said of a tumor cell that is free of host control, or of an episome that replicates independently of the chromosome.

autooxidation. Variant spelling of autoxidation.

autophagic. Self-digesting.

autophagic vacuole. A lysosome enlarged in the process of digesting cellular components.

autoprothrombin I. PROCONVERTIN.

autoprothrombin II. CHRISTMAS FACTOR.

autoprothrombin III. STUART FACTOR.

autoradiogram. The photographic record of a chromatogram that contains radioactively labeled compounds; prepared by exposing a sensitive photographic film to the radioactive radiation by placing it in contact with the chromatogram.

autoradiograph. *See* radioautograph.

autoradiography. *See* radioautography.

autoretardation. The phenomenon that the rate of an endonuclease-catalyzed hydrolysis of a nucleic acid decreases as the high-molecular weight nucleic acid is broken down to smaller fragments.

autosome. Any chromosome that is not a sex (X or Y) chromosome.

autosynthetic cell. A cell-like structure obtained by combining lipid and protein extracts of the brain.

autotroph. A cell or an organism that uses carbon dioxide as its sole carbon source, and that synthesizes all of its carbon-containing biomolecules from carbon dioxide and other small inorganic molecules.

autoxidation. The oxidation of a compound by air alone.

auxiliary pigment. *See* accessory pigment.

auxin. One of a group of plant hormones, such as indoleacetic acid, that promote cell enlargement, chromosomal DNA synthesis, and growth along the longitudinal axis of a plant.

auxocarcinogen. An auxiliary group of atoms in the molecule of a chemical carcinogen that influences the activity of the carcinogenophore.

auxochrome. A group of atoms which, when attached to a molecule containing a chromophore, intensifies the color of the chromophore.

auxotroph. A mutant microorganism that has a block in a metabolic pathway as a result of either the lack of an enzyme or the presence of a defective enzyme. Such mutants require for their growth either the product of the blocked enzymatic reaction or other metabolites that are not required by the wild-type organism.

avalanche. *See* Townsend avalanche.

average. MEAN.

average affinity. AVERAGE INTRINSIC ASSOCIATION CONSTANT.

average deviation. The average of the deviations for a set of measurements regardless of the signs of the individual deviations.

average intrinsic association constant. The value of the association constant for the binding of a given antigen by the corresponding antibodies that is determined for the case when one half of all the antibody sites are occupied by the antigen. *Sym* K_0. *See also* heterogeneity index.

average life. The average length of time that a radioactive atom exists before it disintegrates; the average life is equal to the reciprocal of the decay constant.

average molecular weight. The value of the molecular weight that is determined for a sample consisting of a mixture of molecules. The type of average molecular weight obtained depends on the physical method used in studying the molecules. *See also* number average molecular weight; weight average molecular weight; Z-average molecular weight.

average radius of gyration. The value of the radius of gyration that is based on all the different conformations which a molecule may assume; specifically, $R_G = (\bar{R}^2)^{1/2}$, where R_G is the average radius of gyration and $\bar{R}^2$ is the average of the squares of all possible radii of gyration.

avian. Of, or pertaining to, birds.

avian leucosis virus. An RNA-containing virus that produces leukemia in chickens and that belongs to the group of leukoviruses.

avidin. A protein in raw egg white that combines tightly with the vitamin biotin; when fed to an animal, avidin can produce symptoms of biotin deficiency.

avidity. 1. The tendency of an antibody to bind antigen; measured by the rate of the binding reaction. 2. AFFINITY (2).

avirulent. Not virulent.

avitaminosis. HYPOVITAMINOSIS.

Avogadro's number. 1. The number of molecules in a gram-molecular weight of a compound; denoted by the symbol N and equal to 6.023×10^{23}. 2. The number of atoms in a gram-atomic weight of an element; denoted by the symbol N and equal to 6.023×10^{23}.

AVP. 1. Antiviral protein. 2. Arginine vasopressin.

awu. Atomic weight unit.

axenic animal. GERM-FREE ANIMAL.

axerophthal. VITAMIN A ALDEHYDE.

axerophthol. VITAMIN A.

axial bond. A bond in a molecule having a ring structure that is at right angles to the plane of the ring.

axial ratio. The ratio, for an ellipsoid of revolution, of the axis of revolution to the axis perpendicular to it. The axial ratio is an indication of the overall asymmetry of a macromolecule, since a macromolecule in solution is approximated by an equivalent ellipsoid of revolution.

axial substituent. A substituent attached to an axial bond.

axis of rotational symmetry. An axis of symmetry such that rotation of a body about it will yield one or more structures that are identical to the structure before rotation. The axis is an n-fold axis and denoted C_n if the identical structure is produced by a rotation of $360/n$ degrees, or $1/n$ of a turn.

axis of symmetry. An imaginary axis through a symmetrical body.

axon. The long process of a nerve cell that generally conducts impulses away from the nerve cell body.

axoneme. A bundle of microtubules in cilia and flagella of unicellular eucaryotic organisms.

axoplasm. The cytoplasm of an axon.

8-azaguanine. A purine analogue that retards the growth of some cancers.

azaserine. An antibiotic, either produced by a species of *Streptomyces* or prepared synthetically, that inhibits purine biosynthesis and leads to chromosomal aberrations.

azathioprene. An immunosuppressive drug.

6-azauracil. A uracil analogue.

azeotrope. A mixture of two or more liquids that has a constant boiling point and that is distilled without a change in composition.

azide group. The grouping $-N \overset{\oplus}{=} N \overset{\ominus}{=} N$.

aziridine mutagen. A chemical mutagen that contains the aziridinyl group $N \overset{CH_2}{\underset{CH_2}{\diagup\hspace{-0.5em}\mid}}$ and that functions as an alkylating agent.

azo compound. A compound containing the azo group.

azo dye. A dye that contains the azo group.

azo-dye protein. AZOPROTEIN.

azoferredoxin. IRON PROTEIN.

azo group. The grouping $-N=N-$.

azoic. Without life, particularly in reference to geologic periods antedating life on earth.

azoprotein. A modified protein in which a tyrosine residue has been coupled to an aromatic diazo compound.

B

B. 1. The sum of aspartic acid and asparagine when the amide content is either unknown or unspecified. 2. Boron. 3. 5-Bromouridine.

bacillus (*pl* bacilli). 1. A bacterium having a cylindrically shaped cell; bacilli represent one of the three major forms of eubacteria. 2. A bacterium belonging to the genus *Bacillus*.

bacitracin. A cyclic peptide antibiotic produced by *Bacillus subtilis*.

backbone. The chain structure of a polymer from which the side groups, and/or the side chains, project.

background. 1. The counts of radioactivity registered by a radiation detector in the absence of a radioactive sample; such counts are caused by cosmic radiation, instrument noise, radioactive contamination, and other factors. 2. The appearance of a chromatogram, an electropherogram, or an electron micrograph in areas that are devoid of sample substances.

back mutation. REVERSION (1).

backscattering. *See* backward scattering.

backside displacement. S_N2 MECHANISM.

back titration. The titration of the reagent left after an excess of the reagent has been added to the solution and has been allowed to react with it.

backward flow. The flow of the solvent of a solution of macromolecules that occurs in a closed vessel and that is in a direction opposite to the direction of movement of the macromolecules.

backward flow interface centrifugation. Interface centrifugation, used for cells, bacteria, and viruses, in which the interface is displaced according to the hydrodynamic volume of the particles.

backward scattering. The scattering of radiation in the direction of the beam of radiation and toward the source of the radiation.

bacteremia. The presence of viable bacteria in the blood.

bacteria. *See* bacterium.

bacterial. Of, or pertaining to, bacteria.

bacterial ferredoxin. IRON-SULFUR PROTEIN (c).

bacterial nucleus. NUCLEOID.

bacterial toxin. A toxin of bacterial origin, such as an endo- or an exotoxin.

bacterial-type ferredoxin. IRON-SULFUR PROTEIN (c).

bacterial virus. BACTERIOPHAGE.

bactericidal agent. BACTERICIDE.

bactericide. An agent that kills bacteria.

bacterin. A vaccine consisting of dead pathogenic bacteria.

bacteriochlorophyll. The chlorophyll of photosynthetic bacteria; differs from chlorophyll a of plants in having one of the pyrrole rings in a more reduced form and having a vinyl group replaced by an acetyl group.

bacteriocidal agent. Variant spelling of bactericidal agent.

bacteriocin. A protein produced by one bacterium that is toxic for another bacterium; a bacteriocin differs from an antibiotic in being a protein, having a narrower microbial spectrum, and generally being much more potent.

bacteriocin factor. BACTERIOCINOGEN.

bacteriocinogen. The episome that controls the formation of a bacteriocin.

bacteriology. The science that deals with studies of bacteria.

bacteriolysin. An antibody capable of leading to the dissolution of bacterial cells in the presence of complement.

bacteriolysis. The lysis of bacterial cells.

bacteriolytic. Of, or pertaining to, bacteriolysis.

bacteriophage. A virus that infects bacteria and multiplies in them. *Aka* bacterial virus; phage.

bacteriorhodopsin. A protein molecule that closely resembles the rhodopsin of animals but that is produced by the halophilic bacterium *Halobacterium halobium*.

bacteriostatic. Variant spelling of bacteristatic.

bacteriotropic index. OPSONIC INDEX.

bacteriotropin. IMMUNE OPSONIN.

bacteristasis. The prevention of bacterial growth without killing the bacterial cells.

bacteristat. An agent that prevents the growth of bacteria without killing the bacterial cells.

bacteristatic agent. BACTERISTAT.

bacterium (*pl* bacteria). A minute, unicellular procaryotic organism that is classified as a lower protist.

bactogen. CHEMOSTAT.

baker's yeast. One of several strains of yeast belonging to the species *Saccharomyces cerevisiae*.

BAL. British Anti-Lewisite.

balanced growth. The growth of cells such that all of the cellular components increase by the same factor and that the overall composition of the cells remains constant.

balance study. A study of the overall metabolism of a substance in an organism that is based on a determination of the amount of the substance ingested and of the amount of the same substance, or of its metabolic products, excreted; such a study indicates whether there is a net gain or loss of the substance by the organism.

Balbiani ring. A loop-like structure in polytene chromosomes that is formed by extremely large chromosomal puffs.

ball and stick model. A molecular model in which the atoms are represented by spheres and the bonds by sticks; the bond lengths and the atomic radii are fixed, and the bond angles are correctly indicated for each atom.

band. 1. A zone of macromolecules, such as the zone obtained in density gradient sedimentation, zone electrophoresis, isoelectric focusing, chromatography, or similar techniques. 2. ABSORPTION BAND.

band centrifugation. DENSITY GRADIENT CENTRIFUGATION.

band pass. The range of wavelengths of a radiation that passes through a filter or a similar device.

band width. 1. The width of an absorption band. 2. A range of wavelengths.

Bang method. A method for determining glucose in urine by means of alkaline copper thiocyanate.

bangosome. LIPOSOME.

BAP. 6-Benzylaminopurine.

barbital. Diethylbarbituric acid; a barbiturate.

barbiturate. A sleep-producing drug.

bare lipid membrane. BLACK LIPID MEMBRANE.

Barfoed's test. A colorimetric test for distinguishing mono- from disaccharides by means of a solution of cupric acetate in dilute acetic acid.

barium. An element that is essential to a few species of organisms. Symbol Ba; atomic number 56; atomic weight 137.34; oxidation state +2; most abundant isotope Ba^{138}; a radioactive isotope Ba^{133}, half-life 7.2 years, radiation emitted—gamma rays.

barn. A unit area of the atomic nucleus equal to 10^{-24} cm^2 and used as a measure of the capture cross-section of the nucleus.

baroreceptor. A receptor in the central nervous system that responds to changes in blood pressure.

Barr body. An inactive, condensed X-chromosome in the nuclei of somatic cells of females.

barrier layer cell. PHOTOVOLTAIC CELL.

basal body. KINETOSOME.

basal enzyme. An inducible enzyme that is produced in small amounts in the absence of an inducer.

basal granule. KINETOSOME.

basal level. The low level of concentration at which a basal enzyme is produced by a cell.

basal metabolic rate. The rate of basal metabolism under the following standardized conditions: physical rest, but not sleep; an ambient temperature that does not require energy expenditure for physiological temperature regulation; and a postabsorptive state following a 12-hour fast. The basal metabolic rate may be determined from the energy value of the food intake and the excreted waste products, or from the heat produced by the organism, or from the oxygen consumption by the organism. *Abbr* BMR.

basal metabolism. The level of metabolic activity that is required by an animal for the maintenance of vital functions such as respiration, heart contraction, and kidney and liver function; the maintenance of nonvital functions such as muscular work and digestion is excluded. *See also* basal metabolic rate.

base. 1. Purine. 2. Pyrimidine. 3. Bronsted base. 4. Lewis base. 5. The fixed number, such as 10 or *e*, used in logarithms.

base composition. The relative amounts of the various purines and pyrimidines in a nucleic acid; generally expressed in terms of mole percent.

base line. The line in a chromatogram, an ultracentrifuge pattern, or a similar tracing that corresponds to the solvent rather than to the solution.

basement membrane. The thin, structureless layer that underlies the epithelium and is devoid of cells.

base pair. A pair of hydrogen-bonded bases, a purine and a pyrimidine, that either link two separate polynucleotide strands as in double helical DNA or link parts of the same polynucleotide strand as in the cloverleaf model of transfer RNA.

base pairing. The complementary binding of the bases in a nucleic acid by means of hydrogen bonding. The binding may be between two strands of a double-stranded molecule or between parts of a single-stranded molecule folded back upon itself.

base-pairing rules. The requirements that in a double helical nucleic acid structure adenine must form a base pair with thymine (or uracil) and cytosine must form a base pair with guanine, and vice versa.

base-pair ratio. ASYMMETRY RATIO.

basepiece. *See* supermolecule.

base ratio. One of three concentration ratios for the bases in a nucleic acid that are generally expressed in mole percent: adenine/thymine (or uracil); guanine/cytosine; and purines/pyrimidines.

base sequence. The linear order of the purine and pyrimidine bases, or of their nucleotides, as they occur in a nucleic acid strand.

base stacking. The arrangement of the base pairs in parallel planes in the interior of a double helical nucleic acid structure.

base substitution. The replacement of one base for another in either a nucleotide or a nucleic acid.

basic. 1. Of, or pertaining to, a base. 2. Of, or pertaining to, a solution having a pH greater than 7.0.

basic amino acid. An amino acid that has two amino groups and one carboxyl group.

basic dye. A cationic dye that binds to, and stains, negatively charged macromolecules. *Aka* basic stain.

basophil. A cell that stains with a basic dye.

batch adsorption. A technique for adsorbing a solute from a solution by stirring the solution together with an adsorbent to form a slurry; subsequently the solution is separated from the adsorbent by decantation, filtration, or centrifugation.

batch culture. A culture grown in a given volume of medium. *See also* continuous culture.

bathochromic group. A group of atoms that, when attached to a compound, shifts the absorption of light by the compound to longer wavelengths.

bathochromic shift. A shift in the absorption spectrum of a compound toward longer wavelengths.

BCG. Bacillus Calmette-Guérin; an attenuated strain of *Mycobacterium tuberculosis* used as a vaccine for protection against tuberculosis and leprosy.

B chain. 1. The longer of the two polypeptide chains of insulin that contains 30 amino acids and that is linked to the other chain by two disulfide bonds. 2. The light (L) chain of the immunoglobulins.

B complex. *See* vitamin B complex.

BD-cellulose. Benzoylated diethylaminoethyl-cellulose; a chromatographic support.

Beckmann thermometer. A thermometer with a large bulb and a fine bore that is used for the measurement of small differences in temperature.

bed. The solid support of the column in column chromatography.

Beer's law. The law which relates the absorbance of a solution to its concentration and to the length of the light path through the solution; specifically, $A = \epsilon lc$, where A is the absorbance, l is the length of the light path, c is the concentration, and ϵ is the absorptivity. When l is in centimeters and c in moles per liter, then ϵ is the molar absorptivity. *Aka* Beer-Lambert law.

beet sugar. Sucrose that is isolated from beets.

BEI. Butanol-extractable iodine; the iodine that can be extracted from serum by butanol.

Bellin's hypothesis. COPY-CHOICE HYPOTHESIS.

Bence-Jones protein. An abnormal immunoglobulin that consists only of light chains, generally in the form of dimers, and that is produced by individuals suffering from plasma cell tumors; Bence-Jones protein has unusual thermosolubility properties and is relatively homogeneous.

bending vibration. DEFORMATION VIBRATION.

Benedict's reagent. An aqueous solution of copper sulfate, sodium citrate, and sodium carbonate that is used in Benedict's test for reducing sugars.

benign neoplasm. A harmless, localized, and nonmetastasizing tumor.

Benson model. The model of a biological membrane according to which the proteins are largely globular and are located in the interior of the membrane, and the lipids are intercalated into the folds of the protein chains, with the polar portions of the lipid molecules at the exterior surfaces of the membrane.

bentonite. A clay that consists principally of montmorillonite (aluminum-magnesium-silicate) and that is used as an inhibitor of nucleases.

benzedrine. Tradename for amphetamine.

benzidine test. A test for blood that is based on the formation of a blue compound, benzidine blue, upon treatment of the sample with glacial acetic acid and hydrogen peroxide.

6-benzylaminopurine. A synthetic cytokinin.

benzylpenicillin. *See* penicillin.

beriberi. The disease caused by a deficiency of the vitamin thiamine.

Berkefeld filter. A filter, made of diatomaceous earth, that retains bacteria and that is used for the sterilization of solutions.

Berthelot reaction. A colorimetric test for ammonia in urine that is based on the production of blue indophenol upon treatment of urine with phenol and sodium hypochlorite.

BES. N,N-Bis(2-hydroxyethyl)-2-aminoethanesulfonic acid; used for the preparation of buffers in the pH range of 6.2 to 8.2.

beta. 1. Denoting the second carbon atom next to the carbon atom that carries the principal functional group of the molecule. 2. Denoting a specific configuration of the substituents at the anomeric carbon in the ring structure of carbohydrates. 3. Denoting buffer value. *Sym* β.

beta amylase. The enzyme that catalyzes the

sequential hydrolysis of α $(1 \rightarrow 4)$ glycosidic bonds of starch commencing at the nonreducing end of the starch molecule.

beta carotene. A carotene that is a precursor of vitamin A and that is cleaved in animals to yield two molecules of vitamin A per molecule of beta carotene.

beta chain. One of the two types of polypeptide chains occurring in adult hemoglobin.

beta configuration. The configuration of a polypeptide chain in which the chain is in a partially extended form.

beta decay. The radioactive disintegration of an atomic nucleus that results in the emission of a beta particle.

beta emitter. A radioactive nuclide that decays by emission of a beta particle. A beta emitter is considered to be soft or hard depending on whether the emitted beta particles are of low energy and have a short penetration range, or whether they are of high energy and have a long penetration range.

beta function. *See* Scheraga-Mandelkern equation.

beta galactosidase. *See* β-galactosidase.

betaine. N,N,N-Trimethyl glycine; a compound that serves as a methyl group donor and occurs in plant and animal tissues.

beta keratin. The extended form of keratin, obtainable by stretching alpha keratin, in which the polypeptide chains are in the parallel pleated sheet configuration.

beta lactoglobulin. A protein in milk; the first pure protein for which the complete amino acid composition was determined (1947).

beta lipoprotein. LOW-DENSITY LIPOPROTEIN.

beta method. RAMON METHOD.

beta oxidation. The oxidation of fatty acids in metabolism through successive cycles of reactions, with each operation of the cycle leading to a shortening of the fatty acid by a two-carbon fragment that is removed in the form of acetyl coenzyme A.

beta particle. 1. An electron originating in the atomic nucleus and emitted frequently by radioactive isotopes. Beta particles are considered to be soft or hard depending on whether they are of low energy and have a short penetration range, or whether they are of high energy and have a long penetration range. 2. A glycogen granule in the liver.

beta plateau. The high potential portion of the characteristic curve of a proportional radiation detector at which the count rate is almost independent of the applied voltage and at which the potential is of sufficient magnitude to detect beta particles.

beta ray. A beam of beta particles.

beta-ray spectrometer. An instrument for the analysis of either the energy spectrum or the momentum spectrum of beta rays.

beta threshold. The lowest potential at which beta particles can be detected with a proportional radiation detector.

betatron. An accelerator designed to impart high kinetic energy to electrons by means of electromagnetic induction.

BeV. One billion (10^9) electron volts.

bevatron. An accelerator designed to impart high kinetic energy to protons.

BGG. Bovine gamma globulin.

bi-. 1. Combining form meaning two or twice. 2. Referring to two kinetically important substrates and/or products of an enzymatic reaction; thus a uni bi reaction is a reaction with one substrate and two products.

Bial's reaction. ORCINOL REACTION.

biamperometric titration. An amperometric titration using two like electrodes.

bicine. N,N-Bis(hydroxyethyl)glycine; used for the preparation of buffers in the pH range of 7.7 to 9.1.

bidentate. Designating a ligand that is chelated to a metal ion by means of two donor atoms.

bifluorescence. The variation in the apparent intensity of plane-polarized fluorescence by the rotation of an analyzer through which the fluorescence is observed.

bifunctional antibody. An antibody that has two combining sites for antigen; a divalent antibody.

bifunctional catalyst. A catalyst that can provide both an acidic and a basic catalytic function.

bifunctional feedback. A feedback mechanism that affords control in two directions so that the input of a system is affected both by an increase and by a decrease of the output. An example is a system in which the pH will be adjusted if the pH either rises above or falls below the normal value.

bifunctional reagent. A compound that has two reactive functional groups and that can interact either with two groups of one protein or with one group each of two different proteins.

big insulin. The fraction of free serum insulin that is believed to be identical with proinsulin.

bilayer. A layer that is two molecules thick. *See also* lipid bilayer.

bilayer lipid membrane. BLACK LIPID MEMBRANE.

bilayer model. UNIT MEMBRANE HYPOTHESIS.

bile. The secretion of the liver that aids in the digestion of fats by emulsifying them and that serves to excrete bile pigments, heavy metals, and other waste products of metabolism. *See also* bile salt.

bile acid. A 24-carbon steroid, such as cholic acid or deoxycholic acid, that occurs in the bile in the form of a bile salt.

bile pigment. A degradation product of the heme portion of hemoglobin and other heme proteins; the linear tetrapyrroles, bilirubin and biliverdin, are two examples.

bile salt. A surface-active agent in the bile that aids in the emulsification of fats during digestion and that consists of a bile acid coupled to either glycine or taurine.

biliary. Of, or pertaining to, the bile.

bilin. A colored bile pigment, such as urobilin or stercobilin, that is formed by the oxidation of a colorless bilinogen pigment.

bilinogen. A colorless bile pigment, such as urobilinogen or stercobilinogen, that forms a colored bilin pigment upon oxidation.

bilirubin. A bile pigment. *See also* direct-acting bilirubin; indirect-acting bilirubin.

bilirubin diglucuronide. A conjugated, soluble form of bilirubin that is formed in the liver by the esterification of two molecules of glucuronic acid to two propionic acid residues of bilirubin. *See also* direct-acting bilirubin.

bilirubinemia. HYPERBILIRUBINEMIA.

biliverdin. A bile pigment.

bimolecular lamellar lipid membrane. BLACK LIPID MEMBRANE.

bimolecular layer. BILAYER.

bimolecular leaflet. BILAYER.

bimolecular lipid membrane. BLACK LIPID MEMBRANE.

bimolecular reaction. A chemical reaction in which either two molecules (or other entities) of a single reactant, or one molecule each of two reactants, interact to form products.

binal symmetry. Symmetry in which there are two types of symmetry elements.

binary. Consisting of two parts.

binary complex mechanism. PING PONG MECHANISM.

binary digit. BIT.

binary fission. Asexual division in which a cell divides into two, approximately equal parts.

binder. A substance, such as calcium sulfate, that is mixed with a thin-layer chromatographic adsorbent to increase the mechanical strength of the adsorbent layer.

binding assay. 1. Any method for measuring protein-ligand interactions as in the binding of sugars to periplasmic proteins and in the binding of cyclic AMP to protein kinase. 2. A method for measuring the binding of aminoacyl-tRNA to ribosomes in response to oligoribonucleotides of defined sequences and in the absence of peptide bond formation; based on using labeled aminoacyl-tRNA, collecting the aminoacyl-tRNA-ribosome-oligoribonu-

cleotide complex on Millipore filters, and counting the radioactivity in this complex. *Aka* ribosome binding technique.

binding constant. ASSOCIATION CONSTANT.

binding factor. A protein factor required for the binding of aminoacyl-tRNA to ribosomes.

binding site. The structural part of a macromolecule that directly participates in its specific combination with a ligand. Binding sites are said to be interacting or independent depending on whether the binding of one ligand to one site does, or does not, affect the binding of other ligands to other sites on the same molecule.

binomial coefficient. The coefficient of any term that results from the binomial expansion of $(a + b)^n$.

binomial distribution. A frequency distribution in which the frequencies have the values of binomial coefficients.

binomial nomenclature. A system for naming species of plants and animals in which the name consists of two parts, the first designating the genus, and the second designating the species.

bio-. 1. Combining form meaning life. 2. Combining form meaning a biological or a biochemical system.

bioassay. The measurement of either the activity or the amount of a substance that is based on the use of living cells or living organisms; measurements of infectivity, antibody formation, weight gain, and bacterial growth are examples.

bioautograph. The record obtained when a bioassay is used in conjunction with a chromatographic procedure, as in the case where a paper chromatogram is placed in contact with a bacterial culture on a solid medium.

bioblast. MITOCHONDRION.

biochemical. Of, or pertaining to, biochemistry.

biochemical coupling hypothesis. *See* chemical coupling hypothesis.

biochemical deletion hypothesis. *See* deletion hypothesis.

biochemical energetics. The free energy relationships of biochemical reactions.

biochemical evolution. The evolutionary processes concerned with the formation of biomolecules, cells, cellular structures, metabolic pathways, and other attributes of living cells. *See also* biological evolution; chemical evolution.

biochemical fossil. *See* chemical fossil.

biochemical genetics. MOLECULAR GENETICS.

biochemical inflexibility of tumors. The phenomenon that the control mechanisms of many enzyme systems, such as those of enzyme induction, appear to be frozen in tumor cells.

biochemical lesion. A biochemical alteration, such as the inactivation of an enzyme, that leads to a pathological condition; may be caused by a mutagen, a carcinogen, or other factors.

biochemically deficient mutant. AUXOTROPH.

biochemical marker. A mutable site on a chromosome that, when mutated, leads to a specific enzyme defect which can be detected by biochemical means.

biochemical mutant. AUXOTROPH.

biochemical oxygen demand. The rate at which the oxygen in water is consumed by microorganisms for the oxidation of organic compounds to simple inorganic molecules. *Abbr* BOD.

biochemical predestination theory. The theory that the development of the living cell is determined by, and follows naturally from, the physical-chemical properties of the simplest starting compounds.

biochemical similarity principle. The assumption that the biochemical compounds and processes known to occur ubiquitously in contemporary living systems and to be essential for life, also occurred at the early stages of the development of life and were essential to the origin of life.

biochemistry. The science that deals with the chemistry of living systems and their components.

biochrome. A naturally occurring coloring matter in a plant, an animal, or a microorganism; a pigment.

biochronometry. The science that deals with the temporal organization and the time-keeping mechanisms of biological systems.

biocide. A chemical substance used either to kill or to arrest the growth of living organisms; bactericides, fungicides, and pesticides are examples.

biocytin. ε-Biotinyllysine; a compound formed by linking biotin through its carboxyl group to the ε-amino group of a lysine molecule. Biotin is believed to be similarly linked to a lysine residue in those enzymes which require biotin as a coenzyme.

biod. The dry biomass per unit area of the earth's surface.

biodegradable. Descriptive of a substance that can be decomposed by the enzyme systems of bacteria or other organisms.

bioenergetics. BIOCHEMICAL ENERGETICS.

bioflavonoids. A group of compounds that is designated as vitamin P and that includes the flavanones, the flavones, and the flavonols.

biogel. Trademark for a group of polyacrylamide and agarose gels used in gel filtration.

biogenesis. 1.The synthesis of a substance in a living organism; biosynthesis. 2. The doctrine that living things can come only from other living things.

biogenetic. Of, or pertaining to, biogenesis.

biogenetic law. RECAPITULATION THEORY.

biogenic. 1. Produced by the action of living organisms. 2. Essential to life.

biogenic amine. An amine that is produced by a living organism, particularly a physiologically important amine such as epinephrine, norepinephrine, serotonin, or histamine.

biogeochemistry. The science that deals with the interaction of living organisms with the mineral environment of the earth's crust.

biolith. A geological sediment, such as peat or humus, that consists of residues from organic matter.

biological assay. BIOASSAY.

biological chemistry. BIOCHEMISTRY.

biological clock. The periodicity of either a biological function or a biochemical reaction.

biological evolution. The gradual development of living organisms to their present state; a process that followed the stage of chemical evolution and extended over a period of about 3 billion years.

biological half-life. *See* half-life (2).

biological nitrogen fixation. The conversion of atmospheric nitrogen to ammonia by living organisms.

biological oxidation-reduction. 1. The oxidation-reduction reactions in aerobic cellular respiration whereby nutrients are oxidized in the citric acid cycle, electrons are transported by the electron transport system, and ATP is synthesized through oxidative phosphorylation. 2. Any oxidation-reduction reaction occurring in a living system.

biological oxygen demand. BIOCHEMICAL OXYGEN DEMAND.

biological value. The relative nutritional value of a protein that is based on the amino acid composition of the protein, the digestibility of the protein, and the availability of the digested products. The biological value has been defined in terms of (a) the fraction of absorbed protein nitrogen that is retained in the body; (b) the growth rate of young animals as a function of the dietary level of the protein; (c) the minimal protein concentration required to establish nitrogen balance in adults; and (d) the change in the concentration of essential amino acids in the serum as a function of the dietary level of the protein.

biology. The science that deals with living things; includes botany, zoology, embryology, genetics, morphology, and allied sciences.

bioluminescence. The production of visible light by a living organism. *See also* luciferase.

biolysis. Lysis by biological means.

biomass. The mass of an organism or a group of organisms; variously used in reference to wet mass, dry mass, or mass per unit area. *See also* biod.

biometry. The science that deals with the application of statistics to biological systems.

biomolecule. A molecule, such as a protein or a lipid, that occurs in a living system.

biomonomer. A monomer, such as an amino acid or a nucleotide, that occurs in a living system.

bionomics. ECOLOGY.

bioorthogonal code. An early version of the genetic code based on 24 codons, each containing six nucleotides, such that each codon could undergo two base substitutions and still be recognized as being related to its original form.

biophysics. The science that deals with the physics of living systems and their components.

biopolymer. A polymer, such as a protein or a nucleic acid, that occurs in a living system.

biopterin. A derivative of pteridine, the reduced form of which, dihydrobiopterin, serves as a coenzyme for hydroxylation reactions of amino acids.

bios. A term previously used to denote a growth-promoting substance for yeast; bios I referred to inositol and bios II (or II B) referred to biotin.

biosis. LIFE.

biosphere. The regions of and around the earth that support life; includes the oceans, the upper portion of the land masses, and the lower portion of the atmosphere.

biosterol. VITAMIN A.

biosynthesis. The formation of a substance in a living system.

biosynthetic. Of, or pertaining to, biosynthesis.

biosynthetic pathway. An anabolic pathway.

biotic. Of, or pertaining to, life.

biotin. A vitamin of the vitamin B complex that functions as a coenzyme in carboxylation-decarboxylation reactions.

biotransformation. The metabolic reactions in an organism whereby a foreign chemical, introduced into the organism, is either converted to a different compound or conjugated to a metabolite of the organism.

bireactant reaction. A bimolecular reaction in which two different reactants interact to form products.

birefringence. The phenomenon of a substance possessing two refractive indices depending on the direction along which light passes through the substance.

bispecific antibody. HETEROSPECIFIC ANTIBODY.

bisubstrate reaction. An enzymatic reaction in which two substrates participate.

bit. A binary digit; a single character in a notation using two characters; a basic unit of information content expressed in terms of the probabilities for binary choices such as "yes" and "no" answers.

Bittner's mouse milk factor. MOUSE MAMMARY TUMOR VIRUS.

biuret. A compound formed by the condensation of two molecules of urea.

biuret reaction. A colorimetric reaction for the qualitative and quantitative determination of proteins; based on the production of a purple color upon treatment of biuret, peptides, proteins, or related compounds with copper sulfate in an alkaline solution.

black lipid membrane. An artificially prepared bimolecular lipid membrane that consists of either naturally occurring or synthetic lipids; usually formed in an annular space connecting two compartments filled with an aqueous medium that contains the lipids.

blank. A mixture of reagents that excludes the sample but that is identical to the mixture of reagents with which the sample is treated and that is carried through the same procedures as the sample. *See also* control.

blast cell. A cell with a poorly differentiated, but RNA-rich, cytoplasm that actively synthesizes DNA.

blastoma. A tumor consisting of immature and undifferentiated cells.

blender. A small appliance used to mix, homogenize, or disintegrate liquids and/or solids. *Var sp* blendor.

blender experiment. HERSHEY-CHASE EXPERIMENT.

blind test. A test of a substance or a procedure in which an independent observer records the results without knowing either the identity of the samples or the expected results. *See also* double blind technique.

Blinks effect. CHROMATIC TRANSIENT.

BLM. 1. Bare lipid membrane. 2. Bilayer lipid membrane. 3. Bimolecular lamellar lipid membrane. 4. Bimolecular lipid membrane. 5. Black lipid membrane.

Block. *See* metabolic block; genetic block.

block copolymer. BLOCK NUCLEOTIDE.

blocking antibody. 1. A protective antibody that prevents anaphylaxis by combining with the allergen; it is formed during desensitization and is primarily of the IgG type. 2. An incomplete antibody that actively inhibits agglutination.

blocking group. PROTECTING GROUP.

block nucleotide. A synthetic oligonucleotide of known base sequence. *Aka* block oligonucleotide.

blood. The fluid that circulates through the cardiovascular system and that is composed of a fluid fraction, plasma, and a cellular fraction consisting of erythrocytes, leucocytes, and thrombocytes.

blood-brain barrier. The slow penetration into the brain of substances that are transported by the blood as compared to their more rapid penetration into most other tissues.

blood-cerebrospinal fluid barrier. BLOOD-BRAIN BARRIER.

blood clot. The insoluble network of polymerized fibrin molecules and trapped cells that is formed upon the solidification of blood, most commonly in the course of external bleeding.

blood clotting. The two groups of complex biochemical reactions by which a blood clot is formed. The first group of reactions leads to the formation of the active enzyme thrombin from its inactive precursor, prothrombin; the second group of reactions leads to the thrombin-catalyzed conversion of fibrinogen to the fibrin clot. *Aka* blood coagulation.

blood count. A determination of the number of red and white blood cells per cubic millimeter of blood.

blood group. One of the classes of individuals that belong to a given blood group system.

blood group antigen. BLOOD GROUP SUBSTANCE.

blood group chimera. A chimera containing two different blood types.

blood grouping. BLOOD TYPING.

blood group substance. A genetically determined particulate isoantigen that is attached to the surface of red blood cells and that may be attached to the surface of other cells; related blood group substances represent alternative antigens, specified by allelic genes.

blood group system. A classification of individuals into groups on the basis of their possession, or nonpossession, of specific blood group substances. Some of the blood group systems known are the ABO, Lewis, M-N-S-s, and Rh systems.

blood plasma. *See* plasma.

blood platelet. *See* platelet.

blood serum. *See* serum.

blood sugar. The glucose in the blood.

blood typing. The identification of the blood group substances of an individual.

blood urea nitrogen. The nitrogen of the urea in serum. *Abbr* BUN.

blood vessel. An artery, a vein, or a capillary through which the blood circulates.

blue shift. HYPSOCHROMIC SHIFT.

BMR. Basal metabolic rate.

BND-cellulose. Benzoylated-naphthoylated-diethylaminoethyl-cellulose; a chromatographic support. Also abbreviated BNC.

boat conformation. The arrangement of atoms that resembles the outline of a boat and that is the less stable conformation of the two possible ones for cyclohexane and other six-membered ring systems.

t-**BOC-amino acid.** An amino acid in which the amino group has been protected by attachment of a tertiary butoxycarbonyl group; used in peptide synthesis by the solid phase (Merrifield) method. *Abbr t*-BOC-AA.

BOD. Biochemical oxygen demand.

Bohr effect. The decrease in the oxygen affinity of hemoglobin that is produced either by a decrease in the pH or by an increase in the partial pressure of carbon dioxide. The effect is due to changes in the pK values of ionizing groups in hemoglobin upon the oxygenation and the deoxygenation of the molecule. The effect above pH 6, when oxyhemoglobin is more negatively charged than deoxyhemoglobin, is known as the alkaline Bohr effect; the effect below pH 6, when deoxyhemoglobin is more negatively charged than oxyhemoglobin, is known as the acid Bohr effect.

Bohr magneton. The magnetic dipole moment of a spinning electron.

bolometer. A temperature transducer used for the measurement of minute quantities of radiant heat.

Boltzmann constant. The gas constant divided by Avogadro's number. *Sym* k.

Boltzmann distribution. The most probable distribution of a large number of molecules or particles in a nonuniform field of force at or near equilibrium; the centrifugal force in an ultracentrifuge, and the potential acting on an ion in electrophoresis are examples of such fields of force. The distribution can be expressed as $n_1 = n_2 \, e^{-(E_1-E_2)/kT}$, where n_1 and n_2 are the number of molecule or particles in two locations or in two energy states, E_1 and E_2 are the respective energies, k is the Boltzmann constant, and T is the absolute temperature.

bond. 1. The linkage between two atoms in a molecule. 2. The linkage between two atoms, groups of atoms, ions, or molecules, or between combinations of these.

bond angle. The angle between any two bonds by which an atom is linked to other atoms in the molecule.

bond energy. The energy required to break a chemical bond.

bonding orbital. A molecular orbital in which there is no node of electron density between the bonding atomic nuclei, resulting in a strengthening of the bond between the nuclei. Bonding orbitals are designated sigma (σ) and pi (π) and are generally of lower energy than the antibonding orbitals sigma star (σ^*) and pi star (π^*).

bond length. The length of the bond between two

atoms; equal to the distance between the centers of the nuclei of the two bonded atoms.

bond radius. One half of the bond length

boneblack. An impure charcoal prepared from bone and used as a decolorizing adsorbent.

bone mineral. HYDROXYAPATITE.

bookmark hypothesis. The hypothesis that peptides containing aromatic amino acids serve in vivo as a means of recognizing DNA sequences by having their aromatic amino acid residues become partially inserted between base pairs at certain intercalating sites in the DNA; the different intercalating sites are referred to as pages in a book, and the intercalating peptides are referred to as bookmarks.

booster dose. A dose of antigen, particularly in the form of a vaccine, that is administered to an individual after a priming dose with the intent of stimulating the production of large amounts of antibodies.

booster response. SECONDARY RESPONSE.

boron. An element that is essential to a wide variety of species in one class of organisms. Symbol B; atomic number 5; atomic weight 10.811; oxidation state $+3$; most abundant isotope B^{11}.

boundary. A transition zone, either between solvent and solution or between two different solutions. In analytical ultracentrifugation, a boundary is produced when molecules are sedimented through the solution; in diffusion and in electrophoresis experiments, using a Tiselius apparatus, a boundary is produced by the layering of the solvent over the solution.

boundary sedimentation. Sedimentation in which a boundary is formed, as in sedimentation velocity.

bound enzyme. INSOLUBLE ENZYME.

bound insulin. INSULIN-LIKE ACTIVITY.

Boveri's theory of cancer. See chromosome theory of cancer.

bovine. Of, or pertaining to, cattle.

Bowman-Birk inhibitor. A soybean trypsin inhibitor.

b.p. Boiling point; also abbreviated b.pt.

BPA. Bovine plasma albumin.

B protein. A protein subunit of the enzyme tryptophan synthetase.

Bradshaw test. A test for the presence of Bence-Jones protein in urine that is performed by layering dilute, acidified urine over concentrated hydrochloric acid.

bradykinin. A nonapeptide kinin.

Bragg angle. The angle of incidence, which equals the angle of reflection, in the Bragg equation.

Bragg curve. A plot of specific ionization as a function of either distance or energy.

Bragg equation. An equation relating the angle at which either light rays or x-rays are reflected

from a crystal to the spacing of the atomic planes in the crystal; specifically, $2d \sin \theta = n \lambda$, where θ is both the angle of incidence and the angle of reflection, λ is the wavelength of the radiation, d is the spacing between reflecting planes, and n is an integer.

Bragg law. BRAGG EQUATION.

Bragg peak. The peak of a Bragg curve.

Bragg scattering. The scattering of radiation by a crystal that is described by the Bragg equation.

brain barrier system. BLOOD-BRAIN BARRIER.

brain hormone. A hormone that is produced in the brain of insects and that controls metamorphosis.

brain sparing. The phenomenon that, during starvation, the loss of bulk matter from the brain is smaller than that from other organs.

branched fatty acid. A fatty acid having a branched chain. *See also* ante-iso fatty acid; iso fatty acid.

branched metabolic pathway. A sequence of metabolic reactions that diverges and that can give rise to two or more end products.

branched polymer. A polymer that consists of a main chain to which side chains are attached.

branching. The simultaneous decay of a given type of radioactive atoms by two different modes; a fraction of the atoms decays by one mode and the remaining fraction decays by the other mode.

branching enzyme. An enzyme that catalyzes the synthesis of branch points in a polymer.

branch migration. The process whereby a branch point of two interacting DNA molecules is displaced sequentially along the molecules as a result of their concerted dissociation and reassociation.

branch point. 1. The point in the main chain of a polymer at which either an additional molecule or a second chain is attached. 2. The point at which a sequence of metabolic reactions diverges so that it can give rise to two or more end products.

breakage and reunion model. The model of genetic recombination according to which parts of the parental chromosomes are exchanged as a result of physical breakage of the chromosomes and reunion of the broken fragments. *Aka* breakage hypothesis; break and exchange model.

breakdown potential. The potential at which a Geiger-Mueller tube begins to produce a continuous discharge.

Brei. HOMOGENATE.

Bremsstrahlung. An electromagnetic radiation that is produced when high energy beta particles are decelerated in the electrostatic fields of atomic nuclei.

brewer's yeast. One of several strains of yeast belonging to the species *Saccharomyces cerevisiae.*

bridging atom. An atom that connects two groups in a molecule, such as the oxygen atom that connects two phosphoryl groups in ATP. *Aka* bridge atom.

Briggs-Haldane equation. The rate equation for an enzymatic reaction which is derived on the assumptions that (a) a steady state attains for the enzyme substrate complex, and (b) the velocity of the reaction is an initial velocity, proportional to the concentration of enzyme-substrate complex, so that the reverse reaction from products to enzyme-substrate complex can be neglected. Specifically, $v = V[S]/(K_m + [S])$, where v is the initial velocity of the reaction, V is the maximum velocity, $[S]$ is the substrate concentration, and K_m is the Michaelis constant. *See also* Michaelis-Menten equation.

Briggsian logarithm. *See* logarithm.

Brij. One of a number of polyoxyethylene ethers of higher aliphatic alcohols that are used as nonionic detergents for the solubilization of membrane fractions.

British Anti-Lewisite. The sulfhydryl compound, 2,3-dimercaptopropanol, developed during World War II as a detoxicant for certain war poisons and used subsequently as a chelating agent for heavy metal ions. *Abbr* BAL.

British thermal unit. The amount of heat required to raise the temperature of 1 lb of water by 1°F (From 63 to 64°F). *Abbr* BTU.

brittle diabetes. A disease in which lack of glucose tolerance and sufficiency of insulin activity vary in an unpredictable manner so that glucose and ketone bodies may be present in the urine at one time, and acceptable levels of insulin activity may be present at another time.

broad-beta lipoprotein. FLOATING BETA LIPOPROTEIN.

broad-spectrum antibiotic. An antibiotic, such as chloramphenicol or a tetracycline, that has a wide range of antibacterial activity.

Brodie's solution. A salt solution, used in the Warburg manometer, that has a density of 1.033 g/cc, so that a column 10,000 mm high is equivalent to a pressure of 1 atm; a typical solution contains 46 grams of sodium chloride and 10 grams of sodium cholate per liter of solution.

bromatology. Food science.

bromine. An element that is essential to a wide variety of species in one class of organisms. Symbol Br; atomic number 35; atomic weight 79.909; oxidation states -1, $+1$, $+5$; most abundant isotope Br^{79}; a radioactive isotope Br^{82}, half life 35.4 hours, radiation emitted—beta particles and gamma rays.

5-bromodeoxyuridine. A thymidine analogue used as an antiviral drug. *Abbr* BUDR. *Aka* 5-bromouracildeoxyriboside.

5-bromouracil. A mutagenic pyrimidine analogue that is readily incorporated into DNA in place of thymine to produce transitions. *Abbr* BU.

Bronsted acid. A molecule or an ion that acts as a proton donor.

Bronsted base. A molecule or an ion that acts as a proton acceptor.

Bronsted catalysis law. A quantitative expression of the fact that general acid or base catalysis of a given reaction is primarily a function of the acid or base strength of the catalyst. This may be stated as $log\ k = a\ log\ K + b$, where k is the rate constant of the reaction, a and b are constants, and K is either the dissociation constant in the case of an acid catalyst or the reciprocal of the dissociation constant of the conjugate acid in the case of a base catalyst.

Bronsted-Lowry theory. The theory that describes acids as proton donors and bases as proton acceptors.

Bronsted plot. A plot of the logarithm of the rate constant versus the negative logarithm of the dissociation constant. *See also* Bronsted catalysis law.

bronzed diabetes. A disease characterized by the presence of excessive amounts of glucose in the urine as a result of the deposition of iron in the pancreas, liver, and other organs. *See also* hemochromatosis.

Brownian motion. The random, thermal motion of solute molecules that is due to their continual bombardment by molecules of the solvent. *Aka* Brownian movement.

browning reactions. A group of complex reactions, both enzymatic and nonenzymatic, that occur in various foods upon processing and/or storage; the enzymatic reactions are thought to involve the oxidation of phenolic compounds, and the nonenzymatic reactions are thought to involve caramelization, decomposition of ascorbic acid, and the Maillard reaction.

BrUrd. 5-Bromouridine.

brush border. A striated border that is present on cells that form certain epithelial membranes.

BSA. Bovine serum albumin.

BSV. Bushy stunt virus.

BTU. British thermal unit.

BU. 5-Bromouracil.

BUDR. 5-Bromodeoxyuridine.

bufadienolide. A steroid that is a 24-carbon homologue of a cardenolide and occurs as a cardiac glycoside in plants and toads.

buffer. A solution containing a mixture of a weak acid and its conjugate weak base that is capable of resisting substantial changes in pH upon the addition of small amounts of acidic or basic substances.

buffer capacity. 1. The number of equivalents of either protons or hydroxyl ions that is required to change the pH of 1 liter of a 1M buffer by one pH unit; equal to $(1/m)(dn/dpH)$, where m is the number of moles of buffer, and dpH is the change in pH produced by the addition of dn equivalents of either protons or hydroxyl ions. 2. BUFFER VALUE.

buffer index. BUFFER VALUE.

Buffer value. The number of equivalents of either protons or hydroxyl ions that is required to change the pH of a buffer by one pH unit; equal to dn/dpH, where dpH is the change in pH produced by the addition of dn equivalents of either protons or hydroxyl ions.

building block. A molecule that serves as a structural unit in a biopolymer; a biomonomer.

BUN. Blood urea nitrogen.

Bunsen absorption coefficient. The number of gas volumes that are dissolved by one volume of liquid when the liquid is equilibrated with the gas under 1 atm of pressure; the gas volume is calculated as that volume occupied by the gas at 0°C and 1 atm of pressure. *Aka* Bunsen solubility coefficient.

Bunsen-Roscoe law. RECIPROCITY.

Bunsen solubility coefficient. BUNSEN ABSORPTION COEFFICIENT.

buoyancy factor. The term $1 - \bar{v}\rho$ that appears in equations pertaining to hydrodynamic methods of studying macromolecules, where $\bar{v}$ is the partial specific volume of the solute and ρ is the density of the solution.

buoyant density. The density of a molecule as determined by density gradient sedimentation equilibrium.

buret. A graduated tube with stopcock used for delivery of known volumes of liquid as in a titration. *Var sp* burette.

buried residue. MASKED RESIDUE.

Burkitt's lymphoma. A lymphoma, afflicting children in Africa, from which the Epstein-Barr virus has been isolated.

Burnet's theory. CLONAL SELECTION THEORY.

burst. 1. The rapid presteady-state release of the first product in a ping pong mechanism. 2. The explosion of a phage-infected bacterial cell that is accompanied by the release of phage particles into the medium.

burst size. The average number of phage particles released per infected bacterial cell; equal to the ratio of the final titer to the initial titer in a phage multiplication cycle.

Busch theory. A theory of carcinogenesis that postulates three stages of disease referred to as initiation, promotion, and acceleration. According to this theory, a portion of the DNA is responsible for carcinogenesis and is inhibited in normal cells by combination with a suppressor protein. The carcinogen binds to the cancer DNA-suppressor complex in the initiation stage, releases the DNA in the promotion stage, and allows it to form cancer RNA in the acceleration stage.

bushy stunt virus. TOMATO BUSHY STUNT VIRUS.

butanediol fermentation. BUTYLENE GLYCOL FERMENTATION.

butylene glycol fermentation. The fermentation of glucose, characteristic of *Aerobacter aerogenes* and related forms, that yields primarily butylene glycol and ethanol, and secondarily a number of other products.

butyric-butylic fermentation. The fermentation of glucose that yields butyric acid, *n*-butanol, acetone, and isopropanol in varying proportions.

B vitamin. *See* vitamin B complex.

by-product. A minor product in a chemical reaction.

C

c. 1. Concentration. 2. Curie. 3. Centi.

C. 1. Cytosine. 2. Cytidine. 3. Cysteine. 4. Carbon. 5. Degree centigrade (Celsius). 6. Complement. 7. Coulomb. 8. Heat capacity.

C^{14}. A radioactive isotope of carbon that has a half-life of 5730 years and emits beta particles.

Ca. Calcium.

cachexia. The malnutrition and wasting of bodily tissues that is produced by chronic diseases, such as the drain on host nutrients produced by the proliferation of cancer cells.

cadaverine. A five-carbon polyamine that contains two amino groups.

cadmium. An element that is essential to a wide variety of species in one class of organisms. Symbol Cd; atomic number 48; atomic weight 112.40; oxidation state +2; most abundant isotope Cd^{114}; a radioactive isotope Cd^{109}, half-life 453 days, radiation emitted—gamma rays.

cage. A cavity or enclosed region in the solvent structure into which solute molecules can fit. *See also* clathrate.

Cairns experiment. An experiment that provided evidence for the existence of one replicating fork per molecule of DNA undergoing replication. The experiment consisted of labeling the DNA of growing cells of *Escherichia coli* with tritiated thymine, isolating the DNA by mild procedures, and determining the distribution of label in the DNA by means of radioautography.

cal. Small calorie.

Cal. Large calorie; a kilocalorie, equal to 1000 small calories.

calciferol. A compound that has vitamin D activity and that is obtained by ultraviolet irradiation of ergosterol; denoted vitamin D_2.

calcification. The formation of calcium salt deposits in a tissue.

calcified. Having undergone calcification.

calcitonin. A polypeptide hormone that lowers the level of calcium in the blood and that is secreted by both the thyroid and the parathyroid glands. *Abbr* CT.

calcium. An element that is essential to all plants and animals. Symbol Ca; atomic number 20; atomic weight 40.08; oxidation state +2; most abundant isotope Ca^{40}; a radioactive isotope Ca^{45}, half-life 165 days, radiation emitted—beta particles.

calcium phosphate gel. A gel prepared from calcium chloride and trisodium phosphate and used in the purification of proteins by adsorption chromatography.

calcium pump. The structure and/or the mechanism that mediates the active transport of calcium across a biological membrane.

calculus (*pl* calculi). A hard aggregate or stone that is found in the body and that may consist chiefly of inorganic matter, as in the case of kidney stones, or of organic matter, as in the case of the uric acid stones associated with gout.

calibration. The standardization and graduation of a measuring instrument.

calibration curve. STANDARD CURVE.

callose. A linear homopolysaccharide that occurs in higher plants and that is composed of D-glucose units linked by means of $\beta(1 \rightarrow 3)$ glycosidic bonds.

calomel. Mercurous chloride.

calomel electrode. A reference electrode for pH measurements that contains mercury, mercurous chloride, and a saturated solution of potassium chloride.

caloric intake. The caloric equivalent of the food ingested, calculated on the basis of the energy yield obtainable by complete oxidation of the food.

calorie. A measure of energy equal to the amount of heat required to raise the temperature of 1.0 gram of water by 1°C (from 14.5 to 15.5°C) at a pressure of 1 atm. *Abbr* cal. *Aka* small calorie.

caloric value. The quantity of heat, generally expressed in kilocalories per gram, that is released when a foodstuff is subjected to complete oxidation; the heat of combustion of a foodstuff. *Aka* calorific value.

calorigenesis. The capacity of a substance to generate heat in an organism.

calorigenic. Of, or pertaining to, calorigenesis.

calorigenic action. SPECIFIC DYNAMIC ACTION.

calorimeter. An instrument for measuring the heat that is either absorbed or released by a chemical reaction or by a group of chemical reactions.

calorimetry. The measurement of the heat changes in a chemical reaction or a group of chemical reactions, either in an in vitro system or in an intact organism.

Calvin cycle. A cyclic set of reactions, occurring in chloroplasts, that result in the fixation of carbon dioxide and in its conversion to glucose by means of the ATP and the NADPH formed in the light reaction of photosynthesis. *Aka* Calvin-Bassham cycle; Calvin-Benson cycle.

camera lens. A lens that focuses an image on a photographic plate.

cAMP. Cyclic AMP.

Campbell model. *See* insertion model.

cancer. 1. A disease of multicellular organisms that is characterized by seemingly uncontrolled cellular growth and by the spreading within the organism of apparently abnormal forms of the organism's own cells; cancer cells thus show excessive multiplication, autonomy with respect to the host, and invasiveness (metastasis). 2. A malignant tumor.

cancer biochemistry. The biochemistry of cancer cells. *See also* first law of cancer biochemistry; second law of cancer biochemistry.

cancer gene. ONCOGENE.

cancer-inducing virus. ONCOGENIC VIRUS.

cancerocidal. Capable of killing cancer cells.

cancerogenesis. CARCINOGENESIS.

cancer theory. *See* theory of cancer.

cancroid. 1. *n* A skin tumor of low malignancy. 2. *adj* Cancer-like.

cane sugar. Sucrose isolated from sugar cane.

canine. Of, or pertaining to, dogs.

canonical structure. Any one of the possible resonance structures of a compound.

CAP. 1. Cyclic AMP receptor protein. 2. Catabolite activator protein.

capacitance. An electrical unit equal to the total charge that can be stored in a condenser divided by the potential difference across the plates.

capillarity. The action by which the surface of a liquid, where it is in contact with a solid, is either elevated or depressed as a result of the relative attractions of the molecules of the liquid for each other and for the molecules of the solid.

capillary. 1. *n* A tube or a vessel having a very small diameter. 2. *adj* Of, or pertaining to, a tube or a vessel having a very small diameter.

capillary action. CAPILLARITY.

capillary attraction. The force of adhesion between a solid and a liquid in capillarity.

capillary precipitin test. RING TEST.

capillary viscometer. An instrument for measuring the viscosity of a liquid from the time required for a given volume of the liquid to flow through a capillary. *See also* Ostwald viscometer.

capneic. MICROAEROPHILIC.

capsid. The protein coat, composed of capsomers, that surrounds the nucleic acid of a virus and determines the overall shape of the virus.

capsomer. The morphological unit, one or more of which constitute the viral capsid. The capsomer, in turn, consists of one or more structural units, or monomers. A capsomer that consists of five structural units is known as a pentagonal capsomer, or pentamer, and a capsomer that consists of six structural units is known as a hexagonal capsomer, or hexamer. *Var sp* capsomere.

capsular polysaccharide. A polysaccharide that is a component of a bacterial capsule and that is frequently antigenic.

capsule. A loose gel- or slime-like structure that is rich in polysaccharides and that frequently coats the outer surface of a bacterial cell wall.

capture cross section. The product of the probability that a particle impinging on an atomic nucleus will be captured by that nucleus and the cross-sectional area of the nucleus (in barns).

caramelization. The browning of sugars when they are heated above their melting points.

carbamino compound. A compound formed by the reaction of carbon dioxide with a primary aliphatic amine.

carbaminohemoglobin. The carbamino compound that is formed by the reaction of hemoglobin with carbon dioxide and that represents one of the forms in which carbon dioxide is transported by the blood.

carbamyl group. The acyl group of carbamic acid; the radical H_2NCO-. *Aka* carbamoyl group.

carbamyl phosphate. The high-energy compound $NH_2-COO-PO_3H_2$ that functions in the urea cycle, ammonia fixation, and pyrimidine biosynthesis. *Aka* carbamylating agent.

carbanion. A carbon anion; the species $R_3C:^-$ in which the carbon atom carries an unshared pair of electrons. A carbanion is formed by removal of a group attached to the carbon atom without removing the pair of bonding electrons.

carbene. A neutral organic compound that contains a divalent carbon atom and that is formed by removal of two groups attached to one carbon atom together with one pair of the bonding electrons.

carbocyclic. Of, or pertaining to, an organic compound that has a ring structure consisting only of carbon atoms.

carbohydrase. An enzyme that catalyzes the hydrolysis of carbohydrates.

carbohydrate. An aldehyde or a ketone derivative of a polyhydroxy alcohol that is synthesized by living cells. Carbohydrates may be classified either on the basis of their size into mono-, oligo-, and polysaccharides, or on the basis of their functional group into aldehyde or ketone derivatives.

carbohydrate tolerance test. *See* glucose tolerance test; galactose tolerance test.

carbolic acid. PHENOL (1).

carboligase. The enzyme that catalyzes the formation of acetoin from acetaldehyde and active acetaldehyde.

carbometer. An instrument for measuring the carbon dioxide content of breath.

carbon. An element that is essential to all plants and animals. Symbol C; atomic number 6; atomic weight 12.01115; oxidation states -4, $+2$, $+4$; most abundant isotope C^{12}; the stable isotope C^{13}; a radioactive isotope C^{14}, half-life 5730 years, radiation emitted—beta particles.

carbonaceous. Consisting in part, or entirely, of carbon.

carbon assimilation. CARBON DIOXIDE FIXATION.

carbon chain. A chain of carbon atoms.

carbon clearance test. *See* phagocytic index (2).

carbon clock. The radioactive isotope of carbon, C^{14}, that is used in radiocarbon dating for establishing the age of biological remains.

carbon cycle. 1. The set of reactions whereby photosynthetic organisms reduce carbon dioxide to carbohydrates and heterotrophic organisms oxidize the carbohydrates back to carbon dioxide. 2. CALVIN CYCLE. 3. A series of thermonuclear reactions in the sun believed to be responsible for the energy released by it.

carbon dating. *See* radiocarbon dating.

carbon dioxide assimilation. CARBON DIOXIDE FIXATION.

carbon dioxide capacity of plasma. *See* carbon dioxide combining power of plasma.

carbon dioxide combining power of plasma. The maximum amount of carbon dioxide that 100 ml of plasma can hold in the form of bicarbonate when the plasma is saturated with carbon dioxide at a tension corresponding to the tension of carbon dioxide in normal arterial blood.

carbon dioxide fixation. The photosynthetic conversion of carbon dioxide to carbohydrates. *See also* Calvin cycle.

carbon dioxide transport. The carrying of carbon dioxide by the blood from the tissues to the lungs.

carbon fixation. CARBON DIOXIDE FIXATION.

carbonic anhydrase. The enzyme, located in the erythrocytes, that catalyzes the reversible decomposition of carbonic acid to carbon dioxide and water in the course of respiration.

carbonium ion. A carbon cation; the species R_3C^+ that is formed by the removal of a group attached to a carbon atom together with the pair of bonding electrons.

carbon monoxide hemoglobin. CARBOXYHEMOGLOBIN.

carbon number. EQUIVALENT CHAIN LENGTH.

carbon radical. A radical formed from a compound by removal of a group attached to a carbon atom together with one of the two bonding electrons.

carbon reduction cycle. CALVIN CYCLE.

carbon skeleton. The structure of a molecule considered solely in terms of its carbon atoms.

carboxybiotin. A biotin molecule to which a molecule of carbon dioxide has been attached.

carbonyl group. The grouping $-\overset{\displaystyle O}{\overset{\displaystyle \|}{C}}-$ that occurs in aldehydes and ketones.

carboxyhemoglobin. A hemoglobin derivative in which the sixth coordination position of the iron is occupied by carbon monoxide. *Abbr* HbCO.

carboxylation. The introduction of a molecule of carbon dioxide into an organic compound.

carboxylesterase. An enzyme of low specificity that catalyzes the hydrolysis of esters of carboxylic acids.

carboxyl group. The radical $-COOH$ of an organic acid.

carboxylic acid. An organic compound containing the carboxyl group; an organic acid.

carboxyl terminal. C-TERMINAL.

carboxy-lyase. An enzyme that catalyzes a decarboxylation reaction.

carboxypeptidase. An exopeptidase that catalyzes the hydrolysis of amino acids in a polypeptide chain from the C-terminal. Carboxypeptidase *A* catalyzes the hydrolysis of most amino acids and leads to the sequential degradation of the polypeptide chain from the C-terminal; carboxypeptidase *B* catalyzes only the hydrolysis of C-terminal lysine and C-terminal arginine.

carboxysome. A polyhedral inclusion body, present in some blue-green algae, that contains the enzyme D-ribulose-1,5-diphosphate carboxy-lyase and that functions in the fixation of carbon dioxide by these organisms.

Carbowax. Trademark for polyethylene glycol.

carcinoembryonic antigen. A glycoprotein in blood plasma that, when present at higher than normal concentrations, has been linked to the occurrence of certain cancers. *Abbr* CEA.

carcinogen. A physical or a chemical agent that is capable of producing cancer.

carcinogenesis. The production of cancer.

carcinogenic. Capable of producing cancer.

carcinogenic index. A measure of the activity of a carcinogen; equal to $100A/B$, where A is the number of animals bearing a tumor divided by the number of animals living on the day of appearance of the first tumor, and B is the mean time in days of the appearance of tumors.

carcinogenicity. The capacity to produce cancer.

carcinogenophore. A grouping of atoms in a chemical carcinogen that is primarily responsible for the carcinogenic activity of the molecule. *See also* auxocarcinogen; K region; L region.

carcinoid. A cancer-like tumor of the gastrointestinal tract that grows slowly and rarely metastasizes.

carcinoma. A malignant tumor, derived from epithelial cells, that can occur in a variety of sites such as skin, breast, and liver.

carcinomatosis. A condition in which multiple carcinomas develop simultaneously in different parts of the body as a result of the widespread dissemination of a carcinoma from a primary source.

carcinosis. CARCINOMATOSIS.

cardenolide. A steroid that contains 23 carbon atoms and that is characterized by a 14 β-hydroxyl group and an α,β-unsaturated γ-lactone ring; cardenolides occur as cardiac glycosides in plants and insects.

cardiac. Of, or pertaining to, the heart.

cardiac genins. A collective term for the steroids present in cardiac glycosides; cardenolides and bufadienolides.

cardiac glycoside. One of a group of steroid glycosides, such as oubain and digitalis, that act directly on the heart muscle, increasing the force of systolic contraction and thereby improving cardiac output; they also cause mild vasoconstriction and influence the $Na^+—K^+$ transport across erythrocyte and other cell membranes.

cardiac muscle. The involuntary, striated muscle of the heart.

cardiac puncture. A technique for withdrawing blood from an animal by inserting a syringe directly into the heart.

cardiolipin. The phospholipid hapten, diphosphatidyl glycerol, that is used as an antigen for reacting with the Wasserman antibody in the Wasserman test for syphilis.

cardiovascular. Of, or pertaining to, the heart and the blood vessels.

carnitine. A compound that functions in beta oxidation by transporting fatty acyl groups across the inner mitochondrial membrane.

carnitine barrier. The limited ability of long-chain fatty acids to cross the inner mitochondrial membrane in the form of fatty acyl coenzyme A, as contrasted with their ability to cross the membrane in the form of fatty acyl carnitine.

carnosine. A dipeptide of β-alanine and histidine occurring in vertebrate muscle.

carotene. A hydrocarbon carotenoid.

carotenoid. A polyisoprenoid that may be linear or cyclic and that consists of eight isoprene units (a tetraterpenoid) linked largely in a head-to-tail manner. Carotenoids are water-insoluble pigments that occur in plants and photosynthetic bacteria and that frequently function as accessory pigments in photosynthesis.

carrier. 1. An element or a compound that is added to a sample as an aid in the chemical manipulation of the same, but labeled, element or compound which is present in the sample. 2. Any related or unrelated substance that is added to a sample as an aid in the chemical manipulation of another substance which is present in the sample. 3. A transport agent, generally a protein or an enzyme, that combines with a substance and transports it either across a biological membrane or within a biological fluid. 4. CARRIER GAS.

carrier ampholyte. The ampholyte that forms the pH gradient in isoelectric focusing, as distinct from the sample ampholyte which is fractionated.

carrier culture. A culture, infected with a virus, that nevertheless maintains the multiplication of both the cells and the virus particles. Such a culture can be obtained by making most of the cells of a culture resistant to viral infection as a result of the interferon produced by a small portion of virus-infected cells present in the same culture.

carrier displacement chromatography. Displacement chromatography in which either related or unrelated substances are added to the mixture being chromatographed as an aid in the chemical manipulation of the separated components.

carrier-facilitated diffusion. MEDIATED TRANSPORT.

carrier-free. Descriptive of a radioactive nuclide that is essentially free of its stable isotopes.

carrier gas. The inert gas that functions as the mobile phase in gas chromatography.

carrier protein. 1. A protein that functions as a transport agent. 2. The protein to which a hapten may be conjugated in vitro or to which it may become conjugated in vivo.

carrier state infection. A viral infection in which only a small portion of the cells are infected.

Carr-Price reaction. A colorimetric reaction for the determination of vitamin A that is based on the production of a blue color upon treatment of the sample with antimony trichloride.

cartilage. Connective tissue that consists largely of collagen and chondroitin sulfate.

cascade mechanism. 1. The sequence of successive activation reactions that constitute the process of blood clotting and that achieve a continuous amplification of the initial event due to the fact that the concentrations of the components in the blood increase step by step from the initiating factor to fibrinogen. 2. A sequence of successive activation reactions pertaining to either enzymes or hormones, such as the reaction sequence for the activation of phosphorylase or that for the activation of adrenocorticotropin.

casein. A phosphoprotein that is the principal protein in milk.

CAT. Acronym for a time-averaging computer that is used in nuclear magnetic resonance for increasing the signal-to-noise ratio.

catabolic. Of, or pertaining to, catabolism.

catabolic deletion hypothesis. An early formulation of the deletion hypothesis of cancer, according to which loss of one or more key catabolic enzymes through a deletion mutation in DNA was thought to increase the availability of building blocks for polymers and thereby permit continued cell growth. *Aka* catabolic enzyme deletion hypothesis.

catabolism. 1. The phase of intermediary metabolism that encompasses the degradative and energy-yielding reactions whereby nutrients are metabolized. 2. The cellular breakdown of complex substances and macromolecules to low-molecular weight compounds.

catabolite. Any metabolic intermediate produced in the catabolism of food molecules.

catabolite activator protein. A protein in *Escherichia coli* that binds to, and is activated by, cyclic AMP and that is necessary for the efficient transcription of certain operons which are subject to catabolite repression. *Abbr* CAP.

catabolite gene activator protein. CATABOLITE ACTIVATOR PROTEIN.

catabolite repression. The inhibition of the synthesis of a number of enzymes by a single compound, such as glucose, from a major metabolic pathway; believed to be due to an increase in the concentration of a catabolite produced from the compound in that pathway.

catabolite-sensitive operon. An operon that is inhibited by glucose or its catabolites.

catalase. The hemoprotein enzyme that catalyzes the decomposition of hydrogen peroxide to oxygen and water.

catalysis. The change in the rate of a chemical reaction, generally an increase, that is brought about by the action of a catalyst. *See also* acid-base catalysis; covalent catalysis; etc.

catalyst. A substance that changes the rate of a chemical reaction, generally increasing it. A catalyst remains either unchanged during the reaction or is regenerated in its original form at the end of the reaction. A catalyst functions by changing the activation energy of the rate-determining step, by affecting the orientation of molecules in collision, or by making possible another pathway or mechanism that has a different activation energy.

catalytic. Of, or pertaining to, a catalyst or catalysis.

catalytic amount. The amount of substance that is used in a chemical reaction for catalytic purposes and that is much smaller than the stoichiometric amount of either a reactant or a product; a catalyst, a primer, and a sparker are all used in catalytic amounts.

catalytic center. 1. CATALYTIC SITE. 2. A region within the catalytic site.

catalytic center activity. A measure of enzymatic activity that is equal to the number of molecules of substrate transformed into products per minute per catalytic center of the enzyme. *See also* molar activity.

catalytic constant. MOLECULAR ACTIVITY.

catalytic exchange method. A method for randomly labeling a compound with tritium by dissolving the compound in a tritiated hydroxylic solvent in the presence of metal catalysts and either acid or base.

catalytic reduction method. A method for labeling a compound with tritium in which the compound is dissolved in a nonhydroxylic solvent and double bonds in the compound are reduced by exposure of the solution to tritium gas.

catalytic site. The active site of an enzyme, specifically the active site of a regulatory enzyme as distinct from its allosteric site.

catalytic subunit. The subunit of the regulatory enzyme aspartate transcarbamylase that has enzymatic activity but does not bind the negative effector CTP. *See also* regulatory subunit.

catalyze. To change the rate of a chemical reaction through catalysis.

cataphoresis. 1. The movement of charged particles toward the cathode. 2. ELECTROPHORESIS.

catatoxic steroid. A steroid that protects an animal against a drug by stimulating the activity of drug-metabolizing enzymes.

catch muscle. A muscle, occurring in mollusks, that can remain locked in a contracted form for long periods of time.

catecholamine. A dihydroxyphenylalkylamine derived from tyrosine, such as dopa, dopamine, epinephrine, or norepinephrine.

catechol-O-methyl transferase. An enzyme that functions in the metabolism of epinephrine and norepinephrine. *Abbr* COMT.

catemer. An oligomeric nucleic acid molecule in which complete genomes are held together in an end-to-end manner by either covalent or noncovalent bonds.

catenane. A catenated structure.

catenase. A collective term for an enzyme of either the endo- or the exo-type that catalyzes the cleavage of a polymeric chain; ribonuclease, lysozyme, and carboxypeptidase are examples.

catenated. Interlocked like the links in a chain;

said of interlocking, circular, single-stranded DNA molecules.

catenated dimer. A DNA molecule, present in mitochondria, that consists of two interlocking, circular, single strands.

catharometer. Variant spelling of katharometer.

cathepsin. One of a group of intracellular proteolytic enzymes that occur in most animal tissues.

cathode. The negative electrode by which electrons enter a solution of electrolytes and toward which the cations move in solution.

cathodic. 1. Of, or pertaining to, the cathode. 2. Descriptive of a component that moves toward the cathode in electrophoresis.

cation. A positively charged ion.

cation exchanger. An ion-exchange resin that is negatively charged and that binds cations.

cationic detergent. A surface-active agent in which the surface-active part of the molecule carries a positive charge. *Aka* cationic surface-active agent.

CBG. Cortisol-binding globulin.

CBN. Commission on Biochemical Nomenclature of the International Union of Pure and Applied Chemistry and the International Union of Biochemistry.

CBZ-amino acid. An amino acid in which the amino group has been protected by attachment of a carbobenzoxy group; used in peptide synthesis by the solid phase (Merrifield) method. *Abbr* CBZ-AA.

cc. Cubic centimeter.

CCD. Countercurrent distribution.

cd. Candela; a unit of luminous intensity.

CD. Circular dichroism.

C^{14}-dating. *See* radiocarbon dating.

cDNA. Complementary DNA; a molecule of DNA that is complementary to a molecule of RNA.

CDP. 1. Cytidine diphosphate. 2. Cytidine-5′-diphosphate.

CDPC. CDP-choline.

CDP-choline. A cytidine diphosphate derivative of choline that serves as a donor of a choline residue for the synthesis of certain phosphoglycerides.

Celite. Trademark for a preparation of diatomaceous earth.

cell. 1. The fundamental unit of living organisms; a structure that is capable of independent reproduction and that consists of cytoplasm and a nucleus, or a nuclear zone, surrounded by a cell membrane. 2. An electrical device capable of converting chemical energy into electrical energy, or vice versa; consists of two half-cells, each of which is characterized by a half-reaction. 3. A small container, as that which holds a solution subjected to centrifugation in an analytical ultracentrifuge.

cell-associated virus. The virus that is released into the medium upon the disruption of infected cells which have previously been washed to remove extracellularly adsorbed virus.

cell body. PERIKARYON.

cell coat. The covering of the outer surface of many eucaryotic cells that is rich in glycoproteins and mucopolysaccharides; it plays a role in contact inhibition.

cell count. *See* total cell count; viable cell count.

cell culture. The in vitro growth of either single cells or groups of cells that are not organized into tissues.

cell cycle. The sequence of events occurring in a eucaryotic cell from one mitotic division to the next. The cell cycle is commonly divided into four periods: M, mitosis; G_1, a growth period following mitosis; S, a period following G_1 and characterized by DNA synthesis; and G_2, a second growth period following S.

cell-detaching factor. PENTON.

cell disruption. The breakage of cells.

cell division. The process whereby a parent cell divides, giving rise to two daughter cells.

cell envelope. CELL MEMBRANE.

cell factor. A protein that is produced by cells of solid tumors, occurring in man and animals, and that may be produced in very small quantities by normal cells. The cell factor affects a protein in serum, the serum factor, which leads to proteolysis of fibrin.

cell fractionation. The fractionation of subcellular components.

cell-free amino acid incorporating system. A reconstituted cell-free system for the in vitro study of protein synthesis. It generally consists of ribosomes, messenger RNA (natural or synthetic), transfer RNA, enzymes, amino acids, ATP, GTP, an ATP regenerating system, buffer, and other inorganic and organic compounds.

cell-free extract. A cytoplasmic extract of cells, prepared by rupturing the cells and removing unbroken cells and cell debris, commonly by centrifugation.

cell-free protein synthesis. *See* cell-free amino acid incorporating system.

cell-free system. A system composed of subcellular fractions and/or cell-free extracts, but devoid of intact cells.

cell fusion. HYBRIDIZATION (3).

cell hybridization. HYBRIDIZATION (3).

cell line. A group of cells that are derived from a primary culture and that can be subjected to indefinite serial cultivation.

cell-mediated immunity. CELLULAR IMMUNITY.

cell membrane. The membrane, composed of lipids and proteins, that surrounds a cell; in eucaryotic cells the cell membrane is frequently covered by a cell coat, and in procaryotic and plant cells it is covered by a cell wall.

cellobiose. A disaccharide of glucose in which the glucose molecules are linked by means of a $\beta(1 \to 4)$ glycosidic bond; the repeating unit in cellulose.

cellogel. Gelatinized cellulose acetate; an electrophoretic support.

Cellophane. Trade name for transparent sheets of regenerated cellulose.

cell plate. The structure that is formed between the two daughter nuclei of a dividing plant cell during mitosis and that is the precursor of the cell wall.

cell renewal system. A steady-state normal cell population in an animal in which there is a rapid cell loss that is offset by a rapid replacement of cells.

cell strain. A group of cells of limited transferability that have been derived from either a primary culture or an established cell line by selection and cloning of cells that have specific properties or markers.

cell theory. The theory, proposed by Schleiden and Schwann in 1838, that all animals and plants are composed of cells and products of cells, that cells are the structural and functional units of an organism, and that an organism grows and reproduces by cell division.

cellular. Of, or pertaining to, cells.

cellular immunity. Immunity that is due to cell-bound antibodies, particularly those bound to small lymphocytes; cellular immunity is responsible for such reactions as allograft rejection and delayed-type hypersensitivity.

cellular respiration. *See* respiration (1).

cellulase. An enzyme that catalyzes the hydrolysis of cellulose.

cellulose. A straight chain polysaccharide composed of glucose molecules linked by means of $\beta(1 \to 4)$ glycosidic bonds; the major structural material in the plant world.

cellulose acetate electrophoresis. Zone electrophoresis in which a cellulose acetate sheet is used as the supporting medium.

cell wall. The rigid structure that is external to the cell membrane and that encloses procaryotic and plant cells; the cell wall of procaryotic cells consists primarily of peptidoglycan and that of plant cells consists primarily of cellulose.

Celsius temperature scale. CENTIGRADE TEMPERATURE SCALE.

cementum. The calcified covering of dentine at the submerged portion of a tooth.

Cenozoic era. The most recent geologic time period that began about 63 million years ago and that is characterized by the development of mammals.

center of symmetry. The central point of a sym-

metrical body about which like faces are arranged in opposite pairs.

centi-. Combining form meaning one hundredth and used with metric units of measurement. *Sym* c.

centigrade temperature scale. A temperature scale on which the freezing and boiling points of water at 1 atm of pressure are set at 0 and 100, respectively, and the interval between these two points is divided into 100 deg.

centile. PERCENTILE.

central complex. The transitory form of a complex, composed of an enzyme and one or more compounds, that can undergo, or isomerize to undergo, unimolecular degradation to release either substrates or products.

central dogma. A description of the basic functional relationships between DNA, RNA, and protein. The central dogma states that DNA serves as the template for its own replication and for the transcription of RNA, and that RNA serves as the template for translation into protein. Hence the flow of genetic information is $\overset{\frown}{D}NA \to RNA \to Protein$. *See also* reverse transcriptase.

central metabolic pathway. The metabolic pathway composed of the reactions of the citric acid cycle and some of the reactions of glycolysis. The pathway occupies a central position in metabolism, since it can be used either catabolically for the oxidation of metabolites to carbon dioxide and water, or anabolically for the synthesis and interconversion of metabolites. *Aka* amphibolic pathway.

central nervous system. That part of the nervous system of vertebrates that consists of the brain, the spinal cord, and the nerves originating therefrom. *Abbr* CNS.

centrifugal elutriation. A centrifugal separation technique that is based on the sedimentation of particles through a liquid which flows in a direction opposite to the direction of particle sedimentation. A specially designed rotor is used so that particles which sediment more slowly than the flow velocity of the liquid are washed out from the rotor. The separation process is controlled by varying the flow velocity of the liquid and by varying the rotation rate of the rotor.

centrifugal field. The space within which a centrifugal force is of sufficient strength that its effect can be detected.

centrifugal force. The force exerted on a rotating particle and directed away from the center of rotation; the force increases with increasing distance from the center of rotation.

centrifugation. The process of subjecting either a solution or a suspension to a centrifugal force to separate the components of the solution or the suspension; used for the collection of precipitates, the separation of phases, and the sedimentation of macromolecules. Separation of the components is based on differences in their size, shape, and density.

centrifuge. An instrument capable of generating centrifugal forces by the rotation of a rotor; the rotor holds tubes filled with the solution that is being subjected to centrifugation.

centrifuge cell. *See* cell (3).

centrifuge head. CENTRIFUGE ROTOR.

centrifuge rotor. *See* rotor.

centrifuge tube. The container, constructed of glass, metal, or plastic, that holds the solution that is subjected to centrifugation.

centriole. The central granule in the centrosome.

centripetal force. The force that is exerted on a rotating particle and that is directed toward the center of rotation.

centromere. The junction between the two arms of a chromosome to which the spindle fibers attach during mitosis.

centrosome. A self-replicating cytoplasmic organelle, occurring in animal cells and lower plant cells, that organizes the mitotic apparatus during cell division. *Aka* centrosphere.

cephalin. 1. PHOSPHATIDYL ETHANOL-AMINE. 2. PHOSPHATIDYL SERINE.

cephalin-cholesterol flocculation test. A liver function test that is based on the formation of a flocculant precipitate when serum from individuals with one of several forms of hepatitis is treated with a cephalin-cholesterol suspension.

cephalosporin. An antibiotic, produced by the mold *Cephalosporium*, that resembles penicillin in its action.

cer. Ceramide.

ceramide. An *N*-acyl fatty acid-substituted compound formed from sphinganine, its homologues, its isomers, or its derivatives. *Abbr.* cer. *See also* glycosyl ceramide.

ceramide glycoside. *See* glycosyl ceramide.

ceramide lactoside. *See* cytolipin.

cerebral. Of, or pertaining to, the brain.

cerebrocuprein. A copper-containing protein present in the brain.

cerebroside. A monoglycosyl derivative of a ceramide that generally contains either glucose or galactose and that is abundant in the myelin sheath of nerves and in brain tissue.

cerebrospinal fluid. The fluid that circulates through the subarachnoid spaces of the brain and the spinal cord. *Abbr* CSF.

Cerenkov radiation. A radiation consisting of photons and produced when high energy beta particles pass through either a solid or a liquid medium at speeds greater than that of light in the same medium.

ceride. A wax that is an ester of a long chain fatty acid and a higher aliphatic alcohol.

ceroid. A lipid granule that may be formed in an animal, particularly in the liver, as a result of either the injection of oils rich in unsaturated fatty acids or the ingestion of various experimental diets.

ceruloplasmin. A serum globulin that binds eight atoms of copper per molecule and that serves to transport copper in the blood.

cesium chloride gradient centrifugation. *See* density gradient sedimentation equilibrium.

cevitaminic acid. ASCORBIC ACID.

CF. 1. Citrovorum factor. 2. Complement fixation.

CG. Chorionic gonadotropin.

cGMP. Cyclic GMP.

cgs units. The units of measurement that are based on the centimeter-gram-second system. *See also* SI.

c_H. The constant region of the heavy chains of the immunoglobulins. *See also* c_L.

Ch. Choline.

chain. 1. A group of like atoms linked together in succession. 2. A group of repeating units linked together in succession to form a polymer.

chain conformation. The combined secondary and tertiary structure of either a polypeptide chain or a polynucleotide strand.

chain elongation. *See* elongation.

chain initiation. *See* initiation.

chain isomer. One of two or more isomers that differ from each other in the manner in which the side chains are attached to the main chain and in the lengths of the chains.

chain length. *See* double chain length; triple chain length.

chain propagation. 1. CHAIN ELONGATION. 2. The second stage in a chain reaction.

chain reaction. 1. A series of chemical reactions characterized by initiation, propagation, and termination steps. The reactions of the propagation step are such that each produces a product that can serve as a reactant for a subsequent reaction and the last reaction regenerates a reactant for the first reaction; in this fashion the entire sequence of reactions in the propagation step can be repeated over and over. 2. An autocatalytic reaction, particularly a nuclear one, in which the products react with the reactants to produce more products.

chain termination. *See* termination.

chain termination codon. *See* termination codon.

chair conformation. The arrangement of atoms that resembles the outline of a chair and that is the more stable of two possible conformations for cyclohexane and other six-membered ring systems.

challenge. A dose of antigen, particularly the second or a subsequent dose, that is injected into an animal for the purpose of provoking an immune response.

challenge virus. The virus that is introduced into a host subsequent to, or simultaneously with, the introduction of an interfering virus.

chalone. 1. A chemical substance, such as a hypothalamic hormone, that acts like a hormone in having a target-specific effect but that is not secreted by an endocrine gland. 2. An internal secretion, such as enterogastrone, that has an inhibitory effect.

channel. The interval between the settings of the two discriminators in a scintillation detector that defines the range of pulse intensities that will be recorded by the system. *See also* differential counting.

channels ratio method. A method of correcting for quenching in liquid scintillation counting by using two channels to measure the average energies of pulses of beta particles both before, and after, quenching.

chaotropic agent. A substance that enhances the partitioning of nonpolar molecules from a nonaqueous to an aqueous phase as a result of the disruptive effect that the substance has on the structure of water. Chaotropic agents are generally ions, such as SCN^- and ClO_4^-, that have a large radius, a negative charge, and a low charge density; they are used to solubilize membrane-bound proteins, to alter the secondary and tertiary structure of proteins and nucleic acids, and to increase the solubility of small molecules.

chaotropic series. An arrangement of ions in the order of their effectiveness as chaotropic agents.

characteristic. The whole-number part of a logarithm.

characteristic curve. A plot of the potential applied to a radiation detector versus the count rate.

Chargaff's rules. A set of two quantitative rules that express the base composition of double-stranded, Watson-Crick type DNA: (1) $[A] = [T]$; (2) $[G] = [C]$, where the brackets indicate concentrations of the bases in mole percent and any minor bases, if present, are included in the appropriate concentration terms. Three corollaries follow from these rules: (a) $[A]/[T] = [G]/[C] = 1$; (b) Σ purines $= \Sigma$ pyrimidines; (c) $[A] + [C] = [G] + [T]$, or Σ 6-aminobases $= \Sigma$ 6-ketobases according to the former numbering system of purines and pyrimidines.

charge density. The net electrical charge of a particle per unit surface area of the particle.

charged polar amino acid. A polar amino acid that carries a charge in the intracellular pH range of about 6 to 7.

charged tRNA. A transfer RNA molecule to which an amino acid has been covalently linked; an aminoacyl-tRNA molecule.

charge effect. *See* primary charge effect; secondary charge effect.

charge fluctuation interactions. 1. KIRK-WOOD-SHUMAKER INTERACTIONS. 2. Any interaction between molecules and/or atoms that is due to fluctuating charges; Kirkwood-Shumaker interactions and London dispersion forces are two examples.

charge reversal spectrum. A series of mono- and polyvalent ions in which the ions are arranged in the order of their effective concentrations for reversing the charge of an oppositely charged molecule to which the ion can bind.

charge transfer complex. A complex that may be formed in oxidation-reduction reactions when, as a result of the electron transfer, the electron donor becomes positively charged and the electron acceptor becomes negatively charged so that the two are held together in a complex by electrostatic attraction.

charging. The covalent attachment of an amino acid to a transfer RNA molecule to form an aminoacyl-tRNA molecule.

chase. 1. The effective stoppage of the incorporation of either an isotope or a labeled compound into a substance by the addition of large amounts of either the nonradioactive element or the unlabeled compound; used particularly to stop the incorporation following a pulse. 2. The amount of nonradioactive element or unlabeled compound used to stop the incorporation of an isotope or a labeled compound.

chaulmoogric acid. An unsaturated, cyclic fatty acid that occurs in plants.

Chauvenet's criterion. A criterion for deciding whether or not to reject a measurement that differs greatly from other, identical measurements of the same sample. The criterion states that the measurement should be rejected if the probability of its occurrence is equal to, or less than, $N/2$ where N is the total number of measurements.

ChE. Cholinesterase.

chelate. The ring structure formed by the reaction of two or more groups on a ligand with a metal ion. *Aka* chelate compound.

chelating agent. A compound that can form a chelate with a metal ion.

chelation. The formation of a chelate.

chelator. CHELATING AGENT.

cheluviation. The downward movement of chelate complexes in the soil.

chemical. 1. *n* COMPOUND. 2. *adj* Of, or pertaining to, chemistry.

chemical bond. *See* bond.

chemical coupling hypothesis. A hypothesis of the coupling of ATP synthesis to the electron transport system in oxidative phosphorylation. According to this hypothesis, the transport of electrons leads to the formation of high energy compounds which are then used to phosphorylate ADP to ATP in coupled reactions using the high energy compounds as common intermediates.

chemical element. *See* element.

chemical evolution. The gradual development of the structure and function of biomolecules which includes the synthesis of primitive molecules, their condensation to form primitive polymers, and the self-assembly of these polymers to form large molecular aggregates; a process preceding the stage of biological evolution and extending over a period of about 2 billion years. *See also* Oparin's hypothesis.

chemical equilibrium (*pl* chemical equilibria). The state of a chemical reaction in which there is no more change in the concentrations of the reactants and the products, and the free energy is at a minimum; the rate of the forward reaction is equal to the rate of the reverse reaction so that a small change in one direction is balanced by a small change in the opposite direction.

chemical fossil. A fossil that contains one or more types of organic compounds that were part of the original animal or plant.

chemical interference. The interference that occurs in atomic absorption spectrophotometry if chemical compounds react with the sample and prevent its dissociation into free atoms.

chemical kinetics. The branch of chemistry that deals with the rate behavior of chemical reactions.

chemical messenger. HORMONE.

chemical potential. The partial molar free energy of a substance.

chemical quenching. The quenching that occurs in liquid scintillation counting if some of the energy of the radiation is absorbed either by the sample itself or by other substances in the solution.

chemical race. CHEMOVAR.

chemical reaction. A reaction in which there are changes in the orbital electrons of the reacting atoms as distinct from a nuclear reaction in which there are changes in the atomic nuclei.

chemical score. A measure of the nutritional quality of a protein based on a comparison of its amino acid composition with that of egg, which has a nearly ideal balance of essential amino acids. The amount of each amino acid in the protein is expressed as a percent of the amount of the same amino acid in egg; the lowest value, or score, is given by the essential amino acid that is limiting for growth and is a measure of the nutritional quality of the protein.

chemical shift. The shift in the position of a peak in nuclear magnetic resonance relative to the position of the peak produced by a standard nucleus; it is equal to the difference between the applied magnetic field strength required to produce absorption of energy in a nucleus and the magnetic field strength predicted by the gyromagnetic ratio for an identical nucleus. The chemical shift is due to the fact that each nucleus is in a different part of the molecule and hence experiences a different field, determined by its environment.

chemical thermodynamics. The branch of thermodynamics that deals with chemical compounds and chemical reactions.

chemical transmitter. A compound, such as acetylcholine, that mediates the transmission of a nerve impulse from one nerve cell to another.

chemiluminescence. The production of visible light as a result of a chemical reaction.

chemiosmotic hypothesis. A hypothesis of the coupling of ATP synthesis to the electron transport system in oxidative phosphorylation. According to this hypothesis, the transport of electrons generates an energy-rich proton gradient across the mitochondrial membrane and the proton motive force associated with this gradient then drives the phosphorylation of ADP to ATP.

chemisorption. Sorption that requires strong chemical forces such as those operative in the formation of chemical bonds.

chemistry. The science that deals with the composition, structure, properties, and transformations of substances.

chemoautotroph. 1. A chemotrophic autotroph. 2. LITHOTROPH.

chemoencephalography. The study of the metabolic patterns of the brain as a function of the behavioral experiences of the individual.

chemoheterotroph. 1. A chemotrophic heterotroph. 2. ORGANOTROPH.

chemolithotroph. An organism or a cell that utilizes for its growth (1) oxidation-reduction reactions as a source of energy, (2) inorganic compounds as electron donors for these oxidation-reduction reactions, and (3) carbon dioxide as its source of carbon atoms.

chemoorganotroph. An organism or a cell that utilizes for its growth oxidation-reduction reactions as a source of energy, and organic compounds both as electron donors for these oxidation-reduction reactions and as a source of carbon atoms.

chemoreceptor. A receptor that is stimulated by chemical compounds.

chemostat. An apparatus for maintaining bacteria in the exponential phase of growth over prolonged periods of time. This is achieved by the continuous addition of fresh medium, which is balanced by the continuous removal of the overflow, so that the volume of the growing culture remains constant.

chemosynthetic. Chemotrophic.

chemosynthetic organism. 1. CHEMOLITHOTROPH. 2. CHEMOORGANOTROPH.

chemotactic. Of, or pertaining to, chemotaxis.

chemotactic hormone. A hormone that has a chemotactic effect such as a steroid that causes aggregation of amoeba and slimemolds.

chemotaxin. A substance that is derived from complement and that induces leukocytes to move from an area of lower to one of higher chemotaxin concentration.

chemotaxis. A taxis in which the stimulus is a chemical compound.

chemotaxonomy. The classification of organisms on the basis of the distribution and/or the composition of chemical substances in these organisms; the use of DNA base composition data for the taxonomy of bacteria is an example.

chemotherapeutic agent. A chemical that interferes with the growth of either microorganisms or cancer cells at concentrations at which it is tolerated by the host cells.

chemotherapy. The treatment of a disease by means of chemotherapeutic agents.

chemotroph. An organism or a cell that uses oxidation-reduction reactions as a source of energy.

chemotropism. A tropism in which the stimulus is a chemical compound.

chemovar. A plant that, when grown in one locality, contains one or more different chemical substances as compared to those it contains when grown in a different locality.

chiasma (*pl* chiasmata). The visible connection between chromatids during meiosis.

chick antidermatitis factor. PANTOTHENIC ACID.

chickenpox. VARICELLA.

chimera. An organism that contains a mixture of genetically different cells. *See also* blood group chimera; radiation chimera.

chirality. The necessary condition that allows for a discrimination between two enantiomers; the right- or left-handedness of an asymmetric molecule or of an asymmetric object. Due to their chirality, enantiomeric molecules cannot be brought into coincidence with each other by rotation about axes of symmetry, by reflection in planes of symmetry, or by a combination of these maneuvers.

chi-square method. A statistical test for comparing observed values with expected ones.

chitin. A homopolysaccharide of N-acetyl-D-glucosamine that is a major constituent of the hard, horny exoskeleton of insects and crustaceans.

chitosamine. GLUCOSAMINE.

chlorambucil. A mutagenic, alkylating agent.

chloramphenicol. A broad-spectrum antibiotic, produced by *Streptomyces venezuelae*, that inhibits protein synthesis by attaching to the 70S ribosome, thereby preventing peptide bond formation; chloramphenicol also has immunosuppressive activity. *Abbr* CM.

chloramphenicol particle. A ribosomal subparticle isolated from bacteria in which protein synthesis has been inhibited by chloramphenicol. *Abbr* CM-particle.

chloremia. HYPERCHLOREMIA.

chloride shift. The movement of chloride and hydroxyl ions across the erythrocyte membrane as a result of the movement of bicarbonate ions in the opposite direction; this exchange of ions occurs at both the tissue and the lung level, but the relative directions of ion movement are reversed at the two levels.

chlorine. An element that is essential to several classes of animals and plants. Symbol Cl; atomic number 17; atomic weight 35.453; oxidation states -1, $+1$, $+5$, $+7$; most abundant isotope Cl^{35}; a radioactive isotope Cl^{36}, half-life 3×10^5 years, radiation emitted—beta particles and positrons.

Chlorobium chlorophyll. A chlorophyll occurring in some sulfur bacteria.

chlorocruorin. A hemoglobin-like, respiratory pigment of invertebrates that has a molecular weight of 3.5×10^6 and that contains 190 heme groups per molecule.

p-**chloromercuribenzoic acid.** A reagent that reacts with the sulfhydryl groups of proteins. *Abbr* PCMB.

chloromycetin. CHLORAMPHENICOL.

chlorophyll. The green pigment that occurs in plants and that functions in photosynthesis by absorbing the radiant energy of the sun. The chlorophylls are a group of closely related pigments, structurally related to the porphyrins, but containing magnesium instead of iron. Major chlorophylls of land plants are chlorophyll *a* and *b*, that of some marine organisms is chlorophyll *c*, and that of photosynthetic bacteria is bacteriochlorophyll.

chlorophyllide. A molecule of chlorophyll from which the phytol side chain has been removed by hydrolysis.

chlorophyll unit. PHOTOSYNTHETIC UNIT.

chloroplast. A chlorophyll-containing chromoplast that is the site of photosynthesis in green plants.

cholecalciferol. A compound that has vitamin D activity and that is obtained by ultraviolet irradiation of 7-dehydrocholesterol; designated vitamin D_3.

cholecystokinin. A polypeptide hormone, secreted by the duodenum, that stimulates the secretion of digestive enzymes by the pancreas and that stimulates the contraction of the gall bladder.

cholecystokinin-pancreozymin. CHOLECYSTOKININ.

choleic acid. A specific complex formed between a steroid, particularly a bile acid, and fatty acids, hydrocarbons, or other organic compounds.

choleragen. A toxin of the cholera bacillus that affects the plasma membrane of intestinal cells.

cholestane. The parent ring system of the sterols.

cholesteremia. CHOLESTEROLEMIA.

cholesterol. The principal sterol of vertebrates and a precursor of bile acids and steroid hormones; it is synthesized entirely from acetyl coenzyme A by condensation reactions of isoprene units.

cholesterolemia. The presence of excessive amounts of cholesterol in the blood.

cholesterol ester. An ester formed from cholesterol and a fatty acid.

cholesterol intoxication theory. A theory according to which atherogenesis results from either the ingestion of high-cholesterol fat or the deficiency of certain vitamins.

cholesterolosis. A condition that is characterized by the formation of cholesterol deposits in various organs and tissues, and that is caused by a disturbance in lipid metabolism.

cholesterol oxidase. DESMOLASE.

cholesterosis. CHOLESTEROLOSIS.

cholic acid. The most abundant bile acid in human bile.

choline. A methyl group donor that occurs in some phospholipids and in acetylcholine. It is generally classified with the B vitamins, since it is required in the diet under certain conditions, but it is not a typical vitamin and has no known coenzyme function. *Abbr* Ch.

choline acetyltransferase. The enzyme that catalyzes the reaction in which choline and acetyl coenzyme A are converted to acetylcholine and coenzyme A.

cholinergic. Of, or pertaining to, nerve fibers that release acetylcholine at the nerve endings.

cholinesterase. The enzyme that catalyzes the hydrolysis of acetylcholine and of a variety of other choline esters and that is present in various tissues other than the nervous system. *Aka* cholinesterase II; nonspecific cholinesterase; pseudocholinesterase.

cholinolytic. Descriptive of a pharmacological substance that blocks the action of acetylcholine.

cholinomimetic. Descriptive of a pharmacological substance that imitates the action of acetylcholine.

chondriogene. A plasmagene that is attached to a mitochondrion.

chondriome. A collective term for all of the mitochondria of a cell.

chondriosome. MITOCHONDRION.

chondroitin. A mucopolysaccharide composed of D-glucuronic acid and N-acetyl-D-galactosamine.

chondroitin sulfate. The sulfate ester of chondroitin and a major constituent of bone and cartilage.

chopper. A rotating wheel with alternate silvered and cut-out sections that is placed in the light path of a spectrophotometer, thereby allowing the light beam to pass alternately through the sample solution and through the reference solution.

chorioallantoic membrane. The membrane, used in the assay of viruses, that surrounds the embryo of the chicken and other birds.

chorionic gonadotropin. *See* human chorionic gonadotropin.

chorismic acid. An intermediate in the biosynthesis of the aromatic amino acids; it is located at the branch point from which one pathway leads to tryptophan and another pathway leads to phenylalanine and tyrosine.

Christmas factor. The factor in the intrinsic system of blood clotting that is activated by the plasma thromboplastin antecedent.

ChRNA. Chromosomal RNA.

chromaffin granule. A subcellular organelle that synthesizes, stores, and releases the catecholamines epinephrine and norepinephrine.

chroman. A redox lipid such as tocopherol.

chromatic transient. A sudden short-lived increase or decrease in the rate of a photosynthetic reaction when the wavelength, but not the effective intensity, of the incident light is suddenly changed.

chromatid. One of the two strands that result from the duplication of a chromosome and that are held together by a centromere; the chromatids become separate chromosomes upon division of the centromere.

chromatin. The nuclear material of the chromosomes in higher organisms that consists principally of DNA and histones. *See also* euchromatin; heterochromatin; pseudochromatin.

chromatin body. NUCLEOID.

chromatogram. The visual record of a chromatographic separation, either in the form of the chromatographic support itself, or in the form of a tracing thereof.

chromatographic. Of, or pertaining to, chromatography.

chromatographic resolution. *See* resolution (3).

chromatographic spray. An atomizer used in the spraying of chromatograms for detecting the separated sample spots.

chromatographic support. *See* support.

chromatography. The separation of complex mixtures of molecules that is based on the repetitive distribution of the molecules between a mobile and a stationary phase. The mobile phase may be either a liquid or a gas, and the stationary phase may be either a solid or a solid coated with a liquid. The distribution of the molecules between the two phases is determined by one or more of four basic processes, namely adsorption, gel-filtration, ion-exchange, and partitioning. The operation of these processes, coupled to the movement of the mobile phase, results in a differential migration of the molecules along the stationary phase. *See also* adsorption chromatography; gel-filtration chromatography; ion-exchange chromatography; partition chromatography.

chromatophore. A bacteriochlorophyll-containing chromoplast of photosynthetic bacteria.

chromatophorotropic hormone. MELANOCYTE-STIMULATING HORMONE.

chromatopile. A stack of filter paper disks that have the same diameter and that are compressed within a chromatographic column which is used for preparative-scale separations.

chromatoplate. The plate, covered with a support, that is used in thin-layer chromatography.

chromium. An element that is essential for living organisms but the function of which is unknown. Symbol Cr; atomic number 24; atomic weight 51.996; oxidation states $+2$, $+3$, $+6$; most abundant isotope Cr^{52}; a radioactive isotope Cr^{51}, half-life 27.8 days, radiation emitted—gamma rays.

chromogen. 1. The colorless precursor of a pigment. 2. The colorless parent compound of a dye.

chromogenic. 1. Producing a pigment or a color. 2. Of, or pertaining to, a chromogen.

chromogranin. A soluble protein in chromaffin granules.

chromoisomer. One of two or more isomers that differ from each other in their color.

chromomere. A thickening along a eucaryotic chromosome that results from the local coiling of the chromosome threads.

chromonema (*pl* chromonemata). One of the coiled threads in a eucaryotic chromosome.

chromoneme. The thread of DNA in bacterial cells and in viruses.

chromophobe. A cell that does not stain readily.

chromophore. The group of atoms in a compound that is capable of absorbing light and that is responsible for the color of the compound.

chromoplast. A pigment-containing plastid, such as a chloroplast or a chromatophore, that functions in photosynthesis.

chromoprotein. A conjugated protein in which the nonprotein portion is a pigment or some other chromophoric material.

chromosomal aberration. An abnormality in a chromosome that results from the deletion, the duplication, or the rearrangement of the genetic material; the abnormality is referred to as intrachromosomal or interchromosomal depending on whether it is the result of changes in one or in two chromosomes.

chromosome. 1. A structure in the nucleus of eucaryotic cells that consists of one or more large double-helical DNA molecules that are associated with RNA and histones; the DNA of the chromosome contains the genes and functions in the storage and in the transmission of the genetic information of the organism. 2. The nuclear DNA of eucaryotic cells, the DNA of procaryotic cells, or the DNA of viruses. 3. The RNA of viruses. *See also* genophore.

chromosome break. A break in the structure of a chromosome as that produced by some carcinogenic, alkylating agents.

chromosome map. 1. GENETIC MAP. 2. CYTOGENETIC MAP.

chromosome puff. A localized swelling in a polytene chromosome that represents a region of active RNA synthesis. Very large chromosome puffs form loop-like structures known as Balbiani rings.

chromosome theory of cancer. A theory, proposed by Boveri in 1912, according to which cancer is due to the presence of abnormal chromosomes in the cells as a result of irregularities in mitosis.

chromosorb. A chromatographic adsorbent prepared by the fusion of diatomaceous earth either with, or without, sodium carbonate.

chromotrope. A substance that can appear in two or more different colors depending on the extent to which it is covered with a metachromatic dye.

chronic toxicity test. A toxicity test performed on laboratory animals that requires the administration of a chemical at least once daily for periods of 1 to 2 years.

chronometric method. A method of assaying for the enzyme amylase by measuring the time that is required for the complete hydrolysis of all the starch in the reaction mixture.

chronopotentiometry. An electroanalytical method for studying electrolysis reactions by measuring the potential, as a function of time, at a microelectrode on which is impressed a small constant current.

chronotropic. Affecting the rate of rhythmic movements, especially that of the heart beat.

$C'H_{50}$ unit. UNIT OF COMPLEMENT.

chyle. The lymphatic fluid that is taken up by the lymph vessels from the intestine during digestion and that is characterized by its high content of fat globules.

chylomicron (*pl* chylomicrons; chylomicra). A colloidal fat globule that occurs in blood and lymph and that serves to transport fat from the intestine; a very-low-density lipoprotein, containing about 4% protein, 8% phospholipid, and 88% neutral lipid. Chylomicrons have a density of less than 1.006 g/l, a molecular weight of the order of 10^9 to 10^{10}, and a flotation coefficient above 400S.

chyme. The semifluid mass of partially digested material that is passed from the stomach into the intestine.

chymotrypsin. An endopeptidase that catalyzes the hydrolysis of peptide bonds, principally those in which the carbonyl group is contributed by tryptophan, phenylalanine, or tyrosine.

chymotrypsinogen. The inactive precursor of chymotrypsin.

chymotryptic. Of, or pertaining to, chymotrypsin.

chymotryptic peptides. The peptides obtained by the digestion of a protein with the enzyme chymotrypsin.

Ci. Curie.

cilium (*pl* cilia). A thread-like cellular extension that functions in the locomotion of bacteria and unicellular eucaryotic organisms; cilia are more numerous and shorter than flagella.

ciliary. Of, or pertaining to, cilia.

circadian rhythm. A biological rhythm that has a period of approximately 24 hours.

circular birefringence. The birefringence produced by left and right circularly polarized light.

circular chromatography. A paper chromatographic technique in which the material is allowed to migrate radially; may be carried out by the use of a circular piece of filter paper, to the center of which the sample is applied, and from which a sector is cut out and dipped into the solvent.

circular covalent. Descriptive of a circular polynucleotide strand in which the 3′- and 5′-termini are linked together covalently.

circular dichroism. The dichroism that results from the differences, at a given wavelength, between the extinction coefficients of left and right circularly polarized light when such light is passed through a solution containing a chromophore. Circular dichroism depends on the asymmetry of the electric charge distribution around the chromophore and may either be an intrinsic property of the molecule or be induced in the molecule as a result of perturba-tions in the surroundings. Circular dichroism is used in the study of the secondary structure of macromolecules. *Abbr* CD. *See also* magnetic circular dichroism.

circular DNA. A DNA molecule that has a closed ring-type structure, not necessarily that of a geometric circle.

circular genetic map. The genetic map of a closed ring-type chromosome.

circularly polarized light. Light in which the electric field vectors, at any point on the axis along which the light is being propagated, rotate in a plane perpendicular to that axis. The light is referred to as right or left circularly polarized light, depending on whether the electric field vector rotates in a clockwise or in a counterclockwise direction as seen when looking toward the light source.

circular noncovalent. Descriptive of a circular polynucleotide strand in which the 3′- and 5′-termini are held together by noncovalent bonds.

circular permutation. The formation of different linear segments when the same circle is opened at one or more different points; a principle that is invoked in relating a circular genetic map to the genetic structure of linear DNA molecules.

circulating water bath. A water bath equipped with a pump so that the water can be pumped from the bath to some apparatus and returned from there to the bath.

circulation. The movement of a liquid through a circuit, such as the movement of blood or lymph.

cirrhosis. An inflammatory disease of the liver that is characterized by the replacement of liver cells with fat and fibrous tissue.

cis. 1. Referring to the configuration of a geometrical isomer in which two groups, attached to two carbon atoms linked by a double bond, lie on the same side with respect to the plane of the double bond. 2. Referring to two mutations, particularly those of pseudoalleles, that lie on the same chromosome.

cis effect. The influence of one gene on the expression of another gene that is located on the same chromosome.

cis isomer. *See* cis (1).

cisterna (*pl* cisternae). A cavity in a cell or in an organism that serves as a reservoir.

cis-trans isomers. *See* cis (1); trans (1).

cis-trans test. A complementation test of pseudoalleles.

cistron. The unit of genetic function; a section of the chromosome that codes for a single polypeptide chain; a structural gene.

citrate-activated thrombin. The material obtained by the dissociation of thrombin when thrombin is dissolved in 25% sodium citrate solution.

citrate synthase. The enzyme that catalyzes the first reaction of the citric acid cycle in which acetyl coenzyme A condenses with oxaloacetic acid to yield citric acid and coenzyme A.

citric acid. The symmetrical tricarboxylic acid that is formed in the first reaction of the citric acid cycle in which acetyl coenzyme A condenses with oxaloacetic acid.

citric acid cycle. The cyclic set of reactions that constitutes the core of the central metabolic pathway in most living cells. The citric acid cycle is initiated by the condensation of acetyl coenzyme A with oxaloacetic acid which yields citric acid and coenzyme A. One turn of the citric acid cycle, in conjunction with the operation of the electron transport system and oxidative phosphorylation, achieves the equivalent of the oxidation of one molecule of acetic acid to carbon dioxide and water and of the synthesis of 15 molecules of ATP. *Abbr* TCA cycle.

citrogenase. CITRATE SYNTHASE.

citrovorum factor. Folinic acid; a growth factor for *Leuconostoc citrovorum. Abbr* CF.

citrulline. A nonprotein alpha amino acid that is an intermediate in the urea cycle.

citrullinuria. A genetically inherited metabolic defect in man that is associated with mental retardation and that is characterized by high concentrations of citrulline in the blood and in the urine.

Cl. Chlorine.

c_L. The constant region of the light chains of the immunoglobulins. *See also* c_H.

cladogram. A diagram that depicts the branching sequences in a phylogenetic tree.

classical sedimentation equilibrium. SEDIMENTATION EQUILIBRIUM.

classical thermodynamics. EQUILIBRIUM THERMODYNAMICS.

classification. The systematic arrangement of organisms into groups based on the natural relationships between the organisms. The groups, proceeding from the largest to the smallest, are kingdom, phylum, class, order, family, genus, and species.

clastic reaction. *See* phosphoroclastic reaction; thioclastic reaction.

clathrate. An inclusion complex produced by trapping molecules of one kind in the lattice network formed by molecules of a second kind; frequently refers to the stable complex produced by trapping nonpolar solute molecules in a cage formed by water molecules. *Aka* clathrate compound; clathrate crystal.

Clausius' law. A combined statement of the first and second laws of thermodynamics: the energy of the universe is constant, but its entropy is increasing to a maximum.

clearance. A measure of the efficiency of the kidney in removing a substance from the blood; specifically, $C = UV/P$, where C is the clearance in milliliter of plasma per minute, U and P are the concentrations of the substance in the urine and in the plasma, and V is the flow of urine in milliliter per minute. The clearance is known as either a maximum or a standard clearance depending on whether the flow of urine is more or less than 2 ml/min.

clearance factor lipase. LIPOPROTEIN LIPASE.

clear plaque. A plaque that is produced when all of the cells in the area of the plaque are lysed.

Cleland's reagents. The compounds, dithioerythritol and dithiothreitol, that are used for the protection of sulfhydryl groups against oxidation to disulfides and for the reduction of disulfides to sulfhydryl groups.

Cleland's rules. A set of rules for predicting the type of inhibition of an enzyme-catalyzed reaction or the type of interaction between an enzyme and a cosubstrate or other substance from an inspection of experimental kinetic data. The rules apply to a steady-state mechanism that either contains no random sequences or contains random sequences in rapid equilibrium. The two fundamental rules may be stated as follows: (a) The ordinate intercept of a double reciprocal plot is affected by a substance that associates reversibly with an enzyme form other than the one with which the variable substrate combines. (b) The slope of a double reciprocal plot is affected by a substance that associates with an enzyme form that is the same as, or is connected by a series of reversible steps to, the enzyme with which the variable substrate combines.

climacteric rise. The increase to a maximum in the respiration of ripening fruits that may occur either before or after the removal of the fruit from the plant, depending on the fruit and on the harvesting procedure.

clinical centrifuge. A centrifuge, generally considered to be a small table model, that is capable of generating speeds of approximately 3000 rpm and centrifugal forces of approximately 2000 $\times g$.

clinical chemistry. A branch of chemistry that deals with the qualitative and quantitative determination of chemical substances in man, particularly of substances related to medicine.

Clinistix. Trademark for a group of paper strips impregnated with chemicals and used for semiquantitative determinations of components in urine and/or in blood.

clock-timing. A method of timing used in scintillation counters in which the timing device is not turned off during the interval that is required for the electronic processing of a pulse. *See also* live-timing.

clonal selection theory. A selective theory of antibody formation according to which an antigen selects a particular clone of cells from among a large number of lymphoid cell clones, and then stimulates these cells to proliferate and to synthesize antibodies. Each cell clone is believed to be different and to contain a unique set of genes for specific immunoglobulins so that each clone can synthesize either antibodies having only one type of specificity or, at most, antibodies having a few types of specificity.

clone. A group of cells derived from a single parental cell by asexual reproduction.

cloning. The formation of a clone.

closed chain. RING.

closed circuit system. A system for measurements of indirect calorimetry in which the oxygen consumption, but not the carbon dioxide production, is determined.

closed system. A thermodynamic system that can exchange energy, but not matter, with its surroundings.

close packing. Descriptive of a structure in which nonbonded atoms are surrounded by other nonbonded atoms in such a way that the distances between the atoms are equal, to the extent possible, to the sum of their Van der Waals contact radii.

closure transformation. The transformation of a micellar membrane from one having large spaces between the micelles ("open") to one having small spaces between them ("closed").

clot. *See* blood clot.

clot-promoting factor. HAGEMAN FACTOR.

clot retraction. The shrinking of a blood clot that occurs upon standing and that is accompanied by the expressing of the serum.

clotting time. The time, in minutes, required for blood to clot when it is exposed to air.

cloud chamber. A chamber that contains a supersaturated atmosphere and that is used for observing the tracks produced by ionizing particles; the ions formed by an ionizing particle serve as nuclei for the formation of fog droplets which indicate the path taken by the ionizing particle.

cloverleaf model. A model for the structure of transfer RNA that resembles a cloverleaf and that is based upon the folding of the transfer RNA strand back upon itself so as to permit the formation of a maximum number of intrachain hydrogen bonds; the structure contains four (or five) hydrogen-bonded segments, referred to as arms, to which are attached three (or four) nonhydrogen-bonded segments, referred to as loops.

clupein. A protamine isolated from herring.

cm. Centimeter.

CM. Chloramphenicol.

CMC. Critical micelle concentration.

CM-cellulose. *O*-(Carboxymethyl)-cellulose; a cation exchanger.

CMP. 1. Cytidine monophosphate (cytidylic acid). 2. Cytidine-5′-monophosphate (5′-cytidylic acid).

CM-particle. Chloramphenicol particle.

CM-sephadex. *O*-(Carboxymethyl)-sephadex; a cation exchanger.

CNS. Central Nervous System.

Co. Cobalt.

CoI. Cozymase I.

CoII. Cozymase II.

CoA. Coenzyme A.

coacervate. A polymer-rich phase or droplet that is formed by coacervation and that is believed by some to have been a forerunner of primitive cells.

coacervation. The spontaneous separation of an aqueous solution of a highly hydrated polymer into two phases, one having a relatively high, and the other having a relatively low concentration of the polymer.

coagulase. The enzyme, produced by *Staphylococcus,* that has thrombokinase-like activity and causes citrated, or oxalated, plasma to coagulate. *Aka* coagulating enzyme.

coagulation. The formation of a clot as that formed in blood clotting or in the boiling of egg white; clots may be soft, semisolid, or solid.

coarctation. An increase in the cross-linking, the hardening, and the shrinking of a membrane.

coarse control. The control of biochemical systems that is achieved by the regulation of the amount of an enzyme, as in enzyme induction and enzyme repression.

CoASAc. Acetyl coenzyme A.

CoASH. Coenzyme A.

coat. *See* spore coat; viral coat.

coat protein. A protein of the viral coat.

cobalamine. A form of vitamin B_{12} in which the sixth coordination position of the cobalt atom is not occupied. *Var sp* cobalamin.

cobalt. An element that is essential to several classes of animals and plants. Symbol Co; atomic number 27; atomic weight 58.9332; oxidation states +2, +3; most abundant isotope Co^{59}; a radioactive isotope Co^{60}, half-life 5.26 years, radiation emitted—beta particles and gamma rays.

cobamide. The coenzyme form of vitamin B_{12} in which a 5′-deoxyadenosyl group is attached to the sixth coordination position of the cobalt atom. *Var sp* cobamid.

coboglobin. An artificially prepared hemoglobin or myoglobin molecule in which the iron atom has been replaced by a cobalt atom.

cocarboxylase. THIAMINE PYROPHOSPHATE.

cocarcinogen. An agent that enhances the effect of a carcinogen either by increasing the yield of tumors or by shortening the time required for a tumor to appear.

cocarcinogenesis. The enhancement of the action of one carcinogen by the simultaneous administration of a second carcinogen.

coccus (*pl* cocci). A bacterium having a more or less spherically shaped cell; cocci represent one of the three major forms of eubacteria.

cochromatography. A chromatographic technique for establishing the identity of a compound by applying the compound, together with one or more known compounds, to a chromatographic support.

cocktail. 1. The mixture of reagents required for a cell-free amino acid incorporating system; excludes the messenger RNA, ribosome, and enzyme fractions. 2. The solution of fluors used for liquid scintillation counting.

codase. AMINOACYL-tRNA SYNTHETASE.

code. 1. *n* GENETIC CODE. 2. *v* To direct the incorporation of an amino acid in response to a codon.

codegenerate codon. SYNONYM CODON.

codehydrogenase I. NICOTINAMIDE ADENINE DINUCLEOTIDE.

codehydrogenase II. NICOTINAMIDE ADENINE DINUCLEOTIDE PHOSPHATE.

codeine. An alkaloid narcotic drug that occurs in opium.

codeword. CODON.

codeword triplet. CODON.

coding DNA. That strand of DNA which, in vivo, is transcribed into messenger RNA.

coding ratio. The ratio of the number of nucleotides in a messenger RNA molecule to the number of amino acids in the polypeptide chain that is coded for by the messenger RNA; the number of nucleotides in a codon.

coding strand. CODING DNA.

coding triplet. CODON.

codon. The sequence of three adjacent nucleotides that occurs in messenger RNA and that functions as a coding unit for a specific amino acid in protein synthesis. The codon determines which amino acid will be incorporated into the protein at a particular position in the polypeptide chain. There are 64 codons, 61 of which code for amino acids and 3 of which serve as termination codons.

codon dictionary. *See* dictionary.

codon recognizing site. ANTICODON.

coef. Coefficient; used in the Cleland nomenclature of enzyme kinetics to indicate a factor, composed of one or more rate constants, by which the concentration of a component must be multiplied. Thus (coef A)A may mean, for example, $(k_3 + k_5)$ [A], where [A] is the concentration of component A, and k_3 and k_5 are rate constants.

coefficient of coincidence. COINCIDENCE (2).

coefficient of variation. A measure of the relative dispersion of data in terms of percentages of the mean; specifically, $CV = (\sigma/\bar{X})$ 100, where CV is the coefficient of variation, σ is the standard deviation, and $\bar{X}$ is the observed mean.

coefficient of viscosity. VISCOSITY.

coenocyte. A multinucleate organism that lacks cell walls.

coenzyme. The organic molecule that functions as a cofactor of an enzyme.

coenzyme I. NICOTINAMIDE ADENINE DINUCLEOTIDE.

coenzyme II. NICOTINAMIDE ADENINE DINUCLEOTIDE PHOSPHATE.

coenzyme A. The coenzyme form of the vitamin pantothenic acid that functions in metabolism as a carrier of an acetyl or some other acyl group; the acyl group is linked to the sulfhydryl group of coenzyme A. *Abbr* CoASH; CoA.

coenzyme F. FOLATE COENZYME.

coenzyme Q. One of a group of benzoquinone derivatives that have an isoprenoid side chain of varying length and that function as electron carriers in the electron transport system. Coenzyme Q is structurally related to vitamin K and is frequently considered to represent the coenzyme form of a fat-soluble vitamin. *Abbr* CoQ.

coenzyme R. BIOTIN.

CoF. Coenzyme F.

cofactor. The nonprotein component that may be required by an enzyme for its activity. The cofactor may be either a metal ion (activator) or an organic molecule (coenzyme) and it may be attached either loosely or tightly to the enzyme; a tightly attached cofactor is known as a prosthetic group.

cofactor-requiring mutant. A phage mutant that requires a cofactor for its adsorption to the host cell.

coherent light. Light in which all of the waves are in phase.

coherin. A neurohypophyseal polypeptide that stimulates the coordinate contractions of the intestine.

cohesive end. STICKY END.

Cohn fraction. One of a number of fractions of proteins that are precipitated from plasma when the plasma is treated with ethanol at low temperatures.

coiled coil. SUPERHELIX.

coimmune. Denoting two mutants of the same phage that do not differ in the gene that controls the synthesis of the immunity substance.

coincidence. 1. The occurrence of radioactive events within a span of time that is too short to permit their resolution by a radiation counter. 2. The ratio of the observed number of double crossovers to the theoretical number of double crossovers.

coincidence circuit. An electronic circuit that has two inputs but only one output and that produces an output pulse only if two input pulses arrive either simultaneously or within a known time interval; used in liquid scintillation counting to decrease the level of the dark current due to background counts.

coincidence correction. The correction that is applied in radiation counting for coincidence loss.

coincidence counting. The counting of pulses, produced by radioactive disintegrations, by means of a coincidence circuit.

coincidence loss. The loss of register of pulses as a result of their occurring within too short an interval to permit their resolution by the electronic circuit.

coincidence time. 1. The minimum length of time that must elapse between two events to permit them to be registered as two separate events. 2. The maximum length of time that may separate two pulses and still permit the registration of an output pulse by means of a coincidence circuit.

coion. One of two ions that have charges of the same sign.

coisogenic. Descriptive of animals that are genetically identical except for one or two genetic loci that have been altered by mutation.

cold. Containing no radioactive isotopes.

cold agglutinin. *See* cold hemagglutinin.

cold antibody. An antibody that has a higher titer at lower temperatures.

cold-blooded. POIKILOTHERMIC.

cold hemagglutinin. A hemagglutinin that causes agglutination of red blood cells at lower temperatures but leads to their dispersion at higher temperatures.

cold-sensitive enzyme. An enzyme that has a much lower activity at low temperatures as compared to its activity at room temperature.

cold-sensitive mutant. A mutant that has a higher minimum temperature of growth than the wild-type organism.

cold shock. A sudden chilling.

Col factor. Colicinogenic factor.

colicin. A bacteriocin that is isolated from *Escherichia coli.*

colicinogen. The bacteriocinogen of colicin.

colicinogenic factor. The extrachromosomal DNA element that determines the production of colicin. *Abbr* Col factor.

coliform bacteria. 1. Bacteria belonging to the genera *Escherichia* and *Aerobacter*. 2. A large and diverse group of bacteria that includes *Escherichia coli* and bacteria related to it.

colinear code. A code in which the sequence of the codons in messenger RNA corresponds to the sequence of the amino acids in the polypeptide chain which is coded for by that messenger RNA.

colinearity. The concept that the sequence of the nucleotides in a gene corresponds to the sequence of the amino acids in the polypeptide chain which is specified by that gene. The concept is supported by studies of the enzyme tryptophan synthetase from *Escherichia coli* and of the gene specifying this enzyme; it has been shown that the order and spacings of mutational changes in the gene correspond to the order and spacings of the amino acid substitutions in the enzyme.

coliphage. A phage that infects the bacterium *Escherichia coli.*

collagen. A fibrous scleroprotein that is the major protein of connective tissue and the most abundant protein in higher animals. Collagen forms an unusual triple helix and has an unusual amino acid composition in which glycine, proline, and hydroxyproline together constitute about two thirds of the total amino acid residues. The basic unit of collagen is tropocollagen.

collagenase. An enzyme that catalyzes the hydrolysis of collagen.

collagen helix. The unusual triple helix of collagen in which the polypeptide chains do not have the alpha helical configuration.

collateral sensitivity. The increased sensitivity of an individual to an anticancer drug that results from the individual's resistance to a different anticancer drug.

colligative property. A property of a solution, such as osmotic pressure, that depends on the number of solute particles per unit volume of solution and that does not depend on the size or shape of the particles.

collimating lens. A lens that converts light striking it into a beam of parallel rays.

collimator. A device, composed of either lenses or slits, that is used to convert incident radiation into a narrow beam of parallel rays.

collisional quenching. The energy transfer from an excited molecule to another molecule that occurs when the two molecules approach each other to within the contact distance which they attain during molecular collisions.

colloid. 1. A macromolecule or a particle in which at least one dimension has a length of 10^{-9} to 10^{-6} m. 2. THYROID COLLOID.

colloidal. Of, or pertaining to, colloids.

colloidal dispersion. A colloidal system that consists of a dispersed phase and a dispersion phase and that is thermodynamically unstable and not readily reconstituted after separation of the phases. *See also* colloidal solution; suspension.

colloidal electrolyte. ASSOCIATION COLLOID.

colloidal solution. A true solution that consists of colloidal macromolecules and solvent and that is thermodynamically stable and readily reconstituted after separation of the macromolecules from the solvent. *See also* colloidal dispersion; suspension.

colloid osmotic pressure. The osmotic pressure of a colloidal system that is separated by a membrane which is impermeable to the colloidal particles but is permeable to crystalloids.

colon bacillus. *See Escherichia coli.*

colony. A group of contiguous cells that grow in or upon a solid medium and are derived from a single cell. A bacterial colony may be of smooth or rough morphology depending on whether the cells do or do not possess either a capsule or other surface components.

colorimeter. 1. An optical or a photoelectric instrument for measuring either color differences or color intensities; used for the quantitative determination of compounds in solution by colorimetry. 2. An instrument for the exact matching of two colored solutions.

colorimetry. 1. A method of quantitative analysis in which a compound is determined by a comparison of the color produced by the reaction of a reagent with both standard and test solutions of the compound. 2. A method of quantitative analysis in which a compound is determined by the exact matching of the colors produced by the reaction of a reagent with both a standard and a test solution of the compound.

color quenching. The quenching that occurs in liquid scintillation counting when some of the light that is emitted by the fluor is absorbed by colored components of the sample.

color vision. The capacity to perceive colors that is due to the cones in the retina.

colostral milk. COLOSTRUM.

colostrum. The milk secreted during the first few days after parturition. *Aka* colostral milk.

column. A cylindrical tube that is filled with a chromatographic support and is used in column chromatography.

column chromatography. A chromatographic technique in which the stationary phase consists of a porous solid contained in a cylindrical tube, and the mobile phase percolates through the solid; used primarily for adsorption, gel-filtration, and ion-exchange chromatography.

coma. A state of profound unconsciousness from which the individual cannot be aroused.

comb growth test. A bioassay for androgenic activity that is based on the stimulation of comb growth in capons.

combination code. An early version of the genetic code according to which the nucleotide sequence in messenger RNA was assumed to be random so that all possible sequence permutations of a given triplet could code for the same amino acid.

combination electrode. An electrode that consists of a glass tube into which both a reference electrode and a glass electrode have been incorporated.

combined acidity. The acidity of gastric juice that is due either to protein-bound hydrochloric acid or to acids other than free hydrochloric acid, such as lactic acid and butyric acid. The combined acidity is equal to the difference between the total titratable acidity of gastric juice and the acidity due to free hydrochloric acid.

combining site. *See* antibody combining site.

cometesimal. A body of matter formed from primordial dust; the consolidation of cometesimals is believed to have led to the formation of the planets close to the periphery of the solar system.

comicellization. The solubilization of an insoluble or a slightly soluble compound through the formation of a mixed micelle that consists of the compound and of an amphipathic compound. The process occurs at concentrations of amphipathic compound which are considerably below its critical micelle concentration.

comma-less code. A genetic code in which there are no signals to indicate either the beginning or the end of a codon; in such a code, the displacement of the starting point will lead to the reading of a different set of codons. *Aka* comma-free code.

committed step. A reaction that forms part of a sequence of reactions and that, once it takes place, ensures that all the subsequent reactions in the sequence will also take place. A committed step may be (a) the first reaction catalyzed by a multienzyme system; (b) the first reaction in a biosynthetic pathway; or (c) the reaction at a branch point in a biosynthetic pathway that has no other role than to provide an intermediate for one of the branches of the pathway.

common intermediate principle. The principle that two energetically coupled reactions must proceed by having a common intermediate that transfers the energy from one reaction to the other.

comparative biochemistry. A branch of biochemistry that deals with the nature, the origin, and the control of biochemical differences among organisms.

compartmentation. The unequal distribution of a substance, such as a metabolite or an enzyme, within a cell or within an organism; may refer to the occurrence of the substance in particular structures or to its being a part of a given pool.

compensated acidosis. An acidosis in which the

pH of the blood remains constant due to the effect of mechanisms that counteract the decrease in pH produced initially.

compensated alkalosis. An alkalosis in which the pH of the blood remains constant due to the effect of mechanisms that counteract the increase in pH produced initially.

compensation point. The concentration of carbon dioxide below which, for a given organism, its uptake by photosynthesis is less than its output by respiration.

competence. 1. The physiological state of a bacterial cell that enables it to undergo transformation. 2. The physiological state of a cell that enables it to either recognize an antigen or synthesize antibodies. 3. The physiological state of a part of an embryo that enables it to react to an inductor by determination and differentiation in a specific direction.

competent cell. A cell possessing competence.

competitive inhibition. The inhibition of the activity of an enzyme that is characterized by an increase in the apparent Michaelis constant (substrate concentration required for one half the maximum velocity) and by an increase in the slope of a double reciprocal plot (1/velocity versus 1/substrate concentration) compared to those of the uninhibited reaction; the maximum velocity remains unchanged. *See also* pure competitive inhibition.

competitive inhibitor. An inhibitor that produces competitive inhibition and that generally bears a structural similarity to the substrate of the inhibited enzyme.

competitive protection. The protection of biomolecules against damage from an ionizing radiation that is provided by chemical substances (radical scavengers) which compete with the biomolecules for the harmful free radicals produced by the radiation. *See also* restitutive protection.

competitive protein-binding technique. An assay for a hormone in body fluids that is similar to a radioimmunoassay except that the binding protein is not an antibody but either a plasma, or a cellular, receptor site for the hormone.

complement. A group of serum proteins, present in mammals and many lower animals, that are not immunoglobulins but participate in a variety of immunological reactions. Complement reacts with antigen-antibody complexes and it may, in the presence of cellular particulate antigens, lead to cell lysis. *Sym* C; C′.

complemental air. The volume of air that can be forcibly drawn into the lungs after the normal tidal air has been inspired.

complementarity. The matching up and the mutual adaptation of surfaces in two interacting macromolecules. Complementarity plays a role in such processes as the binding of a substrate to an enzyme, the binding of an antigen to an antibody, and the binding of one nucleic acid strand to another.

complementary. Of, or pertaining to, complementarity.

complementary base sequence. The base sequence in a nucleic acid strand that is related to the base sequence in another strand by the base pairing rules; thus the sequence A-T-G-C in a DNA strand is complementary to the sequence T-A-C-G in a second DNA strand and to the sequence U-A-C-G in an RNA strand.

complementary gene. One of two genes that are similar in their phenotypic effect when they are present separately but which, when they are present together, interact to produce a different phenotypic effect. *Aka* complementary factor.

complementary interaction. The interaction of two genes that leads to phenotypic effects that are different from those produced by either one alone.

complementary strand. A polynucleotide chain that has a complementary base sequence to that in another chain.

complementation. The interaction between two sets of either cellular or viral genes that occurs within the same cell and that permits the cell or the virus to function even though each set of genes carries a mutated and nonfunctional gene. *See also* intergenic complementation; intragenic complementation; in vitro complementation.

complementation map. A genetic map constructed on the basis of complementation experiments.

complementation test. A test for determining whether the mutations of two mutant chromosomes occurred in the same gene so that complementation between the genes is possible; performed by introducing the two mutant chromosomes simultaneously into the same cell.

complement fixation. The binding of complement to an antigen-antibody complex. *Abbr* CF.

complement fixation inhibition test. The inhibition of a complement fixation test, as that produced by the presence of certain haptens or antibodies.

complement fixation test. A test for either antigen or antibody that is based on the binding of complement to the antigen-antibody complex and on the consequent disappearance of complement activity from a mixture of antigen, antibody, and complement.

complete antibody. An antibody that is fully reactive and that gives the ordinary serologic reactions of precipitation and agglutination.

complete antigen. *See* antigen.

complete medium. A minimal medium that is fortified with yeast extract, casein hydrolysate, and the like to permit the growth of auxotrophs.

complete oxidation. The oxidation of organic compounds such that carbon dioxide is the only carbon-containing product; the term may refer either to a single reaction or to a group of reactions.

complete transduction. Transduction in which the DNA from the donor bacterium becomes fully integrated into the chromosome of the recipient bacterium.

complete virion. A fully assembled and infective virus particle; a mature virus.

complex. 1. An aggregate of two or more molecules, particularly of macromolecules, that is formed as a result of their specific interaction. 2. The product formed by the interaction of a metal ion and ligands. *See also* complex ion.

complex I. One of the four complexes derived from electron transport particles that, by itself, can catalyze the oxidation of NADH by coenzyme Q.

complex II. One of the four complexes derived from electron transport particles that, by itself, can catalyze the oxidation of succinate by coenzyme Q.

complex III. One of the four complexes derived from electron transport particles that, by itself, can catalyze the oxidation of reduced coenzyme Q by cytochrome c.

complex IV. One of the four complexes derived from electron transport particles that, by itself, can catalyze the oxidation of reduced cytochrome c by molecular oxygen.

complex hapten. A high-molecular weight hapten that constitutes a separate part of a complete antigen and that gives a visible precipitin reaction with the appropriate antibody.

complex ion. The product that is formed by the interaction of a metal ion and ligands and that carries a charge. *See also* complex (2).

complex lipid. 1. AMPHIPATHIC LIPID. 2. One of a group of diverse lipids. *See also* lipid.

complex locus. The position on a chromosome that is occupied by two or more adjacent genes that code for different polypeptide chains of a multimeric protein.

complex medium. A medium that contains a variety of both known and unknown chemical ingredients.

complex virion. A virus the morphology of which is more intricate than that of either an icosahedral or a helical virus.

component. 1. An independently variable, and chemically distinct, substance. 2. An ingredient of a mixture.

component I. MOLYBDOIRON PROTEIN.

component II. AZOFERREDOXIN.

compound. A substance composed of two or more elements, such that the atoms of the elements are firmly linked together and are present in definite proportions.

compound lipid. COMPLEX LIPID.

compound microscope. A microscope having two or more lenses.

Compton effect. The ejection of an electron from an atom by the impingement on the atom of a high energy photon, such as a photon of x-rays or gamma rays. Part of the energy of the photon is used to eject the electron and to impart kinetic energy to it; the remainder of the energy is emitted as a photon having a lower energy and a longer wavelength than the impinging photon.

Compton recoil electron. The electron ejected from an atom in the Compton effect.

Compton smear. The continuous spectrum of the energies of the photons that are emitted in the Compton effect; a continuous spectrum is obtained, since any fraction of the impinging x-ray or gamma ray energy can be dissipated in this fashion.

computer. An automatic electronic system that can receive a large number of items of information, subject them to specific and often complex calculations, and provide the results either in direct form or in terms of control of other systems. *See also* analog computer; digital computer.

computer interface. The auxiliary equipment used in linking a computer to an apparatus or to an instrument.

COMT. Catechol-*O*-methyl transferase.

comutation. A mutation that occurs in the vicinity of, and simultaneously with, a selected mutation.

concatemer. CATEMER.

concatenate. CATEMER.

concatenation. The formation of catemers.

concave exponential gradient. An exponential density gradient that is formed if the solution introduced into a mixing chamber of constant volume has a lower concentration than the solution initially present in the mixing chamber.

concentrated solution. A solution that contains a large amount of solute.

concentration. The amount of solute in a solution. *See also* formal solution; molal solution; molar solution; osmolal solution; osmolar solution; percent solution; ppb; ppm.

concentration gradient. The change of concentration with distance, as the change of concentration along a density gradient or across a membrane.

concentration of enzymatic activity. A measure of the concentration of an enzyme in solution that is equal to the enzymatic activity divided by the volume of the solution; it is expressed in terms of katals per liter.

concentric cylinder viscometer. An instrument for measuring the viscosity of a liquid by placing the liquid in the space between two concentric cylinders, rotating one cylinder at a constant speed, and measuring the torque exerted on the other cylinder.

concerted acid-base catalysis. Catalysis that consists of the simultaneous action of both acidic and basic catalytic groups.

concerted divalent inhibition. The inhibition of a regulatory enzyme that is produced when two effectors are bound to the enzyme simultaneously, but that is not produced when either effector is bound to the enzyme alone.

concerted feedback inhibition. The feedback inhibition of an enzyme that is produced when two or more end products are present simultaneously, but that is not produced when an end product is present alone.

concerted reaction. A chemical reaction in which a new bond is formed at the same time as, and as a direct consequence of, the breaking of another bond.

concerted transition model. SYMMETRY MODEL.

condensate. 1. The crystalline particles of DNA that are formed during an early stage in the maturation of T-even phages. 2. The liquid obtained by condensation of either a gas or a vapor.

condensation. 1. The linking of two like, or two unlike, molecules with the elimination of either a molecule of water or some other small molecule. 2. POLYMERIZATION. 3. An early stage in the maturation of T-even phages during which the condensate is formed. 4. The transition of either a gas or a vapor to a liquid.

condensation polymer. A polymer formed by the linking of monomers with the elimination of either water molecules or other small molecules.

condensation principle. CONDENSING PRINCIPLE.

condensed conformation. A low-energy conformation of mitochondria that occurs in mitochondrial preparations containing an excess of ADP, and that is characterized by a mitochondrial matrix which is not squeezed together tightly and does not stain heavily.

condensing enzyme. CITRATE SYNTHASE.

condensing principle. A factor that aids in the aggregation of DNA to form a condensate during an early stage in the maturation of T-even phages.

condensing site. PEPTIDYL SITE.

conditional lethal mutant. A phage mutant, the ability of which to grow depends on the condi-

tions; it grows as a normal phage under permissive conditions but does not grow, and thus expresses its lethal mutation, under restrictive conditions.

conditioned vitamin deficiency. A disorder that is caused by an interference with the digestion, absorption, or utilization of a vitamin as distinct from one that is caused by a lack of the vitamin in the diet. *See also* secondary deficiency.

conductance. The property of an electrical circuit that determines the rate at which electrical energy is converted into heat when a given potential is applied across the electrodes; equal to the reciprocal of the electrical resistance.

conduction. The act of conveying either matter or energy. *See also* nerve impulse conduction.

conductivity. 1. The capacity to conduct either electricity or heat. 2. CONDUCTANCE.

conductometry. A method of chemical analysis that is based on measurements of electrical conductivity.

cone. A light receptor in the retina of vertebrates that functions in day and color vision.

cone threshold. The lowest light intensity to which the cones are sensitive.

confidence interval. The range described by the confidence limits.

confidence limits. The upper and lower limits of the range about an experimentally determined mean, such that the true mean will be found in this range with a given degree of probability.

configuration. A unique and fixed spatial arrangement of the atoms in a molecule such that the molecule may be isolated in this stereochemical form. The change from one configuration to another requires the breaking and forming of covalent bonds. *See also* conformation.

confluent growth. The growth of bacterial cells on a solid medium such that the entire surface of the medium is covered by the cells.

confluent lysis. The complete lysis, in a plaque assay, of the entire bacterial lawn.

conformation. A spatial arrangement of the atoms in a molecule that results from the rotation of the atoms about single bonds without a change in the covalent structure of the molecule. Conformation thus refers to a family of structures and not to a single isolatable stereochemical form. The change from one conformation to another does not require the breaking and forming of covalent bonds. *See also* configuration.

conformational analysis. The application of specific rules to determine the conformation of a molecule in which strain will be at a minimum; this conformation is the most stable and the preferred conformation of the molecule.

conformational coupling hypothesis. MECHANOCHEMICAL COUPLING HYPOTHESIS.

conformational isomer. CONFORMER.

conformational map. RAMACHANDRAN PLOT.

conformer. One of two or more isomers that differ from each other in their conformation; any one of the various possible conformations of a molecule.

conformon. A quantized package of energy that, according to the mechanochemical coupling hypothesis, is associated with a localized conformational change in the mitochondrial membrane.

congener. One of a family of related chemical substances, such as derivatives of a compound or elements belonging to the same group in the periodic table.

congenital. Existing at birth.

congenital goiter. FAMILIAL GOITER.

congenital porphyria. A genetically inherited metabolic defect in man that is characterized by the presence of excessive amounts of uroporphyrins in the urine.

conglutination. The agglutination of antigen-antibody-complement complexes by conglutinin.

conglutinin. A protein that is present in normal serum and that causes the agglutination of antigen-antibody-complement complexes; conglutinin is not an antibody. *See also* immunoconglutinin.

conidium (*pl* conidia). An asexual spore of certain fungi. Large and usually multinucleate conidia are known as macroconidia; small and usually uninucleate conidia are known as microconidia.

conjugated acid-base pair. A Bronsted acid and its corresponding Bronsted base; a proton donor and the corresponding proton acceptor.

conjugated antigen. An antigen consisting of a protein that is covalently linked to either a molecule or a group which contains an antigenic determinant.

conjugated enzyme. An enzyme that is a conjugated protein.

conjugated protein. A protein that contains a nonprotein component in addition to the amino acids. The nonprotein component may be either a metal ion or an organic molecule such as a lipid, a carbohydrate, or a nucleic acid. The nonprotein component may be either loosely associated with the protein or bound to it tightly as a prosthetic group.

conjugated redox couple. REDOX COUPLE.

conjugation. 1. The covalent or noncovalent combination of a large molecule, such as a protein or a bile acid, with another molecule. 2.

The alternating sequence of single and double bonds in a molecule. 3. The genetic recombination in bacteria and in other unicellular organisms that resembles sexual reproduction and that entails a transfer of DNA between two cells of opposite mating type which are associated side by side.

conjugon. A genetic element, such as the fertility factor, that is required for bacterial conjugation.

consecutive reactions. A series of two or more reactions in which the product of one reaction serves as a reactant for the next reaction.

cons electrophoresis. ISOTACHOPHORESIS.

conservation equation. An equation that expresses the total concentration of a component in terms of all the various forms in which it occurs. For example, the total concentration of enzyme in a simple enzymatic reaction is equal to the concentration of the free enzyme plus the concentration of the enzyme in the form of the enzyme-substrate complex.

conservative amino acid replacement. CONSERVATIVE SUBSTITUTION.

conservative replication. A mode of replication for double-stranded DNA in which the parental strands do not separate and in which the progeny consists of both original parental duplexes and of newly synthesized duplexes.

conservative substitution. The replacement in a protein of one amino acid by another, chemically similar, amino acid such as the replacement of a polar (nonpolar) amino acid by another polar (nonpolar) amino acid. A conservative substitution is generally expected to lead to either no change or only a small change in the properties of the protein. *See also* radical substitution.

constant region. That part of the immunoglobulin molecule in which virtually no changes in the amino acid sequence are found when immunoglobulins from different sources are compared. The constant region comprises portions of both the light c_L and the heavy c_H chains and does not constitute part of the antibody combining site. *See also* variable region.

constituent concentration. The concentration of a component that takes into account all of the forms in which the component occurs; the total concentration of a macromolecule that exists in two conformational states is an example.

constituent parameter. The concentration average of a parameter, such as a sedimentation coefficient or a diffusion coefficient, for a component that exists in several forms in the solution.

constitutive enzyme. An enzyme that is present in a given cell in nearly constant amounts regardless of the composition of either the tissue or the medium in which the cell is contained.

constitutive mutation. A mutation that results in extensive synthesis of an inducible enzyme in the absence of an inducer and that involves an alteration in either the operator or the regulator gene of the enzyme.

constructive interference. *See* interference (1).

constructive metabolism. ANABOLISM.

contact dermatitis. *See* allergic contact dermatitis.

contact factor. HAGEMAN FACTOR.

contact guidance. The guidance of cells along oriented surfaces.

contact hypersensitivity. The hypersensitivity that is brought about by exposure of the skin to a chemical substance. *See also* allergic contact dermatitis.

contact inhibition. The inhibition of cell growth that occurs in tissue culture when cells of multicellular organisms come into contact with each other. Contact inhibition permits the growth of monocellular layers and prevents the disorderly piling up of cells. The loss of contact inhibition is one of the characteristics of a malignant cell.

contact skin sensitivity. The capacity of an animal organism to respond to a percutaneous application of a chemical sensitizer.

contamination. 1. The mixing of an impurity, such as a heavy metal ion or a radioactive substance, with the sample. 2. An impurity that is present in the sample.

continuity equation. An equation that expresses the conservation of mass during ultracentrifugation on the basis of sedimentation and diffusion.

continuous culture. A culture of cells that is maintained in a growing state over prolonged periods of time. *See also* batch culture; chemostat.

continuous density gradient. A density gradient in which the density changes in an uninterrupted fashion, rather than in a stepwise fashion, from one end of the gradient to the other.

continuous development. A chromatographic technique, used particularly with paper and thin-layer chromatography, in which the solvent is allowed to run continuously over the support.

continuous discharge region. That portion of the characteristic curve of an ionization chamber in which, during gas amplification, there is a continuous discharge in the chamber so that it is no longer usable as a detector.

continuous distribution. A set of experimental data in which the variable being measured can vary continuously and is expressed as a number having one or more decimal places; the weight gain per animal in a group of animals is an example.

continuous emission. The emission of light over a range of wavelengths that is produced in flame photometry when nonionic materials are present in the sample.

continuous flow centrifugation. A preparative-type centrifugation, used for collecting materials from large volumes of liquid, in which a liquid is fed continuously into a rotor, the sediment is accumulated, and the supernatant is continuously withdrawn.

continuous flow electrophoresis. An electrophoretic technique in which the flow of liquid is in a vertical direction and the electric field is at right angles to the direction of liquid flow. The sample is applied continuously at the top of the apparatus and fractions are collected at the bottom of the apparatus at various spacings along the supporting medium.

continuous flow isoelectric focusing. An isoelectric focusing technique in which the flow of liquid is in a vertical direction and the electric field is at right angles to the direction of the liquid flow. The sample is applied continuously at the top of the apparatus and fractions are collected at the bottom of the apparatus at various spacings along the supporting medium.

continuous flow scintillation counter. A liquid scintillation counter designed for the continuous monitoring of radioactivity and used for effluents from amino acid analyzers and gas chromatographs.

continuous flow technique. RAPID FLOW TECHNIQUE.

continuous spectrum. A spectrum in which either the absorption or the emission of radiation covers a range of wavelengths.

continuous variation. *See* method of continuous variation.

contour length. The length of an extended polymer as distinct from the end-to-end distance of the folded polymer.

contracted muscle. A muscle that has undergone contraction.

contractile. Capable of contraction.

contractile protein. A protein, such as actin or myosin, that functions in contractile tissue.

contraction. The shortening of a muscle. *See also* isometric contraction; isotonic contraction.

contributing structure. CANONICAL STRUCTURE.

control. 1. An experiment that serves as a standard of comparison for other experiments; the control is carried out exactly as the other experiments except that it differs from them in one variable, the significance of which can thereby be assessed. *See also* blank. 2. The regulation of a biochemical process. *See also* coarse control; fine control.

controlled atmospheric storage. GAS STORAGE.

controlling gene. A gene, such as a regulator gene, that can turn other genes on or off.

convalescent serum. The serum obtained during convalescence from a disease.

convection. The bulk movement of fluid in which both solvent and solute move together and that is usually due to either density inversions caused by temperature variations or local concentration changes.

conventional animal. An animal raised under ordinary conditions as distinct from one raised in a germ-free environment.

conventional sedimentation equilibrium. SEDIMENTATION EQUILIBRIUM.

convergence theory of cancer. GREENSTEIN HYPOTHESIS.

convergent evolution. An evolutionary pattern in which the lines of development for more recent species come together as a result of independent development from earlier species; such a pattern can be depicted as two or more independent networks, arising from different origins.

conversion. A change in the properties of the host bacterium, such as antigenic character or toxin production, that is brought about by the prophage of that bacterium.

conversion coefficient. The fraction of gamma rays that produce Auger electrons.

conversion electron. The electron emitted from an atom that is undergoing internal conversion.

conversion factor. A number that converts one set of dimensions into another.

conversion stage. That part of the blood clotting process that consists of the conversion of fibrinogen to fibrin under the influence of thrombin.

convertin. The activated form of proconvertin.

converting enzyme. *See* serum converting enzyme.

converting phage. A phage that brings about conversion in its host cell.

convex exponential gradient. An exponential density gradient that is formed if the solution introduced into a mixing chamber of constant volume has a higher concentration than the solution initially present in the mixing chamber.

Conway microdiffusion apparatus. An apparatus for the microchemical analysis of a gas, such as ammonia or carbon dioxide, that can be liberated from a sample by treatment with specific reagents. The apparatus consists of two concentric plates, much like a modified petri dish; the central well contains the sample, and the outer space contains a solution for trapping the gas that will be liberated and that will dif-

fuse away from the sample upon addition of the reagents.

Coombs' reagent. An antiserum that contains antibodies to human immunoglobulins and that is prepared by injecting these immunoglobulins into rabbits.

Coombs' test. A test for demonstrating incomplete antibodies against red blood cells; based on an agglutination reaction in which the incomplete antibodies bind simultaneously to red blood cell antigens and to antibodies against themselves. *See also* direct Coombs' test; indirect Coombs' test.

Coon's method. INDIRECT FLUORESCENT ANTIBODY TECHNIQUE.

cooperative binding. The binding of ligands to interacting sites of a macromolecule so that the binding of one ligand to one binding site affects the binding of subsequent ligands to other binding sites on the same molecule.

cooperative feedback inhibition. The feedback inhibition of an enzyme that is produced by two or more end products such that the inhibition caused by a mixture of two end products present together is greater than that caused by either end product present alone at the same total specific concentration (i.e., the concentration relative to the inhibitor constant).

cooperative hydrogen bonding. The interaction between neighboring hydrogen bonds in a molecule such that the energy required to form these bonds is smaller than the sum of the energies for the individual bonds, and the energy required to break these bonds is greater than the sum of the energies for the individual bonds.

cooperative interactions. *See* cooperative binding; cooperativity.

cooperative kinetics. The kinetics of cooperative binding reactions. *See also* sigmoid kinetics.

cooperativity. 1. The interaction between either identical or different binding sites of a macromolecule so that the binding of a ligand to one site affects the binding of subsequent ligands to other sites on the same molecule. 2. The interaction between neighboring hydrogen bonds in either a protein or a nucleic acid. *See also* cooperative hydrogen bonding.

coordinate covalent bond. A covalent bond formed between two atoms and consisting of two electrons, both of which are donated by only one of the bonded atoms.

coordinated enzymes. The enzymes that are controlled by genes of one operon and that are either induced in coordinate induction or repressed in coordinate repression.

coordinated enzyme synthesis. The synthesis of coordinated enzymes.

coordinate induction. Enzyme induction in which a single inducer brings about the synthesis of a number of inducible enzymes that catalyze a sequence of either consecutive, or related, reactions in which the inducer is generally the first substrate. The structural genes of the coordinated enzymes are contiguous and form part of one operon. *See also* sequential induction.

coordinate regulation. *See* coordinate induction; coordinate repression.

coordinate repression. Enzyme repression in which a single repressor brings about the decreased synthesis of a number of repressible enzymes that catalyze a sequence of either consecutive, or related, reactions in which the repressor is generally the last end product. The structural genes of the coordinated enzymes are contiguous and form part of one operon.

coordination. 1. The formation of a complex between a metal ion and ligands. 2. COORDINATED ENZYME SYNTHESIS.

coordination number. The number of ligands that can be bound to a metal ion to form a complex.

coordination position. The position in the space surrounding a metal ion that can be occupied by a ligand for coordination with the metal ion.

coplanar. Lying in one plane.

copolymer. A polymer formed from two or more different types of monomers that polymerized together.

copper. An element that is essential to all plants and animals. Symbol Cu; atomic number 29; atomic weight 63.54; oxidation states $+1$, $+2$; most abundant isotope Cu^{63}; a radioactive isotope Cu^{64}, half-life 12.8 hours, radiation emitted—beta particles, positrons, and gamma rays.

coprecipitating antibody. An antibody that does not form an antigen-antibody precipitate by itself, but that can be incorporated in an antigen-antibody precipitate under suitable conditions.

coproantibody. An antibody present in feces.

coproporphyrin. The urinary pigment that is derived from coproporphyrinogen. *Abbr* CP.

coproporphyrinogen. An intermediate in the biosynthesis of heme that is derived from uroporphyrinogen. *Abbr* CPG.

copy-choice hypothesis. A hypothesis of the mechanism of genetic recombination according to which the recombinant DNA molecule is synthesized by a system that uses both of the parental DNA molecules as templates, but that copies them in an alternating fashion.

copy error. A mistake in replication.

copy-error mutation. A mutation that results from a mistake in replication.

copy-splice mechanism. A mechanism that describes the genetic control of the synthesis of immunoglobulin light chains in terms of the germ line theory.

CoQ. Coenzyme Q.

CoQH₂. Reduced coenzyme Q.

cor. The bare 15-membered ring of the corrin ring system.

cord factor. A complex disaccharide, composed of trehalose and two molecules of mycolic acid, that occurs in virulent strains of *Mycobacterium tuberculosis* and that is believed to be associated with both the virulence of these organisms and their tendency to grow in cord-like skeins.

core. *See* spore core.

core enzyme. The portion of the enzyme RNA polymerase that consists of an aggregate of five subunits and that possesses catalytic activity, but that requires the attachment of the sigma factor before it can recognize an initiation site of transcription.

core particle. A particle obtained from ribosomes by removal of some of the ribosomal proteins. *Abbr* CP. *See also* intersome.

corepressor. A substance that, together with a repressor, binds to an operator and prevents the synthesis of an enzyme in enzyme repression; the corepressor is generally either a product of the enzymatic reaction or a compound that is structurally similar to the product.

core protein. A structural protein molecule that occurs in complex IV, one of the respiratory assemblies.

Cori coefficient. A measure of the rate of monosaccharide absorption by rat intestine that is expressed as the number of milligrams of monosaccharide absorbed per 100 gram of rat per hour.

Cori cycle. The cyclic group of reactions whereby glycogen is broken down and resynthesized. The sequence begins with the breaking down of muscle glycogen to lactic acid, which is carried by the blood to the liver where it is converted back to glycogen; the liver glycogen, in turn, is then broken down to glucose, which is carried by the blood to the muscle where it is reconverted to glycogen.

Cori ester. Glucose-1-phosphate.

corn sugar. GLUCOSE.

corpuscle. A small particle or body.

corpuscular. Of, or pertaining to, corpuscles.

corpus luteum (*pl* corpora lutea). A yellow progesterone-secreting body that is formed in a ruptured follicle.

corrected absorbance. 1. The absorbance of a solution that has been corrected for the absorbance of either a blank or a reference solution. 2. The absorbance of a solution that is obtained after applying an Allen correction.

correlation. The extent to which two statistical variables vary together; it is measured by the correlation coefficient.

correlation coefficient. A measure of the correlation between two statistical variables; it varies from zero for no correlation to $+1$ or -1 for perfect positive or negative correlation. A correlation coefficient of $+0.3$ (-0.3) means that as one variable increases, the other will increase (decrease) 30% of the time in the long run.

corrinoid. A compound containing the corrin ring system.

corrin ring system. The ring structure of vitamin B_{12} in which a cobalt atom is chelated.

cortex. *See* adrenal cortex; spore cortex.

corticoid. CORTICOSTEROID.

corticosteroid-binding globulin. TRANS-CORTIN.

corticosteroid-binding protein. TRANS-CORTIN.

corticosteroids. The 21-carbon steroid hormones that are derived from the adrenal cortex and the metabolites of these hormones. The two principal groups of corticosteroids are the glucocorticoids and the mineralocorticoids.

corticosterone. A glucocorticoid.

corticotrophin. Variant spelling of corticotropin.

corticotropin. ADRENOCORTICOTROPIC HORMONE.

corticotropin releasing hormone. The hypothalamic hormone that controls the secretion of corticotropin. *Var sp* corticotrophin releasing hormone. *Abbr* CRH. *Aka* corticotropin releasing factor (CRF).

cortin. An acetone extract of the adrenal cortex.

cortisol. The major glucocorticoid in man.

cortisol-binding globulin. TRANSCORTIN.

cortisol-binding protein. TRANSCORTIN.

cortisone. A glucocorticoid.

cosmic rays. The high-energy ionizing radiation that originates outside the earth's atmosphere and that consists primarily of protons and other nuclei.

cosubstrate. A compound that acts somewhat in the capacity of a substrate during an enzymatic reaction, such as a dissociable coenzyme molecule.

cot method. A graphical method for following the renaturation kinetics of denatured and fragmented double-stranded DNA. The fraction of single-stranded molecules that have reassociated is plotted against the cot value. The average length of duplex formed is proportional to the cot value for 50% renaturation. *See also* cot value.

cotransduction. The simultaneous transduction of two or more genes that lie on the same segment of the bacterial DNA that is being transduced.

Cotton effect. The change in sign of the optical rotation in the neighborhood of an absorption band. The effect is due to the circular dichroism of the left and right circularly polarized light components of plane-polarized light. A Cotton effect is referred to as positive or negative depending on whether the optical rotation, with increasing wavelength, passes first through a minimum value and then through a maximum value, or vice versa.

cot value. The product of the concentration of the reassociated DNA and the time as used in the cot method.

Couette viscometer. A concentric cylinder viscometer in which the outer cylinder is rotated at a constant speed and the viscosity is determined from the torque exerted on the inner cylinder.

coulomb. A quantity of electricity equal to a current of 1 A flowing for 1 sec.

Coulomb effect. ION-ION INTERACTION.

Coulombic interactions. The electrostatic interactions that can be described by Coulomb's law.

Coulomb's law. An expression for the electrostatic force F between two point charges; specifically, $F = Q_1Q_2/Dr^2$, where Q_1 and Q_2 are the two point charges, r is the distance between the charges, and D is the dielectric constant of the medium. The force is repulsive if the charges have the same sign, and the force is attractive if the charges have opposite signs. The energy of such interactions is proportional to the reciprocal of r.

coulometer. An instrument for measuring the quantity of electricity.

coulometry. A method of chemical analysis that is based on measurements of the quantity of electricity, in coulombs, which is associated with a quantitative electrode reaction. In constant-current coulometry, the current is kept at a constant level so that the elapsed time is proportional to the total number of coulombs consumed; in constant-potential coulometry, the potential is kept at a constant level and the quantity of electricity consumed is measured with a coulometer.

Coulter counter. A particle counter used for the counting of blood cells and bacteria.

counter. 1. An instrument for indicating, and frequently recording, radioactive radiation; it may include a detector, sample changer, scaler, and printer. Some counters, such as ionization chambers and Geiger-Mueller counters, use the ionization of a gas to measure the radiation; other counters, such as liquid and solid scintillation counters, use the scintillations produced by fluors to measure the radiation. 2. Any instrument for counting, such as a cell counter, a drop counter, etc.

countercurrent diffusion multiplier system. A system for the production of hypertonic urine by certain nephrons in the kidney.

countercurrent distribution. A multistep separation procedure that is based on solubility differences of compounds in two immiscible liquid phases. The compounds are partitioned repeatedly between the two immiscible phases as they "move" along a large number of partition tubes. *Abbr* CCD. *Aka* countercurrent extraction.

countercurrent multiplication mechanism. COUNTERCURRENT DIFFUSION MULTIPLIER SYSTEM.

counter double current distribution. A variation of countercurrent distribution in which the sample is injected continuously into the apparatus and the unwanted components are removed at both ends of the apparatus.

counterelectrophoresis. *See* immunocounterelectrophoresis.

counterflow. The movement of a substance from side *A* of a membrane to side *B*, after equilibrium has been established between the two sides, in response to the addition of a structurally related substance to side *B*. In this process, the substrate moves against its own concentration gradient. The occurrence of counterflow is taken as evidence for the existence of a single carrier which moves both of the substances.

counterflow centrifugation. CENTRIFUGAL ELUTRIATION.

counterimmunoelectrophoresis. A gel electrophoretic technique in which antiserum is placed in one well and antigen in another and an electric field is applied across the two wells. As a result of differences in mobilities and the occurrence of endosmosis, the antigens and antibodies migrate toward each other and form precipitin bands.

counterion. One of two ions that have charges of opposite sign.

counterstain. The staining of either a tissue or a culture with a dye that follows a previous staining with another dye.

counting efficiency. The ratio of the number of registered radioactive counts to the number of actual radioactive disintegrations that occurred during the same time; generally multiplied by 100 to give percent efficiency.

counting loss. COINCIDENCE LOSS.

counting plateau. That portion of the characteristic curve of an ionization chamber that is almost independent of the applied voltage.

counts. A measure of radioactivity that represents the fraction of the radioactive disintegrations that are detected and registered by means of a counter.

coupled layer chromatography. A thin-layer chromatography technique in which a chromatoplate is used, the two halves of which are covered with two different, but adjacent, layers of a chromatographic support.

coupled neutral pump. A coupled pump in which

the movement of one ion across the membrane must be linked to the movement of another ion, of equal valence, in the opposite direction.

coupled pump. A pump for the transport of one solute across a membrane that also drives the transport of a second solute across the same membrane in the opposite direction and in such a fashion that the transport of the second solute is physically dependent on the pump.

coupled reactions. An endergonic and an exergonic reaction that are linked energetically; the endergonic reaction is driven by the exergonic reaction which occurs simultaneously and which shares a common intermediate with the endergonic reaction, such that the overall free energy change for the coupled reactions is negative. *Aka* energetically coupled reactions; energy coupling.

coupled transport. A transport system in which the movement of one solute across the membrane must be linked to the movement of a second solute across the same membrane but in the opposite direction.

coupling. 1. The linking of aerobic respiration, specifically the electron transport system, to the synthesis of ATP. 2. The tendency of linked genes to be inherited together on the same chromosome.

coupling constant. The separation between any two bands of multiple peaks in nuclear magnetic resonance; it is proportional to the magnitude of the spin-spin coupling.

coupling factors. A group of mitochondrial proteins which, when added to specially treated mitochondria, can partially restore the capacity of the mitochondria to synthesize ATP. The mitochondria are treated in such a way that they can carry out electron transport but are incapable of ATP synthesis.

coupling inhibition. UNCOMPETITIVE INHIBITION.

covalent bond. A bond formed between two atoms and consisting of one or more shared pairs of electrons such that one electron in a pair is donated by each of the two bonded atoms. *See also* coordinate covalent bond.

covalent catalysis. Catalysis that requires the formation of a covalent enzyme-substrate-catalyst intermediate.

covalent chromatography. A column chromatographic technique in which a chemical reagent is linked covalently to the column support. When a sample is passed through the column, the reagent reacts with, and binds covalently, the substance of interest. An additional chemical reaction then releases the substance from the support and permits its elution from the column, thereby restoring the initial form of the support.

covalent enzyme-substrate complex. ENZYME-SUBSTRATE COMPOUND.

covalent enzyme-substrate intermediate. SUBSTITUTED ENZYME.

covalent intermediate. 1. A substance formed during covalent catalysis such as the intermediate formed in the transaminase reaction. 2. A covalently linked, high-energy intermediate that, according to the chemical coupling hypothesis, functions in the coupling of oxidative phosphorylation.

covalent labeling. AFFINITY LABELING.

covalently circular. *See* circular covalent.

covalent orbital. An orbital that functions in the bonding of a low-spin complex.

covalent structure analysis. The determination of the covalent bonds that describe the arrangement of monomers in a macromolecule; the bonds that describe the amino acid sequence and the location of disulfide bonds in a protein, or those that describe the nucleotide sequence in a nucleic acid are examples.

covariance. *See* analysis of covariance.

covariation. PLEIOTROPISM.

covirus. One of two or more different viral particles that must be present together for the initiation of infection.

covolume. The difference between the volume of a compound in solution and that given by the sum of its atomic volumes. The additional volume results from the intermolecular forces that set a lower limit to the distance of approach between molecules in a liquid. *Aka* excluded volume.

coxsackievirus. A virus that belongs to the enterovirus subgroup of picornaviruses and that is similar in physical parameters to the polio virus.

cozymase. An early designation of a heat-stable fraction—consisting chiefly of ATP, ADP, AMP, and NAD^+—that was isolated from yeast and participated in the reactions of alcoholic fermentation. Subsequently, cozymase I was used to denote NAD^+ and cozymase II was used to denote $NADP^+$.

c.p. Chemically pure.

CP. 1. Coproporphyrin. 2. Core particle.

CPG. Coproporphyrinogen.

CPK. Creatine phosphokinase.

CPK model. Cory-Pauling-Koltun model; a space-filling molecular model.

cpm. Counts per minute; the number of radioactive counts per minute.

Crabtree effect. The inhibition of oxygen consumption in cellular respiration that is produced by increasing concentrations of glucose. *See also* Pasteur effect.

C-reactive protein. A protein that reacts with the pneumococcal type C polysaccharide and that is present in plasma during some bacterial infections.

creatine. A nitrogenous compound, the phosphorylated form of which, phosphocreatine, is a high-energy compound that serves as a free energy storage compound in muscle.

creatine kinase. The enzyme that catalyzes the conversion of ADP and phosphocreatine to ATP and creatine.

creatine phosphate. *See* phosphocreatine.

creatine phosphokinase. *See* creatine kinase.

creatinine. A cyclic compound, formed from creatine, that represents one of the major forms in which nitrogen is excreted in the urine.

creatinine coefficient. The number of milligrams of creatinine excreted per 24 hours per kilogram of body weight.

creatinuria. The presence of excessive amounts of creatine in the urine.

C region. Constant region.

crenation. The shrinking of red blood cells that occurs when they are placed in a hypertonic solution.

cretinism. A condition of arrested growth that is caused by hypothyroidism.

CRF. Corticotropin releasing factor; *see* corticotropin releasing hormone.

CRH. Corticotropin releasing hormone.

Crick strand. The DNA strand of Watson-Crick type DNA that is not transcribed in vivo. *Abbr* C strand.

Crigler-Najjar syndrome. A genetically inherited metabolic defect in man that is characterized by defective bilirubin metabolism and by jaundice, and that is due to a deficiency of the enzyme uridine diphosphate glucose transferase in the liver.

crisis. A sudden change in the course of an acute disease.

crista (*pl* cristae). An extended infolding of the inner mitochondrial membrane.

cristael. Of, or pertaining to, cristae.

critical group. CRITICAL PAIR.

critical micelle concentration. The concentration of a surface-active compound above which the formation of micelles by this compound becomes appreciable. *Abbr* CMC. *See also* second critical concentration.

critical pair. Two compounds that are not readily separable from each other by their partitioning between two liquid phases, as in countercurrent distribution.

critical temperature. The temperature above which a gas cannot be liquefied by pressure alone; at that temperature the properties of the liquid and of its saturated vapor become indistinguishable.

CRM. Cross-reacting material.

cRNA. Complementary RNA; a molecule of

RNA that is complementary to a molecule of DNA.

cross. *See* genetic cross.

cross absorption. The absorption of either antigens or antibodies by means of the corresponding cross-reacting antibodies or cross-reacting antigens.

cross agglutination test. CROSS-MATCHING.

crossbreeding. OUTBREEDING.

cross-bridge. A short projection from the thick filament of striated muscle; cross-bridges are regularly spaced and link the thick filaments to the thin filaments.

crossed immunoelectrophoresis. An electrophoretic technique in which antigens are first separated by one-dimensional gel electrophoresis; the antigens are then separated in the second dimension by gel electrophoresis, using a gel that contains antibodies and applying an electric field at right angles to the direction of the first separation.

cross-feeding. The phenomenon of two organisms that can grow only in the vicinity of each other or in each other's medium, since each is dependent on the other for an essential growth factor. *See also* syntrophy.

cross induction. The induction of a prophage in a nonirradiated, lysogenic F^- bacterium in response to compounds transferred to the bacterium by conjugation with an ultraviolet-irradiated F^+ cell.

cross infection. The infection of a bacterium by two or more different phage mutants.

crossing-over. The process whereby genetic material is exchanged between chromosomes, leading to gene combinations that are different from those in the parental chromosomes. *See also* breakage and reunion model.

crossing-paper electrophoresis. A paper electrophoretic technique that is used to screen substances for their possible interaction. The various substances are allowed to move in an electric field in two oblique or perpendicular lines. If two substances interact, the line will be deformed since the mobility of the complex will differ from the mobility of at least one of the reactants.

cross-linking. The formation of covalent bonds between chains of polymeric molecules.

cross-matching. A serologic procedure for blood typing in which erythrocytes from a donor of unknown blood type are mixed with the serum of recipients of known blood types.

cross of isocline. The cross-like pattern observed in flow birefringence; the arms of the cross appear dark on a light background.

crossover. 1. The chromosome resulting from crossing over. 2. The individual resulting from crossing over.

crossover method. A method for studying a sequence of oxidation-reduction reactions from the changes produced in the sequence upon the addition of inhibitors. *See also* crossover theorem.

crossover point. 1. The step in a sequence of oxidation-reduction reactions that is being inhibited by the addition of an inhibitor. *See also* crossover theorem. 2. The step in a sequence of metabolic reactions at which the metabolic flux is altered with a resultant change in the concentrations of the remaining reactants. The crossover point is referred to as positive or negative depending on whether the metabolic flux is increased or decreased at the particular step. A positive crossover point results in a decrease of the steady-state levels of the intermediates preceding the crossover point and in an increase of the steady-state levels of the intermediates following the crossover point. These concentration changes are reversed for a negative crossover point.

crossover theorem. The principle that the addition of an inhibitor to a series of oxidation-reduction reactions, such as the electron transport system, will cause the components on the reduced side of the inhibited reaction to become more reduced, and will cause those on the oxidized side to become more oxidized.

crossover unit. A crossover value of 1% between a pair of linked genes.

cross-reacting antibody. An antibody that can combine with antigens that are specific for stimulating the production of different antibodies.

cross-reacting antigen. An antigen that can combine with antibodies which are produced in response to different antigens.

cross-reacting material. A defective protein that is produced by a mutant and that is antigenically similar to the protein produced by the normal wild-type gene. *Abbr* CRM.

cross-reaction. The reaction of an antigen with an antibody that is produced in response to a different antigen; the reaction occurs because of structural similarities between the antigenic determinants of the different antigens. *Aka* reaction of partial identity.

cross-reactivation. The restoration of the activity of a mutant virus, which carries a lethal mutation as a result of previous exposure to a mutagen, by the simultaneous infection of a host cell with both the mutant virus and with one or more active viruses. The mutant virus is activated by a genetic exchange that leads to a replacement of its damaged DNA. *See also* multiplicity reactivation.

cross-sensitization. The immunological sensitization of an organism with an antigen that is different from the antigen which will be used subsequently to trigger an anaphylactic response.

cross-term diffusion coefficient. The diffusion coefficient that a component has when it diffuses in the presence of another component; used in the treatment of diffusion data in a system showing interaction of flows. *See also* main diffusion coefficient.

cross tolerance. The immunological tolerance against one antigen that is produced by the administration of a different, but cross-reacting, antigen.

CRS. Codon-recognizing site.

crude extract. A preparation, derived from biological material, that has not been extensively purified; used particularly to describe a preparation of either homogenized tissue or broken cells from which unbroken cells and cell debris have been removed, commonly by centrifugation.

cryogenic. Of, or pertaining to, low temperatures.

cryoglobulin. A serum globulin that precipitates, gels, or crystallizes upon cooling of either a serum or a solution that contains the globulin.

cryoscope. An instrument for the determination of freezing points.

cryoscopic method. The determination of either molecular weight or osmotic pressure from the depression of the freezing point of a solvent that is produced by the addition of solute.

cryostat. An apparatus for producing and maintaining a controlled low-temperature environment.

cryosublimation. The process whereby water is sublimed from a frozen sample and is collected in a cold trap; cryosublimation refers to the collection of the water, while lyophilization refers to the collection of the dry residue.

cryptic DNA. A DNA of unknown function.

crypticity. 1. The phenomenon that a particulate enzyme has different properties than the same enzyme in soluble form; solubilization of the particulate enzyme requires its detachment from the solid matrix to which it was attached. 2. The phenomenon that an intact cell is unable to use a metabolite because of a deficiency in the transport system that moves the metabolite across the cell membrane; disruption of such a cell permits the utilization of the metabolite by components of the cell. *See also* latency.

cryptic mutant. A mutant cell that can synthesize an inducible enzyme but cannot synthesize a component of the transport system required to move the substrate of that enzyme across the cell membrane.

cryptic prophage. A phage that has suffered a deletion of some of its genes while it was being integrated as a prophage.

cryptoactive. Descriptive of triglycerides that show negligible optical rotation despite the fact that they contain an asymmetric center which results from the attachment of different acyl groups to carbon atoms 1 and 3 of the glycerol.

cryptobiosis. Latent life, as that in a spore.

cryptogenic phage. A phage that can give rise to a cryptic prophage when subjected to ultraviolet curing.

cryptogram. A shorthand presentation of viral properties that consists of four pairs of symbols which indicate the following: type of the nucleic acid/strandedness of the nucleic acid; molecular weight of the nucleic acid/percentage of the nucleic acid in infective particles; outline of the viral particle/outline of the nucleocapsid; kinds of hosts infected/kinds of vectors.

crystal. A solid of definite form that is characterized by geometrically arranged, external plane surfaces as a result of a regularly repeated, internal arrangement of the atoms.

crystal field splitting. The separation of the degenerate *d*-orbitals of a metal ion into orbitals having different energies that is produced by the ligands in a metal ion-ligand complex.

crystal field theory. A description of the way in which the *d*-orbitals of a metal ion are deformed by the electrons of a ligand in a metal ion-ligand complex.

crystal lattice. The three-dimensional arrangement of the atoms in a crystal.

crystallin. The major protein of the lens of the eye that occurs in two forms, denoted α and β, which are present in roughly equal amounts.

crystalline. Of, or pertaining to, crystals.

crystalline fraction. Fc FRAGMENT.

crystallization. The transition of a substance from the molten, the liquid, or the gaseous state to the crystalline state.

crystallographic model. A molecular model, such as a ball and stick or a framework model, in which the bond lengths and the bond angles are clearly indicated.

crystalloid. A noncolloidal low-molecular weight substance.

Cs. Cesium.

CSF. Cerebrospinal fluid.

CSM. An artificially prepared mixture of corn meal, soy flour, and nonfat dry milk that is used as a protein supplement in regions where either a low-protein diet or malnutrition is prevalent.

C strand. Crick strand.

CT. Calcitonin.

C-terminal. The end of a peptide or a polypeptide chain that carries the amino acid which has a free alpha carboxyl group; in representing amino acid sequences, the C-terminal is conventionally placed on the right-hand side. *Aka* C-terminus.

CTP. 1. Cytidine triphosphate. 2. Cytidine-5'-triphosphate.

CTSH. Chorionic thyroid stimulating hormone.

Cu. Copper.

cubic symmetry. A rotational symmetry that is present in three classes of solids of which the prototypes are the tetrahedron, the octahedron, and the icosahedron.

culture. A population of either microbial cells or tissue cells that grows in or on a nutrient medium.

cumulative product feeback inhibition. The inhibition of an enzyme that is produced when the enzyme is inhibited separately and independently by two or more end products.

cuprammonium rayon. Cellulose that has been regenerated from a solution of cuprammonium hydroxide.

cuproprotein. A conjugated protein containing copper as a prosthetic group.

curare. A plant extract, used as an Indian arrow poison, that contains alkaloids which block the transmission of nerve impulses at the neuromuscular junction.

cure. To convert a lysogenic bacterium to a sensitive bacterium that may, upon subsequent infection, be either lysogenized or lysed; commonly achieved either by exposing the bacterial cells to radiation (radiation curing) or by superinfecting them with phage (superinfection curing). *See also* plasmid curing.

curie. 1. A unit of radioactivity equal to 3.7×10^{10} disintegrations per second. 2. The quantity of radioactive nuclide that contains 3.7×10^{10} disintegrations per second. *Sym* C; Ci.

curing. *See* cure; plasmid curing.

Cushing's disease. A disease characterized by an overproduction of adrenocorticotropin and caused by either overactivity or a tumor of the adrenal cortex.

cut and patch repair. EXCISION REPAIR.

cutaneous. Of, or pertaining to, the skin.

cutaneous anaphylaxis. The anaphylactic reaction that is produced in an animal organism by intradermal injections; cutaneous anaphylaxis can be of the active, the passive, or the reverse passive type.

cuvette. A small container for a liquid sample that is to be subjected to optical measurements. Typical cuvettes used in spectrophotometry are rectangular containers, constructed of either pyrex or quartz, that have a light path of 1 cm. Cuvettes, selected to have a specific tolerance with respect to their light transmitting properties, are referred to as matched cuvettes.

CV. Coefficient of variation.

cyanide. The radical CN^- that is a strong poison due to its inhibition of the enzyme cytochrome oxidase at the terminal step of the electron transport system.

cyanocobalamine. Vitamin B_{12}; cobalamine in which cyanide occupies the sixth coordination position of the cobalt atom. *Var sp* cyanocobalamin. *See also* vitamin B_{12}.

cyanogen bromide reaction. The hydrolysis by cyanogen bromide of those peptide bonds in which the carbonyl function is contributed by methionine.

cyanogenic glycoside. A glycoside that is a plant toxin containing a residue of hydrocyanic acid.

cyanoguanidine. DICYANAMIDE.

cyanohemoglobin. A derivative of hemoglobin in which the sixth coordination position of the iron atom is occupied by cyanide.

cyanopsin. A visual pigment in fresh water fish that consists of opsin and retinal$_2$ and that has an absorption maximum at 620 nm.

cyanosis. The bluish coloration of the skin that is caused by insufficient oxygenation of the blood.

cybotactic. Of, or pertaining to, cybotaxis.

cybotaxis. The spatial arrangement of solute molecules in a liquid, particularly of long-chain molecules, such that there is an equilibrium between molecules that have crystal-like orientations and molecules that have random orientations.

cycle. 1. A closed sequence of metabolic reactions, such as the citric acid cycle, in which an end product serves as a reactant for the initiation of the sequence, and in which most of the intermediates serve likewise as both reactants and products. 2. A closed sequence of large scale processes, such as the nitrogen cycle, that describes the nutritional interdependence of plants, animals, and microorganisms.

cyclic. 1. Of, or pertaining to, a cycle. 2. Of, or pertaining to, a ring.

cyclic adenylic acid. ADENOSINE-3',5'-CYCLIC MONOPHOSPHATE.

cyclic AMP. ADENOSINE-3',5'-CYCLIC MONOPHOSPHATE.

cyclic AMP receptor protein. CATABOLITE ACTIVATOR PROTEIN.

cyclic electron flow. The movement of electrons that is limited to photosystem I of chloroplasts and to its associated electron carriers; cyclic electron flow can lead to the synthesis of ATP but does not lead to an accumulation of NADPH.

cyclic GMP. GUANOSINE-3',5'-CYCLIC MONOPHOSPHATE.

cyclic metabolic pathway. CYCLE (1).

cyclic peptide. A peptide that consists of a closed chain and that is devoid of a free alpha amino group and a free alpha carboxyl group.

cyclic photophosphorylation. The synthesis of ATP that is coupled to the cyclic electron flow of photosynthesis.

cyclitol. A cyclic polyhydroxy alcohol.

cyclization. The formation of a ring.

cycloaddition reaction. The formation of a cyclic organic compound by the reaction of two double bonds, located in two separate molecules.

cyclodepsipeptide. *See* depsipeptide.

cycloheximide. An antibiotic, produced by *Streptomyces griseus,* that inhibits protein synthesis by preventing peptide bond formation on 80S ribosomes. *Aka* actidione.

cyclophorase system. A mitochondrial preparation from either kidney or liver that catalyzes the reactions of the citric acid cycle and of beta oxidation.

cyclophosphamide. An immunosuppressive drug.

cycloserine. An antibiotic, produced by *Streptomyces orchidaceus,* that inhibits bacterial cell wall formation and that is used in the treatment of tuberculosis.

cyclotron. An accelerator designed to impart high kinetic energy to particles, such as protons and deuterons, by means of an oscillating electric field and a fixed magnetic field; the particles move in a circular path.

Cyd. Cytidine.

cylindrical axis of symmetry. An axis of rotational symmetry such that $n = \infty \cdot Sym$ C_∞. *See also* axis of rotational symmetry.

cymograph. Variant spelling of kymograph.

CyS. 1. Cysteine. 2. Cysteinyl.

CySH. 1. Cysteine. 2. Cysteinyl.

CySO₃H. Cysteic acid. 2. Cysteyl.

cystathionine. A nonprotein alpha amino acid that is an intermediate in the mammalian biosynthesis of cysteine from methionine.

cysteic acid. A sulfonic acid that is obtained by oxidation of the sulfhydryl group of cysteine to $-SO_3H$. *Sym* $CySO_3H$.

cysteine. An aliphatic, polar alpha amino acid that contains a sulfhydryl group. *Abbr* Cys; CySH; C.

cystic fibrosis. A hereditary disease in man that is characterized by the functional failure of mucus-secreting glands and the resultant presence of an unusual glycoprotein in the mucus which causes the mucus to have an abnormal viscosity.

cystine. The dimer formed from two cysteine residues, linked by means of a disulfide bond.

cystinosis. A genetically inherited metabolic defect in man that is characterized by an inability to utilize cystine.

cystinuria. A genetically inherited metabolic defect in man that is characterized by the presence of excessive amounts of cystine, lysine, and arginine in the urine.

cytidine. The ribonucleoside of cytosine. Cytidine mono-, di-, and triphosphate are abbreviated, respectively, as CMP, CDP, and CTP. The abbreviations refer to the 5′-nucleoside phosphates unless otherwise indicated. *Abbr* Cyd; C.

cytidylic acid. The ribonucleotide of cytosine.

cyto-. Prefix meaning cell.

cytochemistry. The science that deals with the chemistry of cellular components.

cytochrome. A hemoprotein that contains an iron-porphyrin complex as a prosthetic group. Cytochromes function as electron carriers by virtue of the reversible valence change that the heme iron can undergo; they are classified into four groups, designated a, b, c, and d.

cytochrome a. A cytochrome in which the heme prosthetic group contains a formyl side chain; a cytochrome which contains heme A. *Aka* type a cytochrome; class a cytochrome.

cytochrome b. A cytochrome that contains protoheme or a related heme (without a formyl group) as its prosthetic group and in which the prosthetic group is not bound covalently to the protein. *Aka* type b cytochrome; class b cytochrome.

cytochrome c. A cytochrome in which there are covalent linkages between the side chains of the heme and the protein. *Aka* type c cytochrome; class c cytochrome.

cytochrome c′. RHP CYTOCHROME.

cytochrome d. A cytochrome with a tetrapyrrolic chelate of iron as a prosthetic group in which the degree of conjugation of double bonds is less than that in porphyrin; dihydroporphyrin (chlorin) is an example. *Aka* type d cytochrome; class d cytochrome.

cytochrome oxidase. The enzyme that catalyzes the terminal reaction in the electron transport system in which molecular oxygen is reduced to water.

cytochrome P₄₅₀. A cytochrome that functions as an electron carrier in microsomal nonphosphorylating electron transport chains; it plays a role in hydroxylation reactions and is part of the desmolase complex. The cytochrome has an absorption band at 420 nm and its *CO*-derivative has an absorption band at 450 nm.

cytochromoid c. RHP CYTOCHROME.

cytoflav. An impure preparation of riboflavin from heart muscle.

cytogenetic map. A genetic map that shows the location of the genes in a chromosome.

cytogenetics. The science that deals with cellular changes which are related to hereditary phenomena; combines the methods of both cytology and genetics.

cytohemin. Heme A; the prosthetic group of cytochrome a.

cytokinesin. A plant growth-regulating substance that affects cell division and that apparently acts synergistically with auxins.

cytokinesis. The division of the cytoplasm that follows both mitosis and meiosis.

cytokinin. An *N*-substituted derivative of adenine that functions as a plant growth hormone and promotes the division of plant cells in mitosis.

cytolipin. A cytoside; cytolipin H, or ceramide lactoside, can function as a hapten under certain conditions.

cytology. The branch of biology that deals with the origin, the structure, the function, and the life history of cells.

cytolysin. An antibody that can lead to the lysis of cells.

cytolysis. The lysis of cells.

cytolytic. Of, or pertaining to, cytolysis.

cytopathic. Causing either injury or disease to cells.

cytophilic antibody. An antibody that can adhere specifically to macrophages.

cytophotometry. A technique for the quantitative determination of substances by the combined use of microscopy and spectrophotometry; based on measurements of the light absorbed by cellular components that either have or have not been treated with specific stains.

cytoplasm. The protoplasm of a cell, excluding the nucleus or the nuclear zone.

cytoplasmic gene. A nonnuclear gene, such as a gene of mitochondria or chloroplasts.

cytoplasmic inheritance. The non-Mendelian hereditary transmission that depends on replicating cytoplasmic organelles such as mitochondria, viruses, and plastids, rather than on nuclear genes.

cytoribosome. A cytoplasmic ribosome, as distinct from a nuclear, or a mitochondrial, ribosome.

cytoside. A diglycosyl derivative of a ceramide that contains only simple sugars.

cytosine. The pyrimidine 2-oxy-4-aminopyrimidine that occurs in both RNA and DNA.

-cytosis. Combining form meaning an increase in the number of cells.

cytosol. The soluble portion of the cytoplasm that includes dissolved solutes but that excludes the particulate matter.

cytosome. CYTOPLASM.

cytostatic agent. An agent that suppresses cell multiplication and cell growth.

cytotaxis. The attraction or repulsion of cells for one another that leads to the ordering and arranging of new cell structures under the influence of preexisting ones.

cytotoxic. Causing cell death.

cytotoxic anaphylaxis. The anaphylactic reaction that takes place when an animal organism is injected with antibodies which are specific for cell surfaces.

cytotoxic antibody. An antibody that damages antigen-bearing cells, particularly in the presence of complement.

cytotropic anaphylaxis. The anaphylactic reaction that is mediated by cytotropic antibodies.

cytotropic antibody. An antibody that binds to target cells, particularly mast cells, thereby sensitizing the animal for anaphylaxis.

D

d. 1. Deoxy. 2. Dextrorotatory. 3. Deci.

D. 1. D-configuration. 2. Deuterium. 3. Dielectric constant. 4. Translational diffusion coefficient. 5. Aspartic acid. 6. Dihydrouridine.

2,4-D. 2,4-Dichlorophenoxyacetic acid; a synthetic auxin.

$D_{20,w}^0$. Standard diffusion coefficient.

da. Deca.

dactinomycin. ACTINOMYCIN D.

DALA. δ-Aminolevulinic acid.

dalton. A unit of weight that is equal to the mass of one hydrogen atom (1.67×10^{-24} grams) and that is used to express molecular weights.

Danielli-Davson model. UNIT MEMBRANE HYPOTHESIS.

Danielli-Davson-Robertson model. UNIT MEMBRANE HYPOTHESIS.

dansyl amino acid. An amino acid derivative formed by the reaction of 1-dimethylamino-naphthalene-5-sulfonyl chloride with the alpha amino group of an amino acid. Dansyl amino acids are fluorescent and are used for the measurement of amino acids by fluorometry.

dansylation. The introduction of a dansyl group into an organic compound.

Danysz phenomenon. The phenomenon that the extent of formation and dissociation of an antigen-antibody complex depends on whether the antigen is added all at once, or in small doses.

dark adaptation. The time required for the rods in the retina of an animal, previously placed in bright light, to become fully responsive to dim light.

dark current. The current that is obtained in a photoelectric instrument, such as a spectrophotometer or a scintillation counter, in the absence of radiation and that is caused by thermionic emissions of the photomultiplier tube.

dark field illumination. *See* dark field microscope.

dark field microscope. A microscope in which dark field illumination is used so that objects appear bright on a dark background as a result of the light scattered by the objects.

dark reaction. The photosynthetic reaction or reaction sequence that can occur in the absence of light; the fixation of carbon dioxide by plant chloroplasts is an example.

dark reactivation. The enzymatic repair of DNA, previously damaged by exposure to a mutagen, that does not require light and entails an excision-repair mechanism.

dark recovery. DARK REACTIVATION.

dark repair. DARK REACTIVATION.

Darwinian evolution. The cumulative changes, including mutation and selection, that occur through successive generations in organisms that are related by descent. *See also* natural selection.

·d-assay. The construction of a thermal denaturation profile by measurements of the system at the various ambient conditions; the assay measures the extent of the transition. *See also* i-assay.

dative bond. COORDINATE COVALENT BOND.

dative covalent bond. COORDINATE COVALENT BOND.

datum (*pl* data). An experimental finding; a fact; a measurement.

dauermodification. An environmentally induced, phenotypic change that appears to be inherited and survives through one or more generations but that eventually disappears.

daughter. 1. One of the two cells formed from a parent cell by cell division. 2. The DNA molecule or the chromosome formed from either parental molecules or chromosomes by replication. 3. The nuclide formed from a parent nuclide by radioactive decay.

Davie and Ratnoff theory. A theory of blood clotting that is based on a cascade mechanism.

Davis U-tube. A U-tube that contains a porous filter in its lower portion so that bacterial cultures may be placed in either one or both arms of the tube. The filter prevents the passage of bacteria but allows the passage of small, diffusible substances.

day vision. The capacity to see in bright light; due to the cones in the retina.

DCC. Dicyclohexylcarbodiimide.

D-configuration. The relative configuration of a molecule that is based upon its stereochemical relationship to D-glyceraldehyde.

DDT. 1,1,1 - Trichloro - 2,2 - bis(*p*-chlorophenyl) ethane; an insecticide.

deacylase. An enzyme that catalyzes a deacylation reaction.

deacylated tRNA. A transfer RNA molecule from which either a previously bound aminoacyl group or a previously bound peptidyl group has been removed.

deacylation. 1. The removal of an acyl group

from a compound. 2. The formation of acetoacetyl coenzyme A and coenzyme A from two molecules of acetyl coenzyme A.

deadaptation. The changes that occur in an inducible system in the time interval between the point at which the inducer is removed and the point at which synthesis of the inducible enzyme stops.

dead-end complex. An enzyme-inhibitor complex that cannot lead to the formation of products.

dead-end inhibitor. A competitive inhibitor that forms an enzyme-inhibitor complex that cannot react further and that cannot lead to the formation of products until the inhibitor dissociates from the complex.

dead time. COINCIDENCE TIME.

dead-time loss. COINCIDENCE LOSS.

dead vaccine. KILLED VACCINE.

DEAE-cellulose. *O*-(Diethylaminoethyl)-cellulose; an anion exchanger.

DEAE-sephadex. *O*-(Diethylaminoethyl)-sephadex; an anion exchanger.

deamination. The removal of an amino group from an organic compound.

Dean and Webb method. A method for determining the equivalence zone of a precipitin reaction by mixing a constant amount of antiserum with varying dilutions of antigen and taking the tube in which precipitation occurs most rapidly to be indicative of the equivalence zone. *See also* method of optimal proportions; Ramon method.

death phase. The phase of growth of a bacterial culture that follows the stationary phase and during which there is a decrease in the number (or the mass) of the cells.

debranching enzyme. An enzyme that catalyzes the hydrolysis of branch points in a polymer.

de Broglie wavelength. The wavelength associated with a moving particle.

debye. A unit of dipole moment; equal to the dipole moment of two charges, of 4.8×10^{-10} esu each, that are separated by 1 Å.

Debye-Hueckel theory. The theory that relates the activity coefficients of ions in solution to the electrostatic interactions between the ions and to the diameter of the ion atmosphere around each ion.

deca-. Combining form meaning 10 and used with metric units of measurement. *Sym* da.

decade scaler. A scaler that produces one output pulse for every 10 input pulses.

decamer. An oligomer that consists of 10 monomers.

decant. To pour off the liquid layer that is above sedimented material or above a precipitate.

decapitate. To cut off the head.

decapsidate. To remove the viral capsid.

decarboxylase. CARBOXY-LYASE.

decarboxylation. The removal or the loss of a molecule of carbon dioxide from the carboxyl group of an organic compound.

decay. 1. The decomposition of organic matter through the action of microorganisms. 2. RADIOACTIVE DECAY.

decay chain. RADIOACTIVE SERIES.

decay constant. The fraction of radioactive atoms that are decaying per unit time; the constant λ in the equation $N = N_0 e^{-\lambda t}$, where N is the number of radioactive atoms present at time t, N_0 is the number present at time zero, and e is the base of natural logarithms. The term $e^{-\lambda t}$ is known as the decay factor.

decay factor. *See* decay constant.

decay series. RADIOACTIVE SERIES.

deci-. Combining form meaning one tenth and used with metric units of measurement. *Sym* d.

decimal reduction time. The time required for a tenfold reduction in the viability of bacteria that are subjected to a killing procedure.

deciphering of the code. The experimental determination of the nucleotide composition and the nucleotide sequence of the codons of the genetic code.

decline phase. DEATH PHASE.

decoding site. AMINOACYL SITE.

decomplementation. Any process that removes complement activity from a serum; treatment of the serum with heat or antigen-antibody complexes are two examples.

decomposition. The break-up of a chemical substance into two or more simpler substances.

decontamination. The removal of a contamination, particularly a radioactive one.

dedifferentiation. The loss of differentiation, as that due to anaplasia.

deduction. A conclusion arrived at by reasoning from generals to particulars.

deep groove. MAJOR GROOVE.

defective lysogenic strain. A strain of mutant bacterial cells that have incorporated a prophage that cannot replicate upon induction and that cannot give rise to intact phage particles.

defective prophage. A prophage that cannot replicate upon induction and that cannot give rise to intact phage particles.

defective virus. A virus that cannot synthesize one or more of its structural proteins and that can only form intact virus particles in the host cell when it is in the presence of a helper virus. *See also* deficient virus.

defibrinated blood. Blood from which fibrin has been removed.

deficiency. DELETION.

deficiency disease. A disease that results from the deficiency of a nutrient, as that resulting from either a vitamin or a mineral deficiency.

deficient virus. A virus that cannot synthesize

one or more of its functional proteins so that its nucleic acid cannot be replicated autonomously. *See also* defective virus.

defined medium. SYNTHETIC MEDIUM.

deformation vibration. The vibration of a molecule in which there is a change in a bond angle.

deg. Degree.

degeneracy. 1. The existence of two or more synonym codons for a given amino acid. The degeneracy is termed complete or partial depending on whether all, or only some, of all the possible codons code for amino acids; the degeneracy is termed regular if it follows certain rules as distinct from one that is entirely random. 2. The existence of two or more atomic or molecular levels of equal energy; thus an atom may have two or more orbitals of equal energy, and a molecule may exist in two or more conformational states of equal energy.

degenerate. Possessing degeneracy.

degenerate codon. SYNONYM CODON.

degradation. The gradual and stepwise breakdown of a macromolecule to smaller fragments that proceeds by the breaking of covalent bonds.

degree. *See* absolute temperature scale; centigrade temperature scale; Fahrenheit temperature scale.

degree of polymerization. The number of monomers in a polymer.

degrees of freedom. 1. The total number of variables that can be varied arbitrarily without causing the disappearance of a phase. *See also* phase rule. 2. The total number of coordinates along which the velocity of a molecule has either a translational, or a rotational, component. 3. The total number of items—such as observations, deviations, and means—that can vary independently in the light of restrictions imposed on the calculations. *Abbr* d.f.; D/F.

degrowth. The decrease in the mass of an organism that occurs at the end of a period of growth.

dehydrase. DEHYDRATASE.

dehydratase. An enzyme that catalyzes a dehydration reaction.

dehydrated food. Food from which water has been removed.

dehydration. 1. The removal of water from a compound. 2. The loss of body water.

dehydroascorbic acid. The oxidized form of ascorbic acid.

dehydrogenase. An enzyme that catalyzes the removal of hydrogen from a substrate using a compound other than molecular oxygen as an acceptor.

dehydrogenase-type mechanism. SEQUENTIAL MECHANISM.

dehydrogenation. The removal of hydrogen from an organic compound.

deionized water. Water from which ions have been removed; usually done by passing water through an ion-exchange resin, particularly a mixed-bed demineralizer that removes both anions and cations.

deionizer. A device for removing ions from water or from other fluids, usually by means of ion-exchange resins.

delayed-type hypersensitivity. An allergic response that occurs a few hours or a few days after the administration of an antigen. The response depends on the sensitization of certain cells, especially lymphocytes, rather than on circulating antibodies.

deletion. 1. A point mutation in either RNA or DNA in which a single nucleotide is removed from a polynucleotide strand. In double-stranded nucleic acid this also leads to the removal of the complementary nucleotide from the second strand so that an entire base pair is ultimately deleted from the nucleic acid. 2. A chromosomal aberration in which there is a loss of genetic material from the chromosome. The portion lost may be either a nucleotide or a larger fragment, consisting of one or several genes.

deletion hypothesis. A hypothesis according to which cancer is due to the loss of one or more specific enzymes or other proteins. *See also* catabolic deletion hypothesis.

deletion mapping. A method for locating the position of a gene on a genetic map by means of overlapping deletions.

deletion method. A method for isolating specific messenger RNA molecules by hybridizing them with DNA molecules that contain deletions.

deliberate immunization. The purposeful introduction of antigens, antibodies, or lymphocytes into an organism to either stimulate the production of antibodies by the organism or provide the organism with antibodies.

delipidation. The removal of lipid from a biological sample.

delipidized protein. A lipoprotein from which part or all of the lipid has been removed.

deliquescence. The uptake of moisture from the air by a substance under ordinary conditions of temperature and pressure.

delocalized bond. A bond involving more than two atoms.

delocalized orbital. A molecular orbital that is spread over two or more bonding atoms or even over an entire molecule.

delta. 1. Denoting a small difference between two values. *Sym* Δ. 2. Denoting the fourth carbon atom from the carbon atom that carries the principal functional group of the molecule. *Sym* δ.

delta chain. 1. The heavy chain of the IgD immunoglobulins. 2. One of the polypeptide chains of a minor hemoglobin component in normal human adults.

delta ray. A beam of high-energy secondary electrons that have energies above 100 eV.

demineralizer. DEIONIZER.

denaturant. DENATURING AGENT.

denaturation. Any change in the native conformation of a protein or of a nucleic acid other than the breaking of the primary chemical bonds that join either the amino acids or the nucleotides in the chain. Denaturation may involve the breaking of noncovalent bonds such as hydrogen bonds, and the breaking of covalent bonds such as disulfide bonds; it may be partial or complete, reversible or irreversible. Denaturation leads to changes in one or more of the characteristic chemical, biological, or physical properties of the protein or of the nucleic acid.

denatured. Having undergone denaturation.

denaturing agent. A physical or a chemical agent that can bring about denaturation.

dendrite. A short, and usually branched, process of a nerve cell that conducts impulses to the cell body.

dendritic evolution. An evolutionary pattern that, when diagramed, resembles a tree. Such phylogenetic trees are characteristic of the evolutionary patterns of animal species.

denitrification. The formation of molecular nitrogen from nitrate by way of nitrite.

denitrifying bacteria. The bacteria that carry out denitrification.

de novo synthesis. The synthesis of a molecule, particularly a macromolecule, from simple precursors as distinct from its formation by way of anabolism, catabolism, or modification of other macromolecules. Thus the de novo synthesis of a protein refers to its synthesis from the amino acid level and not to (a) its synthesis from the peptide or polypeptide level, (b) its formation by the addition of amino acids to other proteins, or (c) its formation by the breakdown of other proteins.

densitometer. 1. An instrument for measuring either absorbed or reflected light in materials other than solutions. Densitometers are used for the scanning of chromatograms and electropherograms, and for measuring the blackening of photographic films. 2. An instrument for measuring the density or the specific gravity of a gas or of a liquid.

density. 1. Weight per unit volume. 2. The degree of blackening of a photographic film.

density gradient. The change of density with distance; refers particularly to a solution in which there is such a change from one end of the tube or the cell which holds the solution to the other. A density gradient may be set up in an ultracentrifuge cell by virtue of the variation of the centrifugal force with distance from the center of rotation, so that the density increases from the meniscus to the bottom of the cell; solutions of cesium chloride are commonly used for such experiments. A density gradient may also be set up in a centrifuge tube by layering solutions of different densities above each other so that the density increases from the top to the bottom of the tube; solutions of sucrose are commonly used for such experiments.

density gradient centrifugation. The centrifugation of macromolecules through a density gradient for either preparative or analytical purposes.

density gradient sedimentation equilibrium. Density gradient centrifugation that is performed in an analytical ultracentrifuge and that permits the separation of macromolecules which differ only slightly in their densities. The technique involves centrifuging a concentrated salt solution that contains macromolecules, such as a cesium chloride solution, until the salt achieves its equilibrium distribution and thereby produces a density gradient in the cell; the macromolecules band in this density gradient at positions where their densities equal those of the gradient.

density gradient sedimentation velocity. Density gradient centrifugation that is performed in a preparative ultracentrifuge and that is used for the fractionation of macromolecules, coupled with a variety of assay techniques for detection of the various components. The technique involves layering a solution of macromolecules on top of a preformed density gradient and centrifuging the gradient in a swinging bucket rotor. The macromolecules sediment through the gradient as bands and are separated on the basis of their differences in sedimentation rates.

density gradient zonal centrifugation. DENSITY GRADIENT SEDIMENTATION VELOCITY.

-dentate. Combining form indicating the number of groups in a molecule that coordinate with a metal ion to form a complex; used with mono-, bi-, tri-, tetra-, etc.

dentine. The major calcified tissue of teeth; it is covered by either enamel or cementum depending on whether the portion of the tooth is exposed or submerged.

deoxy-. 1. Combining form indicating a compound that contains 2-deoxy-D-ribose. 2. Combining form indicating a compound that contains less oxygen than the parent compound.

deoxyadenosine. The deoxyribonucleoside of adenine.

deoxyadenylic acid. The deoxyribonucleotide of adenine.

deoxycholic acid. A bile acid that is derived from cholic acid by the loss of an oxygen atom. *Abbr* DOC.

deoxycorticosterone. A mineralocorticoid that regulates the excretion and retention of minerals by the kidney, particularly with respect to sodium and potassium. *Abbr* DOC.

deoxycytidine. The deoxyribonucleoside of cytosine.

deoxycytidylic acid. The deoxyribonucleotide of cytosine.

deoxyguanosine. The deoxyribonucleoside of guanine.

deoxyguanylic acid. The deoxyribonucleotide of guanine.

deoxyribonuclease. An endonuclease that catalyzes the hydrolysis of DNA. *Abbr* DNase; DNAase.

deoxyribonuclease I. A deoxyribonuclease that catalyzes the hydrolysis of DNA to mono- and oligonucleotides consisting of, or terminating in, a 5′-nucleotide. *Abbr* DNase I; DNAase I.

deoxyribonuclease II. A deoxyribonuclease that catalyzes the hydrolysis of DNA to mono- and oligonucleotides consisting of, or terminating in, a 3′-nucleotide. *Abbr* DNase II; DNAase II.

deoxyribonucleic acid. The nucleic acid (*abbr* DNA) that constitutes the genetic material in most organisms and that is composed of the genes; together with histones it makes up the chromosomes of higher organisms. DNA is a polynucleotide that is characterized by its content of 2-deoxy-D-ribose and the pyrimidines cytosine and thymine. *See also* DNA forms; Watson-Crick model.

deoxyribonucleoprotein. A conjugated protein that contains DNA as the nonprotein portion. *Abbr* DNP.

deoxyribonucleoside. A nucleoside of 2-deoxy-D-ribose.

deoxyribonucleotide. A nucleotide of 2-deoxy-D-ribose.

deoxyribose. The five-carbon aldose, 2-deoxy-D-ribose, that is the carbohydrate component of deoxyribonucleic acid. *Abbr* dRib; deRib.

deoxyribose nucleic acid. DEOXYRIBONU-CLEIC ACID.

deoxyriboside. A glycoside of deoxyribose.

deoxysugar. A monosaccharide in which one or more hydroxyl groups have been replaced by hydrogen atoms.

deoxythymidine. THYMIDINE.

deoxythymidylic acid. THYMIDYLIC ACID.

deoxyuridine. The deoxyribonucleoside of uracil.

deoxyuridylic acid. The deoxyribonucleotide of uracil.

depancreatize. To remove the pancreas.

dependent form. The phosphorylated form of the enzyme glycogen synthetase that is a regulatory enzyme for which glucose-6-phosphate is a positive effector. *Abbr* D-form.

dependent variable. A quantity that is a mathematical function of one or more independent variables; the value of a dependent variable is fixed once the values for the related independent variables are chosen.

depolarization. The elimination of polarization, as that occurring in a muscle or a nerve membrane upon electrical stimulation.

depolarization fluorescence. *See* fluorescence depolarization.

depolymerization. The degradation of a polymer to oligomers and/or monomers.

depolymerizing enzyme. An enzyme that catalyzes the hydrolysis of a biopolymer to oligomers and/or monomers.

depot fat. The fat that is stored in an organism.

deproteinization. The removal of protein from a biological sample.

depsipeptide. A peptide characterized by its content of both peptide and ester bonds; these peptides are frequently cyclic and are then referred to as cyclodepsipeptides.

depurination. The removal of purines from a nucleic acid.

depyrimidination. The removal of pyrimidines from a nucleic acid.

derepression. Any modification that eliminates the repression of a gene and permits the synthesis of the gene product. Possible modifications include a decrease in the repressor concentration produced by starving the organism of a required nutrient, a reaction of the inducer with the repressor, a mutation of the regulator gene, or a mutation of the operator gene.

deRib. Deoxyribose.

derivative. A compound, usually an organic one, that is obtained by modification of a parent compound as a result of one or more chemical reactions.

derivatize. To synthesize a derivative.

derived carbohydrate. A derivative of a simple sugar, such as a sugar acid or an amino sugar.

derived lipid. A lipid obtained by hydrolysis of a naturally occurring lipid.

derived protein. A product obtained by treatment of a protein with heat, acid, base, enzymes, or other agents. Primary derived proteins, such as proteans and metaproteins, are proteins that have been altered only slightly; secondary derived proteins, such as proteoses and peptones, are proteins that have been altered more extensively.

dermal. Of, or pertaining to, the skin, especially the true skin.

dermatan sulfate. Chondroitin sulfate B.

dermotropic virus. A virus, the target organ of which is the skin.

DES. Diethylstilbestrol.

desalanine insulin. Insulin from which the alanine residue at the carboxyl end of the B chain has been removed by the action of trypsin.

desalting. The removal of inorganic salt ions from a sample; techniques used include electrodialysis, ion-exchange chromatography, gel filtration, and electrophoresis.

desaspidin. A toxic substance occurring in some ferns that is an uncoupler of both oxidative and photosynthetic phosphorylation.

desaturase. An enzyme that catalyzes a desaturation reaction.

desaturation. A reaction, or a reaction sequence, whereby a saturated compound is converted to an unsaturated one.

descending boundary. The electrophoretic boundary that moves downward in one of the arms of a Tiselius electrophoresis cell. *See also* Tiselius apparatus.

descending chromatography. A chromatographic technique in which the mobile phase moves downward along the support.

desensitization. 1. The modification of a regulatory enzyme by either mutation or chemical means that results in an enzyme that has retained its catalytic activity but has lost the capacity to respond to effectors. 2. The attempt to minimize the response of an individual suffering from immediate-type hypersensitivity upon subsequent exposure to an allergen. Common methods include either the repeated injection of small doses of the allergen to form protective blocking antibodies, or the depletion of the individual's tissue stores of histamine and serotonin.

desensitized enzyme. A regulatory enzyme that has been so altered by either mutation or chemical modification that, while it is still catalytically active, it no longer responds to an effector.

desiccant. A substance that absorbs water and that is used to dry air or another substance.

desiccate. To dry by means of a desiccant.

desiccator. A closed container that holds a desiccant and that is evacuated and used for the removal of moisture from a substance and for maintaining the substance in the dry state.

desmoenzyme. PARTICULATE ENZYME.

desmolase. The enzyme complex, composed of a mixed function oxidase and cytochrome P_{450}, that catalyzes the removal of the side chain from cholesterol; cholesterol is thereby converted to pregnenolone, a precursor of the steroid hormones.

desmosome. A thickened zone in the cell membrane of adjoining eucaryotic cells.

desorb. To remove adsorbed molecules from the surface of a solid.

desorption. 1. The removal of adsorbed molecules from the surface of a solid. 2. ELUTION.

desoxy-. DEOXY-.

destain. To remove the excess dye that has not been utilized in staining the materials of interest.

destructive interference. *See* interference (1).

destructive metabolism. CATABOLISM.

detachment. The removal of the prophage from the host bacterial DNA; believed to be a reverse of the insertion process. *See also* insertion model.

detector. 1. A device for detecting the presence of the organic compounds that come off a column in gas chromatography. 2. A device for detecting the presence of radioactive radiation that is given off by a sample in a radiation counter.

detergent. A synthetic, or a naturally occurring, surface-active agent that is used for cleaning.

determinant. 1. EFFECTOR. 2. ANTIGENIC DETERMINANT.

determinate error. An error in a measurement that can be accounted for and that can be avoided, at least in principle; an error due to methodology, instruments, etc.

determination. The establishment of a specific course of development by a part of an embryo that will be pursued regardless of subsequent situations.

deterpenation. The removal of terpenes from essential oils.

detoxification. The enzymatic reactions in an organism whereby foreign compounds, produced within the organism or introduced into it, are converted to less harmful forms and to more readily excretable products; the foreign compounds are either chemically altered or conjugated to normally occurring metabolites of the organism. *Aka* detoxication.

deuridylic acid. An RNA molecule from which uracil residues have been removed by treatment with hydroxylamine.

deuterated. Labeled with deuterium.

deuterium. The stable, heavy isotope of hydrogen that contains one proton and one neutron in the nucleus. *Sym* D.

deuterium exchange. A technique for studying the conformation of a protein by measuring the rate of exchange of the hydrogen (or deuterium) atoms in the protein with the deuterium (or hydrogen) atoms in the medium. The hydrogen (or deuterium) atoms that are in direct contact with the solvent exchange more rapidly than those that are located in the interior of the molecule or those that participate in hydrogen bonding.

deutero-. Combining form indicating a compound that contains deuterium.

deuteron. The deuterium nucleus that consists of one proton and one neutron.

developer. A chemical reducing agent that converts exposed silver halide grains to metallic silver and thereby renders an image visible on a photographic film.

development. 1. The process whereby a mixture, which has been applied to a chromatographic support, is separated into individual components by treatment with a mobile phase. 2. The process whereby an image is rendered visible on a photographic film by means of a developer. 3. The series of orderly changes by which a mature cell, tissue, organ, organ system, or organism comes into existence.

deviation. The difference between a measured value and a reference value, usually the mean.

devolution. Retrograde evolution; degeneration.

dex. Dextrorotatory.

dextran. A branched polysaccharide of D-glucose that serves as a storage material in bacteria and in yeast.

dextrin. One of a group of polysaccharides of intermediate chain length that are formed during the hydrolysis of starch.

dextrogenic amylase. ALPHA AMYLASE.

dextrorotatory. Having the property of rotating the plane of plane-polarized light to the right, or clockwise, as one looks toward the light source. *Abbr* dex; d.

dextrose. D-Glucose; a dextrorotatory monosaccharide.

d.f. Degrees of freedom; also abbreviated D/F.

DF. Dissociation factor.

D-form. Dependent form.

DFP. Diisopropylfluorophosphate.

DFP-peptide. A peptide that contains a serine residue that has been linked to diisopropylfluorophosphate; obtained by treatment of a protein with diisopropylfluorophosphate, followed by partial hydrolysis. DFP-peptides provide information about the amino acid sequence near the active site of those enzymes that possess a serine residue at or near the active site.

DHF. Dihydrofolic acid.

DHU. Dihydrouridine; the nucleoside of dihydrouracil.

DHU arm. The base-paired segment in the clover leaf model of transfer RNA to which the loop, containing dihydrouracil, is attached.

di-. Combining form meaning two or twice.

diabetes. DIABETES MELLITUS.

diabetes innocens. RENAL GLUCOSURIA.

diabetes insipidus. A disease caused by vasopressin insufficiency and characterized by the excretion of large volumes of hypotonic urine.

diabetes mellitus. A disease caused by a deficiency of insulin and characterized by the presence of excessive amounts of glucose in the urine. The insulin deficiency may be caused by inadequate proinsulin production by the pancreas, by an accelerated insulin destruction, or by insulin antagonists and inhibitors.

diabetogenic. Having the capacity to cause or to increase diabetes, specifically diabetes mellitus.

diabetogenic hormone. The hormone hydrocortisone or one of other 11-oxysteroids that are secreted by the adrenal cortex and that stimulate gluconeogenesis.

diagonal electrophoresis. An electrophoretic technique for establishing the position of disulfide bonds in peptides. The peptides, obtained by partial hydrolysis of a protein, are first separated by one-dimensional paper electrophoresis. The paper strip containing the separated peptides is exposed to formic acid vapors which oxidize disulfide bonds to cysteic acid groups. The paper strip is then attached to a second sheet of paper and subjected to electrophoresis at right angles to that of the first separation. The peptides form a diagonal line and any peptide that is not on this line contains cysteic acid groups and was formed by the cleavage of disulfide bonds.

dialysis. The separation of macromolecules from ions and low-molecular weight compounds by means of a semipermeable membrane that is impermeable to colloidal macromolecules but is freely permeable to crystalloids and water.

dialysis equilibrium. *See* equilibrium dialysis.

dialyzate. 1. The solution containing the material that has diffused through a dialysis membrane. 2. The solution containing the material that has not diffused through a dialysis membrane. *Var sp* dialysate.

dialyze. To process by means of dialysis.

dialyzer. An apparatus for performing dialysis that consists of one or more compartments separated by membranes.

diamagnetic. Descriptive of a substance that has paired electrons and has no magnetic dipole moment; when such a substance is placed in a magnetic field, a magnetic dipole is induced in the substance which opposes the applied field and tends to move the substance out of it.

diameters. A measure of optical magnification; a measure of 20,000 diameters means that the diameter of the object has been magnified 20,000 times when viewed through a microscope.

diaminopimelate pathway. A pathway for the biosynthesis of lysine that proceeds by way of diaminopimelic acid and occurs in bacteria and higher plants.

diamond code. An early version of the genetic

code according to which the R groups of the amino acids fit into the diamond-shaped pockets that are present in Watson-Crick type double helical DNA.

diapause. The period of rest, cessation of growth, and decreased level of metabolism that occurs in the life cycle of insects.

diaphorase. One of a group of enzymes that catalyze the reduction of an artificial electron acceptor—such as a dye, ferricyanide, or a quinone—by either reduced nicotinamide adenine dinucleotide or by reduced nicotinamide adenine dinucleotide phosphate. Such enzymes were originally thought to function in the reduction of metabolites in the electron transport system between NADH and the cytochromes, but this need not be the case. One preparation of diaphorase has been shown to be identical with lipoamide dehydrogenase.

diastase. AMYLASE.

diastatic. Of, or pertaining to, amylase.

diastatic index. A measure of amylase activity that is equal to the number of milliliters of 0.1% (w/v) starch solution, the starch of which is hydrolyzed by the enzyme present in 1.0 ml of a sample at 37°C in 30 min.

diastereomer. One of two or more optical isomers of a compound that are not enantiomers. *Aka* diastereoisomer.

diastereotopic. Descriptive of either atoms or groups of atoms in a molecule that bear a diastereomeric relationship to each other. *See also* enantiotopic.

diatomaceous earth. A material that is composed principally of the siliceous skeletons of diatoms and that is used as an aid in filtration and as an adsorbent in column chromatography.

diauxic. Of, or pertaining to, diauxie.

diauxie. The biphasic growth curve that is obtained when an organism is exposed to two substrates that are utilized by the organism in such a fashion that one of the substrates is metabolized by constitutive enzymes while the other substrate is metabolized by inducible enzymes. The substrate requiring the constitutive enzymes is utilized first and represses the induction of the enzymes required for the utilization of the second substrate. Only after the first substrate is used up, can the enzymes be induced by the second substrate which is then metabolized. *Var sp* diauxy.

diazoate. The ion $R—N{=}N—O^-$ formed from a diazonium salt.

diazo compound. A compound containing the diazo group and having the general formula $R_2C{=}\overset{\oplus}{N}{=}\overset{\ominus}{N}$.

diazo group. The grouping $={\overset{\oplus}{N}}{=}{\overset{\ominus}{N}}$.

diazonium compound. DIAZONIUM SALT.

diazonium group. The grouping $—\overset{\oplus}{N}{\equiv}N$.

diazonium salt. The salt of a compound that contains a diazonium group; used specifically for a compound that is a true ionic salt and that is prepared by the reaction of nitrous acid with a primary aromatic amine.

diazotization. The formation of a diazonium salt.

dichroic. Of, or pertaining to, dichroism.

dichroic ratio. The absorbance of plane-polarized light by a polymer in a direction that is parallel to the axis of the polymer divided by the absorbance in a direction that is perpendicular to that axis.

dichroism. The directional effect in the absorption of light that results from the relative orientations of the absorbing chromophoric groups and the direction of polarization of the light. Thus the absorption of light may be limited either to atoms that vibrate in a specific direction or to a component of the light that is polarized in a specific direction.

dictionary. *See* genetic code dictionary.

dictyosome. The Golgi apparatus of plant cells.

dicumarol. An analogue of vitamin K that is used clinically to reduce the blood clotting tendency of an individual.

dicyanamide. A dimer of cyanamide, used as a condensing agent for amino acids.

dicyclohexylcarbodiimide. A compound used as a condensing agent for either amino acids or nucleotides; in this reaction, water is split out between the two condensing molecules and the reagent is converted to dicyclohexylurea. Dicyclohexylcarbodiimide is also an inhibitor of ATPase. *Abbr* DCC.

dielectric. An insulating material that does not conduct an electric current.

dielectric constant. A measure of the polarizability of a medium that is equal to the ratio of the electrostatic force, given by Coulomb's law, in a vacuum to that in the medium. The dielectric constant increases with an increase in the dipole moment of the molecules in the medium. *Sym* D.

dielectric dispersion. The variation of the dielectric constant as a function of the frequency of the electric field.

dielectric increment. The rate of change of the dielectric constant as a function of the solute concentration.

dielectrophoresis. The migration of dipolar molecules toward the region of maximum field strength when placed in an inhomogeneous electric field; a concentration gradient is established when an equilibrium is achieved between dielectrophoretic migration and diffusion.

diesterase. *See* phosphodiesterase.

diet. 1. The eating pattern of an organism. 2. The daily food intake of an organism.

dietary deficiency. Undernutrition that is due to the inadequate intake of one or more essential nutrients even though the diet as a whole may be quantitatively unrestricted.

diethylstilbestrol. A synthetic compound that has high estrogenic activity; a diethyl derivative of 4,4′-dihydroxystilbene (stilbestrol). *Abbr* DES.

difference spectrophotometry. A technique for measuring small changes in absorbance by determining the absorbance of one test solution directly against that of another test solution rather than against that of a reference solution. The two test solutions contain the same solute of interest, and the solute is present in both solutions at the same concentration but under slightly modified conditions. The technique may be used, for example, to measure the absorbance of a solution containing a denatured protein against that of a solution containing the native protein, or to measure the absorbance of a solution containing a protein plus ligand against that of solutions containing the protein and ligand separately.

differential boundary. The boundary formed between two solutions that contain the same components but at different concentrations.

differential centrifugation. The centrifugation of a solution at various speeds so that particles of different sizes can be separated and collected.

differential counting. The counting of pulses in a scintillation counter that fall within a range determined by two preselected levels of intensity. The range of intensities covered is called a window or a channel, and the pulses are selected by means of a pulse height analyzer which consists of two discriminators that reject pulses which are above or below the selected range of intensities.

differential dialysis. The separation of molecules by dialysis through a membrane that has known and specific permeability properties.

differential discrimination. The selection of pulses that takes place in differential counting.

differential flotation centrifugation. Interface centrifugation, used for tissue culture cells, in which particles are separated according to their densities.

differential labeling. A method for investigating the structure of the antibody combining site. The antibody is first allowed to bind a hapten and is then reacted with an unlabeled reagent that binds covalently to regions of the antibody that are not protected by the bound hapten. After removal of the hapten, the antibody is reacted with a labeled compound that reacts with those amino acids that had previously been protected by the hapten.

differential medium. A medium that aids in the identification of bacteria by revealing specific properties of the organism grown on or in it.

differential method. A method for determining the order of a reaction by varying the concentration of each reactant and measuring the resultant rate of the reaction.

differential refractometer. A refractometer for measuring the difference between two refractive indices.

differential scanning calorimetry. A variation of the differential thermal analysis method in which the temperature of both the sample and the reference are maintained either at an equal level or at a fixed differential throughout the analysis. The variation in heat flow to the sample that is required to maintain this level during a transition is then measured. *Abbr* DSC.

differential sedimentation coefficient. The sedimentation coefficient that corresponds to the movement of a differential boundary.

differential thermal analysis. An analytical technique for studying temperature-induced transitions in a sample. The sample and an inert reference are heated or cooled at the same rate, and the difference in temperature between them is recorded. This difference is either zero or constant until a point is reached where a thermal reaction occurs in the sample. The course of this reaction shows up as a peak in a plot of differential temperature versus either time or temperature. The direction of the peak indicates whether the transition was endothermic or exothermic. *Abbr* DTA.

differentiating circuit. An electronic circuit for counting the discontinuous ionizations and the discontinuous electrical currents that are produced in an ionization chamber. *See also* integrating circuit.

differentiation. The process whereby the structures and functions of the cells of a developing organism are progressively changed, thereby giving rise to specialized cells and structures.

diffraction. The modification that radiation undergoes when it passes the edge of an opaque body, passes through small apertures or slits, or is reflected from ruled surfaces or from atomic planes in a crystal.

diffraction grating. A device that serves to separate light into wavelengths and that is used as a monochromator in some spectrophotometers. A diffraction grating consists of a large number of small grooves that are cut into it at such angles that each groove behaves like a prism, so that light is either reflected from, or transmitted through, the grating.

diffraction spots. The spots produced on a photographic film in x-ray diffraction; they are referred to as first, second, third order, etc., corresponding to values of n equal to 1, 2, 3, etc., in the Bragg equation.

diffusate. 1. The material that passes through a dialysis bag. 2. DIALYZATE.

diffusible. Capable of diffusing.

diffusing factor. HYALURONIDASE.

diffusion. 1. The process whereby molecules, as a result of their random Brownian motion, either change their orientation, resulting in rotational diffusion, or move from a region of higher to one of lower concentration, resulting in translational diffusion. 2. TRANSLATIONAL DIFFUSION.

diffusion chamber. A chamber with porous walls that allows the diffusion of dissolved substances but does not allow the passage of cells; may be used for studies of substances released by cells.

diffusion coefficient. *See* rotational diffusion coefficient; translational diffusion coefficient.

diffusion constant. DIFFUSION COEFFICIENT.

diffusion current. The polarographic current caused by the diffusion of ions toward the electrode.

diffusion potential. The membrane potential that arises from the diffusion of ions across the membrane.

DIFP. Diisopropylfluorophosphate.

digest. 1. *n* The mixture of compounds obtained by enzymatic or chemical hydrolysis of macromolecules. 2. *v* To subject macromolecules to enzymatic or chemical hydrolysis either in vivo, as in the digestive tract, or in vitro.

digestion. 1. The process whereby the macromolecules of food are hydrolyzed in the digestive tract to smaller molecules that are absorbed across the intestinal wall and pass into the circulation. 2. Any chemical or enzymatic hydrolysis of macromolecules.

digestive. Of, or pertaining to, digestion.

digestive enzyme. A hydrolytic enzyme that functions in the breakdown of nutrient macromolecules in the digestive tract.

digestive juice. One of the four secretions (salivary, gastric, pancreatic, and intestinal juice) that contain the digestive enzymes. *Aka* digestive fluid.

digestive system. The digestive tract together with the related organs.

digestive tract. The passage in animals that serves for the digestion of food, the absorption of nutrients, and the elimination of waste products.

digital computer. A computer that receives information in the form of discrete and discontinuous data, usually in the form of bits, and processes it by performing a series of arithmetic and logical operations on the data.

digitalis. A plant toxin that is active as a cardiac glycoside.

digitonin. A glycoside used for the fractional precipitation of steroids.

diglyceride. A glyceride formed by the esterification of one glycerol molecule with two fatty acid molecules.

dihedral angle. TORSION ANGLE.

dihydrobiopterin. *See* biopterin.

dihydrofolate reductase. The enzyme that catalyzes the reduction of dihydrofolic acid to tetrahydrofolic acid and that is competitively inhibited by aminopterin and amethopterin.

dihydro-U arm. DHU ARM.

dihydrouracil. The minor base 5,6-dihydrouracil that occurs in transfer RNA; its nucleoside is dihydrouridine and its nucleotide is dihydrouridylic acid. *Abbr* D_iHU.

dihydroxyacetonephosphate. One of the two triose phosphates formed by cleavage of fructose-1,6-diphosphate in glycolysis.

diisopropylfluorophosphate. A reagent that reacts with the hydroxyl group of serine in proteins. The occurrence of this reaction with the enzyme acetylcholinesterase accounts for the reagent being a component of nerve gas. *Abbr* DIFP; DPFP; DFP.

diisopropylphosphofluoridate. DIISOPROPYLFLUOROPHOSPHATE.

dilatation. An enlargement or expansion, as that of a volume. *Aka* dilation.

dilatometry. The measurement of small volume changes, particularly those of liquids, that are produced by either chemical or physical reactions.

dilaudid. A semisynthetic, narcotic drug made by converting morphine to dihydromorphinone.

diluent. The solvent or the solution that is added to another solution for purposes of dilution.

dilute solution. A solution that contains a small amount of solute.

dilution. 1. The lowering of the concentration of a solution. 2. A solution having a lower concentration than another solution.

dilution end point method. A method for determining the antibody titer of an antiserum by titrating a given amount of antigen with various dilutions of the antiserum. The highest dilution of antiserum that produces a detectable precipitin reaction is taken as the antibody titer.

dilution quenching. The quenching in liquid scintillation counting that is caused by the dilution of the fluor with the sample so that the probability of exciting the fluor is decreased.

dilution value of a buffer. The change in the pH of a buffer that occurs when the buffer is diluted with an equal volume of water; the dilution value is positive or negative depending on whether the pH increases or decreases.

dimensional analysis. A check for the validity of

an equation that is made by ascertaining that all the separate terms have the same units.

dimer. 1. An oligomer consisting of two monomers. 2. The condensation product of either two identical or two similar molecules.

dimerization. The formation of a dimer.

dimorphism. The occurrence of two forms. *See also* polymorphism.

dim vision. NIGHT VISION.

dinitrofluorobenzene. SANGER REAGENT.

2,4-**dinitrophenol.** An uncoupler of oxidative phosphorylation. *Abbr* DNP.

dinitrophenyl amino acid. An intensely colored amino acid derivative formed by the reaction of 1-fluoro-2,4-dinitrobenzene with the free alpha amino group of amino acids, peptides, or proteins; used for end group analysis of N-terminal amino acids and for chromatographic detection and quantitative estimation of amino acids. *Abbr* DNP-amino acid. *See also* Sanger reaction.

dinucleotide. 1. A compound consisting of two nucleotides linked by means of a phosphodiester bond. 2. A compound, such as NAD^+ or FAD, that is structurally related to a compound formed from two nucleotides.

dionin. A semisynthetic, narcotic drug made by converting morphine to ethylmorphine.

dioxygenase. OXYGENASE.

DIPF. Diisopropylphosphofluoridate.

diphenylamine reaction. A colorimetric reaction for deoxypentoses that is based on the production of a blue color upon treatment of the sample with an acidic solution of diphenylamine; used for the determination of DNA.

1,3-**diphosphoglyceric acid.** A high-energy compound, the dephosphorylation of which to 3-phosphoglyceric acid leads to the synthesis of ATP from ADP in the second stage of glycolysis.

diphosphopyridine nucleotide. NICOTINAMIDE ADENINE DINUCLEOTIDE.

diphosphoribulose carboxylase. The enzyme that catalyzes the fixation of carbon dioxide by ribulose-1,5-diphosphate in the Calvin cycle.

diphosphothiamine. THIAMINE PYROPHOSPHATE.

dipicolinic acid. The compound, pyridine-2,6-dicarboxylic acid, that is present in large concentrations in spores and that is related to the heat resistance of the spores.

diploid state. The chromosome state in which each of the various chromosomes, except the sex chromosome, is represented twice. *Aka* diploidy.

dipolar ion. A molecule, the dipole of which is due to two or more ionized groups.

dipolar potential. The membrane potential that arises from both the orientation of polar molecules at the surface of the membrane and the ionic double layer of the membrane.

dipole. An atom or a molecule that possesses an asymmetric charge distribution so that the center of its positive charges does not coincide with the center of its negative charges. The charges may arise from ionizations and/or from polar bonds, and the degree of charge separation is measured by the dipole moment.

dipole-dipole interaction. The attractive or repulsive electrical force between two molecules that have permanent dipoles; the energy of such interactions is proportional to r^{-6}, where r is the distance between the dipoles.

dipole flip. A sudden change in the orientation of a dipole.

dipole-induced dipole interaction. The attractive electrical force between a molecule that has a permanent dipole and the dipole that is induced by this molecule in a neighboring atom or molecule. The energy of such interactions is proportional to r^{-6}, where r is the distance between the permanent dipole and the induced dipole.

dipole moment. A measure of the tendency of an atom or a molecule to orient itself in an electric field as a result of the separation of its charges; equal to the product of one of the two separated charges and the distance between the two charges.

diprotic acid. An acid containing two dissociable protons.

direct-acting bilirubin. The water-soluble conjugated form of bilirubin, bilirubin diglucuronide, that gives an immediate reaction with diazotized sulfanilic acid to yield a diazo dye. *See also* indirect-acting bilirubin.

direct calorimetry. A method for determining the basal metabolic rate of an animal from the heat produced by the animal when it is placed in an insulated chamber. *See also* indirect calorimetry.

direct Coombs' test. A Coombs' test in which the red blood cells have been coated with antibody in vivo. *See also* indirect Coombs' test.

direct effect. The change, such as an excitation or an ionization, that is produced in a molecule as a result of its direct interaction with radiation. *See also* indirect effect.

direct fluorescent antibody technique. A fluorescent antibody technique in which either the antigen or the antibody of interest is reacted directly with the corresponding antibody or antigen that has been labeled with a fluorochrome. *See also* indirect fluorescent antibody technique.

direct isotope dilution analysis. A technique for determining the amount of an unlabeled compound by the addition of a known amount of the same, but labeled, compound. The mixture of labeled and unlabeled compounds is then isolated and its specific activity is determined.

direct oxidation pathway. HEXOSE MONOPHOSPHATE SHUNT.

direct plot. A plot of x versus y, where x and y are two variables.

disaccharide. A carbohydrate composed of two monosaccharide units linked by means of a glycosidic bond.

disassembly. The stepwise removal of ribosomal proteins from ribosomes, as that achieved with concentrated salt solutions.

disassociation. DISSOCIATION (2).

disc gel electrophoresis. An electrophoretic technique in which discontinuities of pH, ionic strength, buffer composition, and gel concentration are purposely built into the gel system. Used particularly for a zone electrophoretic technique that permits high resolution analysis of small samples and that is performed in a polyacrylamide gel; the gel consists of three portions which differ in their composition and in their pH and which are loaded in three stages into open-ended cylindrical glass tubes. *Var sp* disk gel electrophoresis.

Dische reaction. DIPHENYLAMINE REACTION.

discontinuous density gradient. A density gradient in which the density changes in a stepwise fashion from one end of the gradient to the other.

discontinuous distribution. A set of experimental data in which the variable being measured cannot vary continuously and is expressed as a whole number; the number of amino acids per protein in a group of proteins is an example.

discriminator. An electronic device that is capable of either rejecting or accepting a pulse depending on the intensity of the pulse. *See also* differential counting; integral counting.

discriminator hypothesis. The hypothesis that there is one site that is identical in a given group of isoacceptor transfer RNA molecules, and that this site is a factor, though not necessarily the only one, in the recognition of the isoacceptor transfer RNA molecules by the corresponding aminoacyl-tRNA synthetase.

disease. A disturbance in the structure and/or the function of an organ, a tissue, or an organism, as that produced by either a viral or a bacterial infection.

disease of regulation. A disease that is characterized by the abnormality of an otherwise normal bodily function and that is caused by a disturbance of the equilibria that control the bodily function; high blood pressure is an example.

disinfectant. A chemical substance used for disinfection.

disinfection. The destruction of viable and harmful bacteria, or the destruction of the toxins and the vectors of these bacteria.

disintegration. The spontaneous transformation of a radioactive nuclide into another nuclide, generally with the emission of radioactive radiation.

disintegration constant. DECAY CONSTANT.

dismutation reaction. A chemical reaction in which a single compound serves as both an oxidizing and a reducing agent and gives rise to two or more compounds by gain and loss of electrons. An example is the conversion of two molecules of pyruvate plus a molecule of water to one molecule each of lactate, acetate, and carbon dioxide. *See also* disproportionation reaction.

dispensable enzyme. NONESSENTIAL ENZYME.

dispensable gene. NONESSENTIAL GENE.

dispersed phase. The solute particles of a colloidal dispersion.

dispersion. *See* dispersion medium; colloidal dispersion.

dispersion effect. DISPERSION FORCES.

dispersion forces. The attractive electrical forces between atoms and/or nonpolar molecules that result from the formation of small, transient, induced dipoles. The motions of the electrons in one atom or molecule lead to the formation of a small, instantaneous, and transient dipole that induces a corresponding dipole in a neighboring atom or molecule. The energy of such interactions is proportional to r^{-6}, where r is the distance between the instantaneous dipoles. *Aka* London dispersion forces.

dispersion medium. The solvent of a colloidal dispersion.

dispersive replication. A mode of replication for double-stranded DNA in which the parental strands are broken into segments that are incorporated equally, together with newly formed segments, into the two daughter DNA molecules.

displacement analysis. *See* displacement chromatography; displacement electrophoresis.

displacement chromatography. A chromatographic technique, useful for preparative separations, in which a compound is applied to a chromatographic column and is then displaced from the column by elution with a solution containing a second compound.

displacement development. DISPLACEMENT CHROMATOGRAPHY.

displacement electrophoresis. ISOTACHOPHORESIS.

displacement reaction. A chemical reaction in which a group is displaced from a carbon atom by the attack of a nucleophile. *See also* S_N1 mechanism; S_N2 mechanism.

displacer. The eluent used in displacement analysis.

disproportionation reaction. A chemical reaction in which a single compound gives rise to two or more different compounds; an example is the reaction $2AB \rightarrow AB_2 + A$, where AB is a copolymer. *See also* dismutation reaction.

disseminated infection. A viral or a bacterial infection in which the infecting agent spreads from the site of entry to other parts of the body.

dissimilation. CATABOLISM.

dissociation. 1. The breakdown of a compound into smaller components, such as the dissociation of an acid to a proton and an anion. 2. The breakdown of a molecular complex, particularly one of macromolecules, such as the dissociation of an enzyme-substrate complex. 3. The appearance of a novel bacterial colony as a result of mutation and selection.

dissociation constant. The equilibrium constant for dissociation (1, 2).

dissociation factor. *See* ribosome dissociation factor.

dissymmetric. Descriptive of a molecule that lacks overall molecular symmetry but which possesses elements of symmetry within the molecule.

dissymmetry constant. FRICTIONAL RATIO.

dissymmetry of scattering. The ratio of the intensity of light scattered by a solution at the angle θ to that scattered at the angle $180 - \theta$.

dissymmetry ratio. 1. The ratio of the intensity of light scattered by a solution at 45° to that scattered at 135°. 2. ASYMMETRY RATIO.

distal. Remote from a particular location or from a point of attachment.

distillate. The liquid produced by the condensation of its vapor during distillation.

distillation. The process whereby a liquid is purified by boiling and the vapors are condensed and collected.

distilled water. Water that is purified and collected by distillation.

distorted bond model. A model for the structure of water according to which the water structure is an altered ice lattice that is produced by the bending and the distorting, but not the breaking, of the hydrogen bonds in ice.

distribution coefficient. PARTITION COEFFICIENT.

distribution equation. One of a set of equations that describe the distribution of an enzyme among various possible forms; such forms may include EI or EP, where E is the enzyme, I is an inhibitor, and P is a product. Each equation provides an expression for a ratio such as $[EI]/[E_t]$ or $[EP]/[E_t]$ in terms of kinetic constants, equilibrium constants, and concentrations of components; $[E_t]$ is the total enzyme concentration.

distribution isotherm. PARTITION COEFFICIENT.

distribution law. PARTITION LAW.

distribution potential. The membrane potential that arises from the unequal distribution of ions on both sides of the membrane.

distribution ratio. PARTITION COEFFICIENT.

disulfide. A compound containing a disulfide bond.

disulfide bond. The covalent bond formed between two sulfur atoms, particularly that formed in peptides and proteins between two sulfhydryl groups of two cysteine residues.

disulfide bridge. DISULFIDE BOND.

disulfide interchange. A chemical reaction in which there is an interchange between the groups attached to two or more disulfide bonds, as in the reaction $R_1-S-S-R_2 + R_3-S-S-R_4 \rightleftarrows R_1-S-S-R_3 + R_2-S-S-R_4$.

disulfide link. DISULFIDE BOND.

DIT. Diiodotyrosine.

dithioerythritol. *See* Cleland's reagent.

dithiothreitol. *See* Cleland's reagent.

diuresis. An increase in the excretion of urine.

diuretic. 1. *n* An agent that increases the excretion of urine. 2. *adj* Of, or pertaining to, diuresis.

diurnal. 1. Pertaining to the daytime. 2. Occurring daily.

divergent evolution. An evolutionary pattern in which the lines of development for more recent species branch out from earlier species; such a pattern can be depicted as one network branching out from a common origin.

Dixon plot. A graphical method for determining the inhibitor constant of either competitive or noncompetitive inhibition of an enzymatic reaction; consists of plotting the reciprocal of the velocity of the reaction as a function of the inhibitor concentration.

dizygotic twins. Twins that are formed from two separate eggs which were fertilized by separate sperms in the same maternal organism.

d,l-pair. A pair of enantiomers.

DME. Dropping mercury electrode.

DMSO. Dimethylsulfoxide.

DNA. Deoxyribonucleic acid.

DNA-agar technique. A technique for measuring the extent of hybridization between nucleic acid molecules. Nucleic acid fragments from one source are trapped in an agar gel and are allowed to hybridize with radioactively labeled nucleic acid fragments from a second source. The amount of radioactivity in the agar is then determined.

DNA-arrest mutant. A phage mutant that initiates the synthesis of DNA in normal fashion but stops the synthesis soon thereafter.

DNAase. Deoxyribonuclease.

DNA-cellulose chromatography. Affinity chromatography in which the column support consists of cellulose to which DNA molecules have been adsorbed noncovalently.

DNA complexity. The total number of base pairs that occur in nonrepeating sequences of a given DNA genome.

DNA-delay mutant. A phage mutant that initiates the synthesis of DNA after a delay of several minutes.

DNA-dependent DNA polymerase. *See* DNA polymerase.

DNA-dependent RNA polymerase. *See* RNA polymerase.

DNA duplex. A double-stranded DNA molecule; a double helix of DNA.

DNA duplicase. DNA-DEPENDENT DNA POLYMERASE.

DNA forms. The configurations of DNA that are a function of the relative humidity and the type of positive counterions present. The A, B, and C forms of DNA are stable, respectively, at intermediate, high, and low relative humidities and have 11, 10, and 9.3 nucleotide pairs per turn of double helix. The B form is considered to be the biologically important form.

DNA groove. *See* major groove; minor groove.

DNA ligase. The enzyme that catalyzes the joining together of two single-stranded DNA segments which may either be parts of the same duplex or be parts of different duplexes.

DNA-like RNA. An RNA molecule that resembles DNA in its overall base composition and base ratios; the RNA is generally rich in adenine and uracil and has an adenine/uracil ratio of approximately 1.0.

DNA methylase. An enzyme that catalyzes the methylation of the bases in DNA; the methylation occurs generally subsequent to, rather than prior to, the incorporation of the bases into the polynucleotide strand.

DNA-negative mutant. A phage mutant that is unable to initiate the synthesis of DNA.

DNA phage. A DNA-containing phage.

DNA polymerase. An enzyme that catalyzes the synthesis of DNA from the deoxyribonucleoside triphosphates, using either single or double-stranded DNA as a template; referred to as DNA-dependent DNA polymerase to distinguish it from RNA-dependent DNA polymerase which uses an RNA template for the synthesis of DNA.

DNA polymerase I. A DNA-dependent DNA polymerase, originally thought to function in the replication of DNA but now believed to function in the repair-synthesis of DNA. *Aka* Kornberg enzyme.

DNA polymerase II. A DNA-dependent DNA polymerase that is sensitive to sulfhydryl reagents and that is believed to function in the replication of DNA.

DNA polymerase III. A DNA-dependent DNA polymerase that is sensitive to sulfhydryl reagents and that is believed to function in the replication of DNA. Under certain assay conditions the enzyme behaves as a reverse transcriptase and utilizes polyriboadenylic acid preferentially.

DNA puff. CHROMOSOME PUFF.

DNA repair. *See* cut and patch repair; excision repair; patch and cut repair.

DNA replication. *See* replication (1); conservative replication; dispersive replication; end-to-end conservative replication; semiconservative replication.

DNA restriction enzyme. *See* restriction enzyme.

DNA-RNA hybrid. A double helix composed of a single strand of DNA which is hydrogen-bonded to a single strand of a complementary RNA.

DNase. Deoxyribonuclease.

DNA transcriptase. DNA-DEPENDENT RNA POLYMERASE.

DNA virus. A DNA-containing virus.

DNFB. Dinitrofluorobenzene.

DNP. 1. 2,4-Dinitrophenyl group. 2. Deoxyribonucleoprotein. 3. 2,4-Dinitrophenol.

DNP-amino acid. Dinitrophenyl amino acid.

Dns-amino acid. Dansyl amino acid.

DOC. 1. Deoxycorticosterone. 2. Deoxycholate.

DOCA. Deoxycorticosterone acetate.

doctrine of monomorphism. The belief that microorganisms show constancy with respect to their morphological form and their physiological function. *See also* doctrine of pleomorphism.

doctrine of original antigenic sin. The phenomenon that antibodies, formed in a secondary response that is elicited by an immunogen related to, but not identical with, the immunogen that elicited the primary response, react more strongly with the primary, than with the secondary, immunogen.

doctrine of pleomorphism. The belief that microorganisms have great capacity for variation with respect to their morphological form and their physiological function. *See also* doctrine of monomorphism.

doctrine of uniformitarianism. The assumption that the chemical and physical laws have remained unchanged from the time of the formation of the earth, throughout the development of life, and up to the present time.

dominant. 1. DOMINANT GENE. 2. The trait produced by a dominant gene.

dominant gene. A gene that partially or entirely suppresses the expression of another, allelic gene.

Donnan equilibrium. GIBBS-DONNAN EQUILIBRIUM.

Donnan potential. The membrane potential that arises from the unequal distribution of ions which is produced by the establishment of a Gibbs-Donnan equilibrium.

Donnan term. A measure of the ionic distributions attained in a Gibbs-Donnan equilibrium; equal to the difference between the sum of the concentrations of the diffusible ions on that side of the membrane that contains nondiffusible ions, and the sum of the concentrations of the diffusible ions on the other side.

donor site. PEPTIDYL SITE.

DOPA. 3,4-Dihydroxyphenylalanine; an intermediate both in the conversion of tyrosine to melanin pigments and in the synthesis of epinephrine and norepinephrine from tyrosine.

dormancy. The inactive state of a spore.

dormant infection. A bacterial infection that does not produce overt disease symptoms but in which the bacteria can be detected by the use of proper techniques. *See also* latent infection.

Dorn effect. SEDIMENTATION POTENTIAL.

dosage. 1. A specified dose. 2. The administration of a dose.

dose. A measured quantity of a chemical, a microorganism, a virus, or radiation that is administered to an organism.

dose-action curve. DOSE-RESPONSE CURVE.

dose-effect curve. DOSE-RESPONSE CURVE.

dose fractionation. The exposure of an object or an organism to small doses of radiation, administered at regular intervals.

dose meter. DOSIMETER.

dose rate. The dose of radiation received per unit time.

dose-response curve. A curve that describes the relationship between the dose that is administered to organisms and either the number or the percentage of organisms that show a response. Examples of kinds of doses are those of carcinogens, drugs, poisons, viruses, and radiation; examples of kinds of responses are tumor development, recovery from a disease, and death. *Abbr* DRC.

dosimeter. An instrument for measuring the cumulative dose of radiation to which an organism or an object has been exposed. *Aka* dose meter.

dosimetry. The measurement of doses of radiation, microorganisms, viruses, or chemicals.

double beam in space spectrophotometer. A spectrophotometer in which the original light beam is split into two beams that pass through two identical optical systems at the same time; one beam passes through the sample solution while the other passes through a reference solution.

double beam in time spectrophotometer. A spectrophotometer in which the original light beam is made to pass, at any given time, either through the sample solution or through a reference solution; commonly achieved by means of a light chopper.

double-beam spectrophotometer. A spectrophotometer in which a light beam passes through both the sample solution and a reference solution so that a direct measurement of the difference in absorbance between the two solutions can be made. *See also* double beam in space spectrophotometer; double beam in time spectrophotometer.

double-blind technique. A technique for testing either a substance or a procedure in which neither the subject nor the experimenter know the identity of the samples and the expected results; used in the testing of drugs, vaccines, etc., particularly in conjunction with the use of placebos. *See also* blind test.

double bond. A covalent bond that consists of two pairs of electrons, shared by two atoms.

double carbon dioxide fixation. The conversion of two molecules of carbon dioxide to organic compounds by means of two different reactions, as that occurring in *Chromatium*.

double-chain length. A crystalline form of glycerides in which two acyl groups form a unit structure.

double diffusion. A technique of immunodiffusion in which both the antigen and the antibody are allowed to diffuse through a gel; can be used for either one-dimensional or two-dimensional diffusion.

double-displacement mechanism. PING PONG MECHANISM.

double helix. *See* Watson-Crick model.

double isotope dilution analysis. A technique for determining the amount of a labeled compound of unknown specific activity by treating two aliquots of the labeled compound with different and known amounts of the same, but unlabeled, compound. The mixtures of labeled and unlabeled compounds are then isolated and their specific activities are measured.

double-label experiment. An experiment based on the use of either one compound labeled with two different isotopes, or two compounds. each labeled with a different isotope.

double layer. *See* ionic double layer.

double lysogeny. The phenomenon in which a bacterium is infected twice with the same temperate phage, the two prophages of which are inserted in tandem into the host bacterial DNA.

double-reciprocal plot. A plot of $1/x$ versus $1/y$, where x and y are two variables; a Lineweaver-Burke plot is an example.

double refraction. BIREFRINGENCE.

double refraction of flow. FLOW BIREFRINGENCE.

double resonance. SPIN DECOUPLING.

double-stranded. Descriptive of a nucleic acid molecule that consists of two polynucleotide chains. *Abbr* ds.

doublet. 1. A sequence of two adjacent nucleotides in a polynucleotide strand. 2. A double peak, as that obtained in nuclear magnetic resonance.

doublet code. A genetic code in which an amino acid is specified by two adjacent nucleotides in messenger RNA; thought to have been a forerunner of the present genetic code.

double-well technique. An immunoelectrophoretic procedure for determining the common precipitin arcs produced by two mixtures.

doubling dilution. Serial dilution such that the dilution in each tube is twice that in the preceding tube.

doubling time. The observed time required for a cell population to double in either the number of cells or the cell mass; it is equal to the generation time only if all the cells in the population are capable of doubling, have the same generation time, and do not undergo lysis. *Sym* t_D.

doubly lysogenic strain. A bacterial strain in which the cells carry two prophages per genome.

doughnut model. A model of the conformational changes produced in a cell membrane by the action of complement. According to this model, complement brings about the formation of a rigid, stable channel that stretches across the phospholipid bilayer of the membrane; the outside of this channel is hydrophobic, but the hollow inside core is hydrophilic and permits the free passage of water molecules and ions. *See also* leaky patch model.

Dowex. Trademark for a group of ion-exchange resins.

downhill reaction. EXERGONIC REACTION.

Down's syndrome. MONGOLISM.

DPFP. Diisopropylfluorophosphate.

dpm. Disintegrations per minute; the number of radioactive disintegrations per minute.

DPN$^+$. Diphosphopyridine nucleotide. *Aka* NAD$^+$.

DPNH. Reduced diphosphopyridine nucleotide. *Aka* NADH.

dps. Disintegrations per second; the number of radioactive disintegrations per second.

DPT. Diphosphothiamine.

drag effect. SOLVENT DRAG.

Draper's law. FIRST LAW OF PHOTOCHEMISTRY.

DRC. Dose-response curve.

Dreiding model. A framework model.

dRib. Deoxyribose.

Drierite. A brand of anhydrous calcium sulfate used as a desiccant.

drift. *See* genetic drift.

driving force. The source of free energy and/or the conditions and factors that are responsible for causing a reaction to proceed in a given direction; frequently refers to the mechanism by which the free energy of activation of the transition state is lowered.

D-RNA. DNA-like RNA; also abbreviated d-RNA.

droplet sedimentation. The phenomenon that may occur in density gradient sedimentation velocity when a solution of macromolecules is layered over a density gradient that contains a highly diffusible low-molecular weight solute; the unequal diffusion of this solute and of the macromolecules may lead to a density inversion and to convection at the interface between the solution of the density gradient and that of the macromolecules.

dropping mercury electrode. An electrode, used in polarography, in which mercury drops from a reservoir through a capillary into the solution that is being studied; the capillary is mounted vertically below the reservoir. *Abbr* DME.

Drosophila. The genus of the fruit fly; an organism used in genetic research.

Drude equation. An equation that describes the variation of optical rotation with wavelength; specifically, $[m'] = a_0\lambda_0^2/(\lambda^2 - \lambda_0^2)$, where $[m']$ is the reduced mean residue rotation, λ is the wavelength, and a_0 and λ_0 are constants.

Drude term. The expression $a_0\lambda_0^2/(\lambda^2 - \lambda_0^2)$ that is part of the Drude equation and that is also the first term in the Moffit-Yang equation.

drug-detoxication enzyme. DRUG-METABOLIZING ENZYME.

drug-induced apnea. A genetically inherited metabolic defect in man that is due to a deficiency of the enzyme pseudocholinesterase.

drug-metabolizing enzyme. A mammalian enzyme that acts on drugs and other foreign chemicals but which is not known to act on any metabolite normally occurring within the same organism. *See also* normal enzyme.

drug resistance. The relative resistance of mutant microorganisms to the action of drugs.

dry box. A moisture-free enclosure.

dry ice. Solid carbon dioxide.

drying agent. DESICCANT.

drying oil. A highly unsaturated oil that tends to undergo autoxidation.

dry weight. The weight of a sample from which the water has been removed.

ds. Double-stranded.

DSC. Differential scanning calorimetry.

dsDNA. Double-stranded DNA.

D-site. Donor site.

dsRNA. Double-stranded RNA.

DTA. Differential thermal analysis.

DTNB. 5,5′-Dithiobis (2-nitrobenzoic acid); Ellman's reagent.

dual-bed chromatography. COUPLED-LAYER CHROMATOGRAPHY.

dual-effect mutant. A mutant possessing a polar mutation; a polarity mutant.

ductless gland. ENDOCRINE GLAND.

Du Nouy ring tensiometer. An instrument for measuring surface tension and interfacial tension by a determination of the force required to detach a platinum ring from a surface.

duocrinin. The gastrointestinal hormone that controls the secretion from Brunner's gland, located in the duodenum.

duplex. 1. *n* A double helix. 2. *adj* Double helical.

duplex DNA. The double helix of the Watson-Crick model of DNA.

duplicase. DNA-DEPENDENT DNA POLYMERASE.

duplicate gene. One of the multiple copies of a gene, produced by gene duplication.

duplication. A chromosomal aberration in which a chromosome bears two identical groups, each composed of one or several genes.

Duponol. Trademark for a group of detergents composed of sulfate esters of alcohols that are derived from long-chain fatty acids.

dwarfism. A condition of being undersized that results from the premature arrest of skeletal growth and that may be caused by insufficient secretion of growth hormone.

dye. A compound that strongly absorbs light in the visible region and that can be firmly attached to a surface as a result of chemical and/or physical interactions.

dye-sensitized photooxidation. The oxidation of a biologically important molecule that occurs in the absence of oxygen but in the presence of a photosensitizing dye and an appropriate electron and/or hydrogen acceptor. *See also* photosensitization.

dynamic capacitor electrometer. VIBRATING REED ELECTROMETER.

dynamic equilibrium. 1. EQUILIBRIUM. 2. STEADY STATE.

dynamic osmometer. An osmometer, the operation of which is based on the application of hydrostatic pressure sufficient to just prevent osmosis from occurring.

dynamic viscosity. The viscosity of a liquid that has not been corrected for the density of the liquid. *See also* kinematic viscosity.

dyne. A unit of force equal to the force which, when applied to a mass of one gram, will impart to it an acceleration of 1 cm/sec^2.

dynein. A protein in cilia.

dysfunction. An impairment of normal function; a malfunction.

dysgammaglobulinemia. A condition in which there is an imbalance in the relative amounts of the various types of immunoglobulins in an individual.

dysgenic. Detrimental to the genetic qualities of a race or a breed. *See also* eugenic.

dysmyelination. The formation of a myelin that has a faulty structure.

dystrophy. Defective nutrition.

E

e. 1. The base of natural logarithms; a constant, equal to 2.7183. . . . 2. Electron.

E. 1. Enzyme. 2. Energy. 3. Extinction. 4. Reduction potential. 5. Glutamic acid.

E_0. *See* standard electrode potential.

E^0. *See* standard electrode potential.

E_0'. *See* standard electrode potential.

$E_{1cm}^{1\%}$. The extinction coefficient of a 1% solution, the absorbance of which is measured in a cuvette having a light path of 1 cm.

EAA. Essential amino acids.

Eadie-Hofstee plot. A single reciprocal plot of the Michaelis-Menten equation in which either v/S is plotted versus v, or S/v is plotted versus S; v is the velocity of the reaction and S is the substrate concentration. *Aka* Eadie plot.

early enzyme. A virus-specific enzyme that is transcribed from an early gene.

early gene. A viral gene that is transcribed early after the infection of a host cell by the virus. *Aka* early function gene.

early protein. A virus-specific protein that is transcribed from an early gene.

early RNA. A virus-specific RNA that is synthesized by RNA polymerase shortly after the infection of the host cell by the virus.

earthy group. A group of compounds that have high melting points and that are believed to have occurred in the original gas-dust of the solar nebula. *See also* gaseous group; icy group.

EBV. Epstein-Barr virus.

EC. Electron capture.

E.C. Enzyme commission; in the listing of an enzyme, the abbreviation is followed by four numbers indicating the classification of the enzyme according to main division, subclass, sub-sub class, and serial number in the sub-sub class.

ECC. End carbon chain.

eccrine gland. A sweat gland.

ecdysone. A steroid hormone that influences molting in insects.

ECF. Extracellular fluid.

ECG. Electrocardiogram.

echovirus. A virus that belongs to the enterovirus subgroup of picornaviruses and that is similar to the polio virus in its physical parameters.

ECL. Equivalent chain length.

eclipse. 1. The time interval between the infection of a bacterial cell by a phage and the appearance of intracellular infective phage particles. 2. The time interval in bacterial transformation during which the transforming DNA cannot be extracted from the cell in a form that retains its activity for transformation.

eclipsed conformation. The conformation of a molecule in which, in a Newman projection, a large number of the atoms are either partially or completely concealed from view by other atoms. In such a conformation, interatomic distances are relatively small and interatomic interactions are maximized; as a result, an eclipsed conformation is less stable than a staggered one.

E. coli. *Escherichia coli.*

ecological system. ECOSYSTEM.

ecology. The study of the interrelationships between organisms and the interrelationships between organisms and their external environments.

economic species. A desirable species; a species, such as humans, that uses chemicals to eliminate other, undesirable species.

economic toxicology. The branch of toxicology that deals with man's use of chemicals to selectively affect tissue function or to selectively eliminate uneconomic species.

ecosystem. The system of interrelated organisms and nonliving components in a particular environment.

ECTEOLA-cellulose. Epichlorhydrin-triethanolamine cellulose; an anion exchanger.

ectochemistry. The chemistry of the surface of either a cell or an organism.

ectocrine. ECTOHORMONE.

ectohormone. A chemical, such as a pheromone, that is produced by one organism and that exerts an effect on another organism.

-ectomy. Combining form meaning surgical removal.

ectopic protein. A protein produced by a neoplasm that is derived from a tissue that is not normally engaged in the synthesis of that protein.

ectoplasm. The layer of the cytoplasm near the periphery of the cell that is more rigid than that in the interior of the cell. *See also* endoplasm.

ED_{50}. Median effective dose.

eddy diffusion. The irregularity in the diffusion of solute molecules that occurs in a porous chromatographic support. The phenomenon is due to the fact that (a) the pathlength for some solute molecules is either shorter or longer than that for the bulk of the molecules; and (b) the rate of solvent flow varies in different regions of the porous support.

eddy migration. The irregularity in the electrophoretic mobility of solute molecules that occurs in a porous support. The phenomenon is due to the fact that (a) the pathlength for some solute molecules is either shorter or longer than that for the bulk of the molecules; and (b) the rate of solvent flow varies in different regions of the porous support.

edema. An abnormal accumulation of fluids in the tissues.

Edman degradation. A method for the stepwise removal of amino acids from peptides and proteins that allows a determination of both the N-terminal amino acid and the amino acid sequence. The method is based on the reaction of the Edman reagent, phenylisothiocyanate, with the free alpha amino group of amino acids, peptides, or proteins, and on the removal of the N-terminal amino acid in the form of a phenylthiohydantoin. These steps are then carried out repetitively.

ED pathway. Entner-Doudoroff pathway.

EDTA. 1. Ethylenedinitrolotetraacetic acid. 2. Ethylenediaminetetraacetic acid.

EEG. Electroencephalogram.

EF. 1. Extrinsic factor. 2. Elongation factor.

EFA. Essential fatty acids.

effective half-life. The half-life of a radioactive isotope in a biological system that is equal to the product of the biological half-life and the radioactive half-life.

effective lethal phase. The stage in the development of an organism carrying a lethal gene at which that gene generally brings about the death of the organism.

effective mean residue rotation. REDUCED MEAN RESIDUE ROTATION.

effector. A metabolite that, when bound to an allosteric enzyme, alters the catalytic activity of the enzyme. The effector generally alters either the Michaelis constant of the enzyme or the maximum velocity of the reaction. An effector that functions as an activator and leads to an increase in the binding of the substrate and other effectors is known as a positive effector; an effector that functions as an inhibitor and leads to a decrease in the binding of the substrate and other effectors is known as a negative effector. An effector may likewise bind to a nonenzymatic, allosteric protein and lead to a change in the properties of the protein.

efficiency of infection. The ratio of the number of viral particles in a group of animals, cell cultures, or other test units to the number of infective test units produced.

efficiency of plating. 1. The proportion of animal cells that give rise to colonies. 2. The ratio of the plaque count in a plaque assay

under a given set of conditions to that under standard conditions.

efflorescence. The giving up of water by a substance when it is exposed to air under ordinary conditions of temperature, pressure, and moisture.

effluent. ELUATE.

efflux. Outward flow, as that out of a cell.

egest. To discharge material, such as waste, from either a cell or a body.

EGF. Epidermal growth factor.

egg white injury. The condition of biotin deficiency in man or animals that results from an excessive intake of raw egg white; the condition is due to the protein avidin that is present in egg white and that combines tightly with biotin.

egg white injury factor. BIOTIN.

Ehrlich ascites tumor. A tumor, derived from a mouse carcinoma, that grows in the peritoneal cavity and that has been kept alive in tissue culture.

Ehrlich reaction. A colorimetric reaction for tryptophan and other compounds containing the indole ring; based on the production of a red color upon treatment of the sample with *p*-dimethylaminobenzaldehyde and concentrated hydrochloric acid.

Ehrlich's reagent. 1. A reagent that contains *p*-dimethylaminobenzaldehyde and that gives a red color with porphobilinogen, δ-aminolevulinic acid, and related compounds. 2. A reagent that contains diazotized sulfanilic acid and that gives a blue color with bilirubin in the Van den Bergh reaction.

Ehrlich's receptor theory. An early selective theory of antibody formation that was formulated by Ehrlich in 1900. The theory proposed that cells were covered with antibody-like receptors that contained haptophore side chains with which the antigens combined. The receptors were then thought to be liberated from the cells and to enter the blood stream as circulating antibodies. *Aka* Ehrlich's side chain theory.

EHTP. Equivalent height of a theoretical plate.

EI. Enzyme-inhibitor complex.

einstein. A mole of photons; Avogadro's number of photons.

Einstein law of photochemical equivalence. FIRST LAW OF PHOTOCHEMISTRY.

Einstein-Sutherland equation. An equation relating Brownian motion and diffusion; specifically, $D = RT/Nf$, where D is the diffusion coefficient, R is the gas constant, T is the absolute temperature, N is Avogadro's number, and f is the frictional coefficient.

EIS. Enzyme-inhibitor-substrate complex.

elaidinization. The cis-trans isomerization of mono- or polyunsaturated fatty acids. The term is derived from the isomerization of oleic acid (cis) to elaidic acid (trans).

elastase. An enzyme that catalyzes the hydrolysis of the peptide bonds of elastin and other peptide bonds which are formed by neutral amino acids.

elastin. A major scleroprotein of connective tissue, especially of the elastic tissue of tendons and arteries.

elastomer. A natural or a synthetic polymer that possesses rubber-like properties.

elective enrichment. ENRICHMENT CULTURE.

elective theory. SELECTIVE THEORY.

electric. 1. Of, or pertaining to, electricity. 2. Pertaining to the motion, emission, and behavior of currents of free electrons in passive elements such as wires, resistors, and inductors. *Aka* electrical.

electric birefringence. The birefringence produced by molecules that have become oriented as a result of the application of an electric field. *Aka* electrical birefringence.

electric dichroism. The dichroism that occurs when polarized light is absorbed by molecules, the orientation of which is affected by the direction of an applied electric field. *Aka* electrical dichroism.

electric dipole. DIPOLE.

electric double layer. IONIC DOUBLE LAYER.

electric field. The space surrounding electrical charges in which a mechanical force will be exerted on a charge introduced into it.

electrochemical potential. The free energy change for the transport of a charged solute either up or down a concentration gradient; equal to the sum of the free energy changes due to the concentration gradient and due to the electrical potential.

electrochemistry. The branch of chemistry that deals with the interrelationships and the transformations of chemical and electrical energy.

electrochromatogram. ELECTROPHEROGRAM.

electrochromatography. 1. Any electrophoretic procedure in which the separation of solute particles does not depend solely on the mobilities of the particles, but is also significantly affected by the sorptive interactions between the solute particles and the solid support. 2. ZONE ELECTROPHORESIS. 3. CONTINUOUS FLOW ELECTROPHORESIS.

electrochromatophoresis. ZONE ELECTROPHORESIS.

electrocortin. ALDOSTERONE.

electrode. A device, frequently a wire or a plate, by which an electric current passes into, or out of, an electric cell, a solution, an apparatus, or a body.

electrodecantation. A technique for the separation and fractionation of proteins by means of electrodialysis. During this process, the protein molecules move toward the electrodes and tend to form zones of high protein concentration at those portions of the retaining membrane that are near the electrodes. These zones are gravitationally unstable and tend to move downward in the container, carrying the protein molecules along. Thus in the absence of stirring, there is a gradual decrease in the concentration of protein molecules in the top layers of the solution, and a gradual increase in their concentration in the lower layers.

electrode potential. 1. A measure of the tendency of an oxidation-reduction half-reaction to occur by either loss or gain of electrons. The electrode potential measuring electron loss is known as an oxidation potential and that measuring electron gain is known as a reduction potential; the two are numerically equal, but of opposite sign, and are denoted E. The electrode potential at pH 7.0 is the biochemical electrode potential and is denoted E'. *See also* standard electrode potential. 2. The potential across a chemically reactive electrode that participates either in the formation of ions from atoms or in the formation of atoms from ions.

electrodialysis. A technique for the removal of ions from a solution in which dialysis of the solution is carried out with the simultaneous application of an electric field across the dialysis bag.

electrodiffusion. The spreading of a spot or a zone in electrophoresis as a result of diffusion; may occur when an electrophoretic component is present as a number of rapidly interconvertible forms that have different mobilities.

electroencephalogram. The recording of the potential changes of the brain. *Abbr* EEG.

electroendosmosis. ELECTROOSMOSIS.

electrofocusing. ISOELECTRIC FOCUSING.

electrogenic pump. A pump that generates a gradient of electrochemical potential in the course of its operation, since the movement of one ion across the membrane does not have to be linked to the movement of another ion in the opposite direction.

electrokinetic phenomenon. One of four phenomena that describe either the electrical forces produced by the relative motion of solids and liquids, or the relative motion of solids and liquids produced by electrical forces. The four phenomena are electrophoresis, electroosmosis, streaming potential, and sedimentation potential.

electrokinetic potential. ZETA POTENTIAL.

electrolysis. The decomposition of a substance by the action of an electric current.

electrolyte. A substance that dissociates, partly

or entirely, into two or more ions in water; solutions of electrolytes conduct an electric current.

electrolyte balance. The reactions and factors involved in maintaining a constant internal environment in the body with respect to the distribution of electrolytes between the various fluid compartments.

electrolytic. Of, or pertaining to, electrolytes or electrolysis.

electrolytic desalting. The removal of ions from a solution by means of electrolysis.

electromagnetic spectrum. The entire range of either the wavelengths or the frequencies of electromagnetic radiations, from the shortest cosmic rays to the longest radio waves.

electromechanochemical coupling hypothesis. A modified form of the mechanochemical coupling hypothesis. *Abbr* EMC hypothesis.

electrometer. An instrument for measuring electrical potential, electrical charge, or electrical current.

electrometric titration. A titration in which electromotive force is measured as a function of titrant added.

electromigration. 1. ZONE ELECTRO-PHORESIS. 2. ISOTACHOPHORESIS.

electromotive force. The directly measurable electrical energy that can be derived from an electrical cell composed of two half-cells. *Abbr* EMF.

electromyogram. The recording of the potential changes of a muscle. *Abbr* EMG.

electromyography. The study of the electrical properties of muscle by means of recordings of its action potential.

electron. 1. The elementary particle of nature that has a charge of minus one and a mass of 0.000549 amu (9×10^{-28} grams). 2. The elementary particle of nature that has a mass of 0.000549 amu and a charge of either plus one or minus one. *Sym* e.

electron acceptor. A substance that is being reduced; an oxidant; an oxidizing agent.

electron affinity. The tendency of a substance to accept electrons and to function as an oxidizing agent.

electron capture. A mode of radioactive decay in which an orbital electron is attracted to the nucleus of an atom and combines with a proton in that nucleus. The electron is generally derived from the K-shell of the atom and its combination with the proton leads to the formation of a neutron and to the emission of energy in the form of x-rays. *Abbr* EC.

electron carrier. A substance that serves as a donor and acceptor of either electrons or electrons and protons in an electron transport system. *See also* electron transfer protein.

electron configuration. The arrangement of the electrons about an atomic nucleus.

electron dense. Descriptive of a substance, such as a heavy metal ion or concentrated macromolecular matter, that has the capacity to scatter impinging electrons and to prevent their passage through it.

electron-dense label. An atom or a compound that is electron dense and that is frequently used as a marker in electron microscopy. *See also* ferritin-labeled antibody.

electron-density map. The three-dimensional representation of the structure of a molecule that is based on x-ray diffraction data and that is prepared by superimposing layers which correspond to electron densities of various planes in the molecule.

electron diffraction. The diffraction of a beam of electrons that results from the wavelike nature of the electrons.

electron donor. A substance that is being oxidized; a reductant; a reducing agent.

electronegative. 1. Describing the tendency of an atom or a group of atoms to gain electrons. 2. Having a negative charge; having an excess of electrons.

electron equivalent. A measure of reducing power that is equivalent to one electron.

electron-exchange resin. A resin that contains groups which are capable of undergoing reversible oxidation and reduction; such resins can be used as insoluble oxidizing and reducing agents in column chromatography.

electron hole. The energy level, in either the ground state or a deexcited state, of an atom or a molecule that has lost an electron and that has a great tendency to recapture an electron. *See also* hole.

electronic. Pertaining to the motion, emission, and behavior of currents of free electrons in vacuum tubes, gas tubes, semiconductors, and superconductors.

electronic transition. The transition of electrons of an atom or a molecule from one atomic orbital to another. Electronic transitions require large amounts of energy and can be induced by visible light, ultraviolet light, and x-rays.

electron magnetic resonance. *See* electron paramagnetic resonance; electron spin resonance.

electron micrograph. The photographic record of a pattern observed in an electron microscope.

electron microscope. A microscope in which beams of electrons are focused by means of magnetic fields onto a fluorescent screen or onto a photographic film. The electron microscope has great resolving power because of the short wavelengths that are associated with electrons. *Aka* transmission electron microscope.

electron microscope radioautography. The use of electron microscopy in conjunction with radioautography.

electron-nuclear double resonance. A technique for determining proton hyperfine splitting of organic free radicals in solution. *Abbr* ENDOR.

electron pair bond. COVALENT BOND.

electron paramagnetic resonance. A method for studying the interaction of unpaired electrons in a substance with the environment of these electrons. A substance containing unpaired electrons has a permanent magnetic moment (i.e., paramagnetic) as a result of the magnetic properties of its spinning electrons and will tend to orient itself in an applied magnetic field; the magnetic field of the unpaired electron may be either parallel or antiparallel to that of the applied field. A transition from one state to the other is associated with an energy difference and occurs when the applied field is of sufficient strength and the transition is accompanied by absorption of electromagnetic radiation; the relative magnitude of this radiation and that of the applied magnetic field are interpreted in terms of the electron's interaction with its environment. *Abbr* EPR. *See also* electron spin resonance.

electron pressure. The tendency of a substance to donate electrons and to function as a reducing agent.

electron sink. An electronegative atom or a group of atoms that captures an electron from other components of the system.

electron spin resonance. The common name for electron paramagnetic resonance when it is applied to compounds that are characterized by g-values close to that of the free electron, such as most organic free radicals. *Abbr* ESR.

electron transfer chain. ELECTRON TRANSPORT SYSTEM.

electron transfer flavoprotein. An electron carrier that serves as a link between the reduced form of fatty acyl coenzyme A dehydrogenase and the electron transport system. *Abbr* ETF.

electron transfer protein. A protein that serves as a donor and acceptor of either electrons or electrons and protons in oxidation-reduction reactions. Six types of electron transfer proteins have been identified: flavoproteins, proteins containing reducible disulfide groups, cytochromes, iron-sulfur proteins, cuproproteins, and molybdoproteins.

electron transfer system. ELECTRON TRANSPORT SYSTEM.

electron transport chain. ELECTRON TRANSPORT SYSTEM.

electron transport particle. 1. RESPIRATORY ASSEMBLY. 2. SUPERMOLECULE.

electron transport system. 1. The group of bio-logical oxidation-reduction substances that are present in mitochondria and that act sequentially in the transport of either electrons or electrons and protons. The electrons and protons are abstracted from metabolites in glycolysis, the citric acid cycle, beta oxidation, and other metabolic reactions. Each oxidation-reduction substance, or electron carrier, is capable of oxidizing a preceding one in the sequence and the oxidation proceeds from a metabolite to molecular oxygen as the ultimate oxidizing agent. The complete sequence is metabolite-NAD^+-FAD-CoQ-cytochromes-oxygen and the reduction potential of the electron carriers becomes progressively more positive in this direction. The free energy change corresponding to the potential difference between metabolite and oxygen is utilized for the synthesis of ATP from ADP, a reaction that is coupled to the electron transport system at various sites. 2. Any group of electron carriers such as that functioning in photosynthesis. *Abbr* ETS.

electron trap. ELECTRON SINK.

electron volt. A unit of energy equal to the energy acquired by an electron in passing through a potential gradient of 1 V.

electroosmosis. The flow of a conducting liquid through a porous medium that results from the application of an electrical potential; an electrokinetic phenomenon that is the reverse of streaming potential.

electropherogram. The record of a zone electrophoresis pattern, either in the form of the electrophoretic support itself or in the form of a tracing thereof.

electrophile. An atom or a group of atoms that is electron pair seeking.

electrophilic. Of, or pertaining to, either an electrophile or a reaction in which an electrophile participates.

electrophilic catalysis. Catalysis in which the catalyst abstracts a pair of electrons from a reactant.

electrophilic displacement. A chemical reaction in which an electrophilic group attacks and displaces a susceptible group in a compound and then binds covalently to the compound at that site. *Aka* electrophilic substitution.

electrophorese. To cause a compound or a mixture of compounds to move by electrophoresis.

electrophoresis. The movement of charged particles through a stationary liquid under the influence of an electric field. Electrophoresis is a powerful tool for the separation of particles and for both preparative and analytical studies of macromolecules. The particles are separated primarily on the basis of their charge and to a lesser extent on the basis of their size and shape. Moving boundary or free electrophoresis

electrophoretic. refers to electrophoresis performed in solution, while zone electrophoresis refers to electrophoresis performed in a porous medium.

electrophoretic. Of, or pertaining to, electrophoresis.

electrophoretic effect. The decrease in the electrophoretic mobility of a macromolecule that is caused by the movement of counterions and solvent molecules in a direction opposite to that of the macromolecule.

electrophoretic mobility. The velocity with which a charged particle moves in electrophoresis divided by the electric field strength that is applied across the support; generally expressed in units of $cm^2/(sec)(V)$.

electrophoretic retardation. ELECTRO-PHORETIC EFFECT.

electrophorogram. Variant spelling of electropherogram.

electroplax. A flat plate in the electric organ of some fish that, when stacked, allows for the generation of a large potential difference.

electropositive. 1. Describing the tendency of an atom or a group of atoms to lose electrons. 2. Having a positive charge; having a deficiency of electrons.

electrosorptive spreading. The spreading of a spot or a zone in electrophoresis that occurs if the moving particle is strongly adsorbed to the support.

electrostatic. Pertaining to electrical charges that are not in motion.

electrostatic bond. IONIC BOND.

electrostatic interactions. The attractive and repulsive electrical forces between atoms, and/or groups of atoms, and/or molecules that are caused both by the presence of ionized species and by the electropositive and electronegative properties of the atoms. *See also* ion-ion interaction.

electrostriction. The decrease that occurs in the volume occupied by a charged molecule in water as compared to the volume occupied by an uncharged molecule that has the same empirical formula as the charged one. The difference in volume is due to the strong attraction of water molecules to, and their resultant compression and close packing around, the charged groups of the solute molecule.

electroviscous effect. The dependence of the viscosity of a polymer on the electrical charge of the polymer.

element. A fundamental form of matter that has special properties and that is not decomposable by ordinary chemical means; it consists of atoms that are of one type and that have the same atomic number.

elementary analysis. The quantitative chemical analysis of the relative amounts of the different elements in a compound.

elementary particle. 1. SUPERMOLECULE. 2. RESPIRATORY ASSEMBLY.

elements of symmetry. The centers, axes, and planes of symmetry of either a molecule or a body.

eliminase. PECTATE LYASE.

elimination reaction. A chemical reaction in which there is a decrease in the number of groups attached to carbon atoms so that the molecule becomes more unsaturated.

ellipsoid of revolution. A geometrical solid formed by the rotation of an ellipse about one of its axes. The ellipsoid of revolution is called oblate or prolate depending on whether the rotation occurred about the minor or the major axis of the ellipse. *See also* equivalent ellipsoid of revolution.

elliptically polarized light. Light that is composed of left and right circularly polarized light components of unequal amplitude.

ellipticity. 1. The ratio of the axes of an ellipse. 2. MOLAR ELLIPTICITY.

Ellman's reagent. 5,5′-Dithiobis(2-nitrobenzoic acid); a reagent for the determination of sulfhydryl groups in proteins.

elongation. The stage in the polymerization of amino acids during protein synthesis that covers all of the steps between the initiation and the termination of the polypeptide chain. *See also* elongation cycle.

elongation cycle. The set of repetitive reactions that occur during the elongation stage of protein synthesis. These reactions are (a) the attachment of the incoming aminoacyl-tRNA to the aminoacyl site on the ribosome; (b) the peptidyl transferase-catalyzed formation of a peptide bond between the incoming amino acid and the growing polypeptide chain; and (c) the translocase-catalyzed shifting of the peptidyl-tRNA from the aminoacyl site to the peptidyl site on the ribosome with a simultaneous shift of the messenger RNA by one codon.

elongation factor. One of at least two protein factors that function in the elongation of polypeptide chains during protein synthesis and that are designated either as T and G, or as I and II. Elongation factor T is required for the binding of aminoacyl-tRNA to the ribosome and is dissociable into a stable and an unstable subunit referred to as T_s and T_u; elongation factor G is the translocase.

eluant. 1. Variant spelling of eluent. 2. The solute that is being separated on, and eluted from, a chromatographic column.

eluate. The liquid that is collected by elution from a chromatographic column.

eluent. The liquid that is used for the elution of substances from a chromatographic column.

eluent strength. The solvent adsorption energy

per unit surface area of a fully activated adsorbent.

eluotropic series. A group of solvents arranged in the order of their relative eluting strength for a given chromatographic adsorbent.

elute. To remove and collect a solute from the stationary phase in chromatography by means of elution with a solvent.

eluting agent. ELUENT.

elution. The process whereby a solute is removed and collected from the stationary phase in chromatography by passage of a solvent over the chromatographic support.

elution analysis. ELUTION CHROMATOGRAPHY.

elution band. A zone of separated sample components that is obtained in column chromatography.

elution centrifuge. A centrifuge in which a cylindrical holder, containing a separation column, revolves about the center of the apparatus and simultaneously rotates about its own axis; both feed and return tubes for the solutions are led through the center of the holder.

elution chromatography. A chromatographic technique in which the sample is applied to a column and the sample components are separated into bands or zones that can be eluted and collected. *Aka* elution development.

elution profile. A plot of some property of a column eluate, such as absorbance or radioactivity, as a function of either the eluate fraction number or the cumulative volume of the collected eluate.

elution time. RETENTION TIME.

elution volume. The volume of eluate collected in column chromatography from the time of sample application to the time at which a given component is eluted at maximal concentration.

elutriation. 1. The separation of suspended particles according to their size by washing, decantation, and settling. 2. CENTRIFUGAL ELUTRIATION.

Embden-Meyerhof-Parnas pathway. GLYCOLYSIS.

Embden-Meyerhof pathway. GLYCOLYSIS.

embryo. 1. The unborn or unhatched young vertebrate during the early stages of its development. *See also* fetus. 2. The rudimentary plant within a seed.

embryology. The science that deals with the development of an organism.

embryonic induction. MORPHOGENIC INDUCTION.

EMC hypothesis. Electromechanochemical coupling hypothesis.

emergency hormone. CATECHOLAMINE.

emergency hypothesis. The hypothesis that the catecholamines constitute the principal regulatory mechanism of an animal in an emergency

situation and that this mechanism allows the animal to mobilize to meet physical or emotional challenges.

Emerson enhancement effect. *See* enhancement effect.

EMF. 1. Electromotive force. 2. Erythrocyte maturation factor.

EMG. Electromyogram.

-emia. 1. Combining form meaning the presence of a substance in blood. 2. Combining form meaning the presence of excessive amounts of a substance in blood.

emission. The discharge of energy in the form of radiation.

emission spectrum. A plot of the emission of electromagnetic radiation by a molecule as a function of either the wavelength or the frequency of the radiation.

emit. To discharge energy in the form of radiation.

EMP pathway. Embden-Meyerhof-Parnas pathway.

emphore. A protein that is not an enzyme, but that specifically binds a ligand and thereby involves the ligand in biological activity.

empirical formula. The chemical formula of a compound that indicates the number and the types of the different atoms in the compound but not the manner in which they are linked together.

empty capsid. The viral capsid without the nucleic acid which it normally encloses.

emulsification. The formation of an emulsion.

emulsifying agent. A substance, such as a surface-active agent, that stabilizes an emulsion by coating the droplets of the dispersed phase, thereby preventing their coalescence. *Aka* emulsifier.

emulsion. A colloidal dispersion of one liquid in another, immiscible or partially miscible, liquid.

emulsion fractionation. A chromatographic technique in which the stationary phase is the surface of droplets and the mobile phase is the liquid that flows between the droplets. *See also* foam fractionation.

enamel. The calcified covering of dentine at the exposed portion of a tooth.

enamine. An organic compound that contains the grouping $-HN-\underset{|}{C}=\underset{|}{C}-$.

enantiomer. One of two optical isomers of a compound that are nonsuperimposable mirror images of each other.

enantiomorph. ENANTIOMER.

enantiotopic. Descriptive of either the atoms or the groups of atoms in a molecule that bear an enantiomeric relationship to each other; the two identical substituents of a meso carbon atom are an example of enantiotopic groups. *See also* diastereotopic.

encapsidate. To enclose the viral nucleic acid with a capsid. *Aka* encapside.

encapsulated. Having a capsule.

encephalovirus. ARBOVIRUS.

end absorption. The intense absorbance of saturated compounds at and below wavelengths of 200 nm.

end carbon chain. The number of carbon atoms in a fatty acid methyl ester molecule from the terminal methyl group to the center of the double bond nearest this methyl group. The length of the end carbon chain is important in determining the retention of fatty acid methyl esters in gas chromatography. *Abbr* ECC.

endemic. Of, or pertaining to, a disease that has a low, and more or less constant, incidence in a particular locality or region.

endergonic reaction. A chemical reaction that requires the input of energy; an uphill reaction with a positive free energy change.

end-group analysis. The determination of both the type and the number of the terminal groups in a polymer; used as an assessment of purity and for calculations of minimum molecular weights. The Sanger reaction is used for end-group analysis of proteins and exhaustive methylation is used for end-group analysis of carbohydrates.

endo-. Combining form meaning within or internal.

endo conformation. The conformation of the furanose ring in which the protruding carbon atom 2 or 3 is on the same side as the protruding carbon atom 5.

endocrine gland. A ductless gland of internal secretion that produces one or more hormones which are secreted directly into the circulation.

endocrine system. The endocrine glands.

endocrinology. The science that deals with the structure, the function, and the products of the endocrine glands and of other specialized secretory cells.

endocytosis. The process whereby cells take up fluids and particles by pinocytosis and phagocytosis.

endoenzyme. An enzyme that is released inside the cell.

endoergic reaction. ENDERGONIC REACTION.

endogenote. The genetic complement of the recipient cell that combines with a genetic fragment from a donor cell in bacterial transformation. *See also* exogenote.

endogenous. Originating within the organism.

endogenous metabolism. The level of metabolism in the absence of added nutrients.

endolysin. PHAGE LYSOZYME.

endometrium. The mucous membrane that lines the uterus.

endonuclease. An enzyme that catalyzes the hydrolysis of a polynucleotide strand at internal positions of the strand, rather than at the ends.

endopeptidase. An enzyme that catalyzes the hydrolysis of a polypeptide chain at internal positions of the chain, rather than at the ends.

endoplasm. The portion of the cytoplasm in the interior of the cell that is more fluid than that near the periphery of the cell. *See also* ectoplasm.

endoplasmic reticulum. The cytoplasmic network that consists of cisternae, vesicles, and tubules and that is frequently continuous with the cell membrane and the nuclear membrane. The endoplasmic reticulum has ribosomes attached to it and serves as a transport system for the proteins synthesized on the adhering ribosomes. *Abbr* ER.

ENDOR. Electron-nuclear double resonance.

endoskeleton. The internal skeleton of an organism, such as the bony skeleton of vertebrates.

endosmosis. The osmotic movement of fluid toward the inside of a cell or a vessel.

endospore. A spore formed within a cell, as that formed in a bacterium.

endosymbiotic infection. A viral infection in which the infected cells multiply for several generations even though they continue to release virus particles. The phenomenon applies particularly to animal viruses and is due to differences in the rates of cell and virus multiplication rather than to lysogeny.

endothermic reaction. A chemical reaction that requires the input of heat; a reaction with a positive enthalpy change.

endotoxin. A toxic lipopolysaccharide that is released from the cell wall of gram-negative bacteria upon destruction of the cell by autolysis or by other means.

end plate. The base plate of a T-even phage. *See also* T-even phage.

end point. The experimental point in a titration that corresponds to the theoretical equivalence point.

end-point method. A virus assay in which a fixed volume of a serially-diluted virus sample is inoculated into a number of animals, cell cultures, or other test units.

end-point mutation. A mutation that is expressed in an organism after a period of growth following exposure of the organisms to a mutagen.

end product. The chemical compound that represents a final substance formed in a sequence of chemical reactions.

end-product inhibition. FEEDBACK INHIBITION.

end-product repression. COORDINATE REPRESSION.

end-to-end conservative replication. A mode of replication for double-stranded DNA in which the parental molecule breaks into two halves so that each daughter molecule consists of one half of a parental duplex and one half of a newly synthesized duplex.

end-to-end distance. The distance between the ends of a folded polymer as distinct from the contour length of the extended polymer. *See also* root-mean-square end-to-end distance.

end window counter. A Geiger-Mueller counter in which either a thin window or a membrane separates the sample from the detector.

enediol. An organic compound that contains two hydroxyl groups attached to the two carbon atoms of a double bond.

energized conformation. ORTHODOX CONFORMATION.

energy. The capacity to do work. *Sym* E.

energy barrier. 1. The difference between the bond energy at which a molecule dissociates and the energy of the ground state of the molecule. 2. The difference between the ground state energy of the activated complex of a reaction and the sum of the ground state energies of the reactants.

energy charge. A measure of the availability of high energy phosphate bonds in the ATP-ADP-AMP system. It is equal to the expression $([ATP] + \frac{1}{2}[ADP])/([AMP] + [ADP] + [ATP])$ and has a value of 1.0 if all of the adenine nucleotide is present as ATP, and a value of zero if all of the adenine nucleotide is present as AMP. The energy charge is the mole fraction of ATP or its equivalent in the total adenylate pool; $\frac{1}{2}$ADP is equivalent to 1ATP because of the interconversion between ADP and ATP which is catalyzed by the enzyme myokinase. *See also* phosphate potential.

energy coupling. 1. The synthesis of ATP that is linked to the operation of the electron transport system. 2. COUPLED REACTIONS.

energy diagram. 1. POTENTIAL ENERGY DIAGRAM. 2. The diagrammatic representation of the energy content of various states such as those of (a) reactants, activated complex, and products in a reaction, (b) the nuclear energy levels of an atom, and (c) the electronic energy levels of an atom.

energy fluence. The energy carried by the photons in photon fluence; the energy fluence rate is the energy carried by the photons in photon fluence rate.

energy of activation. *See* activation energy.

energy-poor compound. LOW-ENERGY COMPOUND.

energy-regenerating system. ATP REGENERATING SYSTEM.

energy-rich compound. HIGH-ENERGY COMPOUND.

energy sink. A molecule or a group of atoms in a molecule that readily accepts energy transferred to it from other components of the system.

energy transduction. *See* transduction (2).

energy transfer. The transfer of excitation energy from one chromophore or one molecule to another by a radiationless process. The energy may then be dissipated in a number of ways, such as through fluorescence. Energy transfer is strongly dependent on the distance between the chromophores and is useful for studying structural relationships among groups of atoms in a macromolecule.

energy well. *See* potential energy well.

enhancement. 1. The prolongation of the life of a transplant in a recipient by injection of killed tissue from the donor into the recipient prior to implantation of the transplant. 2. The increase in the rate of growth of a tumor that occurs in an animal which has been immunized with antigens of the tumor.

enhancement effect. The increase in photosynthetic efficiency (quantum yield) of chloroplasts that occurs when light of longer wavelengths is supplemented with light of shorter wavelengths. The longer wavelengths are those above 680 nm; the shorter wavelengths are those below 600 nm.

enol. The tautomer of a ketone in which the carbonyl group has been converted to an alcoholic hydroxyl group that is attached to a double bond.

enolase. The enzyme that catalyzes the formation of the high-energy compound, phosphoenolpyruvic acid, from 2-phosphoglyceric acid in glycolysis.

enol-keto tautomerism. *See* keto-enol tautomerism.

enriched food. Food to which nutrients have been added after naturally occurring ones have been removed; enrichment does not replace all of the lost nutrients.

enriched medium. MAXIMAL MEDIUM.

enrichment. 1. PURIFICATION (2). 2. The selective growth of bacteria by means of an enrichment culture. 3. The increase in the concentration of a stable isotope above its natural abundance.

enrichment culture. A culture used for the selection of specific bacterial strains from among a mixture; such a culture favors the growth of the desired bacteria under the conditions used.

enrichment medium. A liquid selective medium of such a composition that it favors the growth of a specific bacterial strain over others in a mixed bacterial population.

entatic. Pertaining to a chemical bond that is strained.

entatic site hypothesis. The hypothesis that in reactions catalyzed by metalloenzymes, a metal ion-enzyme complex is formed prior to the formation of an enzyme-substrate or enzyme-inhibitor complex. The metal ion-enzyme complex is believed to be characterized by distorted geometries of the metal coordination sites so that the enzyme is in a state of tension, and the metal ion is in a state that approximates its transition state for the particular reaction. Such an activated metal ion-enzyme complex would serve to lower the activation energy of the reaction.

enteric. Of, or pertaining to, the intestine.

enteric virus. A virus, the target organ of which is the intestine.

enterocrinine. A gastrointestinal hormone, present in the intestinal mucosa, that controls the secretion of intestinal juice.

enterogastrone. A gastrointestinal hormone, secreted by the duodenum, that inhibits gastric motility and gastric secretion.

enterohepatic circulation. The circulatory system that connects the intestine and the liver by way of the bile and the portal blood. The system transports, for example, cholesterol and bile acids; these are excreted from the liver to the intestine by way of the bile and are reabsorbed from the intestine and returned to the liver by way of the portal blood.

enterokinase. A proteolytic enzyme, secreted by the intestine, that catalyzes the activation of trypsinogen to trypsin.

enterotoxin. An exotoxin that is released by staphylococci and that induces nausea, cramps, diarrhea, and vomiting.

enterovirus. 1. A virus that belongs to a subgroup of picornaviruses which includes polio virus, coxsackievirus, and echovirus; enteroviruses infect the gastrointestinal tract of man and some also cause acute respiratory infections. 2. ENTERIC VIRUS.

enthalpy. Heat content; the thermodynamic function H in $H = E + PV$, where E is the internal energy of the system, P is the pressure exerted on the system, and V is the volume of the system.

enthalpy change. The difference between the enthalpy of formation of the products and that of the reactants in a chemical reaction. *Sym* ΔH.

Entner-Doudoroff pathway. A pathway for the conversion of glucose to glyceraldehyde-3-phosphate that occurs in some bacteria. *Abbr* ED pathway.

entomology. The study of insects.

entropic doom. The terminal equilibrium state of the universe at which the free energy is at a minimum and the entropy is at a maximum; this state is predicted on the basis of the first and second laws of thermodynamics which imply that the universe is progressing toward a state of increased randomness and disorder.

entropic strain. The concept that a good substrate, when bound to an enzyme, is restricted in its rotation and is in a conformation which approximates its conformation in the activated complex. A poor substrate, on the other hand, is considered to be able to rotate more freely about critical bonds so that it will achieve the correct conformation only occasionally and at random.

entropy. The thermodynamic function that is a measure of that part of the energy of a system that cannot perform useful work; the degree of randomness or disorder of a system. *Sym* S.

entropy change. The difference between the entropy of the products and that of the reactants in a chemical reaction. *Sym* ΔS.

entropy compensation. The increase in entropy that may accompany the formation of an enzyme-substrate complex as a result of conformational changes of the enzyme.

entropy stabilization. The stabilization of a structure due to an increase in entropy associated with the formation of the structure.

entropy unit. A unit of entropy equal to 1 cal/(deg)(mole). *Abbr* EU.

entry site. AMINOACYL SITE.

enucleated. Without a nucleus.

envelope. 1. The two nuclear membranes of a eucaryotic cell. 2. The membrane surrounding some viruses and consisting of lipid, carbohydrate, and protein. 3. The bacterial cell membrane, cell wall, and capsule. 4. A specific conformation of the furanose ring of monosaccharides.

enveloped nucleocapsid. A nucleocapsid surrounded by a membrane.

Enz. Enzyme.

enzymatic. Of, or pertaining to, enzymes.

enzymatic activity. 1. The catalytic activity of an enzyme. 2. The rate of reaction of a substrate that may be attributed to catalysis by an enzyme and that is expressed in terms of katals.

enzymatic reversion. DEADAPTATION.

enzyme. A protein molecule, produced by living cells, that functions as a catalyst of biochemical reactions. The number and the types of reactions catalyzed by an enzyme are determined by the specificity of the enzyme. Enzymes are classified into the six main divisions of oxidoreductases, transferases, hydrolases, lyases, isomerases, and ligases. *Abbr* Enz; E. *See also* enzyme classification.

enzyme I. A soluble bacterial enzyme that is part of the phosphotransferase system for the transport of sugars across the cell membrane. The enzyme catalyzes the reaction P-enolpyruvate + HPr → pyruvate + P-HPr, where *HPr* is a heat-stable, low molecular weight protein, and *P* designates phosphate.

enzyme II. A membrane-bound bacterial enzyme that is part of the phosphotransferase system for the transport of sugars across the cell membrane. The enzyme catalyzes the reaction *P-HPr* + sugar → sugar-*P* + *HPr*, where *HPr* is a heat-stable, low molecular weight protein, and *P* designates phosphate. The enzyme is responsible for the specificity of the transport with respect to the sugar and functions in some systems in conjunction with another protein (factor III).

enzyme adaptation. ENZYME INDUCTION.

enzyme assay. The measurement of enzymatic activity that is based on a determination of either the rate or the extent of the formation of a product or the disappearance of a reactant.

enzyme cascade. CASCADE MECHANISM.

enzyme classification. The systematic arrangement and the naming of enzymes that is based on the 1972 recommendations of the Enzyme Commission of the International Union of Biochemistry. Each enzyme is denoted by a number composed of four figures. The first figure denotes one of the six main divisions: oxido-reductases, transferases, hydrolases, lyases, isomerases, ligases. The second figure denotes the subclass and the third figure denotes the sub-subclass. The last figure denotes the serial number of the enzyme in its sub-subclass. The enzyme number is preceded by the abbreviation E.C.

enzyme commission. A special commission of the International Union of Biochemistry that made recommendations for the classification and naming of enzymes and for the definitions of the mathematical constants used in enzymology. The recommendations were first published in 1964 and were published in revised form in 1972. *Abbr* E.C.

enzyme complex. MULTIENZYME SYSTEM.

enzyme concentration. *See* concentration of enzymatic activity.

enzyme conservation equation. *See* conservation equation.

enzyme deletion hypothesis. *See* deletion hypothesis.

enzyme electrode. An electrode that incorporates an enzyme into its design and that is specific for measuring the concentration of a reactant or a product of the reaction catalyzed by the enzyme. The enzyme is frequently trapped within a gel matrix around the electrode or trapped within a liquid film in a cellophane membrane.

enzyme fractionation. Protein fractionation applied to an enzyme preparation.

enzyme induction. The process whereby an inducible enzyme is synthesized in response to an in-

ducer. The inducer combines with a repressor and thereby prevents the blocking of an operator by the repressor. The operator controls the structural gene of the enzyme and the active, unblocked operator permits the transcription of that gene.

enzyme-inhibitor complex. The complex that consists of an enzyme and an inhibitor which is bound either to the catalytic site of the enzyme or to some other site on the enzyme. *Abbr* EI.

enzyme kinetics. The kinetics of enzyme-catalyzed reactions.

enzyme labeling. A method for locating antigens or antibodies in tissues; based on binding the antigen or the antibody to an enzyme and then determining the location of the enzyme in the tissues by making use of the known properties of the enzyme.

enzyme multiplicity. The occurrence of two or more forms of the same enzyme, all of which catalyze the same reaction.

enzyme multiplicity feedback inhibition. Feedback inhibition in which two or more forms of an enzyme, all of which catalyze the same reaction, are inhibited to different degrees and by different end products.

enzyme pH electrode. An enzyme electrode that incorporates in its design a conventional glass electrode which is sensitive to hydrogen ions.

enzyme repression. The process whereby the synthesis of a repressible enzyme is decreased in response to either a repressor or a repressor-corepressor complex. The repressor or the repressor-corepressor complex binds to and blocks an operator and thereby prevents the transcription of the structural gene of the enzyme which is controlled by that operator.

enzyme-specific electrode. ENZYME ELECTRODE.

enzyme specificity. *See* specificity (1).

enzyme-substrate complex. The complex that consists of an enzyme and the substrate that is bound to the enzyme at its catalytic site. *Abbr* ES.

enzyme-substrate compound. The enzyme-substrate complex in which the substrate is covalently linked to the enzyme.

enzyme system. MULTIENZYME SYSTEM.

enzyme unit. The amount of enzyme that, under defined conditions, will catalyze the transformation of 1 μmole of substrate per minute, or, where more than one bond of each substrate molecule is attacked, 1 μeq of the group concerned per minute. *Sym* EU; U. *See also* katal.

enzyme variant. *See* variant (1); multiple forms of an enzyme; isozyme.

enzymic. ENZYMATIC.

enzymoelectrophoresis. An electrophoretic technique for the detection and determination of isozymes by the combined use of electrophoresis and an enzyme assay. The technique is used specifically for the isozymes of lactate dehydrogenase which are first separated by gel electrophoresis and are then assayed by covering the gel with another gel which contains substrate and coenzyme; the paired gels are then scanned spectrophotometrically.

enzymology. The study of enzymes and enzyme-catalyzed reactions.

enzymolysis. Hydrolysis by means of enzymes.

eobiogenesis. The first instance of the formation of living matter from nonliving matter.

eobiont. A primitive prototype of a living cell.

Eoff process. The formation of glycerol from dihydroxyacetone phosphate by yeast under alkaline conditions.

eosome. A fundamental molecule of the ribosome; a molecule of ribosomal RNA or a molecule of ribosomal protein.

EP. Enzyme-product complex.

epidemic. Of, or pertaining to, a disease that has a widespread incidence which is significantly above the endemic level.

epidemiology. The study of diseases in populations.

epidermal growth factor. A low-molecular weight protein, isolated from the submaxillary glands of mice, that stimulates proliferation and keratinization of epidermal tissue in several species. *Abbr* EGF.

epigenesis. The concept that an organism develops through the appearance and diversification of structures and functions that are not present in the egg. *See also* preformation.

epigenetics. The study of the mechanisms that result in the expression of the phenotypic effects of genes during differentiation and development.

epimer. One of two optical isomers of a compound that differ from each other only in the configuration about one asymmetric carbon atom.

epimerase. An enzyme that catalyzes the interconversion between two optical isomers, each of which contains only one asymmetric center. *See also* racemase.

epimeric carbon. The asymmetric carbon atom of two optical isomers that differ from each other only in the configuration about this carbon atom.

epimerization. The formation of one epimer from another.

epinephrine. A catecholamine hormone that is secreted by the adrenal medulla. Epinephrine raises the level of blood sugar by increasing the breakdown of glycogen, stimulates the mobilization of free fatty acids, and has various effects on the cardiovascular system and on muscular tissue.

episome. A genetic element that can replicate either autonomously in the cytoplasm or together with the chromosome into which it becomes integrated, and that can shift back and forth between these two states. A temperate phage and the sex factor are two examples.

epistasis. The suppression of the expression of one gene by another, nonallelic gene.

epistatic gene. A gene, the expression of which suppresses or reduces the expression of another, nonallelic gene.

epithelial cell. A cell of the epithelium.

epithelioma. A malignant tumor derived from epithelial cells.

epithelium. The sheet of cells, consisting of one or more layers, that covers surfaces and lines tubes of animal tissue.

epitope. ANTIGENIC DETERMINANT.

epitype. A family of related epitopes.

EPR. Electron paramagnetic resonance.

epsilon. Denoting the fifth carbon atom from the carbon atom that carries the principal functional group of the molecule. *Sym* ϵ.

epsilon chain. The heavy chain of the IgE immunoglobulins.

Epstein-Barr virus. A herpesvirus that has been implicated in the etiology of infectious mononucleosis and of Burkitt's lymphoma.

eq. Equivalent.

equation of state. An equation that relates the volume, pressure, temperature, and mass of a substance.

equator. The line in an x-ray diffraction pattern that passes through the zero point and that is perpendicular to the axis of the fiber that is being studied. *Aka* zero layer line.

equatorial bond. A bond in a molecule having a ring structure that is roughly in the plane of the ring.

equatorial reflection. An x-ray diffraction spot that lies on the equator.

equatorial substituent. A substituent attached to an equatorial bond.

equilibrium (*pl* equilibria). The state of a system in which no further change is occurring and in which the free energy is at a minimum. At equilibrium, the rate of the forward reaction is equal to the rate of the reverse reaction so that a small change in one direction is balanced by a small change in the opposite direction.

equilibrium banding. DENSITY GRADIENT SEDIMENTATION EQUILIBRIUM.

equilibrium constant. The constant K that is characteristic of a given chemical reaction at a specified temperature and that is based on the activities of all of the reactants and all of the products at equilibrium. *Sym* K; Keq. *See also* apparent equilibrium constant.

equilibrium dialysis. A method for measuring the binding of low-molecular weight ligands to macromolecules. The macromolecules are placed inside a dialysis bag, the ligands are placed outside the bag, and dialysis is allowed to proceed until equilibrium is established. From the known concentration of macromolecules, the expected distribution of ligand across the dialysis membrane in the absence of binding, and the initial and final concentrations of ligand on either side of the dialysis membrane, it is possible to calculate the average number of ligand molecules that are bound per macromolecule. The method permits a determination of the number of binding sites per macromolecule for a given ligand and a determination of the intrinsic association constant which governs that binding.

equilibrium potential. The membrane potential that arises from the differences in the concentrations of ions across a membrane and that exactly balances the tendency of these ions to diffuse from the more concentrated to the more dilute solution.

equilibrium thermodynamics. The branch of thermodynamics that deals with changes between equilibrium states of closed systems; the change from one equilibrium state to another may be either a reversible or an irreversible process.

equimolar. Containing an equal number of moles and hence an equal number of molecules.

equine. Of, or pertaining to, horses.

equivalence point. 1. The point in a titration where a chemically equivalent amount of a compound has been added to the compound being titrated. 2. EQUIVALENCE ZONE.

equivalence zone. A zone in the precipitin curve of the antigen-antibody reaction in which maximum precipitation of the antigen-antibody complex occurs.

equivalent. GRAM-EQUIVALENT WEIGHT.

equivalent chain length. The number of carbon atoms, generally a nonintegral number, that is derived for an organic compound on the basis of its retention time in gas chromatography. The number is obtained by interpolation of a semilogarithmic plot of retention time versus the number of carbon atoms in the chain for a series of saturated, straight-chain organic compounds. *Abbr* ECL.

equivalent ellipsoid of revolution. The ellipsoid of revolution that has the same volume as the actual hydrodynamic unit in the solution; the hydrodynamic unit consists of the macromolecule together with tightly bound solvent.

equivalent height of a theoretical plate. *See* height equivalent to a theoretical plate.

equivalent thickness. The thickness, in centimeters, of an absorbing material that is equivalent to a thickness of 1 cm of air with respect to absorption of alpha particles, multiplied by the density of the absorbing material in grams per cubic centimeter.

ER. Endoplasmic reticulum.

erg. A unit of work that is equal to the work done when a force of 1 dyne acts through a distance of 1 cm.

ergastoplasm. ROUGH-SURFACED ENDOPLASMIC RETICULUM.

ergocalciferol. CALCIFEROL.

ergosome. POLYSOME.

ergosterol. A sterol that yields calciferol upon irradiation with ultraviolet light.

erosion model. RELIC MODEL.

error. The difference between a measured value of a quantity and either the true or the expected value.

error curve. NORMAL DISTRIBUTION.

error function. NORMAL DISTRIBUTION.

error theory. A theory of aging according to which aging is due to the occurrence of errors in the biosynthesis of proteins; the accumulation of the partially active, or of the inactive, proteins that are produced in this fashion then leads to the death of cells.

erythrocin. ERYTHROMYCIN.

erythro configuration. The configuration of a compound in which two asymmetric carbon atoms have identical substituents on the same side, as is the case in erythrose.

erythrocruorin. A hemoglobin-like respiratory pigment of invertebrates that has a molecular weight of about 4×10^5 to 6.7×10^6 and contains 30 to 400 heme groups per molecule.

erythrocyte. A mature red blood cell that is no longer engaged in the synthesis of hemoglobin and that derives its energy primarily from anaerobic glycolysis and from the phosphogluconate pathway.

erythrocyte ghost. *See* ghost (1).

erythrocyte maturation factor. VITAMIN B_{12}.

erythrocytosis. POLYCYTHEMIA.

erythromycin. A macrolide antibiotic produced by *Streptomyces erythreus.*

erythropoiesis. The formation of red blood cells.

erythropoietic. Of, or pertaining to, erythropoiesis.

erythropoietic porphyria. A congenital porphyria that is caused by an excessive formation of heme precursors in the developing red blood cells of the bone marrow.

erythropoietin. A glycoprotein, formed in the kidney, that has hormone-like properties and that leads to an increase in the formation of erythrocytes.

erythrose. A four-carbon aldose that is an intermediate in the phosphogluconate pathway.

ES. Enzyme-substrate complex.

Escherichia coli. A bacterium that is normally present in the intestine and that is widely used in biochemical and genetic research. *Abbr E. coli. Aka* colon bacillus.

ESI. Enzyme-substrate-inhibitor complex.

ESR. 1. Electron spin resonance. 2. Erythrocyte sedimentation rate.

essential amino acid. An amino acid that is required by an organism for normal growth and functioning, but that cannot be synthesized by the organism; such an amino acid must be obtained through the diet. *Abbr* EAA.

essential amino acid index. The chemical score of the essential amino acids of a protein. *Abbr* EAA index.

essential enzyme. An enzyme without which a cell or an organism cannot grow or survive.

essential fatty acid. A fatty acid that is required by an organism for normal growth and functioning, but that cannot be synthesized by the organism; such a fatty acid must be obtained through the diet. *Abbr* EFA.

essential gene. A gene that codes for an essential enzyme.

essential metabolite. A metabolite required for the growth of cells or the growth of an organism.

essential nutrient. A nutrient that is required by an organism for normal growth and functioning, but that cannot be synthesized by the organism; such a nutrient must be obtained through the diet.

essential oil. An oil that is produced by a plant and that has a characteristic odor and flavor. Essential oils are rich in terpenes and in oxygenated compounds such as alcohols, aldehydes, ketones, acids, and esters; the oxygenated compounds are most responsible for the odors and flavors of the essential oils.

established cell line. A group of cells that can be subcultured and that multiply serially whether for short or long periods; refers to either the primary or the permanent cell strain. *Aka* established cell strain.

established tissue culture. A long-term tissue culture.

established tumor. A tumor that has been transplanted and that has been allowed to grow in the new host.

ester. An organic compound formed from an alcohol and an acid by splitting out a molecule of water; the water is formed from the *H* of the alcohol and from the *OH* of the acid.

esterase. An enzyme that catalyzes the hydrolysis of an ester.

esterification. The formation of an ester.

ester interchange. An interesterification reaction in which two esters react with each other to produce two new esters.

esterolytic protease. A proteolytic enzyme that catalyzes the hydrolysis of ester bonds in appropriate substrates at a faster rate than it

catalyzes the hydrolysis of natural peptide bonds.

ester value. The number of milligrams of potassium hydroxide that are required to saponify the esters in one gram of a fat or an oil; identical to the saponification number if the sample does not contain free fatty acids.

estolide. An intermolecular lactone of hydroxy fatty acids.

17-β-estradiol. A steroid hormone, secreted by the ovaries, that is a major female sex hormone and that is responsible for the regulation of feminine characteristics and for the regulation of the menstrual-ovulatory cycle.

estrane. The parent ring system of the estrogens.

estrogen. An 18-carbon steroid that is a female sex hormone or one of its metabolites; the major estrogens are estrone and 17-β-estradiol. *See also* female sex hormone.

estrone. A major female sex hormone.

estrous cycle. The sequence of endocrine-related events from the beginning of one estrus to the beginning of the next.

estrus. 1. The period of reproductive activity in an animal. 2. ESTROUS CYCLE.

Et. Ethyl group.

ETF. Electron-transfer flavoprotein.

ether. 1. An organic compound derived from an alcohol by replacing the hydrogen of the hydroxyl group with an organic radical. 2. Diethyl ether; ethyl ether.

ethereal sulfate. One of a group of phenolic, sulfur-containing compounds that are excreted in the urine.

etherification. The formation of an ether.

ethionine. An amino acid analogue that can be incorporated into protein during protein synthesis.

ethylenediaminetetraacetic acid. ETHYLENEDINITROLOTETRAACETIC ACID.

ethylenedinitrolotetraacetic acid. A chelating agent. *Abbr* EDTA.

ethylenimine. An aziridine mutagen.

ethylmethane sulfonate. A chemical mutagen.

etiology. The study of causes, specifically the causes of disease.

ETP. Electron transport particle.

ETS. Electron transport system.

ETS particle. Electron transport system particle; *see* electron transport particle.

EU. 1. Enzyme unit. 2. Entropy unit.

eubacterium (*pl* eubacteria). A typical, or "true," bacterium; eubacteria are classified either as cocci, bacilli, and spirilla on the basis of their shape, or as gram-positive and gram-negative on the basis of their staining properties.

eucaryon. The nucleus of a eucaryotic cell.

eucaryote. A higher organism that consists of cells that possess a true nucleus; the nucleus is surrounded by a nuclear membrane and contains the genetic material within multiple chromosomes. *See also* procaryote.

eucaryotic. Of, or pertaining to, eucaryotes.

euchromatin. The diffuse chromatin that is active in RNA synthesis and that stains weakly during interphase, but stains strongly during metaphase.

eugenic. Related to, or aimed at improving, the genetic qualities of a race or a breed. *See also* dysgenic.

eugenics. The science that deals with the improvement of the genetic qualities of a race or a breed through alteration of its genetic makeup.

Euglena. A genus of unicellular photosynthetic flagellates that are used for genetic studies and that are classified as either green algae or protozoa.

euglobulin. A water-insoluble globulin.

eukaryon. Variant spelling of eucaryon.

eukaryote. Variant spelling of eucaryote.

euploid state. The chromosome state in which the number of chromosomes is an exact multiple of the basic number in the genome. *Aka* euploidy.

euthenics. The science that deals with the improvement of the human race through control and alteration of its physical, biological, and social environments.

eutrophication. The process whereby a body of water becomes deficient in oxygen.

eutrophic lake. A shallow lake, having a depth of 10 m or less, that is murky due to dense growth of planktonic algae and has a high rate of nutrient supply in relation to its volume of water. In such a lake, both plant biomass and productivity are high and the bottom layers of the lake frequently contain low concentrations of dissolved oxygen at the end of the summer. *See also* mesotrophic lake; oligotrophic lake.

eV. Electron volt.

even-carbon fatty acid. EVEN-NUMBERED FATTY ACID.

evenly labeled. UNIFORMLY LABELED.

even-numbered fatty acid. A fatty acid molecule that has an even number of carbon atoms.

eversion theory. A theory of aging according to which aging is due to changes in the structures of macromolecules.

everted sac technique. A technique for studying the absorption of substances across the intestinal wall by means of in vitro experiments performed on small segments of the intestine, turned inside out.

evocation. The process whereby a morphogenetic effect is brought about by an evocator.

evocator. A chemical substance that is mor-

phogenetically active and that is emitted by an organizer.

evolution. 1. DARWINIAN EVOLUTION. 2. The process of continuous change by which something complex develops from something simpler. 3. The liberation of a gas during a chemical reaction.

evolutionary tree. PHYLOGENETIC TREE.

evolve. To undergo evolution.

excision. 1. The enzymatic removal of a nucleotide segment, particularly a thymine dimer, from a nucleic acid strand. 2. DETACHMENT.

excision repair. A repair mechanism for damaged DNA. First, the damaged nucleotide segment of a DNA strand is removed by two nuclease-catalyzed incisions on either side of the damaged segment. Second, the correct segment is synthesized by means of DNA polymerase with the second strand of the DNA serving as a template. Last, the newly synthesized segment is joined to the existing strand by means of DNA ligase. *Aka* cut and patch repair. *See also* patch and cut repair.

excitability. The capacity of living matter to respond to an external stimulus.

excitation. 1. The transition of an atom or a molecule from a lower to a higher energy level as that brought about by the raising of electrons from lower to higher energy orbitals. 2. The process whereby an external stimulus brings about changes in living matter, such as the changes in a muscle fiber that are initiated by a nerve impulse.

excited state. Any state of a nucleus, an atom, or a molecule that is of a higher energy level than that of the ground state.

exciton. A quantum of excitation energy; a photon.

excluded volume. 1. COVOLUME. 2. The volume of solvent that is required to elute a component in gel filtration when that component is neither adsorbed to, nor permeates into, the gel bed.

exclusion chromatography. GEL FILTRATION CHROMATOGRAPHY.

exclusion diagram. RAMACHANDRAN PLOT.

exclusion limit. The molecular weight of the largest molecules of a particular shape that can be effectively fractionated by a specific gel in gel filtration. Molecules of similar shape but higher molecular weights will not penetrate the gel particles and will move right through the column without being fractionated.

exclusion reaction. The prevention of further phage infection of a phage-infected bacterium that is brought about by strengthening the cell envelope when the phage-infected cell is heated.

exergonic reaction. A chemical reaction that is

accompanied by a release of energy; a downhill reaction with a negative free energy change.

exchange diffusion. The passive transport of two solutes across a biological membrane such that one solute moves in one direction while an equimolar amount of the second solute moves in the opposite direction.

excitatory autacoid. HORMONE.

exhaustion theory. A theory of aging according to which aging is due to the exhaustion of an essential nutrient.

exhaustive methylation. Maximal methylation; in carbohydrate chemistry this refers to the formation of a methyl ether at every free hydroxyl group on the carbohydrate. The hydrolysis of an exhaustively methylated carbohydrate, followed by separation, identification, and quantification of the components, is used as an aid in determining the structure of the carbohydrate.

exo-. Combining form meaning outside or external.

exobiology. The study of extraterrestrial life.

exo conformation. The conformation of the furanose ring in which the protruding carbon atom 2 or 3 is on the opposite side as the protruding carbon atom 5.

exocrine gland. A gland of external secretion that discharges its secretion by means of a duct.

exocytosis. The process whereby fluids and particles are discharged from a cell by a reversal of endocytosis.

exoenzyme. An enzyme that is released outside the cell.

exoergic reaction. EXERGONIC REACTION.

exogenote. The genetic fragment of the donor cell that is transferred to a recipient cell in bacterial transformation. *See also* endogenote; exosome.

exogenous. Originating outside the organism.

exogenous metabolism. The level of metabolism due to added nutrients.

exonuclease. An enzyme that catalyzes the sequential hydrolysis of nucleotides from one end of the polynucleotide strand.

exopeptidase. An enzyme that catalyzes the sequential hydrolysis of amino acids from one end of the polypeptide chain.

exophthalmic goiter. GRAVE'S DISEASE.

exophthalmos-producing substance. An uncharacterized substance, thought to be a hormone that functions in the production of bulging eyes.

exoskeleton. A protective external covering, such as the scales of fish or the horny structure of insects and crustaceans.

exosmosis. The osmotic movement of fluid toward the outside of a cell or a vessel.

exosome. A genetic fragment that is transferred to a recipient cell in bacterial transformation and that is not readily integrated into the recipient chromosome but can remain unintegrated and can replicate, be transcribed, and express biochemical function in this state. *See also* endogenote.

exospore. A spore formed by budding, as that generally formed by a fungus.

exosporium. A loose outer layer that frequently covers a spore.

exothermic reaction. A chemical reaction that is accompanied by a release of heat; a reaction with a negative enthalpy change.

exotoxin. A toxic protein that is discharged from a bacterial cell. Exotoxins are generally produced by gram-positive bacteria and can be demonstrated in a bacterial culture in which no appreciable autolysis has occurred.

experimental error. The error that results from the inability to reproduce precisely the experimental conditions in an otherwise carefully and accurately performed analysis.

expire. To exhale.

exponential. A function obtained by raising the constant e to a power; the exponential of x is e^x, where e is the base of natural logarithms and x is a variable.

exponential decay. The mode of radioactive decay that can be described by the equation $N = N_0 e^{-\lambda t}$, where N is the number of radioactive atoms present at time t, N_0 is the number of radioactive atoms originally present, e is the base of natural logarithms, and λ is the decay constant.

exponential density gradient. A density gradient in which the density increases exponentially from one end of the gradient to the other. It is described by the equation $c = a + b\, e^{-dV}$, where V is the volume of the gradient, c is the concentration, e is the base of natural logarithms, and a, b, and d are constants. *See also* concave exponential gradient; convex exponential gradient.

exponential growth. The growth of cells in which the number of cells (or the cell mass) increases exponentially and the growth at any time is proportional to the number of cells (or the cell mass) present.

exponential growth rate constant. The reciprocal of the doubling time, expressed as the number of generations per hour.

exponential phase. The phase of growth of a bacterial culture in which the number of cells (or the cell mass) increases exponentially so that a plot of the logarithm of the number of cells (or the cell mass) as a function of time yields a straight line.

exponential survival curve. A survival curve that indicates an exponential loss of active units as a function of increasing dose. Such data will yield a straight line when they are replotted in terms of the logarithm of the surviving fraction as a function of the dose.

exposure. The rate of irradiation per unit area, perpendicular to the beam of radiation, and multiplied by the time interval of irradiation.

exposure dose. The intensity of a dose that is based on the ionizations produced by the radiation in air; generally expressed in terms of roentgens. *See also* radiation absorbed dose.

exsanguinate. To drain of blood.

extant. Living or existing at the present time, as opposed to extinct.

extensin. A protein, rich in hydroxyproline, that is attached to cellulose fibrils in plant cell walls.

extensive property. A property of a system, such as volume or energy, that can be defined only by specifying the precise amounts of all of the substances involved. *See also* intensive property.

external indicator. An indicator that is outside the titration vessel and to which is added a drop of the liquid that is being titrated.

external monooxygenase. A monooxygenase in which the cosubstrate that incorporates the second oxygen atom is not itself a product of the reaction.

external quenching. The stoppage of secondary and subsequent ionizations in an ionization detector that is caused by a momentary reduction in the applied potential.

external respiration. *See* respiration (1).

external-sample scintillation counter. A scintillation counter in which radiation from an external source interacts with a solid fluor that is coupled to a photomultiplier.

external standard. A standard that is treated separately from the sample.

external suppression. INTERGENIC SUPPRESSION.

extinct. Not living or existing at the present time, as opposed to extant.

extinction. ABSORBANCE.

extinction angle. The angle between the plane of polarization and the cross of isocline in flow birefringence.

extinction coefficient. ABSORPTIVITY.

extinction coefficient ϵ **(P).** A molar absorptivity used for nucleotides and nucleic acids, the concentration of which is expressed in terms of moles of phosphorus per liter; specifically, $\epsilon (P) = A/cd$, where $\epsilon (P)$ is the extinction coefficient, A is the absorbance, c is the concentration in terms of gram-atoms of phosphorus per liter, and d is the length of the lightpath in centimeters.

extinction dilution. The dilution of a bacterial or a viral sample to such an extent that the diluted preparation is no longer infectious.

extinction point. The critical oxygen tension for a green plant below which aerobic fermentation occurs as part of the metabolism of the plant, and above which aerobic respiration occurs exclusively.

extra-. Prefix meaning outside.

extra arm. The base-paired helical nucleic acid segment that is of variable length and that occurs in the cloverleaf model of some transfer RNA molecules.

extracellular enzyme. EXOENZYME.

extracellular titer. The titer of phage particles that is obtained after the infected cells are removed by centrifugation; a measure of the phage particles that have been released by the host cells. *See also* intracellular titer.

extrachromosomal inheritance. CYTOPLASMIC INHERITANCE.

extracistronic suppression. INTERGENIC SUPPRESSION.

extract. 1. CRUDE EXTRACT. 2. A preparation that serves as a source of vitamins and coenzymes for microbiological media and that is prepared from meat, yeast, or other materials by destruction of the cells.

extraction. The removal of a substance from a solid or a liquid mixture by dissolving it in a solvent.

extraction ratio. The fraction of a substance that is reabsorbed by the kidneys from the glomerular filtrate.

extragenic suppression. INTERGENIC SUPPRESSION.

extranuclear inheritance. CYTOPLASMIC INHERITANCE.

extrapolation. The process of extending a graph from a region containing experimental data to one that is devoid of data.

extrapolation number. The extrapolated value for a multitarget survival curve that may or may not correspond to the actual number of sensitive targets per irradiated unit.

extrinsic blood coagulation. EXTRINSIC SYSTEM.

extrinsic Cotton effect. A Cotton effect that is caused by a small molecule which is bound to the protein, and not by the protein itself.

extrinsic factor. VITAMIN B_{12}.

extrinsic fluorescence. Fluorescence that is caused by a small molecule which is bound to the protein, and not by the protein itself. *See also* autofluorescence.

extrinsic heterogeneity. The heterogeneity of antibodies that results from extrinsic factors, such as the large number of antigenic determinants in any particular antigen.

extrinsic system. The series of reactions in blood clotting that are initiated by the release of thromboplastin from the blood platelets of injured tissue.

extrinsic thromboplastin. THROMBO-PLASTIN.

extrude. To remove material by extrusion.

extrusion. The process of expelling or thrusting out, as the expelling of an adsorbent from a chromatographic column.

F

f. 1. Frictional coefficient. 2. Femto.

f₀. The frictional coefficient of a sphere that has the same volume as the macromolecule being studied.

F. 1. Degree Fahrenheit. 2. Fertility factor. 3. Faraday. 4. Farad. 5. Fluorine. 6. Fick unit. 7. Force. 8. Phenylalanine. 9. Folic acid.

F⁺. A male, or donor, bacterial strain.

F⁻. A female, or recipient, bacterial strain.

FA. 1. Fatty acid. 2. Folic acid. 3. Filtrable agent.

Fab fragment. The fragment of the IgG immunoglobulin molecule that is obtained by treating the molecule with the enzyme papain. The fragment consists of an intact light chain and the Fd fragment of one heavy chain, held together by means of a disulfide bond. Two Fab fragments are obtained per IgG molecule, and each fragment contains one antibody combining site.

Fab′ fragment. The fragment of the IgG immunoglobulin molecule that is obtained by treating the molecule with the enzyme pepsin, followed by reduction. The fragment consists of an intact light chain and the Fd′ fragment of one heavy chain, held together by means of a disulfide bond. Two Fab′ fragments are obtained per IgG molecule, and each fragment contains one antibody combining site.

F(ab′)₂ fragment. The fragment of the IgG immunoglobulin molecule that is obtained by treating the molecule with the enzyme pepsin without subsequent reduction. The fragment is a dimer of two Fab′ fragments, held together by means of two disulfide bonds. The F(ab′)₂ fragment contains both of the antibody combining sites of the molecule.

Fabry's disease. A genetically inherited metabolic defect in man that is associated with kidney failure and that is due to a deficiency of the enzyme trihexosyl ceramide α-galactosylhydrolase.

facilitated diffusion. MEDIATED TRANSPORT.

F-actin. The polymerized, fibrous form of actin that consists of a double helix of G-actin monomers.

factor. A component that is not yet completely identified; frequently the term is retained after the factor has been fully identified.

factor I. FIBRINOGEN.

factor II. PROTHROMBIN.

factor III. 1. THROMBOPLASTIN. 2. A protein that functions in conjunction with enzyme II in some systems.

factor IV. The calcium ions that function in blood clotting.

factor V. PROACCELERIN.

factor VI. ACCELERIN.

factor VII. PROCONVERTIN.

factor VIII. ANTIHEMOPHILIC FACTOR.

factor IX. CHRISTMAS FACTOR.

factor X. STUART FACTOR.

factor XI. PLASMA THROMBOPLASTIN ANTECEDENT.

factor XII. HAGEMAN FACTOR.

factor XIII. FIBRINASE.

factor F. INITIATION FACTOR.

factor G. TRANSLOCASE (2).

factorial. A function of a positive integer n that is denoted $n!$ and that is equal to the product of all the integers between 1 and n; thus $5! = 5 \times 4 \times 3 \times 2 \times 1 = 120$.

factor IF. INITIATION FACTOR.

factor R. RELEASE FACTOR.

factor T. ELONGATION FACTOR T.

factor theory. The theory according to which blood clotting proceeds by a cascade mechanism, involving a large number of components.

factor X. 1. VITAMIN B_{12}. 2. BIOTIN.

factor Y. PYRIDOXINE.

factory. A cytoplasmic region in poxvirus-infected animal cells that is actively engaged in the replication of the viral DNA.

facultative. Capable of living under more than one set of conditions.

facultative aerobe. FACULTATIVE ANAEROBE.

facultative anaerobe. An organism or a cell that can grow either in the absence, or in the presence, of molecular oxygen.

facultative water excretion. The urinary excretion of water that is greater than that required for the elimination of waste products.

FAD. Flavin adenine dinucleotide.

FADH₂. Reduced flavin adenine dinucleotide.

F agent. FERTILITY FACTOR.

Fahrenheit temperature scale. A temperature scale on which the freezing and boiling points of water at 1 atm of pressure are set at 32 and 212, respectively, and the interval between these two points is divided into 180 deg.

fall curve. The decrease in color intensity of a

sample solution as a function of time; used in reference to determinations with an autoanalyzer.

falling sphere viscometer. An instrument for measuring the viscosity of a solution from the time required for a sphere to fall through a known column height of the solution.

fallout. The radioactive substances that are produced by nuclear explosions and that fall through the atmosphere onto the earth's surface.

familial goiter. A genetically inherited metabolic defect in man that is characterized by an excessive loss of iodinated tyrosines from the thyroid gland and that is due to a deficiency of the enzyme iodotyrosine dehalogenase.

family antigen. An antigen that is common to a group of viruses that constitute a family.

Fanconi's syndrome. A genetically inherited metabolic defect in man that is characterized by an increased excretion of amino acids.

farad. A unit of electrical capacitance, equal to the capacitance of a capacitor that requires 1 C to raise its potential by 1 V. *Abbr* fd.

faraday. The quantity of electricity that is transported per gram-equivalent weight of an ion; 96,500 C/eq; 23,060 cal/(V) (eq). *Sym* F.

Faraday effect. The exhibition of optical rotatory power by an optically inactive substance that is placed in a magnetic field.

farnoquinone. Vitamin K_2.

Farr test. A radioimmunoassay for determining the antigen binding capacity of an antiserum. The antiserum is treated with radioactively labeled antigen, and the labeled antigen-antibody complex is precipitated with 50% saturated ammonium sulfate; the radioactivity of the precipitate is then determined.

fast hemoglobin. A hemoglobin that, after electrophoresis, is located closer to the anode than is normal, adult hemoglobin.

fast reaction. A reaction, or a step in a reaction sequence, that has a very large rate constant and that proceeds very rapidly; special techniques are required for analysis of such reactions.

fast-sweep polarography. A sensitive polarographic technique in which a hanging mercury drop electrode is used.

fat. 1. NEUTRAL FAT. 2. The oily and greasy material of adipose tissue.

fat body. A structure that contains storage fat and that occurs in insects, reptiles, and amphibians.

fat-soluble A. A fraction of fat-soluble vitamins prepared from egg yolk.

fat-soluble vitamin. One of a diverse group of vitamins—including vitamins A, D, E, and

K—that is soluble in organic solvents and insoluble in water.

fat solvent. A nonpolar organic solvent, such as chloroform, that will extract lipids from tissues.

fat-splitting. The hydrolysis of fats to free fatty acids that is produced by water at elevated temperatures.

fatty acid. A long-chain carboxylic acid that occurs in lipids; may be branched or unbranched, and saturated or unsaturated.

fatty acid activating enzyme. FATTY ACID THIOKINASE.

fatty acid activation. The conversion of a fatty acid to a fatty acyl coenzyme A ester which is the first step in the reactions of beta oxidation. The fatty acyl coenzyme A ester can be formed in a reaction catalyzed by a thiokinase or in a reaction catalyzed by a thiophorase.

fatty acid:CoA ligase (AMP). FATTY ACID THIOKINASE.

fatty acid oxidation. 1. BETA OXIDATION. 2. Any set of reactions whereby a fatty acid is oxidized in metabolism.

fatty acid synthetase complex. A cytoplasmic, multienzyme complex that catalyzes a cyclic set of reactions whereby a fatty acid is synthesized from one molecule of coenzyme A and successive molecules of malonyl coenzyme A; the complex consists of six enzymes and the acyl carrier protein.

fatty acid thiokinase. The enzyme that catalyzes the formation of a fatty acyl coenzyme A ester from the fatty acid and coenzyme A in fatty acid activation.

fatty acyl group. The acyl group of a fatty acid; the grouping $R-\overset{\overset{\displaystyle O}{\|}}{C}-$.

fatty degeneration. The degeneration of a tissue due to the formation of fat globules in the cytoplasm of the affected cells; the fat of these globules is derived from the cells themselves. *See also* fatty infiltration.

fatty infiltration. The degeneration of a tissue due to the formation of fat globules in the cytoplasm of the affected cells; the fat of these globules is derived from outside the cells. *See also* fatty degeneration.

fatty liver. A liver that is characterized by fatty degeneration and/or fatty infiltration and that may develop due to various conditions such as diabetes, chemical poisoning of the liver, or a deficiency of lipotropic agents.

Favism. A severe type of hemolytic anemia that occurs in individuals suffering from a hereditary glucose-6-phosphate dehydrogenase deficiency upon eating broad beans.

FCCP. Carbonylcyanide *p*-trifluoromethoxy phenylhydrazone; an uncoupling agent.

Fc fragment. The fragment of the IgG immunoglobulin molecule that is obtained by treating the molecule with the enzyme papain. The fragment consists of two heavy chain fragments joined by means of two disulfide bonds.

fd. Farad.

Fd. Ferredoxin.

FDA. Food and Drug Administration; an agency of the U.S. Public Health Service.

Fd fragment. The fragment of the IgG immunoglobulin molecule that is obtained by treating the molecule with the enzyme papain, followed by reduction. The fragment consists of that portion of the heavy chain that is joined to an intact light chain in the Fab fragment. Two Fd fragments are obtained per molecule of IgG.

Fd′ fragment. The fragment of the IgG immunoglobulin molecule that is obtained by treating the molecule with the enzyme pepsin, followed by reduction. The fragment consists of that portion of the heavy chain that is joined to an intact light chain in the Fab′ fragment. Two Fd′ fragments are obtained per molecule of IgG.

FDNB. 1-Fluoro-2,4-dinitrobenzene; the Sanger reagent.

FDP. Fructose-1,6-diphosphate.

F-duction. SEXDUCTION.

Fe. Iron.

feedback. 1. A process in which part of the output of a system is returned to the input of the same system in such a fashion that it affects the subsequent output by the system. *See also* negative feedback; positive feedback. 2. That part of the output of a system that is returned to the input in a feedback process.

feedback deletion hypothesis. A modification of the deletion hypothesis of cancer according to which cancer results from the loss of a repression mechanism that restrains DNA synthesis in normal cells. The loss of the repression mechanism may be due to a number of factors such as lack of repressor synthesis or segregation of the repressor in a part of the cell.

feedback inhibition. A negative feedback mechanism in which a distal product of an enzymatic reaction inhibits the activity of an enzyme that functions in the synthesis of this product.

feedback loop. The cyclic system of components that participate in a feedback mechanism.

feedback repression. A negative feedback mechanism in which a metabolite regulates the amount of an enzyme through enzyme repression.

feeder cell. An irradiated cell that is metabolically active but that cannot multiply; such cells are used to provide growth factors for unirradiated cells.

feeder layer. A layer of feeder cells that is used in tissue culture.

feeder pathway. A metabolic pathway that provides metabolites for another, major pathway.

Fehling's test. A test for reducing sugars that is based on the reduction of cupric ions by a reducing sugar when an alkaline solution of the sugar is treated with copper sulfate.

Felderstruktur. The structure of muscle fibers that is characterized by an incomplete separation of the thick myofibrils.

feline. Of, or pertaining to, cats.

female hormone. Any one of the three related compounds: estradiol, estrone, and estriol.

female sex hormone. An estrogen that affects the estrous cycle, the reproductive cycle, and the development of secondary sex characteristics in the female; the female sex hormones are produced primarily by the ovaries and the placenta.

FeMo protein. MOLYBDOIRON PROTEIN.

femto-. Combining form meaning 10^{-15} and used with metric units of measurement. *Sym* f.

F episome. FERTILITY FACTOR.

Fe protein. IRON PROTEIN.

ferment. 1. *n* An early term for enzyme. 2. *v* To process by means of fermentation.

fermentation. 1. The energy-yielding, metabolic breakdown of organic compounds by microorganisms that generally proceeds under anaerobic conditions and with the evolution of gas. 2. The energy-yielding, metabolic breakdown of organic compounds in an organism that proceeds in the absence of molecular oxygen and with the use of organic compounds both as oxidizing agents and as oxidizable substrates. *See also* anaerobic respiration.

fermentor. An apparatus for the growth of microorganisms in liquid media and for the study of microbial metabolism, including fermentation.

ferredoxin. A nonheme iron-containing protein that has a low reduction potential and that serves as an early electron acceptor in both photosynthesis and nitrogen fixation. *Abbr* Fd.

ferredoxin-reducing substance. A flavoprotein electron carrier, possibly a bound form of ferredoxin, that is the immediate acceptor for the electrons from pigment P_{700} in photosystem I of chloroplasts. *Abbr* FRS.

ferric. Designating iron that has a valence of three; Fe^{+++}; Fe(III).

ferrichrome. A cyclic hexapeptide, composed of three residues each of glycine and hydroxyornithine, that is complexed with a ferric atom and occurs in fungi.

ferricytochrome. A cytochrome in which the iron is in the ferric form.

ferriheme. A heme in which the iron is in the ferric form.

ferrihemochrome. A hemochrome in which the iron is in the ferric form.

ferriprotoporphyrin. A protoheme in which the iron is in the ferric form.

ferritin. A conjugated and electron-dense protein that is composed of a protein shell, apoferritin, which surrounds four discrete micelles of ferric hydroxide-phosphate salts. Ferritin functions in the absorption of iron through the intestinal mucosa and serves as a storage protein for iron in the liver, the spleen, and other animal tissues. *See also* ferritin-labeled antibody.

ferritin-labeled antibody. An antibody to which ferritin has been linked covalently; the electron-dense ferritin provides a suitable label for localizing the antibody in electron microscope preparations. Antigens or other proteins may be labeled with ferritin in an analogous manner.

ferrochelatase. HEME SYNTHETASE.

ferrocytochrome. A cytochrome in which the iron is in the ferrous form.

ferroflavoprotein. A complex flavoprotein that contains iron in the form of a heme or in some other form, in addition to containing either FMN or FAD.

ferroheme. A heme in which the iron is in the ferrous form.

ferrohemochrome. A hemochrome in which the iron is in the ferrous form.

ferrous. Designating iron that has a valence of two; Fe^{++}; $Fe(II)$.

ferroprotoporphyrin. A protoheme in which the iron is in the ferrous form; ferroprotoporphyrin IX is the prosthetic group of hemoglobin, myoglobin, catalase, peroxidase, and cytochrome b.

fertility factor. The bacterial episome that enables a bacterium to function as a male in bacterial conjugation; the male bacterium serves as a donor of DNA and produces an F-pilus through which the DNA is transferred to the recipient female cell. *Sym* F.

fertility vitamin. VITAMIN E.

fetal. Of, or pertaining to, the fetus.

fetal hemoglobin. The hemoglobin that occurs during the development of the fetus and that diminishes rapidly after birth; consists of two alpha and two gamma chains. *Sym* HbF.

fetus. The unborn or unhatched young vertebrate during the later stages of its development. *See also* embryo.

Feulgen reaction. A staining reaction that is specific for DNA and that is based on converting the DNA to apurinic acid and treating the apurinic acid with Schiff's reagent for aldehydes.

fever blisters. HERPES SIMPLEX.

f/f$_0$. Frictional ratio; f is the frictional coefficient of the molecule being studied and f_0 is the frictional coefficient for a sphere that has the same volume as the molecule.

FFA. Free fatty acids.

F factor. FERTILITY FACTOR.

F′ factor. An augmented F factor; a fertility factor that has an additional chromosomal segment attached to it.

(F$_1$ + F$_2$) fragment. HEAVY MEROMYOSIN.

F$_1$ fragment. The globular head portion of the myosin molecule.

F$_2$ fragment. The central tail fragment of the myosin molecule.

F$_3$ fragment. LIGHT MEROMYOSIN.

FGAR. Formylglycinamide ribonucleotide; an intermediate in the biosynthesis of purines.

F-genote strain. F′-STRAIN.

FH$_2$. Dihydrofolic acid.

FH$_4$. Tetrahydrofolic acid.

fiber. A threadlike structure, as that of a muscle or a nerve, that consists of bundles of fibrils.

fiber diagram. The x-ray diffraction pattern that a fibrous material yields when it is analyzed by the rotating crystal method.

fiber optics probe. A flexible probe consisting of fine glass or of fine plastic fibers that are optically aligned to transmit an image and/or to transmit light.

fibril. A fine threadlike structure, bundles of which constitute a fiber; the myofibrils of muscle and the neurofibrils of a nerve are two examples.

fibril ghost. A myofibril that has lost its myosin.

Fibrillenstruktur. The structure of muscle fibers that is characterized by a relatively uniform distribution of well separated myofibrils.

fibrillin. A fibrous protein believed to contribute to the structural integrity of myofibrils.

fibrils long spacing. An artificially prepared assembly of collagen molecules that is characterized by having long periodicities of about 2500 Å when examined in the electron microscope; produced from neutral solutions of collagen in the presence of either glycoproteins or excess mucopolysaccharides. *Abbr* FLS. *See also* segment long spacing.

fibrin. The protein that is formed from fibrinogen by the action of thrombin and that is polymerized to form the blood clot; the monomeric, soluble form is denoted fibrin-s, and the polymerized, insoluble form is denoted fibrin-i. *See also* hard clot; soft clot.

fibrinase. The enzyme that catalyzes the cross-linking of fibrin molecules in blood clotting so that a hard clot is formed; the enzyme catalyzes joining of the γ-carboxyl groups of glutamic acid residues to the ε-amino groups of lysine residues.

fibrin monomer. The monomeric fibrin that is

formed from fibrinogen by the action of the enzyme thrombin.

fibrinogen. The serum protein that gives rise to two fibrinopeptides and fibrin when it is acted upon by the enzyme thrombin during blood clotting; the fibrin is subsequently polymerized to form the blood clot.

fibrinokinase. The enzyme that catalyzes the conversion of plasminogen to plasmin.

fibrinolysin. PLASMIN.

fibrinopeptide. One of two peptides, denoted *A* and *B*, that are removed from fibrinogen during its conversion to fibrin by the action of thrombin.

fibrin polymer. The polymerized fibrin molecules that form the basis of the blood clot.

fibrin stabilizing factor. FIBRINASE.

fibroblast. A cell that occurs in connective tissue and that functions in fiber formation.

fibroin. *See* silk fibroin.

fibroma. A benign tumor derived from fibrous connective tissue.

fibrous. Consisting either largely, or entirely, of fibers.

fibrous actin. F-ACTIN.

fibrous lamina. A thick filamentous layer that reinforces the inner membrane of the nucleus of many cells.

fibrous long spacing. FIBRILS LONG SPACING.

fibrous protein. A protein in which the polypeptide chains are either extended or coiled in one dimension. The polypeptide chains in such a protein are held together largely by interchain hydrogen bonds and form sheets or fibers. Fibrous proteins serve principally as structural proteins. *See also* globular protein.

ficin. An endopeptidase of broad specificity.

Fick's first law. The law that relates the flow of material in translational diffusion to the concentration gradient and that may be expressed as $dm/dt = -DA (dc/dx)$, where dm is the mass transferred in time dt through a cross-sectional area A, D is the diffusion coefficient, and dc/dx is the concentration gradient. The negative sign indicates that the transfer of material occurs from a region of higher to one of lower concentration.

Fick's second law. The law that relates the flow of material in translational diffusion to the concentration gradient and to the change of this gradient with time. The law may be expressed as a differential equation, and the equation can be solved for the concentration gradient; this yields a normal distribution which describes the diffusion pattern.

Fick unit. A unit of the diffusion coefficient equal to 10^{-7} cm²/sec. *Sym* F.

ficoll. A synthetic water-soluble copolymer of sucrose and epichlorhydrin that has a weight

average molecular weight of about 400,000 and that is used for the preparation of density gradients.

FID. Flame ionization detector.

fidelity. The degree to which replication, transcription, or translation proceeds without errors.

field effect. ELECTROSTATIC INTERACTIONS.

figure of merit. The liquid scintillation counting that is performed at the highest possible efficiency and with the lowest possible background count.

filament. 1. A very thin, threadlike structure. 2. MYOFILAMENT.

filamentous phage. *See* minute phage.

film badge. A photographic film holder that is worn by an individual and that is used for approximate measurements of the radiation to which the individual has been exposed.

film diffusion. The rapid diffusion of ions in an ion-exchange resin such that the exchange rate is controlled by the speed at which the ions diffuse through the solution surrounding the resin particles.

filter. 1. A porous material used for the collection of either a precipitate or suspended matter. 2. A light-absorbing material that transmits only selected wavelengths of light.

filterable agent. An early designation of a phage that mediates transduction in *Salmonella*.

filterable virus. An early term for virus.

filter fluorometer. A fluorometer in which filters are used to select the desired excitation and emission wavelengths.

filter paper chromatography. *See* paper chromatography.

filter photometer. A photometer in which filters are used for the isolation of bandwidths.

filtrable. Variant spelling of filterable.

filtrate. The liquid that has passed through a filter.

filtration enrichment. A method for the isolation of fungal auxotrophs by growing either wild-type or mutagenized fungal spores on a minimal medium. Since on this medium only normal spores will germinate and develop mycelia, they can be removed by filtration, and the remaining auxotrophic spores can then be supplied with an enriched medium to permit their germination and growth.

filtration fraction. The fraction of plasma that is filtered through the glomeruli of the kidney and that is equal to the glomerular filtration rate divided by the renal plasma flow; frequently taken as being equal to the ratio of inulin clearance to *p*-aminohippuric acid clearance.

fine control. The control of biochemical systems that is achieved by regulating the activity of an enzyme, as in the case of allosteric enzymes.

fine structure. The splitting of a spectral peak into a number of peaks, as that which occurs in nuclear magnetic resonance and in electron paramagnetic resonance.

fine-structure genetic mapping. The determination of the relative positions of the mutable sites on a chromosome in terms of intervals of decreasing size and, ultimately, in terms of nucleotides.

fingerprint. 1. A pattern of spots that is obtained when a partial hydrolysate of either a protein or a peptide is subjected to paper chromatography in one dimension and to paper electrophoresis in the second dimension. The term fingerprint is likewise applied to a similar two-dimensional map of a hydrolysate of either a nucleic acid or a nucleotide, as well as to maps obtained by modified procedures, involving other support media and/or other separation techniques. 2. The infrared absorption spectrum of a molecule.

fingerprint region. The middle region of the infrared spectrum that is of most use for determinations of molecular structure because both group frequency bands and skeletal bands occur in this region.

first law of cancer biochemistry. The principle, enunciated by Warburg in 1930, that cancer cells carry out glycolysis virtually universally, whether under aerobic or anaerobic conditions. The principle has been restated to mean that shifting of the metabolism of a cell to the anaerobic state is the main biochemical difference between a tumor and a normal cell.

first law of photochemistry. The law that light must be absorbed by a molecule before that molecule can undergo a photochemical reaction. *Aka* Draper's law; Grotthus-Draper law.

first law of thermodynamics. The principle of conservation of energy: the total energy of a system plus its surroundings is constant and is independent of any transformations that they may undergo; energy can neither be created nor destroyed by chemical means.

first-order kinetics. The kinetics of a first-order reaction.

first-order reaction. A chemical reaction in which the velocity of the reaction is proportional to the concentration of one reactant.

first-set rejection. The sequence of events that leads to the rejection of an initial transplant by an unprimed individual.

first-step-transfer DNA. A DNA fraction of T5 phage that is injected into the bacterial host during the first few minutes of infection and that controls the breakdown of the host DNA and the injection of the remainder of the phage DNA. *Abbr* FST-DNA.

Fischer formula. 1. A straight chain, planar projection of carbohydrates. *Aka* Fischer projec-

tion. 2. A simplified, planar formulation of porphyrins.

Fischer-MacDonald degradation. A degradative procedure for aldoses whereby the sugar is converted to the next lower aldose; based on the removal of the anomeric carbon by treatment with ethyl mercaptan.

Fiske-SubbaRow method. A colorimetric procedure for the determination of phosphate in biological materials that is based on the production of a blue color by treatment of the sample with ammonium molybdate and 1-amino-2-naphthol-4-sulfonic acid.

fixation. 1. The chemical reactions whereby an atmospheric gas is converted to either an inorganic or an organic compound. *See also* carbon dioxide fixation; nitrogen fixation. 2. The preparation of tissues for cytological or histological study by converting cellular substances to insoluble components with as little alteration of the original biological structures as possible.

fixative. A protein-denaturing solution that is used for the fixation of biological tissues.

fixed virus. A virus that is obtained by passage through an organism or by cultivation in tissue culture. *See also* street virus.

fixer. A chemical reagent that removes the unexposed and unreduced silver halide grains from a photographic film.

flagellar. Of, or pertaining to, flagella.

flagellin. The monomeric protein of flagella.

flagellum (*pl* flagella). A threadlike, cellular extension that functions in the locomotion of bacterial cells and of unicellular eucaryotic organisms; flagella are longer and less numerous than cilia.

flame emission spectrophotometer. A spectrophotometer used for flame photometry.

flame ionization detector. An ionization detector that is used in gas chromatography for the detection of organic compounds and that converts these compounds into ions by means of a flame. *Abbr* FID.

flame photometry. The determination of elements in solution from the emission spectrum that is produced when the solution is sprayed into a flame. The electrons of the atoms are excited by the flame and emit the excitation energy as light when they fall back to lower energy levels.

flash evaporator. An apparatus for the removal of solvent from a solution; the solvent is evaporated from a thin film of solution, formed by the rotation of a round-bottom flask that is submerged in a water bath.

flash photolysis. A technique for studying short-lived primary or subsequent photochemical intermediates; based on increasing the concentration of these intermediates by irradiating the sample with light pulses of short duration and great intensity. The light transmitted by the sample is then separated into its component wavelengths by means of a spectrograph and allowed to expose a photographic film.

flat-bed chromatography. Chromatography, such as paper or thin-layer chromatography, in which a flat layer of chromatographic support is used.

flat-bed electrophoresis. Electrophoresis, such as paper or thin-layer electrophoresis, in which a flat layer of electrophoretic support is used.

flat spectrum counting. A counting method, used in liquid scintillation, that minimizes the effects of quenching.

flavanone. *See* bioflavonoid.

flavin adenine dinucleotide. The flavin nucleotide, riboflavin adenosine diphosphate, which is a coenzyme form of the vitamin riboflavin and which functions in dehydrogenation reactions catalyzed by flavoproteins; abbreviated as FAD in its oxidized form and as $FADH_2$ in its reduced form.

flavin-linked dehydrogenase. A dehydrogenase that contains a flavin nucleotide as a prosthetic group.

flavin mononucleotide. The flavin nucleotide, riboflavin-5′-monophosphate, which is a coenzyme form of the vitamin riboflavin and which functions in dehydrogenation reactions catalyzed by flavoproteins; abbreviated as FMN in its oxidized form and as $FMNH_2$ in its reduced form.

flavin nucleotide. A collective term for flavin mononucleotide and flavin adenine dinucleotide.

flavodoxin. A protein that is similar to ferredoxin and that is apparently capable of replacing ferredoxin in nitrogen fixation; it differs from ferredoxin in that it contains FMN instead of iron.

flavone. *See* bioflavonoid.

flavonoid. One of a group of aromatic, oxygen-containing, heterocyclic compounds that are widely distributed among higher plants. Many of the flavonoids are intensely colored and constitute plant pigments; others have been credited with vitamin P activity.

flavonol. *See* bioflavonoid.

flavoprotein. A conjugated protein in which the nonprotein portion is a flavin nucleotide, such as the dehydrogenases that use either FMN or FAD as a coenzyme. FMN is linked noncovalently to the protein, while FAD may be linked either covalently or noncovalently. Some flavoproteins are more complex and may also contain either metal ions or heme in addition. *Abbr* FP.

flexible active site. An active site that can un-

dergo conformational changes in the course of a reaction. *See also* induced-fit theory.

flexible polymer. A polymer that can assume a number of different conformations of essentially identical energy but that is not a free-draining polymer.

flickering cluster model. A model for the structure of water according to which the water structure results primarily from cooperative-type hydrogen bonding between the water molecules, such that short-lived clusters of extensively hydrogen-bonded molecules form and break repeatedly.

flip-flop. 1. An inside-outside transition as that in which a molecule moves from the inner monolayer to the outer monolayer of a bilayer. 2. A mechanism for an enzyme that shows half-site reactivity. *See also* half of the sites phenomenon.

floating beta lipoprotein. A lipoprotein that occurs in some conditions of hyperlipoproteinemia and that has an unusually low density due to its high content of triglycerides.

flocculation. The precipitation of finely divided particles in the form of fleecy masses.

flocculation reaction. The flocculation that is obtained in a precipitin reaction when soluble antigen-antibody complexes are formed in both the antigen excess zone and the antibody excess zone.

Florisil. Trademark for a magnesium silicate adsorbent that is used in the column chromatography of lipids.

Flory-Huggins lattice theory. A theory used in calculating the chemical potential of relatively concentrated polymer solutions.

flotation. The movement of molecules in solution under the influence of a centrifugal field and toward the center of rotation. Such systems lead to the formation of inverted peaks in the schlieren optical system, and the peaks rise from the bottom of the cell toward the meniscus as centrifugation proceeds.

flotation coefficient. The analogue of the sedimentation coefficient for molecules that undergo flotation. *Sym* s_f. *See also* standard flotation coefficient.

flow birefringence. The birefringence that is caused by the orientation of asymmetric molecules in a solution that is subjected to flow and shear; used for determining rotational diffusion coefficients, and commonly measured by placing the solution between two concentric cylinders, of which the outer one is being rotated, and by passing polarized light through the solution.

flow dichroism. The dichroism that results from the orientation of asymmetric macromolecules when a non-Newtonian liquid, which contains the macromolecules, is subjected to shear.

flow method. *See* rapid flow technique; stopped flow technique.

flow potential. STREAMING POTENTIAL.

flow quenching. A rapid flow technique in which the enzyme and the substrate are mixed in the usual manner, but the mixture then flows into a second mixing chamber rather than into an observation cell. The enzymatic reaction is stopped rapidly in the second chamber by mixing a chemical quenching reagent with the enzyme and the substrate. A reactant or a product of the reaction is then determined by any convenient method.

flowsheet. A schematic outline of the steps in the synthesis of a compound or in the isolation of a subcellular fraction.

FLS. Fibrils long spacing.

fluctuation test. A statistical analysis for proving that selective variants, such as phage- or drug-resistant mutants, arise spontaneously and not as a result of exposure to the phage, drug, or other agent.

fluctuation theory. 1. A theory, proposed by Linderstrom-Lang, according to which proteins fluctuate continuously between a large number of closely related conformational states. *Aka* motility model. 2. A theory of light scattering by solutions that is based on the fluctuations in the concentrations of solute and solvent molecules in small volume elements of the solution.

fluence. *See* photon fluence; energy fluence.

fluid compartment. The total amount of fluid in the body that either is located in a particular type of tissue or has a particular composition. The two major fluid compartments are the intracellular fluid and the extracellular fluid; the extracellular fluid consists of several subcompartments such as interstitial fluid, cerebrospinal fluid, blood plasma, and lymph.

fluidity. The reciprocal of viscosity.

fluid mosaic model. LIPID-GLOBULAR PROTEIN MOSAIC MODEL.

fluor. 1. A liquid or a solid that is used in scintillation counters and that emits a flash of light when it is excited by radioactive or other radiation. 2. FLUOROCHROME.

fluoresce. To exhibit fluorescence.

fluorescein. A fluorescent compound used in the fluorescent antibody technique.

fluorescence. The emission of radiation by an excited molecule that occurs as a result of an electronic transition whereby the molecule returns from the excited state to the ground state. The emitted radiation is usually of a different and longer wavelength than that of the exciting radiation, and the time interval between excitation and emission is of the order of 10^{-9} to 10^{-7} sec. *See also* phosphorescence.

fluorescence depolarization. Fluorescence in which the exciting light is polarized and the emitted light is partially depolarized; the degree of depolarization of the emitted light is then measured.

fluorescence enhancement. A method for studying the binding of fluorescent ligands to antibodies by determining the difference between the fluorescence of the free ligands and that of the antibody-bound ligands.

fluorescence microscope. A microscope in which structures are illuminated with ultraviolet light and are made visible by fluorescence.

fluorescence polarization. Fluorescence in which the exciting light is not polarized and the emitted light is partially polarized; the degree of polarization of the emitted light is then measured.

fluorescence quenching. A method for studying the binding of haptens to antibodies. The aromatic amino acids of the antibody are first excited and caused to fluoresce; subsequently the fluorescence is decreased when the antibody is allowed to combine with the hapten and an energy transfer takes place from the excited antibody to the hapten. The method can likewise be used to study binding reactions with other proteins.

fluorescent antibody. An antibody that is covalently linked to a fluorescent dye, such as fluorescein or rhodamine, and that has retained its immunochemical activity.

fluorescent antibody technique. A technique for locating either antigens or antibodies in a microscopic preparation of cells or tissues by treating the preparation with the corresponding fluorescent antibodies or fluorescent antigens. *See also* direct fluorescent antibody technique; indirect fluorescent antibody technique; anticomplement fluorescent antibody technique.

fluorescent antigen. An antigen that is covalently linked to a fluorescent dye, such as fluorescein or rhodamine, and that has retained its immunochemical activity.

fluorescent screen. A plate coated with a material, such as calcium tungstate or zinc sulfide, which fluoresces upon irradiation.

fluoridation. The addition of fluoride to water supplies in an attempt to decrease dental caries; the final fluoride concentration is usually 1 mg/l.

fluorimeter. Variant spelling of fluorometer.

fluorimetry. Variant spelling of fluorometry.

fluorine. An element that is essential to a wide variety of species in one class of organisms. Symbol F; atomic number 9; atomic weight 18.9984; oxidation state -1; most abundant isotope F^{19}.

fluorochrome. A substance that, when irradiated with light of a certain wavelength, emits light of a longer wavelength; a fluorescent compound, particularly one used to stain biological specimens.

1-fluoro-2,4-dinitrobenzene. *See* Sanger reaction.

fluorography. *See* solid scintillation fluorography.

fluorometer. An instrument for the measurement of fluorescence that contains both a light source for supplying the excitation energy and a light detector for measuring the emission energy; filter fluorometers and spectrofluorometers are the two basic types.

fluorometry. The measurement of fluorescence that may include a determination of one or more of the following: (a) the concentration of a fluorescent compound; (b) the relative efficiencies of various exciting wavelengths to cause fluorescence; (c) the relative intensities of various wavelengths in the emitted fluorescent light; and (d) the probability that an absorbed photon will generate an emitted photon in fluorescence.

fluorophenylalanine. An amino acid analogue of phenylalanine that can be incorporated into protein during protein synthesis.

fluorophore. A potentially fluorescent group of atoms in a molecule.

5-fluorouracil. A pyrimidine analogue that is used in cancer chemotherapy.

flux. 1. The metabolic rate with respect to a particular substrate in a given tissue; equal to AV/K_m, where A is the substrate concentration in the tissue, V is the maximum velocity, and K_m is the Michaelis constant. 2. The rate of flow of either matter or radiation; equal to the number of particles or of photons that pass through a unit area per unit time.

fMet-tRNA. *N*-Formylmethionyl tRNA.

FMN. Flavin mononucleotide.

FMNH$_2$. Reduced flavin mononucleotide.

foam. The colloidal dispersion of a gas in a liquid.

foam fractionation. A chromatographic technique in which the stationary phase is the surface of bubbles and the mobile phase is the liquid flowing between the bubbles. *See also* emulsion fractionation.

folacin. 1. A generic descriptor for all folates and related compounds that exhibit qualitatively the biological activity of tetrahydropteroylglutamic acid. 2. FOLIC ACID.

folate. A generic descriptor for the family of compounds that contain the pteroic acid nucleus.

folate coenzyme. Tetrahydrofolic acid or one of its derivatives.

Folch method. A method for the isolation of lipids from either tissues or fluids by extraction with chloroform/methanol/water mixtures.

folic acid. Pteroylglutamic acid; a widely distributed vitamin of the vitamin B complex. The coenzyme forms of folic acid are derivatives of tetrahydrofolic acid and they function in the metabolism of one-carbon fragments. *Abbr* F.

folic acid coenzyme. FOLATE COENZYME.

folic acid conjugate. One of a group of folic acid derivatives that contain from two to seven glutamyl residues per molecule.

folic acid reductase. The enzyme tetrahydrofolate dehydrogenase.

Folin-Ciocalteau reaction. A colorimetric reaction for tyrosine that is used for the quantitative determination of proteins; based on the production of a blue color upon treatment of the sample with a complex phosphomolybdotungstic acid reagent.

folinic acid. N^5-Formyltetrahydrofolic acid; a reduced and formylated derivative of folic acid that is more stable to air oxidation than is the parent compound.

Folin method. LOWRY METHOD.

Folin reaction. A colorimetric reaction for amino acids that is based on the production of a red color by treatment of an alkaline solution of the sample with 1,2-naphthoquinone-4-sulfonate.

Folin-Wu method. One of a group of analytical procedures for the determination of glucose or other components in blood. In the case of glucose determination, the proteins are precipitated with tungstic acid and the protein-free filtrate is heated with an alkaline copper tartrate solution, followed by treatment with phosphomolybdic acid.

follicle-stimulating hormone. The gonadotropic protein hormone, secreted by the anterior lobe of the pituitary gland, that stimulates the growth of ovarian follicles in the female and spermatogenesis in the male. *Abbr* FSH.

follicle-stimulating hormone releasing hormone. The hypothalamic hormone that controls the secretion of the follicle-stimulating hormone. *Abbr* FSHRH; FRH. *Aka* follicle-stimulating hormone releasing factor (FSHRF; FRF).

food additive. A substance that is added to, and not naturally present in, a food; includes those substances that are added in the preparation of a fortified food exclusive of sugar, salt, and vinegar.

food chain. A sequence of organisms that feed one upon the other in succession and that provide for the transfer of food energy from the simpler to the more complex organisms.

food web. A system of interlocking food chains.

Forbe's disease. GLYCOGEN STORAGE DISEASE TYPE III.

forbidden clone hypothesis. The hypothesis of autoimmunity according to which the normal tolerance of an animal to self antigens is due to the death, in fetal life, of the clones which are responsible for the synthesis of the corresponding autoantibodies. If, as a result of a mutation, such normally forbidden clones reemerge during the adult life of the animal, they would then lead to the synthesis of autoantibodies and produce autoimmunity.

forbidden transition. A transition between energy states of either an atom or a molecule that is forbidden on the basis of the quantum mechanical selection rules; in practice this means that the transition occurs at a negligibly small rate.

forced diffusion. The diffusion that takes place when free diffusion is modified by the application of an external force such as an electrical or a centrifugal force.

force-feeding. The feeding of experimental animals by forcing food into them; frequently performed by insertion of a tube through the nose into the stomach.

forescattering. *See* forward scattering.

formal electrode potential. The electrode potential of a solution that contains equal concentrations of both the oxidized and the reduced forms of the substance of interest, and that contains all other substances at specified concentrations. *Aka* formal potential.

formalin. An aqueous 37% (w/v) solution of formaldehyde that is used as a fixative in microscopy and as a reagent for reacting with the amino groups of proteins and nucleic acids.

formality. The concentration of a solution expressed in terms of the number of formula weights of solute in 1 liter of solution.

formal solution. A solution that contains one formula weight of solute per liter of solution.

formation constant. The equilibrium constant for the formation of a metal ion-ligand complex.

formazan. A sugar derivative formed by the reaction of a carbohydrate phenylosazone with a diazo compound.

formed elements. A collective term for the erythrocytes, leucocytes, and thrombocytes of blood.

formic fermentation. MIXED ACID FERMENTATION.

formimino group. The grouping $HN{=}CH{-}$.

formol titration. The titration of an amino acid, peptide, or protein in the presence of formaldehyde. The formaldehyde reacts with, and lowers the pK of, the amino groups so that the region in the titration curve in which the amino groups are being titrated can be identified.

formula weight. The sum of the atomic weights, expressed in grams, in the formula of a compound; identical to the molecular weight for those substances that exist as true molecules.

formylation. The introduction of a formyl group into an organic compound.

formyl group. The grouping $-CH{=}O$ that is derived from formic acid.

N-formylmethionine. A formylated form of methionine that, when bound to a specific transfer RNA molecule, serves as the initiating aminoacyl transfer RNA in the translation of bacterial systems. *See also* N-acetylserine.

N-formylmethionyl-tRNA. The initiating aminoacyl-transfer RNA molecule in the translation of bacterial systems. *Abbr* fMet-tRNA.

N^5-formyltetrahydrofolic acid. FOLINIC ACID.

Forssman antigen. A heterogenetic antigen located in the red blood cells and in the tissue cells of various organisms.

Forssman hapten. The nonprotein portion of the Forssman antigen that is believed to be an oligohexosyl ceramide.

fortified food. Food to which nutrients have been added over and above those occurring naturally; may refer either to the increase in concentration of a naturally occurring nutrient, or to the addition of a different nutrient.

forward mutation. A mutation that leads to an altered phenotype which will differ from the wild type.

forward scattering. The scattering of radiation in the direction of the beam of radiation and away from the source of the radiation.

Fouchet's test. A test for bilirubin in urine that is based on the production of a green color by treatment of urine with ferric chloride and trichloroacetic acid.

Fould's rules. A set of six general principles that describe the progression of tumors in animals: (1) Different tumors in the same animal progress independently; (2) Progression occurs independently in different characters of the same tumor; (3) Progression occurs in latent tumor cells and in tumor cells in which growth is arrested; (4) Progression is continuous or discontinuous and proceeds gradually or by abrupt changes; (5) Progression follows one of alternate paths of development; (6) Progression does not always reach an endpoint within the lifetime of the host. *See also* tumor progression.

Fourier synthesis. The process of computing the form of a function from the values of its coefficients in a Fourier series (an infinite series of sine and cosine terms); used in deriving the electron density distribution of a molecule from x-ray diffraction data.

FP. Flavoprotein.

F6P. Fructose-6-phosphate.

F-pilus (*pl* F-pili). A hollow tube that is formed by a male bacterium and that serves for the transfer of its DNA to a female bacterium during conjugation.

fraction. 1. A preparation derived from a biological source and composed of one or more components. 2. A discrete portion of material obtained by fractionation.

fractional centrifugation. DIFFERENTIAL CENTRIFUGATION.

fractional distillation. The slow distillation of a liquid that permits the collection of fractions which distill at different temperatures.

fractional electrical transport. An electrochemical technique for determining whether an "active principle," for which an assay is available, is a strong acid, a strong base, an amphoteric compound, a nonelectrolyte, or a combination of these. The technique is based on subjecting dilute solutions of the "active principle" to electrolysis under high potentials and in cells that can be cut into a number of compartments in which the principle can then be assayed.

fractional precipitation. The stepwise separation of substances, particularly macromolecules, from a solution; based on precipitating the substances in the order of their increasing solubilities by changing the ionic strength, the pH, the dielectric constant, etc., of the solution.

fractional sterilization. The sterilization of a material by means of short, intermittent periods of heating that are separated by periods of incubation. Spores are allowed to germinate during the incubation period and are then killed by the subsequent heating period. The method is used for materials that cannot tolerate the long exposure to the high temperatures that are used in ordinary sterilization procedures.

fractional turnover rate. The reciprocal of the turnover time.

fraction-antibody binding. Fab FRAGMENT.

fractionated dose. A dose of radiation that is administered in the form of a series of short exposures.

fractionation. The separation of sample material into discrete portions.

fraction collector. An automatic device for the collection of consecutive portions of a liquid, as those obtained from a chromatographic column or from a density gradient. The collection may be based on time, volume, weight, or number of drops.

fraction-crystalline. Fc FRAGMENT.

fragment length mapping. A technique that aids in determining the base sequence of a nucleic acid molecule that is synthesized in vitro by a nuclease-free polymerase. The incubation of the polymerase with substrate for varying lengths of time results in the synthesis of polynucleotide fragments of different size, all of which contain the same 5′-terminal. The fragments are separated according to their size by means of polyacrylamide gel electrophoresis, and a

specific oligonucleotide is then located in the nucleic acid by determining the minimum fragment length that is required to ensure the presence of the oligonucleotide.

fragmentation reaction. A chemical reaction in which there is a cleavage of one or more carbon-to-carbon bonds, leading to a breakage in the carbon skeleton of the molecule.

frame-shift mutation. A mutation that leads to an alteration in the normal relationship between nucleotide readout from a nucleic acid and amino acid sequence in the corresponding protein. A point mutation that results in a deletion or an insertion of a nucleotide constitutes a frame-shift mutation; in this case, the codons will be unchanged up to the point of the mutation and will specify correct amino acids, but all subsequent codons will be altered and will specify incorrect amino acids.

framework model. A molecular model in which the atoms are represented by their nuclei and by their bonds and in which the bond angles have their actual values.

Franck-Condon principle. The principle that a change in the electronic configuration of a molecule as a result of the emision or the absorption of a photon requires about one hundredth the length of time that is required for a vibration of the molecule; consequently, the atomic nuclei do not alter their positions appreciably during the emission or the absorption of a photon.

Frank-Evans iceberg. *See* iceberg.

fraudulent DNA. A DNA molecule into which purine and/or pyrimidine analogues have been incorporated that do not occur naturally.

fraudulent nucleotide. A nucleotide into which purine and/or pyrimidine analogues have been incorporated that do not occur naturally.

frayed end. The terminal region of a double-stranded nucleic acid molecule in which the bases are not perfectly complementary.

free diffusion. The diffusion across a boundary in a solution or that in a gas phase as opposed to the diffusion across a membrane or some other porous medium.

free-draining polymer. A polymer in solution that is coiled and flexible in such a fashion that the solvent held within the polymer is free to travel at its own velocity rather than at the velocity of the polymer; consequently, a free-draining polymer is not well approximated by a solid particle in solution.

free electrophoresis. MOVING BOUNDARY ELECTROPHORESIS.

free energy. That component of the total energy of a system that can do work under conditions of constant temperature and pressure; known as the Gibbs free energy (G) and expressed by the thermodynamic function $G = H - TS$, where H

is the enthalpy, T is the absolute temperature, and S is the entropy. *See also* Helmholtz free energy.

free energy change. The difference between the free energy of formation of the products and that of the reactants in a chemical reaction; denoted ΔG. The free energy change for a reaction at pH 7.0 is the biochemical free energy change and is denoted $\Delta G'$. *See also* standard free energy change.

free fatty acids. Nonesterified fatty acids. *Abbr* FFA.

free insulin. IMMUNOREACTIVE IN-SULIN.

free radical. An atom or a group of atoms that has an odd number of electrons; an atom or a group of atoms containing an unpaired electron.

free rotation. The sterically unhindered rotation of an atom or a group of atoms about a single bond.

freeze-blowing. A modification of the freeze-clamp technique that is used for the preparation of brain specimens; entails the rapid removal and rapid freezing of brain tissue from conscious animals.

freeze-clamp technique. A technique for determining the concentration and the compartmentation of metabolites under conditions that closely resemble those of the in vivo state; performed by abruptly stopping the metabolic reactions in a tissue specimen by pressing the tissue into a thin, frozen wafer between two aluminum blocks, previously cooled in liquid nitrogen, and then analyzing the tissue for the metabolites.

freeze-cleaving. FREEZE-FRACTURING.

freeze-drying. LYOPHILIZATION.

freeze-etching. A modification of the freeze-fracturing technique in which the specimen is suspended in water prior to freezing, rather than in glycerol or in some other cryptoprotective agent. The cleaved surfaces can then be etched by a brief heating of the specimen to $-100°C$ in a vacuum; this removes some of the ice bound to the surfaces in the specimen.

freeze-fracturing. A technique used to prepare specimens for electron microscopy. The specimen is frozen and then fractured with a knife edge to yield cleavage surfaces; the surfaces are then replicated by metal casting and are examined microscopically after the original cellular material has been digested away.

freeze-stop technique. FREEZE-CLAMP TECHNIQUE.

freeze-substitution. The dehydrating of tissue specimens by freezing them in a cold organic solvent such as propane or isopentane.

freeze-thawing. The disruption of cells by repeated freezing and thawing of a cell suspension.

freezing. 1. A restriction in the rotation of a molecule that enhances its binding to another molecule, as in the freezing of a substrate that binds to an enzyme, or in the freezing of ribosomes that bind to messenger RNA. 2. The transition of a liquid to a solid that is produced by a lowering of the temperature.

freezing-point osmometer. An osmometer, the operation of which is based on the lowering of the freezing point of a solvent by the addition of a solute.

French press. An apparatus used for the disruption of cells and for the preparation of cell-free extracts. The cells, in suspension, are subjected to high pressures by means of a piston and the suspension is then forced through a small orifice. The sudden decrease in pressure leads to an explosion of the cells. Some cell breakage is also caused by the shearing forces.

frequency. The number of times that an event recurs per unit time, such as the number of vibrations per second. For electromagnetic radiations, the frequency is equal to the speed of light divided by the wavelength of the radiation.

frequency distribution. A systematic graphical arrangement of statistical data that shows both the classes into which a variable is divided and the frequencies of these classes.

frequency of recombination. A measure of the distances between loci on a genetic map that is equal to the number of recombinants divided by the total number of progeny.

Freund's adjuvant. An adjuvant that is used as either a complete or an incomplete adjuvant: complete Freund's adjuvant consists of mineral oil, detergent, and dead mycobacteria; incomplete Freund's adjuvant consists of mineral oil and detergent.

FRF. Follicle-stimulating hormone releasing factor; *see* follicle-stimulating hormone releasing hormone.

FRH. Follicle-stimulating hormone releasing hormone.

frictional coefficient. *See* rotational frictional coefficient; translational frictional coefficient.

frictional force. The force that is exerted on a particle in solution as a result of friction; it is equal to the product of the velocity of the particle and its frictional coefficient.

frictional ratio. The ratio of the experimentally determined translational frictional coefficient of a macromolecule to the translational frictional coefficient of a sphere that has the same molecular weight as the macromolecule. The frictional ratio is commonly divided into two factors which measure the contribution of hydration and the contribution of molecular asymmetry. *See also* Oncley equation.

Friend leukemia virus. A mouse leukemia virus that belongs to the group of leukoviruses.

frontal analysis. *See* frontal chromatography.

frontal chromatography. A column chromatographic technique in which the sample is passed continuously into the column and only the component that emerges first from the column can be obtained in pure form.

frontside attack. The mechanism of a chemical displacement reaction that proceeds with retention of configuration.

frozen accident theory. A theory of the evolution of the genetic code according to which the code evolved by chance until the current codon assignments had been developed; once this was achieved, the code was essentially prevented from any further evolution because of the deleterious, or even lethal, consequences that this would have for the organism. *See also* lethal mutation model.

frozen replica method. FREEZE-ETCHING.

FRS. Ferredoxin-reducing substance.

Fru. Fructose.

fructan. A homopolysaccharide of fructose that occurs in plants.

fructose. A six-carbon ketose that, together with glucose, makes up a molecule of sucrose. *Abbr* Fru.

fructose-1,6-diphosphate. A metabolite that is cleaved in glycolysis to two triose phosphates, glyceraldehyde-3-phosphate and dihydroxyacetone phosphate. *Abbr* FDP.

fructose intolerance. A genetically inherited metabolic defect in man that is due to a deficiency of the enzyme fructose-1-phosphate aldolase.

fructose-6-phosphate. A metabolite that is phosphorylated in glycolysis to fructose-1,6-diphosphate.

fructosuria. A genetically inherited metabolic defect in man that is characterized by the presence of excessive amounts of fructose in the urine and that is due to a deficiency of the enzyme fructokinase.

fruit fly. A fly of the genus *Drosophila*, an organism used for genetic research.

fruit sugar. FRUCTOSE.

FSF. Fibrin-stabilizing factor.

FSH. Follicle-stimulating hormone.

FSHRF. Follicle-stimulating hormone releasing factor; *See* follicle-stimulating hormone releasing hormone.

FSHRH. Follicle-stimulating hormone releasing hormone.

FST-DNA. First-step-transfer DNA.

F′-strain. A bacterial strain in which an augmented fertility factor has become incorporated into the bacterial chromosome. *See also* F′ factor.

F test. A statistical test for comparing the variances of two sets of results by means of their F value (the ratio of the two variances).

fucose. A deoxysugar that occurs in some bacterial cell walls.

fucosidosis. A genetically inherited metabolic defect in man that is characterized by cerebral degeneration, muscle spasticity, and an accumulation of *H*-isoantigen; a lipid storage disease that is due to a deficiency of the enzyme alpha fucosidase.

Fuller's earth. Any clay that has adequate decolorizing capacity and that contains aluminum magnesium silicate.

fulvic acid. The complex mixture of acid-soluble and alkali-soluble substances that are extracted from the organic matter of soil. *See also* humic acid; humin.

fumarase. The enzyme that catalyzes the reversible hydration of fumaric acid to malic acid in the citric acid cycle.

fumarate pathway. A catabolic pathway whereby either phenylalanine or tyrosine is converted to fumaric acid, which then feeds into the citric acid cycle.

fumaric acid. The unsaturated, dicarboxylic acid that is formed from succinic acid in the citric acid cycle.

functional death. The death of an organism that results from the inability of a gene, or a group of genes, to carry out their functions.

functional group. A reactive atom or a group of atoms in a molecule that has specific properties; aldehyde, ketone, amino, hydroxyl, carboxyl, and sulfhydryl groups are some examples.

functional group isomer. One of two or more isomers that have the same molecular composition but differ from each other in the type of functional group that they contain.

function of state. *See* state function.

fungal. Of, or pertaining to, fungi.

fungicide. An agent that kills fungi.

fungus (*pl* fungi). A plant protist that is nonphotosynthetic and that is devoid of chlorophyll; fungi generally contain a mycelium and are frequently coenocytic.

furan. A heterocyclic compound, the structure of which resembles the ring structure of the furanoses.

furanose. A monosaccharide having a five-membered ring structure.

furanoside. A glycoside of a furanose.

fused ring. A ring that has two or more atoms in common with another ring.

fusidic acid. A steroid antibiotic that inhibits protein synthesis by blocking the GTP-dependent reactions that are catalyzed by the enzyme translocase.

futile cycle. A set of two metabolic reactions that accomplishes nothing except waste the free energy difference of the two reactions. For example: the reaction *glucose + ATP = glucose-6-phosphate + ADP* together with the reaction *glucose-6-phosphate + H₂O = glucose + P_i* leads only to the net reaction of *ATP = ADP + P_i*.

fuzzy coat. CELL COAT.

F value. *See* F test.

Fv fragment. The N-terminal portion of the Fab fragment of the immunoglobulins; it consists of the variable portions of one heavy and one light chain.

G

g. 1. Gram. 2. Gravity; used to describe centrifugal forces. *See also* relative centrifugal force. 3. g-value.

G. 1. Guanine. 2. Guanosine. 3. Glycine. 4. Gibbs free energy. 5. Glucose. 6. Gauss. 7. G-value.

G$_1$. *See* cell cycle.

G$_2$. *See* cell cycle.

GA. 1. Glyceric acid. 2. Glutamic acid.

GABA. γ-Aminobutyric acid.

G-actin. The monomeric, water-soluble, and globular form of actin.

gal. Gallon.

Gal. Galactose.

galactin. PROLACTIN.

galactocerebroside. A galactose-containing cerebroside.

galactolipid. A galactose-containing lipid.

galactosamine. The amino sugar of galactose that occurs in glycolipids and in chondroitin sulfate.

galactose. A six-carbon aldose. *Abbr* Gal.

galactosemia. A genetically inherited metabolic defect in humans that is characterized by the inability of an infant to metabolize the galactose that is derived from the lactose in milk; due to a deficiency of the enzyme galactose-1-phosphate uridyl transferase.

galactose tolerance test. A liver function test that measures the ability of the liver to remove galactose from the blood stream and to convert it to glycogen; performed in a similar manner as a glucose tolerance test.

β-galactosidase. The inducible enzyme that catalyzes the hydrolysis of β-galactosides.

Galactostat. Tradename for a galactose oxidase reagent.

galactosuria. The presence of excessive amounts of galactose in the urine.

galactozymase. The enzyme system responsible for the inducible fermentation of galactose in *Escherichia coli*; consists of galactokinase, galactose-1-phosphate uridyl transferase, and UDP-glucose epimerase.

GalNAc. *N*-Acetylgalactosamine.

gamete. A mature, haploid germ cell.

gametocyte. A cell that is destined to develop into a gamete; an oocyte or a spermatocyte.

gametogenesis. The formation of gametes; oogenesis or spermatogenesis.

gamma. 1. *n* A microgram. 2. *adj* Denoting the third carbon atom from the carbon atom that carries the principal functional group of the molecule. *Sym* γ.

gamma aminobutyrate bypass. *See* γ-aminobutyrate bypass.

gamma aminobutyric acid. *See* γ-aminobutyric acid.

gamma chain. 1. The heavy chain of the IgG immunoglobulins. 2. One of the two types of polypeptide chains present in fetal hemoglobin.

gamma globulin. 1. A protein that belongs to a specific fraction of serum proteins. 2. IMMUNOGLOBULIN.

gamma ray. An electromagnetic radiation of short wavelengths that is frequently emitted by radioactive isotopes and that consists of photons which originate in the nucleus of an atom.

gamma ray spectrometer. An instrument for measuring the distribution of the energies of gamma rays.

ganglioside. 1. A ceramide oligoglycoside that contains simple sugars and acylated neuraminic acids, and that may also contain amino sugars. 2. Any cerebroside containing neuraminic acid.

gangliosidosis. *See* generalized gangliosidosis; Tay-Sachs disease.

gap filament. A very thin filament seen in electron micrographs in the gap between the A-band and the I-band of striated muscle; believed to be composed of accidental sarcomere constituents that appear as a result of muscle stretching.

GAR. Glycinamide ribonucleotide; an intermediate in the biosynthesis of purines.

gargoylism. A genetically inherited metabolic defect in man that is characterized by skeletal deformities and mental deficiency and that is due to a defect in mucopolysaccharide storage, resulting in the presence of excessive amounts of chondroitin sulfate in the urine. *See also* Hunter's syndrome; Hurler's syndrome.

gas amplification. 1. The process whereby the ions formed in an ionization chamber produce more ions from the gas molecules in the chamber and thereby increase the electrical current that is measured. The magnitude of the current produced depends on the applied potential; as the potential is increased, the current rises above a saturation current, passes through a proportional region, a limited proportional region, a Geiger-Mueller region, and ultimately reaches a level at which there is a continuous discharge in the chamber. 2. The ratio of the actual charge collected at the electrode of an ionization chamber to the charge produced in the chamber by the initial radiation.

gas chromatogram. The tracing of a gas chromatographic separation; a plot of detector response versus either time or volume of carrier gas. Each component, or mixture of components, is represented by a peak in a differential chromatogram and by a step in an integral chromatogram.

gas chromatography. 1. Column partition chromatography in which the stationary phase is a solid and the mobile phase is a gas. *Abbr* GSC. 2. Column partition chromatography in which the stationary phase is an inert carrier, coated with an essentially nonvolatile liquid, and the mobile phase is an inert gas which sweeps volatile compounds through the column. *Abbr* GLC.

gas chromatography-mass spectrometry. The combination of gas chromatographic and mass spectrometric techniques in which components are separated by a gas chromatography column and are then fed into a mass spectrometer. *See also* mass fragmentography.

gas constant. The physical constant that is derived from the ideal gas law, $PV = nRT$, where R is the gas constant [1.987 cal/(deg) (mole)], P is the pressure, V is the volume, T is the absolute temperature, and n is the number of moles.

gaseous exposure method. WILZBACH METHOD.

gaseous group. A group of compounds that have very low melting points and that are believed to have occurred in the original gas-dust of the solar nebula. *See also* earthy group; icy group.

gas flow counter. A radiation counter in which the ionization detector must be flushed continually with the counting gas, used for gas amplification, to eliminate air leakage into the detector.

gas hold-up. HOLD-UP VOLUME.

gas ionization. The formation of ion pairs from gases that are subjected to an ionizing radiation.

gas liquid chromatography. *See* gas chromatography (2).

gasometry. The measurement of gas volumes and/or gas pressures.

gas solid chromatography. *See* gas chromatography (1).

gas storage. The storage of fruit at decreased oxygen levels and at increased carbon dioxide levels compared to those in air; used in an attempt to prolong the storage life of the fruit.

gastric. Of, or pertaining to, the stomach.

gastric juice. The digestive juice that consists of the secretion from the stomach and that contains hydrochloric acid and pepsinogen. *Aka* gastric fluid.

gastricsin. A proteolytic enzyme present in gastric juice.

gastrin. A peptide hormone of 17 amino acids that is produced by the gastric mucosa and that stimulates the secretion of gastric juice.

gastrointestinal. Of, or pertaining to, the stomach and the intestine.

gate. 1. A structure in a biological membrane through which passive transport takes place. 2. A cut-off level for pulses in scintillation counting. *See also* differential counting; integral counting.

Gaucher's disease. A genetically inherited metabolic defect in man that is characterized by an accumulation of cerebrosides in tissues and by an enlargement of the spleen and the liver; due to a deficiency of the enzyme glucocerebrosidase.

gauss. A unit of magnetic field strength.

Gauss error function. NORMAL DISTRIBUTION.

Gaussian chain. The chain of a polymer that consists of a number of stiff segments which are linked by joints, the positions of which are random. The wormlike coil model can be considered to be a limiting case of a Gaussian chain.

Gaussian distribution. NORMAL DISTRIBUTION.

Gaussian error curve. NORMAL DISTRIBUTION.

GC. Gas chromatography.

GC-MS. Gas chromatography-mass spectrometry.

GC-rich nucleus. A region in a double helical nucleic acid structure that is rich in guanine and cytosine; GC-rich nuclei are denatured more slowly than regions that are rich in either adenine and thymine, or in adenine and uracil.

GDH. 1. Glucose dehydrogenase. 2. Glycerophosphate dehydrogenase. 3. Glutamate dehydrogenase.

GDP. 1. Guanosine diphoshate. 2. Guanosine-5′-diphosphate.

GDPM. Guanosine diphosphate mannose.

GEF. Gel electrofocusing.

gegenion. COUNTERION.

Geiger-Mueller counter. A radiation counter, of the ionization chamber type, that is designed for operation in the Geiger-Mueller region. *Aka* Geiger counter.

Geiger-Mueller plateau. The counting plateau obtained with a Geiger-Mueller counter.

Geiger-Mueller region. That portion of the characteristic curve of an ionization chamber in which, during gas amplification, maximum gas amplification is obtained, and the collected charge is independent of the size of the initial charge produced by the radiation.

gel. A solid colloidal dispersion consisting of a network of particles and a solvent that is immobilized in this network.

Gelarose. Trademark for a group of agarose gels that are used in gel filtration chromatography.

gelatin. *See* parent gelatin.

gelation. Gel formation; the transition from a sol to a gel.

gel chromatography. GEL FILTRATION CHROMATOGRAPHY.

gel electrofocusing. Isoelectric focusing in which a gel is used as the supporting medium for the pH gradient; gel electrofocusing is faster and requires smaller amounts of sample and reagents than isoelectric focusing in a density gradient. *Abbr* GEF.

gel electrophoresis. Zone electrophoresis in which a gel is used as the supporting medium.

gel exclusion chromatography. GEL FILTRATION CHROMATOGRAPHY.

gel filtration chromatography. 1. A column chromatographic technique in which the stationary phase consists of gel particles of controlled size and porosity, as those prepared from polymeric carbohydrates. Molecules are fractionated on such a column on the basis of their size and their rates of diffusion into the gel particles; smaller molecules diffuse more rapidly into the gel and move more slowly through the column than larger ones. 2. Molecular sieving in which aqueous systems are used. *Abbr* GFC. *Aka* gel filtration.

gel immunofiltration. *See* immuno-gel filtration.

gel osmometer. An osmometer, the operation of which is based on changes in the dimensions of a gel.

gel permeation chromatography. 1. GEL FILTRATION CHROMATOGRAPHY. 2. Molecular sieving in which nonaqueous systems are used. *Abbr* GPC.

gem-. Combining form meaning geminal.

geminal. Referring to two substituents on the same carbon atom. *Abbr* gem.

gene. The unit of heredity that occupies a specific locus on the chromosome. The gene may be a functional unit (cistron; structural gene), a mutational unit (muton), or a recombinational unit (recon).

gene amplification. The selective replication of specific genes; frequently occurs during oogenesis for the genes which code for ribosomal RNA.

gene conversion. The asymmetrical segregation of genes during replication that leads to an apparent conversion of one gene into another.

gene dosage. The number of times that a particular gene occurs in the nucleus of a cell.

gene duplication. The process whereby one or more genes from one chromosome are integrated by crossing-over into a second chromosome that already carries the same genes. This results in a duplication, or in higher states of multiplication, of the material present in the chromosome. Gene duplication may be partial or virtually complete, and is considered to be responsible for the similarities in amino acid sequences among groups of proteins, such as the immunoglobulins, the hemoglobins, and the haptoglobins.

gene frequency. A measure of the proportion of an allele in a population; equal to the number of loci at which a particular allele is found divided by the total number of loci at which it could occur.

gene hypothesis. The hypothesis that, in the development of life, nucleic acids were formed first and proteins arose later; based in part on the attributes of life shown by nucleic acids in that they are able to replicate, code for protein, and undergo mutation.

gene linkage. *See* linkage.

gene pool. The sum total of all the genes in a population of sexually reproducing organisms.

general acid-base catalysis. The catalysis in solution in which the catalysts are various acidic and/or basic species that serve as proton donors and/or proton acceptors. *See also* specific acid-base catalysis.

general adaptation syndrome. The sequence of reactions that are initiated by an increased secretion from the adrenal cortex in response to a stress and that allow an animal to adapt to the stress.

generalized anaphylaxis. Systemic anaphylaxis.

generalized gangliosidosis. A genetically inherited metabolic defect in man that is associated with mental retardation and that is characterized by an accumulation of gangliosides and by an enlargement of the liver; due to a deficiency of a beta galactosidase.

generalized transduction. Bacterial transduction in which any of the genes of the donor bacterium may become transduced.

generally labeled. Designating a compound that is randomly labeled at various positions in the molecule; usually refers to tritiated compounds. *Sym* G.

generation cycle. CELL CYCLE.

generation time. The time required by a cell for the completion of one cycle of growth. *Sym* T_g. *See also* doubling time.

gene redundancy. The presence of two or more copies of the same gene in a chromosome.

generic. Of, or pertaining to, a genus.

genesis. The origin or the coming into existence of anything; the formation or the production of anything.

gene substitution. The replacement of one allele by another allele of the same gene.

gene therapy. The use of genetic engineering in the combat of disease.

genetic. Of, or pertaining to, genetics.

genetic block. A metabolic block that results from a mutation.

genetic code. The specification of a sequence of amino acids in a protein by a sequence of nucleotides in a nucleic acid; the set of codons that specify the amino acids and carry the information for protein synthesis.

genetic code dictionary. The set of 64 codons, 61 of which code for amino acids commonly occurring in proteins, and 3 of which are termination codons.

genetic complementation. *See* complementation.

genetic cross. 1. The mating of two organisms that results in the formation of genetic recombinants. 2. The production of progeny containing genotype of two or more parents, as in the simultaneous infection of bacteria with several types of phages. 3. The progeny derived from two or more parents by mating or by other means.

genetic disease. A hereditary disease that arises from an abnormality in the genetic makeup of an organism, as that caused by the presence of a mutant gene. *See also* inborn error of metabolism.

genetic drift. The random change of gene frequency due to chance fluctuations, particularly that occurring in small populations.

genetic engineering. The application of molecular genetics to the purposeful alteration of the hereditary makeup, particularly that of man.

genetic equilibrium. The condition in which a given gene frequency remains constant throughout successive generations.

genetic expression. The phenotypic aspect of a gene; the active function of a gene.

genetic fine structure. The location of mutable sites on a chromosome as determined by fine-structure genetic mapping.

genetic information. The hereditary information contained in a sequence of nucleotides in either chromosomal DNA or chromosomal RNA.

genetic linkage. *See* linkage.

genetic locus. *See* locus.

genetic map. The linear arrangement of mutable sites on a chromosome that is deduced from genetic recombination experiments.

genetic marker. A mutable site on a chromosome that, when mutated, leads to gross and visible changes in the organism. *See also* biochemical marker.

genetic material. The chromosomal nucleic acid, predominantly DNA but at times RNA, that carries the information for the synthesis of proteins and for the synthesis of other nucleic acids.

genetic recombination. *See* recombination.

genetic reversion. *See* reversion (1).

genetics. The science that deals with heredity.

genetotrophic principle. The principle that the nutritional requirements of an organism are determined by its genetic makeup.

genic. Of, or pertaining to, genes.

genome. A complete single set of the genetic material of a cell or of an organism; the complete set of genes in a gamete.

genopathy. A pathological condition that results from a genetic defect.

genophore. The nucleic acid, devoid of histones, that is the physical counterpart in procaryotic cells and viruses to the chromosomes in eucaryotic cells.

genotype. 1. The genetic makeup of an organism; the totality of the genes of an organism. 2. A group of organisms that have an identical genetic makeup. *See also* phenotype.

genotypic. Of, or pertaining to, genotype.

genotypic adaptation. The preferential growth of a genotypically varied organism.

genotypic variation. A rare mutation that involves only a few organisms in a population and that leads to a new genotype.

gentobiose. A disaccharide of D-glucose in which the glucose molecules are linked by means of a β $(1 \rightarrow 6)$ glycosidic bond.

genus (*pl* genera). A taxonomic group that includes one or more closely related species.

geobiochemistry. BIOGEOCHEMISTRY.

geometrical isomer. One of two isomers that differ from each other in the configuration of the groups attached to two carbon atoms that are linked by a double bond; the groups may be on the same side (cis; syn) with respect to the plane of the double bond, or they may be on opposite sides (trans; anti).

geotaxis. A taxis in which the stimulus is gravity.

geotropism. A tropism in which the stimulus is gravity.

geriatrics. The branch of medicine that deals with the problems and the pathological aspects of old age and aging people. *See also* gerontology.

germ cell. A reproductive cell; a cell that can be fertilized when it is mature and that can reproduce the organism; an ovum or a spermatozoon, or any of their antecedents. *See also* somatic cell.

germ-free animal. An animal reared in a bacteria-free environment.

germicidal agent. An agent that kills microorganisms.

germicide. GERMICIDAL AGENT.

germination. 1. The overall process—consisting

of activation, initiation, and outgrowth—whereby a spore is converted to a vegetative cell. 2. The second stage in the conversion of a spore to a vegetative cell that is characterized by the rehydration of the spore and by the loss of dipicolinic acid and glycopeptides. *Aka* initiation. 3. The beginning of growth of a seed or of a reproductive body after a period of dormancy.

germ line. The gametes and their antecedents.

germ-line theory. A theory of the origin of the genes that code for the variable regions of antibody molecules and that allow for the great diversity of antibodies. According to this theory all cells, including lymphocytes, have the same set of genes as those in the germ cells from which the individual arose. All the genes for all the antibodies that an individual can make are, therefore, already present in the fertilized egg and are transmitted through the germ line.

germ plasm. The genetic material in the germ cells; the sum total of the genes transmitted to the offspring through the germ cells.

gerontology. The science that deals with the problems and the pathological aspects of old age and aging people. *See also* geriatrics.

gestagen. PROGESTIN.

gestation. Pregnancy.

GeV. Giga (10^9) electron volts.

G-factor. GUANOSINE TRIPHOSPHATASE.

GFC. Gel filtration chromatography.

GFR. Glomerular filtration rate.

GGE. Gradient gel electrophoresis.

GH. Growth hormone.

GHIF. *See* growth hormone regulatory hormone.

GHIH. *See* growth hormone regulatory hormone.

ghost. 1. An erythrocyte that has lost some or all of its cytoplasmic content; prepared by allowing the erythrocyte to swell in distilled water so that its permeability is increased and the cytoplasmic material leaks out. 2. A T-even phage that has lost some or all of its DNA; prepared by subjecting the phage to an osmotic shock and digesting the DNA of the ruptured phage with deoxyribonuclease. 3. A spheroplast that has lost essentially all of its intracellular material; prepared by growing gram-negative bacteria in the presence of penicillin or by digesting the cell wall of gram-negative bacteria with lysozyme.

ghost peak. An unexpected peak that is present in a gas chromatogram and that is usually due to a contaminant.

GHRF. *See* growth hormone regulatory hormone.

GHRH. *See* growth hormone regulatory hormone.

GHRIF. *See* growth hormone regulatory hormone.

GHRIH. *See* growth hormone regulatory hormone.

giant chromosome. POLYTENE CHROMOSOME.

gibberellic acid. The parent compound of the gibberellins.

gibberellin. One of a group of plant hormones that consist of five fused rings and that stimulate the growth of leaves and shoots.

Gibbs-Donnan equilibrium. The unequal distribution of diffusible ions that is established at equilibrium on the two sides of a membrane if one side contains a nondiffusible ion. The unequal distribution of the diffusible ions becomes more pronounced with an increase in the concentration and/or the charge of the nondiffusible ion; it becomes less pronounced with increasing concentrations of the diffusible ions added initially to that side of the membrane that does not contain the nondiffusible ion. *Aka* Gibbs-Donnan effect; Gibbs-Donnan membrane equilibrium.

Gibbs-Duhem equation. An equation that relates the chemical potential of different components in a system. At constant temperature and pressure the equation can be written as $dG = \sum_i \mu_i dn_i + \sum_i n_i d\mu_i$, where G is the free energy, and μ_i and n_i are the chemical potential and the number of moles of component i. Since $dG = \sum_i \mu_i dn_i$, the equation simplifies to $\sum_i n_i d\mu_i = 0$

Gibbs free energy. *See* free energy.

Gibbs phase rule. PHASE RULE.

GIF. *See* growth hormone regulatory hormone.

giga-. Combining form meaning one billion (10^9) and used with metric units of measurement. *Sym* G.

gigantism. ACROMEGALY.

GIH. *See* growth hormone regulatory hormone.

Gilbert's disease. CRIGLER-NAJJAR SYNDROME.

Girard's reagent. One of a group of reagents used to extract ketosteroids from urine; reagent T is trimethylammonium acetyl hydrazide chloride, and reagent P is pyridinium acetyl hydrazide chloride.

GK. Glycerokinase.

gland. A cell or an organ that produces specific substances which are secreted outside the cell or the organ, either directly or through ducts, and which are used in other parts of the body or are eliminated from the body.

glandular. Of, or pertaining to, glands.

glass electrode. An electrode that has a thin glass membrane incorporated into its design and that is sensitive to the hydrogen ion concentration of solutions; it is used for the measurement of pH.

Glc. Glucose.

GLC. Gas liquid chromatography.

GlcA. Gluconic acid.

GlcN. Glucosamine.

GlcNAc. *N*-Acetylglucosamine.

GlcUA. Glucuronic acid.

GLDH. Glutamate dehydrogenase.

gliadin. A seed protein of wheat.

Gln. 1. Glutamine. 2. Glutaminyl.

globin. The polypeptide chain that is associated with an iron-porphyrin group in both hemoglobin and myoglobin.

globinometer. An instrument for measuring the amount of oxyhemoglobin in blood.

globin zinc insulin. The zinc salt of a mixture of globin and insulin that is less soluble than insulin alone. *See also* NPH insulin.

globoid leukodystrophy. KRABBE'S DISEASE.

globoside. A ceramide oligoglycoside that contains simple sugars, amino sugars, and *N*-acetyl amino sugars.

globular. Spherical.

globular actin. G-ACTIN.

globular protein. A protein in which the polypeptide chain (or chains) is coiled in three dimensions to form a more or less globular molecule. The polypeptide chain is held together through intrachain bonds such as hydrogen bonds, hydrophobic bonds, electrostatic bonds, and disulfide bonds, and varying lengths of the polypeptide chain may be in helical configuration. Globular proteins have diverse functions and occur in the form of enzymes, storage proteins, transport proteins, and so on. *See also* fibrous protein.

globulin. A globular and simple protein that is either insoluble or sparingly soluble in water, is soluble in dilute salt solutions, and is precipitated by ammonium sulfate at 50% saturation.

glomerular filtrate. The filtrate, free of cells and colloidal particles, that is produced from blood by the glomeruli of the kidney.

glomerular filtration rate. The rate at which the glomerular filtrate is produced; a measure of the efficiency of the kidney that is generally expressed in terms of the clearance of a substance, such as inulin, which is metabolically inert, freely filterable, and neither absorbed nor secreted by the kidney. *Abbr* GFR.

glomerulus (*pl* glomeruli). *See* nephron.

glove box. A sealed box of glass or plastic that has two or more gloves fitted into its sides; allows for the safe manipulation of the contents of the box without breaking the atmospheric seal. *Aka* glove bag.

Glu. 1. Glutamic acid. 2. Glutamyl.

GluA. Glucuronic acid.

glucagon. A polypeptide hormone that antagonizes the action of insulin and leads to an increase in the level of blood sugar by stimulating glycogenolysis; it is secreted by the Islets of Langerhans in the pancreas.

glucalogue. A monosaccharide analogue of glucose, such as deoxyglucose, methylglucose, or galactose.

glucan. A polysaccharide composed of glucose units.

glucocerebroside. A glucose-containing cerebroside.

glucocorticoid. A 21-carbon steroid that is secreted by the adrenal cortex and that acts primarily on carbohydrate, lipid, and protein metabolism. Glucocorticoids include corticosterone, cortisone, and cortisol, and lead to protein catabolism, gluconeogenesis from the amino acids thus formed, lipid mobilization, and an increase in ketone bodies. In addition, glucocorticoids also have antiallergic and antiinflammatory effects.

glucocorticosteroid. GLUCOCORTICOID.

glucogenesis. The synthesis of glucose from precursors other than glycogen.

glucogenic. Of, or pertaining to, glucogenesis.

glucokinase. An enzyme that catalyzes the phosphorylation of glucose to glucose-6-phosphate.

glucolipid. A glucose-containing lipid.

gluconeogenesis. The synthesis of glucose from nonglucose precursors such as amino acids, intermediates of glycolysis, or intermediates of the citric acid cycle.

gluconeogenic. Of, or pertaining to, gluconeogenesis.

glucosamine. An amino sugar of glucose that is a component of chitin and occurs in vertebrate tissues. *Abbr* GlcN.

glucose. The six-carbon aldose that is the major sugar in the blood and a key intermediate in metabolism. *Abbr* Glc; G.

glucose effect. The catabolite repression produced by glucose or by its catabolites.

glucose-6-phosphate dehydrogenase. The pyridine-linked dehydrogenase that catalyzes the first reaction of the hexose monophosphate shunt whereby glucose-6-phosphate is oxidized to 6-phosphoglucono-δ-lactone. *Abbr* G6PDH; G6PD.

glucose-6-phosphate dehydrogenase deficiency. A genetically inherited metabolic defect in man that is due to a deficiency of the enzyme glucose-6-phosphate dehydrogenase. Afflicted individuals show primaquine sensitivity, develop Favism upon eating broad beans, and appear to be protected from malaria transmitted by *Plasmodium falciparum*.

glucose tolerance test. A measure of the rate at which glucose is metabolized; used as a screening test for diabetes. A dose of glucose is administered to a fasting individual and the blood glucose concentration is then determined as a function of time. In normal individuals, the glucose concentration rises to a maximum within about 30 minutes and drops back to the initial level within about 2 hours. In diabetic individuals, the glucose concentration rises to a higher value and does not drop back as rapidly as is the case for normal individuals.

glucoside. A glycoside of glucose.

Glucostat. Tradename for a glucose oxidase reagent.

glucosuria. The presence of excessive amounts of glucose in the urine.

glucosylation. The introduction of a glucose residue into an organic compound.

glucosyl group. A glucose residue that is linked to another group or molecule by means of a glycosidic bond.

glucuronate pathway. A metabolic pathway for the conversion of glucose to xylulose-5-phosphate that is operative in higher plants, mammals, crustaceans, and yeast; it is apparently not a major pathway for the oxidation of glucose. The pathway serves to provide vitamin C in plants and in those animals capable of synthesizing this vitamin. *Aka* glucuronate-gulonate pathway; glucuronate-xylulose cycle.

glucuronic acid. A sugar acid of glucose. *Abbr* GluA; GlcUA.

glucuronide. A compound formed by linking glucuronic acid to another compound by means of a glycosidic bond; many toxic compounds are detoxified by being converted to a glucuronide and are then excreted in this form.

glumitocin. A peptide hormone, secreted by the posterior lobe of the pituitary gland, that is related in structure and function to oxytocin and occurs in some fish.

GluNH$_2$. 1. Glutamine. 2. Glutaminyl.

glutamic acid. An aliphatic, acidic, and polar alpha amino acid. *Abbr* Glu; E; GA.

glutamic semialdehyde. A derivative of glutamic acid in which only one of the two carboxyl groups has been converted to an aldehyde group.

glutamine. An aliphatic, polar alpha amino acid; the amide of glutamic acid. *Abbr* Gln; GluNH$_2$; Q.

glutathione. A widely distributed tripeptide, γ-glutamyl-cysteinyl-glycine, that serves as a coenzyme for some enzymes and is thought to function as an antioxidant in protecting the sulfhydryl groups of enzymes and other proteins. Reduced glutathione is synonymous with glutathione and is abbreviated GSH; oxidized glutathione is a dimer of two glutathione molecules, linked by means of a disulfide bond, and is abbreviated GSSG.

glutelin. A simple, globular protein of plant origin that is insoluble in water, alcohol, or salt solutions, but is soluble in dilute solutions of acids or bases.

gluten. The principal protein of wheat; the properties of gluten permit the formation of leavened bread.

Glx. The sum of glutamic acid and glutamine when the amide content is either unknown or unspecified.

Gly. 1. Glycine. 2. Glycyl.

glycan. POLYSACCHARIDE.

glycemia. The presence of glucose in the blood.

glyceraldehyde. A three-carbon aldose, a phosphorylated derivative of which is an intermediate in glycolysis; serves as a reference compound for the assignment of D and L configurations to amino acids, carbohydrates, and related compounds.

glycerate pathway. An anaplerotic reaction sequence whereby glyoxylate is converted to 3-phosphoglyceric acid.

glyceride. An ester of glycerol and one to three molecules of a fatty acid; a neutral fat. Depending on the number of fatty acid molecules esterified, the product is called mono-, di-, or triglyceride.

glycerin. GLYCEROL.

glycero-. Combining form denoting a compound having one asymmetric carbon atom.

glycerol. A three-carbon trihydroxy alcohol that occurs in many lipids; phosphorylated derivatives of glycerol are intermediates in glycolysis.

glycerol fermentation. The formation of small amounts of glycerol during alcoholic fermentation.

glycerolipid. A lipid derived from glycerol.

glycerol phosphate shuttle. A shuttle involving glycerol-3-phosphate, dihydroxyacetone phosphate, and glycerol phosphate dehydrogenase.

glycerol phosphatide. PHOSPHOGLYCERIDE.

glycerophosphoric acid. PHOSPHOGLYCERIC ACID.

glycine. The simplest alpha amino acid; it can be classified as an aliphatic and polar amino acid. *Abbr* Gly; G. *Aka* aminoacetic acid; glycocoll.

glycitol. An acyclic polyhydroxy alcohol.

glycocalyx. CELL COAT.

glycocholic acid. The compound formed by the conjugation of cholic acid and glycine; the salt of glycocholic acid is one of the bile salts.

glycocyamine. GUANIDOACETIC ACID.

glycogen. A highly branched homopolysaccharide of D-glucose units that is the major form of storage carbohydrate in animals.

glycogenesis. The synthesis of glycogen; the anabolism of glycogen.

glycogen granule. A cytoplasmic storage particle of glycogen that also contains proteins and enzymes that function in the synthesis and breakdown of glycogen.

glycogenic. Of, or pertaining to, glycogenesis.

glycogenic amino acid. An amino acid that can serve as a precursor of pyruvic acid, glucose, and glycogen in metabolism.

glycogenolysis. The breakdown of glycogen; the catabolism of glycogen.

glycogenosis (*pl* glycogenoses). GLYCOGEN STORAGE DISEASE.

glycogen phosphorylase. The enzyme that catalyzes the successive hydrolytic removal of glucose residues, in the form of glucose-1-phosphate, from the nonreducing end of glycogen; this reaction is the first step for the utilization of glycogen in glycolysis.

glycogen storage disease. One of a group of genetically inherited metabolic defects in man that are characterized by an abnormal accumulation of liver and tissue glycogen and that are due to deficiencies of enzymes which function in glycogen metabolism. The enzymatic deficiencies and the synonymous names of the eight known types of glycogen storage diseases are as follows: type I—glucose-6-phosphatase; von Gierke's disease; type II—α-1,4-glucosidase; Pompe's disease; type III—amylo-1,6-glucosidase; Cori's disease; type IV—amylo (1,4 $\rightarrow$ 1,6) transglucosylase; Andersen's disease; type V—muscle glycogen phosphorylase; McArdle's disease; type VI—liver glycogen phosphorylase; Hers' disease; type VII—muscle phosphofructokinase; type VIII—liver phosphorylase kinase.

glycogen synthetase. The enzyme that catalyzes the synthesis of the straight chains of glycogen from UDP-glucose. Two forms of glycogen synthetase exist, a dependent form, denoted *D*, and an independent form, denoted *I*. *See also* dependent form; independent form.

glycolaldehyde group. The grouping $CH_2OH—CO—$.

glycolipid. Any lipid containing a carbohydrate.

glycolysis. 1. The anaerobic degradation of carbohydrates whereby a molecule of glucose is converted by a series of steps to two molecules of lactic acid. The overall reaction sequence yields a limited amount of ATP and is commonly divided into two stages. Stage I refers to the ATP-requiring reactions whereby glucose is converted to glyceraldehyde-3-phosphate; stage II refers to the ATP-yielding reactions whereby glyceraldehyde-3-phosphate is converted to lactic acid. *Aka* anaerobic glycolysis. 2. The sequence of reactions from glucose to pyruvic acid that is common to carbohydrate catabolism under both aerobic and anaerobic conditions. *Aka* aerobic glycolysis.

glycolytic. Of, or pertaining to, glycolysis.

glycolytic pathway. Glycolysis.

glycone. The carbohydrate portion of a glycoside.

glyconeogenesis. The synthesis of glycogen from noncarbohydrate precursors, such as fat or protein.

glycopeptide. A peptide that is linked covalently to a carbohydrate, as that which is polymerized to form the peptidoglycan of baterial cell walls.

glycophosphoglyceride. A compound composed of phosphatidic acid, the phosphate group of which is linked to a sugar.

glycophosphosphingolipid. A phosphosphingolipid that is a derivative of a ceramide and that contains both sugars and phosphate esters.

glycoprotein. A conjugated protein in which the nonprotein portion is a carbohydrate that is linked covalently to the protein and that contains less than 4% hexosamine.

glycoside. A mixed acetal (or ketal) derived from the cyclic hemiacetal (or hemiketal) form of an aldose (or a ketose); a compound formed by replacing the hydrogen of the hydroxyl group of the anomeric carbon of the carbohydrate with an alkyl or an aryl radical.

glycosidic bond. The bond between the anomeric carbon of a carbohydrate and some other group or molecule. *Aka* glycosidic link.

glycosphingolipid. 1. A mono-, oligo-, or polyglycoside of a ceramide; ceramide monoglycosides include cerebrosides and sulfatides, and ceramide oligoglycosides include cytosides, globosides, and gangliosides. *See also* glycosyl ceramide. 2. Any lipid that contains sphinganine, its homologues, isomers, or derivatives, and one or more sugar molecules.

glycosuria. The presence of excessive amounts of glucose and/or other sugars in the urine.

glycosyl ceramide. A carbohydrate-containing derivative of a ceramide. Monoglycosyl ceramides contain one carbohydrate group per molecule and include cerebrosides and sulfatides; oligoglycosyl ceramides contain 2 to 10 carbohydrate groups per molecule and include cytosides, globosides, and gangliosides; polyglycosyl ceramides contain more than 10 carbohydrate groups per molecule. Mono-, oligo-, and polyglycosyl ceramides are also known as either ceramide mono-, ceramide oligo-, and ceramide polyglycosides, or as ceramide mono-, ceramide oligo-, and ceramide polysaccharides. *See also* glycosphingolipid.

glycosyl glyceride. A glycerolipid that is a glycoside of a diacylglycerol.

glycosyl group. A sugar residue that is linked to another group or molecule by means of a glycosidic bond.

glyoxalase. The enzyme that catalyzes the conversion of methyl glyoxal to lactic acid.

glyoxalate. The ionized form of glyoxal; the compound $O{=}CH—COO^-$. *Var sp* glyoxylate.

glyoxylate. Variant spelling of glyoxalate.

glyoxylate bypass. A set of two enzymatic reactions whereby isocitric acid is converted to malic acid; the operation of these reactions results in a modified citric acid cycle, known as the glyoxylate cycle.

glyoxylate cycle. A modified citric acid cycle in which the reaction sequence between isocitric acid and malic acid is altered and acetate is used both as a source of energy and as a source of intermediates. The cycle occurs in some plants and in some microorganisms, and requires the input of two molecules of acetyl coenzyme A; it leads to the synthesis of a molecule of succinic acid which is used for the synthesis of carbohydrates and other cell components. *Aka* glyoxylate shunt.

glyoxylic acid reaction. HOPKINS-COLE REACTION.

glyoxysome. A cytoplasmic organelle of plants that contains key enzymes of the glyoxylate cycle.

gm. Gram.

G-M counter. Geiger-Mueller counter.

Gm group. A group of allotypic antigenic sites on the gamma chain of human IgG immunoglobulins.

GMP. 1. Guanosine monophosphate (guanylic acid). 2. Guanosine-5'-monophosphate (5'-guanylic acid).

G myeloma protein. An abnormal IgG immunoglobulin that is produced by individuals suffering from multiple myeloma.

gnotobiosis. The rearing of gnotobiotic animals.

gnotobiota. The known microfauna and microflora of a gnotobiotic animal.

gnotobiotic animal. 1. GERM-FREE ANIMAL. 2. A germ-free animal that has purposely been infected with one or more known bacterial species.

goiter. An abnormal enlargement of the thyroid gland.

goitrogenic. Causing goiter.

goitrogenic glycoside. A plant toxin that is a glycoside and that causes hyperthyroidism.

Golgi apparatus. A cytoplasmic organelle, composed of cisternae and vesicles, that functions in the collection and in the subsequent secretion of substances synthesized by the cell; an example of such substances are the proteins that are synthesized by the ribosomes attached to the endoplasmic reticulum. *Aka* Golgi body; Golgi complex; Golgi material.

gonad. A sex gland; an ovary or a testis.

gonadotrophic hormone. Variant spelling of gonadotropic hormone.

gonadotropic hormone. GONADOTROPIN.

gonadotropin. A hormone that stimulates the gonads; gonadotropins are secreted by the posterior lobe of the pituitary gland, the placenta, and the endometrium. *Var sp* gonadotrophin.

GOT. Glutamate-oxaloacetate transaminase. *Aka* aspartate transaminase.

gout. A metabolic disease that is characterized by an increase in the concentration of uric acid in the serum and by its precipitation as sodium urate in various tissues of the body.

Gouy interferometer. An interferometer in which constructive and destructive interference produce a series of horizontal light and dark fringes.

GP. Glycerophosphate. *Aka* glycerol-1-phosphate; α-glycerophosphate.

G1P. Glucose-1-phosphate.

G3P. Glyceraldehyde-3-phosphate.

G6P. Glucose-6-phosphate.

GPC. Gel permeation chromatography.

G6PD. Glucose-6-phosphate dehydrogenase.

GPDH. Glyceraldehyde phosphate dehydrogenase.

G6PDH. Glucose-6-phosphate dehydrogenase.

GPT. Glutamate-pyruvate transaminase.

gradient. The change in the value of a property per unit distance in a specified direction.

gradient curve. The plot of refractive index gradient, which is proportional to the concentration gradient, versus distance; obtained with the schlieren optical system.

gradient elution. A column chromatographic technique in which the composition of the eluent is changed continuously, usually with respect to ionic strength and/or pH.

gradient-flow method. A method for studying the kinetics of enzymatic reactions by means of a flow system and a gradient, such as a linear substrate gradient.

gradient gel electrophoresis. An electrophoretic technique in which the particles move across a concentration gradient by passing through a gel of progressively decreasing pore size. *Abbr* GGE.

gradient layer. A layer, the composition of which forms a gradient; used in thin-layer chromatography and thin-layer electrophoresis.

gradient mixer. A device for preparing density gradients.

gradient sievorptive chromatography. Any sievorptive chromatographic technique in which there is a gradient of small molecules across the chromatographic support.

graduated. Divided into equal units by means of lines, such as the scale of a thermometer.

Graffi leukemia virus. A mouse leukemia virus that belongs to the group of leukoviruses.

graft. TRANSPLANT.

graft copolymer. A synthetically produced copolymer in which one type of polymer forms a long branch of a second polymer, as in the branching of a vinyl polymer from a cellulose chain.

graft-versus-host reaction. The disease in the recipient of a transplant that is caused by the transfer of immunocompetent cells together with the transplant from the donor; as a result, the immunocompetent cells of the transplant form antibodies to the tissue antigens of the recipient.

gram. A metric unit of weight; the mass of 1 cm^3 of water at 4°C. *Abbr* g; gm.

gram-atom. GRAM-ATOMIC WEIGHT.

gram-atomic weight. The atomic weight expressed in grams; the weight of an element in grams that is numerically equal to its atomic weight.

gram-equivalent weight. The weight of a substance in grams that will either release or combine with 1 gram of hydrogen or 8 grams of oxygen.

gramicidin-S. An ionophorous antibiotic that also acts as an uncoupler of oxidative phosphorylation.

gram-mole. MOLE.

gram-molecular weight. MOLE.

gram-negative. Designating a bacterium that does not retain the initial gram stain but retains the counterstain. Gram-negative bacteria possess a relatively thin cell wall that is not readily digested by the enzyme lysozyme, and in which the peptidoglycan layer is covered with lipopolysaccharide.

gram-positive. Designating a bacterium that retains the initial gram stain and is not stained by the counterstain. Gram-positive bacteria generally possess a relatively thick and rigid cell wall that is readily digested by the enzyme lysozyme, and that consists of a layer of peptidoglycan.

gram stain. A set of two stains that are used to stain bacteria; the staining depends on the composition and the structure of the bacterial cell wall. *See also* gram-negative; gram-positive.

granule. A small grain or particle, such as a starch or a glycogen granule.

granum (*pl* grana). A stack of thylakoid disks in a chloroplast.

grape sugar. GLUCOSE.

GRAS. Acronym for "Generally recognized as safe," a category of food additives established by the Food and Drug Administration.

grating. *See* diffraction grating.

gratuitous induction. Enzyme induction in which the synthesis of an inducible enzyme is brought about in response to compounds other than the natural substrates of the enzyme.

Grave's disease. A pathological condition caused by thyroid hyperfunction and characterized by protruding eyeballs and by goiter.

gravimetric analysis. A method of chemical analysis that is based on separating the substance of interest from other components and weighing either the purified substance or one of its derivatives.

greasy spots. The matching, nonpolar surfaces of subunits that aid in the association of the subunits to form an oligomeric protein.

greater membrane. The components that are external to the unit membrane structure, particularly the carbohydrate-rich layer that is present on the exterior of many cells.

Greenstein hypothesis. A hypothesis of cancer according to which tumors are characterized by a general convergence of enzyme patterns which leads to a biochemical uniformity of tumor tissues.

GRF. 1. Growth hormone releasing factor. *See* growth hormone regulatory hormone. 2. Growth factor for guinea pigs.

GRH. *See* growth hormone regulatory hormone.

grid. 1. A two-dimensional network of uniformly spaced horizontal and vertical lines. 2. A screen for mounting specimens in electron microscopy.

GRIF. *See* growth hormone regulatory hormone.

GRIH. *See* growth hormone regulatory hormone.

Gross leukemia virus. A mouse leukemia virus that belongs to the group of leukoviruses.

Grotthus-Draper law. FIRST LAW OF PHOTOCHEMISTRY.

ground state. The normal, unexcited, lowest-energy level of a nucleus, an atom, or a molecule.

ground substance. The gel-like and mucopolysaccharide-containing matrix of extracellular space.

group frequency band. An infrared absorption band that is characteristic of a particular chemical group in a molecule, such as the $C{=}O$, $C{=}C$, $C{-}H$, or $C{=}N$ group. *See also* skeletal band.

group specificity. The selectivity of an enzyme that allows it to catalyze a reaction with a group of related substrates.

group transfer potential. The free energy change for the hydrolysis reaction in which a given group of atoms is removed from a compound.

group transfer reaction. A reaction, other than an oxidation-reduction reaction, in which a functional group is transferred from one molecule to another.

group translocation process. A process whereby a group of atoms, such as a phosphoryl group, is moved across a membrane by being bound successively to a number of different substrates.

growing point. REPLICATING FORK.

growth curve. A plot of either the number of cells or the cell mass of a growing culture as a function of time.

growth factor. 1. Any factor, such as a mineral, a vitamin, or a hormone, that promotes the growth of an organism. 2. A specific substance that must be present in a growth medium to permit cell multiplication.

growth hormone. The protein hormone that is secreted by the anterior lobe of the pituitary gland, stimulates body growth, and affects many aspects of metabolism. *Abbr* GH.

growth hormone regulatory hormone. One of two hypothalamic hormones (or factors) that, respectively, stimulate or inhibit the release of growth hormone from the pituitary gland. The growth hormone releasing hormone (or factor) is abbreviated variously as GRH (GRF), GHRH (GHRF), or SRH (SRF) where S stands for somatotropin. The growth hormone release-inhibiting hormone (or factor) is abbreviated variously as GIH (GIF), GHIH (GHIF), GRIH (GRIF), GHRIH (GHRIF), SIH (SIF), or SRIH (SRIF).

growth hormone release-inhibiting hormone. *See* growth hormone regulatory hormone.

growth hormone releasing hormone. *See* growth hormone regulatory hormone.

growth rate constant. The relative increase in either the number of cells or the cell mass per unit time; specifically, $dN/dt = \alpha N$, where α is the growth rate constant, and dN is the increase in the number of cells or the cell mass in the time dt.

GSC. Gas solid chromatography.

GSH. Glutathione.

GSSG. Oxidized glutathione; *see* glutathione.

GTP. 1. Guanosine triphosphate. 2. Guanosine-5′-triphosphate.

GTPase. Guanosine triphosphatase.

guanidinium group. GUANIDO GROUP.

guanidino acetic acid. An intermediate in the biosynthesis of creatine that is formed by transfer of the guanido group from arginine to glycine.

guanidino group. GUANIDO GROUP.

guanido group. The basic grouping NH_2—C—NH— that is present in the amino
$$\|$$
$$NH$$
acid arginine.

guanine. The purine, 2-amino-6-oxypurine, that occurs in both RNA and DNA. *Abbr* G.

guanosine. The ribonucleoside of guanine. Guanosine mono-, di-, and triphosphate are abbreviated, respectively, as GMP, GDP, and GTP. The abbreviations refer to the 5′-nucleoside phosphates unless otherwise indicated. *Abbr* Guo; G.

guanosine-3′, 5′-cyclic monophosphate. A cyclic nucleotide, commonly called cyclic GMP, that is formed from GTP in a reaction catalyzed by the enzyme guanylate cyclase. Cyclic GMP is present intracellulary at very low concentrations and is believed to function as an antagonist of cyclic AMP in systems composed of opposing reactions that are controlled in both directions, such as muscle contraction-muscle relaxation, and glycogen synthesis-glycogen breakdown. *Abbr* cGMP. *Aka* cyclic guanylic acid.

guanosine triphosphatase. The enzyme that catalyzes the hydrolysis of GTP, as in the process of peptide bond formation during protein synthesis. *Abbr* GTPase.

guanosine-5′-triphosphate. A high-energy compound that is required for peptide bond formation in protein synthesis. *Abbr* GTP.

guanylate cyclase. The enzyme that catalyzes the formation of guanosine-3′,5′-cyclic monophosphate from guanosine-5′-triphosphate.

guanylic acid. The ribonucleotide of guanine.

gum. An excretion of certain plants that usually contains polysaccharides composed of glucuronic acid, galactose, arabinose, and, at times, other sugars.

gum arabic. The gum produced by trees of the genus *Acacia*.

Guo. Guanosine.

g-value. A factor used in electron paramagnetic resonance to relate the frequency of the absorbed radiation to the strength of the applied magnetic field. The magnitude of this factor is a measure of the extent to which an electron interacts with other electrons or nuclei.

G-value. A measure of the sensitivity of a compound to undergo a reaction subsequent to and as a result of exposure to an ionizing radiation; equal to the number of molecules sensitized by the radiation per 100 eV absorbed.

gyratory shaker. A shaker that provides a rotational motion.

gyromagnetic coefficient. The ratio of the magnetic moment of a nucleus to its angular momentum spin; used in nuclear magnetic resonance.

gyromagnetic ratio. A constant that is characteristic of an individual atomic nucleus and that is related to the energy which must be absorbed by this nucleus before it can undergo a transition. *See also* nuclear magnetic resonance.

H

h. 1. Hour. 2. Planck's constant; 1.58×10^{-34} cal/sec. 3. Hecto.

H. 1. Hydrogen. 2. Enthalpy. 3. Histidine. 4. Henry. 5. Magnetic field strength.

H³. Tritium; the radioactive isotope of hydrogen that has a half-life of 12.5 years and is a weak beta emitter.

HA. Hydroxyapatite.

Hageman factor. The initiating factor of the intrinsic system of blood clotting.

Haldane-Oparin hypothesis. The hypothesis that the origin of life on the earth was preceded by a very long period of abiogenic evolution. During this period simple organic compounds were formed first, primarily from gases in the atmosphere, and more complex compounds were then formed from them by a variety of reactions, occurring primarily in the seas. These reactions and compounds then gave rise to macromolecules which ultimately assembled to form structures that were the forerunners of living cells.

Haldane relationship. An expression that relates the equilibrium constants of an enzymatic reaction to the kinetic constants of the forward and the reverse reactions. For a simple, one-substrate enzymatic reaction the Haldane relationship is $K_{eq} = V_f K_{m_r} / V_r K_{m_f}$, where K_{eq} is the equilibrium constant, V is the maximum velocity, and K_m is the Michaelis constant; the subscripts f and r refer to the forward and the reverse reaction, respectively. *Aka* Haldane equation; Haldane.

half-band width. The width of an absorption band at the point at which the absorption equals one half of the maximum absorption.

half-cell. That part of an electrical cell that consists of one half-reaction.

half-life. 1. The time required for one half of either the mass or the number of atoms of a radioactive substance to undergo radioactive decay. *Aka* radioactive half-life. 2. The time required for one half of the mass of a substance to be either metabolized or excreted by an organism. *Aka* biological half-life. 3. The time required for one half of the mass of a reactant to undergo chemical reaction. For a first order reaction $t_{1/2} = 0.693/k$, where $t_{1/2}$ is the half-life and k is the rate constant.

half of the sites phenomenon. The phenomenon, observed with a number of regulatory enzymes, in which the reaction at one site of the enzyme prevents the reaction at a second site so that, at any given time, only one half of the potential sites of the enzyme participate in a reaction.

half-period. HALF-TIME.

half-reaction. A reaction in which there is either a gain or a loss of electrons; a half-reaction in which there is a gain of electrons can take place only in the presence of a second half-reaction in which there is a corresponding loss of electrons, and vice versa.

half-thickness. HALF-VALUE LAYER.

half-time. The time required to achieve one half of the maximum of a reaction.

half-time of exchange. The time required for one half of the exchangeable atoms to be exchanged in a reaction that involves the exchange of atoms.

half-value dose. The dose of radiation or toxic compound that produces deaths in 50% of the cells, or loss of infectivity in 50% of the virus particles, in a test group within a specified time; analogous to the median lethal dose for animals.

half-value layer. The thickness of an absorbing material that reduces the intensity of a beam of incident radiation to one half of its original intensity. *Abbr* HVL.

half-wave potential. The polarographic potential at which one half of the maximum current is obtained.

halide. A binary compound formed from a halogen and either a metal or an organic radical; a fluoride, a chloride, a bromide, or an iodide.

hallucinogenic drug. A substance that, if taken in appropriate doses, produces distortion of perception, vivid images, or hallucinations. *Aka* hallucinogen.

halogen. An element of group VIIA in the periodic table that consists of the elements fluorine F, chlorine Cl, bromine Br, iodine I, and astatine At.

halogenation. The introduction of a halogen into an organic compound.

halogen quenching. The quenching of an ionization detector that occurs when a halogen gas is added to the mixture used for gas amplification. *See also* internal quenching.

halophile. An organism that grows only in solutions of either moderate or high salt concentration.

Hamilton syringe. A finely tooled syringe for the delivery of volumes in the microliter range.

Hammett equation. One of two equations that

describe the effect of meta and para substituents of an aromatic compound on either the rate or the equilibrium constant of a reaction. Specifically, $log\ k - log\ k° = \rho\sigma$ or $pK - pK° = \rho\sigma$, where k and $k°$ are the rate constants for the substituted and for the unsubstituted compound, and pK and $pK°$ are the negative logarithms of the corresponding equilibrium constants; ρ is a constant that characterizes the reaction with respect to its sensitivity to electron supply at the reaction site and that is independent of the substituent; and σ is a constant that characterizes a substituent with respect to its electron-withdrawing power and that is independent of the nature of the reaction.

handedness. The property of a helix of being either right-handed or left-handed.

Hanes plot. A graphical treatment of enzyme kinetics data in which S/v is plotted as a function of S, where S is the substrate concentration, and v is the velocity of the reaction.

hanging drop technique. A technique for the microscopic examination of live microorganisms that avoids possible distortion of the organisms which may be caused by drying or fixing; performed by suspending the microorganisms in a drop of fluid on a concave microscope slide.

hanging mercury drop electrode. An electrode used for fast-sweep polarography.

hanging strip electrophoresis. An electrophoretic technique in which paper strips are used that are suspended from a central support in the form of an inverted V, with each end of the paper dipping into a buffer compartment.

H-antigen. 1. A flagellar antigen of bacteria. 2. A precursor of the A and B antigens of the human ABO blood group system.

Hanus iodine number. An iodine number determined by the use of a solution of iodine in glacial acetic acid, with iodine bromide serving as an accelerator of the reaction.

haploid number. The fundamental number of chromosomes that comprise a single set; the number of chromosomes in a genome; the gametic chromosome number. *Sym* N.

haploid state. The chromosome state in which each type of chromosome is represented only once. *Aka* haploidy.

hapten. A substance that can react selectively with antibodies of the appropriate specificity but that stimulates the production of these antibodies in an animal only when it is coupled to a carrier.

haptenic antigen. HAPTEN.

haptenic group. The hapten plus the amino acid through which it is covalently linked to a protein carrier; the haptenic group may constitute either all, or part of, an antigenic determinant.

hapten inhibition test. A test that measures the extent to which a hapten inhibits an antigen-antibody reaction; performed by allowing the hapten to bind to and block the combining sites of the appropriate antibody, and then adding antigen that is directed against the same antibody.

hapten protection. The shielding of the combining sites of an antibody by allowing a hapten to bind to them. The reaction may be used to specifically label the combining site of an antibody by reacting the antibody-hapten complex first with an unlabeled reagent, and, after removal of the hapten, with the same, but labeled, reagent.

haptoglobin. A globulin-containing mucoprotein that binds free hemoglobin, derived from destroyed erythrocytes, and transports it to the reticuloendothelial system for catabolism.

hard clot. The final, insoluble blood clot that is formed by the cross-linking of fibrin molecules in the presence of calcium ions and the enzyme fibrinase.

Harden-Young ester. FRUCTOSE-1,6-DI-PHOSPHATE.

hard soap. A sodium salt of a long-chain fatty acid; hard soaps are less water-soluble than soft soaps.

hardware. The physical equipment used with computers; the electronic, magnetic, and mechanical items such as cabinets, tubes, transistors, and wires.

hard water. Water that contains appreciable concentrations of calcium, magnesium, and iron ions; these ions form insoluble soaps that are ineffective as surface-active agents.

hard x-rays. High-frequency x-rays that have short wavelengths and great penetrating power.

Harris-Ray test. A test for vitamin C that is based on the titrimetric reduction of the dye 2,6-dichlorophenol indophenol.

Hartnup's disease. A genetically inherited metabolic defect in man that is associated with mental retardation and that is due to a deficiency of the enzyme tryptophan pyrrolase.

harvest. The collection of bacterial cells at a particular stage of growth.

Hatch-Slack-Kortschak pathway. An alternative pathway to the Calvin cycle of photosynthesis in which phosphoenolpyruvate is carboxylated to yield oxaloacetate as the first product; the cycle is believed to be operative in a number of tropical grasses and in certain dicotyledons. *Abbr* HSK pathway.

Haworth projection. A representation of carbohydrates in which the ring structures are drawn as regular hexagons or pentagons, in a plane that is perpendicular to the plane of the paper; the attached atoms or groups of atoms are indicated as being either above or below the plane of the ring. *Aka* Haworth formula.

hazard. The likelihood of toxic injury to a living system by a harmful chemical under the circumstances of its intended use.

Hb. Hemoglobin; also abbreviated as HHb. Related compounds are abbreviated as follows: HbA-adult hemoglobin; HbCO-carbon monoxide hemoglobin; HbF-fetal hemoglobin; HbO_2, $HHbO_2$-oxyhemoglobin; and HbS-sickle cell hemoglobin.

H band. H ZONE.

HCG. Human chorionic gonadotropin.

H chain. 1. One of the two types of polypeptide chains of lactate dehydrogenase isozymes; denoted H, since the tetramer of H chains is found predominantly in heart tissue. 2. HEAVY CHAIN.

HCR. Host-cell reactivation.

HD. High-density lipoprotein.

HDL. High-density lipoprotein.

HDP. Hexose diphosphate.

head. 1. The hexagonal, DNA-containing structure of a T-even phage. 2. The activated portion of a condensing unit. 3. The globular portion of the myosin molecule. 4. The 5′-phosphate end of an oligo- or a polynucleotide strand. 5. ROTOR.

headpiece. *See* supermolecule.

head-to-head condensation. The condensation of two molecules by way of their activated ends, as in the condensation of two molecules of acetaldehyde to form acetoin.

head-to-tail condensation. The condensation of two molecules by way of the active end of one molecule and the passive end of the other molecule. The condensation of isoprene units in the biosynthesis of cholesterol, and the polymerization of amino acids in the formation of peptides and proteins are two examples.

headward growth. The polymerization mechanism in which the passive tail of a monomer adds to the activated head of a chain, thereby making its own head the receptor for the next addition of monomer. *See also* head-to-tail condensation.

heat capacity. The quantity of heat required to raise the temperature of a given amount of substance by 1°C.

heat labile. Descriptive of a molecule that loses its activity upon heating to moderate temperatures of about 50°C.

heat labile citrovorum factor. 10-Formyltetrahydrofolic acid. *Abbr* HLCF.

heat stable. Descriptive of a molecule that retains its activity upon heating to moderate temperatures of about 50°C.

heavy. Labeled with a heavy isotope.

heavy atom method. ISOMORPHOUS REPLACEMENT.

heavy chain. One of two polypeptide chains that are linked to two light chains to form the immunoglobulin molecule. The molecular weight of a light chain is about 25,000 and that of a heavy chain is about 50,000. The heavy chains of the IgG, IgA, IgM, and IgD immunoglobulins are denoted, respectively, as γ, α, μ, and δ chains.

heavy hydrogen. DEUTERIUM.

heavy isotope. An isotope that contains a larger number of neutrons in the nucleus than the more frequently occurring common isotope of that element.

heavy label. A heavy isotope that is generally introduced into a molecule to facilitate its separation from identical molecules containing the more frequently occurring isotope.

heavy meromyosin. The fragment of the myosin molecule that consists of the globular head and a portion of the tail of the molecule. *Abbr* H-meromyosin; HMM. *Aka* ($F_1 + F_2$) fragment.

heavy metal contamination. The presence of trace amounts of heavy metals, such as lead and zinc, in the water and/or in the chemicals that are used in the preparation of media and other solutions.

heavy ribosome. 1. A ribosome labeled with a heavy isotope. 2. POLYSOME.

heavy strand. 1. A polynucleotide chain labeled with a heavy isotope. 2. The naturally occurring polynucleotide chain of a duplex that has a greater density than the complementary chain.

heavy water. Deuterium oxide; D_2O.

hecto-. Combining form meaning one hundred and used with metric units of measurement. *Sym* h.

Hehner number. The percentage of nonvolatile, water-insoluble fatty acids in a fat. *Aka* Hehner value.

height equivalent to a theoretical plate. The length of a gas chromatographic column, some other chromatographic column, or a distillation column divided by its efficiency in terms of theoretical plates; the length of a column over which the separation effected is equivalent to that of one theoretical plate. *Abbr* HETP; EHTP. *See also* theoretical plate.

Heisenberg principle. UNCERTAINTY PRINCIPLE.

Heitler-London theory. VALENCE BOND THEORY.

HeLa cells. Cells that have been derived from a human carcinoma of the cervix and that have been maintained in tissue culture since 1953.

helical. Of, or pertaining to, a helix.

helical content. The proportion of hydrogen-bonded base pairs in a nucleic acid, or the proportion of hydrogen-bonded amino acid residues in a protein.

helical cross. The cross-like x-ray diffraction

pattern that is obtained with fibrous helical material such as DNA.

helical virion. A virus, such as tobacco mosaic virus, in which the capsid is a helix that forms a hollow cylinder.

helicity. Helical structure.

heliotropism. PHOTOTROPISM.

helix. A coiled structure, or spiral, that is described by the thread of a bolt or the turns of a tubular spring; the curve that is traced on the surface of a cylinder by the rotation of a point that cuts the elements of the cylinder at a constant oblique angle. A helix is said to be left-handed or right-handed depending on whether it corresponds to the thread of a left-handed or a right-handed bolt. *See also* alpha helix; collagen helix; Watson-Crick model.

helix-breaking amino acid. An amino acid that, wherever it occurs in the polypeptide chain, interrupts the alpha helical structure and creates a bend in the chain. This is always the case for proline and hydroxyproline and appears to occur occasionally for glutamic acid, aspartic acid, and histidine.

helix-coil transition. The transition of a polymer from a helical configuration to that of a random coil, as in the denaturation of helical proteins and double-stranded DNA.

Heller's test. A qualitative test for protein that is based on the formation of a white precipitate at the interface between concentrated nitric acid and a test solution which is layered above it.

Helmholtz free energy. That component of the total energy of a system that can do work under conditions of constant temperature and volume; expressed by the thermodynamic function $A = E - TS$, where A is the Helmholtz free energy, E is the internal energy, T is the absolute temperature, and S is the entropy.

helper phage. *See* helper virus.

helper virus. A virus that, by infecting a cell and supplying a missing product, allows the simultaneous infection of that cell by a defective virus.

hemadsorption. The adsorption of a substance to red blood cells.

hemagglutination. The agglutination of red blood cells.

hemagglutination inhibition. The inhibition of hemagglutination; used for assaying hemagglutinating viruses by adding antiviral antibodies to a mixture of virus particles and red blood cells. *Abbr* HI.

hemagglutinin. An agglutinin of red blood cells.

hematin. FERRIHEME.

hematinometer. HEMOGLOBINOMETER.

hematocrit. 1. The relative volume of blood occupied by the erythrocytes and expressed in cubic centimeters per 100 cc of blood. 2. An apparatus for measuring the relative volume of cells and plasma in blood, usually by means of centrifugation.

hematology. The science that deals with blood and blood-forming organs.

hematolysis. HEMOLYSIS.

hematopoiesis. HEMOPOIESIS.

hematoside. One of a class of monosialoganglio-sides; a compound having the structure: ceramide-glucose-galactose-*N*-acetylneuraminic acid.

hematuria. The presence of either blood or red blood cells in the urine.

heme. 1. Any tetrapyrrolic chelate of iron, generally an iron porphyrin complex, in which four coordination positions of iron are occupied. 2. A protoheme; an iron-porphyrin complex that has a protoporphyrin nucleus, specifically one containing protoporphyrin IX which is the oxygen-binding portion of the hemoglobin molecule. Heme A (protoheme A) is the prosthetic group of cytochrome a; heme C (protoheme C) is the prosthetic group of cytochrome c.

heme-heme interactions. The cooperative interactions between the hemes of the subunits of hemoglobin with respect to the binding of oxygen.

hemel. An aziridine mutagen.

heme protein. *See* hemoprotein.

hemerythrin. A nonheme, iron-containing, respiratory pigment of sipunculid worms.

heme synthetase. The enzyme that catalyzes the incorporation of iron into protoporphyrin.

hemiacetal. A compound formed by either an inter- or an intramolecular reaction between an aldehyde group and an alcohol group.

hemicellulose. A polymer of D-xylose that contains side chains of other sugars and that serves to cement plant cellulose fibers together.

hemichrome. FERRIHEMOCHROME.

hemichromogen. FERRIHEMOCHROME.

hemiketal. A compound formed by either an inter- or an intramolecular reaction between a ketone group and an alcohol group.

hemin. FERRIHEME.

hemizygous gene. A gene that is present only once among the chromosomes of a cell, such as a sex-linked gene.

hemochromatosis. A disease that is caused by excessive iron absorption from the intestine and by excessive iron deposition in various organs; frequently associated with glucosuria. *See also* bronzed diabetes.

hemochrome. A low-spin compound of heme in which the fifth and sixth coordination positions of the iron are occupied by strong-field ligands.

hemochromogen. FERROHEMOCHROME.

hemoconcentration. An increase in the

concentration of red blood cells in the blood; a decrease in the concentration of plasma.

hemocyanin. A copper-containing respiratory pigment of mollusks and crustaceans.

hemodialysis. The dialysis of blood by means of a semipermeable membrane.

hemodialyzer. An artificial kidney; an apparatus that circulates blood from and back into the body by passing it through a series of semipermeable membranes which remove waste products in a manner analogous to the operation of the kidney.

hemodilution. A decrease in the concentration of red blood cells in the blood; an increase in the concentration of plasma.

hemoflavoprotein. A heme-containing flavoprotein.

hemoglobin. The oxygen-transporting protein of the blood that consists of four polypeptide chains, two alpha and two beta chains, each one surrounding a heme group. Hemoglobin occurs in several normal forms such as adult and fetal hemoglobin, and in various abnormal forms, such as sickle cell hemoglobin and hemoglobin C. *Abbr* Hb; HHb. *See also* Hb.

hemoglobinemia. The presence of free hemoglobin in blood plasma.

hemoglobinometer. An instrument for the visual or the photoelectric measurement of the hemoglobin content of blood.

hemoglobinopathy. A genetically inherited metabolic defect in man that is characterized by the presence of a structurally altered hemoglobin.

hemoglobinuria. The presence of either free hemoglobin or closely related pigments in the urine.

hemoglobin variant. *See* variant (1).

hemolymph. The circulatory fluid of various invertebrates that is functionally comparable to the blood and lymph of vertebrates.

hemolysate. The suspension of lysed cells obtained upon hemolysis.

hemolysin. 1. An antibody that, in the presence of complement, causes hemolysis. 2. A substance, such as the bacterial toxin streptolysin, that causes hemolysis.

hemolysis. The lysis of red blood cells.

hemolytic. Of, or pertaining to, hemolysis.

hemolytic anemia. An anemia that is characterized by an excessive destruction of red blood cells; several hemolytic anemias are genetically inherited metabolic defects due to a deficiency of a glycolytic enzyme.

hemolytic antibody. HEMOLYSIN (1).

hemolytic plaque technique. *See* plaque technique.

hemolytic system. The mixture of red blood cells and antibodies against them that is used in the complement fixation test.

hemolyze. To lyse red blood cells.

hemophilia. A genetically inherited, sex-linked metabolic defect in man that is characterized by prolonged clotting times and that is caused by a deficiency of the antihemophilic factor in blood. Hemophilia A is caused by a deficiency of antihemophilic factor A, and hemophilia B is caused by a deficiency of antihemophilic factor B.

hemopoiesis. The formation of red blood cells.

hemopoietic. Of, or pertaining to, hemopoiesis.

hemopoietine. ERYTHROPOIETIN.

hemoprotein. A conjugated protein that contains a heme as a prosthetic group.

hemorrhagic disease. The pathological condition that is characterized by bleeding and by long clotting times, and that is caused by a deficiency of vitamin K.

hemorrhagic disease of the newborn. A temporary hemorrhagic disease that occurs in an infant and that persists only until a bacterial flora is established in the infant's intestine.

hemosiderin. A protein that serves to store iron in the form of ferric hydroxide in the liver, the spleen, and the bone marrow.

hempa. An aziridine mutagen.

Henderson-Hasselbalch equation. An equation that relates the pH of a solution to the dissociation constant of a weak acid, and to the concentrations of the proton donor form HA and the conjugated proton acceptor form A^- of the acid. Specifically, $pH = pK_a + log [A^-]/[HA]$, where pK_a is the negative logarithm of the dissociation constant and brackets indicate molar concentrations.

Henle's loop. *See* nephron.

Henri equation. An equation, analogous to that obtained by integration of the Michaelis-Menten equation, that is applicable to an enzymatic reaction if the decrease in reaction rate as a function of time is due solely to the decrease in enzyme saturation as the concentration of the substrate decreases. The equation was reported by Victor Henri in 1902, prior to the derivation of the Michaelis-Menten equation.

henry. A unit of inductance, equal to the inductance in which an induced electromotive force of 1 V is produced when the inducing current is changed at the rate of 1 A/sec. *Sym* H.

Henry's function. The product of the radius of a macromolecule and its reciprocal ion-at-mosphere radius.

Henry's law. The law that the solubility of a gas in a liquid, at constant temperature, is proportional to the partial pressure of the gas.

heparin. A naturally occurring polysaccharide composed of D-glucuronic acid-2-sulfate and D-glucosamine-2,6-disulfate; heparin has anticoagulant activity and acts by preventing the conversion of prothrombin to thrombin.

hepatectomy. The surgical removal of the liver.

hepatic. Of, or pertaining to, the liver.

hepatic porphyria. A porphyria that is characterized by the formation of excessive amounts of heme precursors in the liver.

hepatitis. An inflammation of the liver, usually caused by a viral infection but sometimes caused by a toxic compound.

hepatocyte. A liver cell.

hepatoflavin. An impure preparation of riboflavin from liver.

hepatoma. A carcinoma of liver cells.

HEPES. *N*-2-Hydroxyethylpiperazine-*N*′-2-ethanesulfonic acid; used for the preparation of buffers in the pH range of 6.8 to 8.2.

hepta-. Combining form meaning seven.

heptamer. An oligomer that consists of seven monomers.

heptose. A monosaccharide that has seven carbon atoms.

herbicide. A chemical that kills herbs, especially weeds.

hereditary code. *See* genetic code.

hereditary disease. *See* genetic disease.

hereditary material. *See* genetic material.

heroin. A semisynthetic narcotic drug, made by converting morphine to diacetylmorphine.

herpes simplex virus. A virus that belongs to the herpesvirus group and that infects man and a variety of animals.

herpesvirus. An enveloped, icosahedral virus that contains double-stranded DNA and that infects man and lower animals.

herpes zoster. The recurrent form of the disease that is produced by varicella virus in a host that was previously infected by the virus.

Hers' disease. GLYCOGEN STORAGE DISEASE TYPE VI.

Hershey-Chase experiment. An experiment that demonstrates that, during phage infection, the phage DNA is injected into, and the phage protein coat remains outside of, the bacterial cell. The experiment is performed by infecting *Escherichia coli* cells with P^{32} or S^{35}-labeled T-even phages, followed by removal of the empty phage heads and the phage tails by shearing in a blender.

hertz. A unit of frequency; one cycle per second.

het. A partially heterozygous phage.

heteroallelic. Pertaining to alleles of a gene that have mutations at different sites.

heteroantibody. An antibody to a heteroantigen.

heteroantigen. An antigen that is immunogenic in a given animal species and that is either produced by synthetic reactions or is derived from plants, microorganisms, or other animal species.

heteroatom. An atom in a ring structure that is not a carbon atom.

heterocatalytic. Pertaining to the catalysis of the reaction of one substance by another, different substance, as distinct from autocatalytic.

heterochromatin. The condensed chromatin that is not very active in RNA synthesis and that stains strongly during interphase.

heterochromosome. A chromosome that consists primarily of heterochromatin.

heterocyclic. Of, or pertaining to, an organic compound that has a ring structure which consists of carbon atoms and one or more non-carbon atoms.

heterocyclic atom. HETEROATOM.

heterocytotropic antibody. A cytotropic antibody that binds to and sensitizes target cells of a species that differs from the one in which the antibody was produced.

heterodimer. A protein composed of two nonidentical polypeptide chains.

heterodisperse. Consisting of macromolecules that differ greatly in their size.

heteroduplex. A double-stranded nucleic acid molecule in which the two strands are complementary except for one or more regions in which they are not complementary. A double-stranded phage DNA molecule that is formed by hybridizing one strand from a wild-type phage with one strand from a mutant of this phage is an example.

heteroenzyme. One of two or more isodynamic enzymes derived from different sources, such as alcohol dehydrogenase from yeast and alcohol dehydrogenase from liver. *See also* isodynamic enzyme; isozyme.

heterofermentative lactic acid bacteria. Lactic acid bacteria that produce in fermentation less than 1.8 moles of lactic acid per mole of glucose; in addition to lactic acid, these organisms produce ethanol, acetate, glycerol, mannitol, and carbon dioxide. *See also* homofermentative lactic acid bacteria.

heterogeneic. XENOGENEIC.

heterogeneity. 1. The state of a preparation of macromolecules in which the macromolecules differ with respect to size, charge, structure, or other properties. 2. The state of a system in which there are two or more phases.

heterogeneity index. A measure of antibody heterogeneity that is based on the assumption of an essentially random distribution of the binding affinity of antibody combining sites and hapten molecules. It can be expressed as $r/n = (K_0c)^a/[1 + (K_0c)^a]$, where a is the heterogeneity index, n is the antibody valence, r is the average number of hapten molecules bound per molecule of antibody, c is the concentration of free hapten molecules, and K_0 is the average intrinsic association constant (the association constant for that concentration of hapten at which half of the antibody sites are occupied by hapten molecules). The equation can be rewritten as $log\,[r/(n-r)] = a\,log\,c + a\,log\,K_0$, and a can be determined from a plot of $log\,[r/(n-r)]$ versus $log\,c$. The heterogeneity index varies from 0 to 1; a value of 1 indicates homogeneous antibody. *See also* Hill equation; Hill plot.

heterogeneous. Of, or pertaining to, heterogeneity.

heterogeneous catalysis. Catalysis in a system that consists of two or more phases.

heterogenetic. Widely distributed in many species.

heterogenetic antigen. An antigen that is produced by several, phylogenetically unrelated species.

heterograft. A transplant from one individual to another individual of a different species.

heteroimmune phage. One of two phages that are sensitive to different immunity substances.

heteroimmune superinfection. The superinfection of a bacterium with a heteroimmune phage, as in the infection of a lysogenic bacterium with a phage that is insensitive to the immunity substance of the prophage.

heteroimmunity. 1. The immune reactions in which the antigens and the antibodies are derived from different animal species. 2. The immune reactions in which the antigens are derived from plants, microorganisms, or synthetic reactions.

heterokaryon. A fungus cell that contains two or more nuclei of different genotypes. *Var sp* heterocaryon.

heterolactic fermentation. The fermentation of glucose that yields lactic acid as well as other products. *See also* heterofermentative lactic acid bacteria.

heterologous. 1. Pertaining to subcellular fractions, such as transfer RNA, ribosomes, and enzymes, that have been isolated from different species. 2. Pertaining to an antigen and an antibody (or an antiserum) when the antibody has been produced in response to the administration of a different antigen. 3. Pertaining to an antigen, an antibody, or an antiserum when the antigens and antibodies are derived from different species. 4. Pertaining to genetically dissimilar individuals of the same species; xenogeneic.

heterologous graft. HETEROGRAFT.

heterologous interference. The viral interference induced by either an active or a suitably inactivated virus against a virus of a different taxonomic group.

heterolysis. The cleavage of a chemical bond between two atoms in which both of the two electrons associated with the bond move to one of the atoms; a reaction of the type $R:X = R^- : + X^+$.

heterophagic. Other-digesting, as in the digestion by lysosomes of exogenous material taken up by the cell.

heterophile antibody. An antibody that reacts with antigens of more than one species.

heterophile antigen. HETEROGENETIC ANTIGEN.

heteroploid state. The chromosome state in which the number of chromosomes differs from that of the characteristic diploid (or haploid) state. *Aka* heteroploidy.

heteropolar. AMPHIPATHIC.

heteropolar bond. POLAR BOND.

heteropolymer. A polymer composed of two or more types of monomers.

heteropolysaccharide. A polysaccharide composed of two or more types of monosaccharides.

heteropycnosis. The variation of the degree of condensation of either different chromosomes or of parts of the same chromosome; used as the basis for classifying chromatin into euchromatin and heterochromatin.

heteroreactivation. The reactivation of a poxvirus by another poxvirus of a different immunological subgroup. *See also* reactivation (2).

heteroside. A glycoside, the "aglycone" portion of which is not a carbohydrate. *See also* holoside.

heterosomal aberration. An interchromosomal aberration. *See also* chromosomal aberration.

heterospecific antibody. An artificially produced antibody that has one combining site which is specific for one antigen and a second combining site which is specific for a different antigen.

heterothallic. Descriptive of organisms, such as certain fungi and algae, that exist in the form of two mating types so that only gametes from strains of opposite mating type can fuse to form zygotes.

heterotherm. POIKILOTHERM.

heterotroph. A cell or an organism that requires a variety of carbon-containing compounds from animals and plants as its source of carbon, and that synthesizes all of its carbon-containing biomolecules from these compounds and from small inorganic molecules.

heterotropic. Pertaining to a regulatory enzyme in which the effector is a metabolite other than the substrate of the enzyme.

heterovalent resonance. Resonance in which the various resonance structures do not have the same number of chemical bonds.

heterozygosity. The state of having different alleles at one or more loci in the homologous chromosomes.

heterozygote. A zygote that carries different alleles at one or more loci in the homologous chromosomes and that does not breed true; a cell or an organism that carries two different alleles of the same gene.

heterozygous. Of, or pertaining to, a heterozygote.

HETP. Height equivalent to a theoretical plate.

heuristic process. A problem-solving process in which solutions are discovered by evaluating the progress made toward the final solution as by means of a controlled trial and error method, exploratory methods, the sequencing of investigations, etc. The process is valuable for stimulating further research, calculations, and the like, even though individual steps may remain unproven or be incapable of proof. *See also* algorithm; stochastic process.

hexa-. Combining form meaning six.

hexagonal capsomer. *See* capsomer.

hexamer. 1. An oligomer that consists of six monomers. 2. HEXAGONAL CAPSOMER.

hexokinase. The enzyme that catalyzes the conversion of glucose to glucose-6-phosphate in the first step of glycolysis.

hexon. The capsomer of adenoviruses that is surrounded by six other capsomers in the intact virus particle.

hexon antigen. An antigen of the hollow capsomer of adenoviruses.

hexosamine. An amino sugar of a six-carbon monosaccharide.

hexosan. A polysaccharide of hexoses.

hexose. A monosaccharide that has six carbon atoms.

hexose monophosphate oxidative pathway. The formation of ribose-5-phosphate from glucose-6-phosphate in the initial reactions of the hexose monophosphate shunt.

hexose monophosphate pathway. HEXOSE MONOPHOSPHATE SHUNT.

hexose monophosphate shunt. The metabolic pathway that requires the input of six molecules of glucose-6-phosphate and leads to the complete oxidation of one molecule of glucose-6-phosphate to carbon dioxide, water, and phosphate. The pathway functions to generate reducing power in the form of NADPH, allows for the interconversion of monosaccharides, and is linked to the fixation of carbon dioxide in photosynthesis. *Abbr* HMS; HMP shunt.

hexose phosphoketolase pathway. A catabolic pathway of glucose that is related to the hexose monophosphate shunt and that occurs in some bacteria.

hexuronic acid. ASCORBIC ACID.

Hfr strain. HIGH-FREQUENCY OF RECOMBINATION STRAIN.

HFT. HIGH-FREQUENCY TRANSDUCTION.

HFT lysate. High-frequency transduction lysate;

a lysate prepared by induction of a prophage that possesses an unusually high transducing power.

Hg. Mercury.

HGF. Hyperglycemic-glycogenolytic factor.

HGG. Human gamma globulin.

HGH. Human growth hormone.

HI. Hemagglutination inhibition.

5-HIAA. 5-Hydroxyindoleacetic acid.

high-angle x-ray diffraction. LARGE ANGLE X-RAY DIFFRACTION.

high-density lipoprotein. A lipoprotein that contains approximately 30% protein, 30% neutral lipids, and 40% phospholipids; these lipoproteins have molecular weights of the order of 100,000 to 400,000, and a density of about 1.062 to 1.21 g/l. *Abbr* HDL; HD.

high-efficiency liquid chromatography. HIGH-PRESSURE LIQUID CHROMATOGRAPHY.

high-energy bond. A covalent bond, the hydrolysis of which under standard conditions yields a large amount of free energy. The term refers to the large negative free energy change associated with the hydrolysis reaction, and not to the bond energy. High-energy bonds are commonly denoted by a squiggle ($\sim$).

high-energy compound. A compound that, upon hydrolysis under standard conditions, yields a large amount of free energy. High-energy compounds are frequently those in which a phosphate group is removed by hydrolysis and the free energy change of this reaction is of the order of 7 kcal/mole or more. *See also* high-energy bond.

high-energy phosphate donor. A high-energy compound that can function as a phosphoryl group donor to a low-energy phosphate acceptor by way of the ADP-ATP phosphoryl group carrier system. *Aka* high-energy phosphate compound.

high-frequency of recombination strain. A bacterial strain that has a high frequency of recombination due to the fact that the episomal fertility factor has become incorporated into the bacterial chromosome. *Abbr* Hfr strain.

high-frequency transduction. Transduction in which the phages that are capable of transducing constitute a high proportion of the total phage population. *Abbr* HFT.

high-lipid lipoprotein. LOW-DENSITY LIPOPROTEIN.

high-performance liquid chromatography. HIGH-PRESSURE LIQUID CHROMATOGRAPHY.

high polymer. A polymer of very high molecular weight, such as a polymer occurring in plastics, rubber, fibers, or human tissues.

high-potential iron-sulfur protein. IRON-SULFUR PROTEIN (c).

high-pressure liquid chromatography. A column chromatographic technique that is rapid and provides high resolution; it can be used with the various modes of liquid chromatography such as gel filtration, adsorption, partition, and ion-exchange. The apparatus consists of a column, a hydraulic system, a detector, and a recorder. The liquid is forced in series and under pressure through a column which is maintained at a constant temperature, through a detector, and then into a fraction collector. *Abbr* HPLC.

high-resistance-leak method. A method for amplifying the ion current produced in an ionization chamber.

high-sensitivity liquid chromatography. HIGH-PRESSURE LIQUID CHROMATOGRAPHY.

high-speed liquid chromatography. HIGH-PRESSURE LIQUID CHROMATOGRAPHY.

high-speed sedimentation equilibrium. MENISCUS DEPLETION SEDIMENTATION EQUILIBRIUM.

high spin. The state of a complex in which there is a maximum of unpaired electrons; referred to as a state of essentially ionic bonding and ascribed to certain hemoproteins.

high-temperature-short-time method. The pasteurization of material by heating it at 71.7°C for 15 sec. *Abbr* HTST method.

high-voltage electrophoresis. An electrophoretic technique in which the applied electric field is greater than 20 V/cm and in which separations are achieved in short times; useful for the separation of low-molecular weight compounds such as amino acids, peptides, and nucleotides.

Hill coefficient. The coefficient n_H in the Hill equation; it is also the slope in a Hill plot. *See also* Hill equation; Hill plot.

Hill equation. An equation that describes the binding of ligands to a protein; common forms of the equation are: (i) $r/n = K'[S]^{n_H}/(1 + K'[S]^{n_H})$; (ii) $r/(n - r) = K'[S]^{n_H}$; (iii) $log [r/(n - r)] = n_H log [S] + log K'$; (iv) $Y_s = [S]^{n_H}/(K + [S]^{n_H})$; (v) $log [Y_s/(1 - Y_s)] = n_H log [S] - log K$, where r is the number of moles of ligand S bound per mole of total protein, n is the number of binding sites of a given type per molecule of protein, K' is the intrinsic association constant for this type of site, Y_s is the fraction of total binding sites occupied by ligand S ($Y_s = r/n$), K is the intrinsic dissociation constant ($K = 1/K'$), and n_H is the interaction factor for the sites which varies from 1, for noninteracting sites, to n, for highly cooperative binding.

Hill plot. A graphical representation of binding data based on the Hill equation that is used for determining intrinsic association constants and for determining the number of binding sites of a given type per molecule of protein; consists of a plot of $log [r/(n - r)]$ as a function of $log [S]$ or a plot of $log [Y_s/(1 - Y_s)]$ as a function of log [S]. *See also* Hill equation.

Hill reaction. The light reaction in photosynthesis that is carried out in the presence of an artificial electron acceptor; the reaction: $2H_2O + A = 2 AH_2 + O_2$, where A is the electron acceptor.

Hill reagent. The artificial electron acceptor in the Hill reaction.

hinge point. The position in an analytical ultracentrifuge cell at which the concentration, during sedimentation equilibrium, is equal to the initial concentration of the solution.

hinge region. 1. That portion of the IgG immunoglobulin molecule that is adjacent to the two disulfide bonds which link the two heavy chains together; the hinge region is near the sites of action of papain and pepsin and is believed to be a flexible region that permits the molecule to open up in the form of a Y. 2. A region in the predominantly helical rod section of the myosin molecule that is less stable than the rest of the molecule, and that is more susceptible to proteolytic attack; the trypsin-sensitive region of myosin.

hippuric acid. *N*-Benzoylglycine; the form in which benzoic acid is detoxified and excreted in the urine.

His. 1. Histidine. 2. Histidyl.

H-isoantigen. A pentahexosyl ceramide that accumulates in individuals suffering from fucosidosis.

His operon. The operon in *Salmonella typhimirium* that consists of the genes which code for the 10 enzymes involved in the biosynthesis of histidine.

histamine. A pharmacologically active mediator of the allergic response that causes vasodilation, increased capillary permeability, and contraction of smooth muscle. Histamine is formed by decarboxylation of histidine and is widely distributed in mammalian tissues, particularly in mast cells from which it is released during the allergic response.

histidine. A heterocyclic, basic, and polar alpha amino acid that contains the imidazole ring system. *Abbr* His; H.

histidinemia. A genetically inherited metabolic defect in man that is characterized by elevated blood and urine levels of histidine, and that is due to a deficiency of the enzyme histidase that functions in histidine catabolism.

histochemistry. The science that deals with the chemical constitution and the chemical changes of tissues and cells by combining the techniques of biochemistry and histology.

histocompatibility. Immunological tolerance to transplanted tissue.

histocompatibility antigen. A tissue antigen of the donor of an allograft that induces transplantation immunity in the recipient of the transplant.

histocompatibility gene. A gene that codes for a histocompatibility antigen.

histogram. A graphical representation of a frequency distribution by means of rectangles, the widths of which represent the class interval and the heights of which represent the frequencies of the different classes.

histohematin. CYTOCHROME.

histoincompatibility. Immunological intolerance to transplanted tissue that results from antigenic differences between the donor and the recipient of a transplant.

histological chemistry. See histochemistry.

histology. The branch of anatomy that deals with the structure and the properties of tissues, as examined by staining and microscopy.

histolysis. The destruction of tissues.

histone. A basic, globular, and simple protein that is characterized by its high content of arginine and lysine. Histones are found in association with nucleic acids in the nuclei of many eucaryotic cells; they are classified into four major groups, denoted I-IV, on the basis of their arginine and lysine content.

histoplasmin. A crude and sterile filtrate, derived from mycelia of the fungus *Histoplasma capsulatum,* that is used for intradermal injections as a test for delayed-type skin hypersensitivity.

hit theory. See target theory.

hive. The inflammation of human skin produced during cutaneous anaphylaxis.

HL. One of four chloroform-soluble fractions from liver (HL-1 through HL-4) that contain growth factors for *Lactobacillus helveticus* and *Lactobacillus lactis* and that may be related to folic acid and lipoic acid.

HLCF. Heat-labile citrovorum factor.

HMC. 5-Hydroxymethylcytosine; a minor base.

H-meromyosin. Heavy meromyosin.

HMG. 1. β-Hydroxy-β-methyl glutarate. 2. Human menopausal gonadotropin.

HMM. Heavy meromyosin.

HMP. 1. Hexose monophosphate pathway. 2. Hexose monophosphate.

HMP shunt. Hexose monophosphate shunt.

HMS. Hexose monophosphate shunt.

HnRNA. Heterogeneous (heterodisperse), high-molecular weight DNA-like RNA that occurs in the nuclei of eucaryotic cells; believed to be degraded within the nucleus and to serve as a precursor of messenger RNA.

Hodge scheme. A schematic summary of various enzymatic and nonenzymatic mechanisms of the browning reactions.

Hofmeister series. LYOTROPIC SERIES.

holandric. Designating a trait that appears only in males.

holandric gene. A gene that is located only on the Y-chromosome and that appears only in males.

holdback carrier. A nonradioactive compound that is added to a sample to prevent either the coprecipitation or the adsorption of a soluble radioactive compound. *Aka* holdback agent.

hold-up volume. The gas chromatographic retention volume of a nonadsorbed component; the volume of mobile phase that must be eluted before a component, which is not retarded by the stationary phase, can be eluted.

hole. 1. An energy level that is not occupied by a particle even though adjacent energy levels are filled. 2. An unoccupied position in a crystal, a metal, or a liquid where there should normally be either an electron or an atom.

hollow fiber technique. A technique for dialyzing, desalting, concentrating, and fractionating solutions of macromolecules; entails the use of bundles of semipermeable, hollow-bore fibers that function as molecular sieves and that have pores of controlled dimensions.

holocrine gland. A gland that produces a secretion that consists of altered secretory cells of the gland itself.

holoenzyme. An entire conjugated enzyme that consists of a protein component, or apoenzyme, and of a nonprotein component, which may be either a coenzyme or an activator.

hologynic. Designating a trait that appears only in females.

holophytic nutrition. A mode of nutrition, as that of green plants, which requires only inorganic compounds for growth and maintenance of the organism.

holoside. A glycoside whose "aglycone" portion is another carbohydrate; an oligosaccharide. *See also* heteroside.

holothurin. ANIMAL SAPONIN.

holozoic nutrition. A mode of nutrition, as that of animals, which requires organic compounds for growth and maintenance of the organism.

homeomorph. One of a group of possible and reasonable reaction mechanisms that obey the same rate equation.

homeostasis. 1. The constancy of the internal environment of an organism; the steady state with respect to functions, tissues, and fluids of the organism. 2. The processes involved in the regulation and maintenance of the internal environment of an organism.

homeostatic. Of, or pertaining to, homeostasis.

homeothermic. Variant spelling of homoiothermic.

homoallelic. Pertaining to alleles of a gene that have mutations at identical sites.

homoamino acid. An amino acid that has an additional CH_2 group compared to another amino acid; an amino acid homologue.

homocellular transport. The transport across the cell membrane that moves material into or out of the cell. *See also* intracellular transport; transcellular transport.

homochromatography. A chromatographic technique in which a group of compounds are separated by development with a solvent that contains either the same or related compounds; applied, for example, to the paper chromatographic separation of a mixture of labeled nucleotides from a hydrolysate of *Escherichia coli* 5S RNA by development with a solvent containing a mixture of unlabeled nucleotides from a hydrolysate of yeast RNA.

homocopolymer. A copolymer in which individual chains are composed entirely of one type of monomer, as in poly dA:dT.

homocysteine. A homologue of cysteine that contains one CH_2 group more than cysteine.

homocystinuria. A genetically inherited metabolic defect in man that is associated with mental retardation and that is characterized by the presence of excessive amounts of homocystine in the urine; due to a deficiency of the enzyme cystathionine synthetase which functions in the metabolism of cysteine.

homocytotropic antibody. A cytotropic antibody that binds to and sensitizes target cells in the same species in which it was produced. *See also* reagin.

homodimer. A protein that is composed of two identical polypeptide chains.

homofermentative lactic acid bacteria. Lactic acid bacteria that produce in fermentation 1.8 to 2.0 moles of lactic acid per mole of glucose. *See also* heterofermentative lactic acid bacteria.

homogenate. The suspension prepared by homogenization of tissues, cells, or cellular components.

homogeneity. 1. The state of a preparation of macromolecules of a single type (e.g., of one enzyme) in which the macromolecules are identical with respect to size, charge, structure, and all other properties. 2. The state of a system in which there is only one phase.

homogeneous. Of, or pertaining to, homogeneity.

homogeneous catalysis. Catalysis in a system that consists of one phase.

homogenization. The disruption of tissues, cells, or cellular components and their reduction to particles of small size so that a relatively uniform suspension is obtained.

homogenizer. A device, frequently a tube with a finely tooled pestle, that is used for homogenization.

homograft. ALLOGRAFT.

homoimmune phage. One of two phages that are sensitive to the same immunity substance.

homoiothermic. Descriptive of an organism that has a nearly constant body temperature irrespective of the temperature of its environment. *Aka* warm-blooded.

homokaryon. A fungus cell that contains two or more nuclei of only one genotype. *Var sp* homocaryon.

homolactic fermentation. The fermentation of glucose that yields lactic acid as the sole product. *See also* homofermentative lactic acid bacteria.

homologous. 1. Pertaining to subcellular fractions such as tRNA, ribosomes, and enzymes that have been isolated from the same species; isologous. 2. Pertaining to genetically dissimilar individuals of the same species; allogeneic. 3. Pertaining to proteins or nucleic acids from different species that have identical or similar functions, such as the hemoglobins of various vertebrate species. 4. Pertaining to two proteins that show sequence homology. 5. Pertaining to an antigen and an antibody (or an antiserum) when the antibody has been produced in response to the administration of that antigen. 6. Pertaining to an antigen and an antibody (or an antiserum) when the antigens and the antibodies are derived from the same species; isologous. 7. Pertaining to chemical compounds that are members of a homologous series.

homologous chromosomes. Chromosomes that occur in pairs, with one each derived from the male and female parent, and that contain the same linear gene sequences so that each gene is present in duplicate.

homologous disease. GRAFT-VERSUS-HOST REACTION.

homologous enzyme variants. Enzyme variants that have similar molecular structures and catalytic properties. *See also* homologous (3).

homologous hapten. A hapten that closely resembles the haptenic group of the immunogen.

homologous interference. The inhibition of the multiplication of one virus by an active or a suitably inactivated different virus which belongs to the same taxonomic group.

homologous series. A series of organic compounds in which each member differs from the preceding one by having one additional CH_2 group.

homologue. 1. A member of a homologous series. 2. One of a pair of homologous chromosomes. *Var sp* homolog.

homolysis. The cleavage of a chemical bond between two atoms in which one of the two electrons associated with the bond moves to one atom and the second electron moves to the other atom; a reaction of the type $R:X = R\cdot + X\cdot$.

homophilic bonding. The attraction of like groups for each other, as that of nonpolar groups in a hydrophobic bond or that of polar groups in an electrostatic bond.

homopolar bond. NONPOLAR BOND (1).

homopolymer. A polymer composed of only one type of monomer.

homopolynucleotide. A polynucleotide composed of only one type of nucleotide.

homopolypeptide. A polypeptide composed of only one type of amino acid.

homopolysaccharide. A polysaccharide composed of only one type of monosaccharide.

homoreactant antibody. An autoantibody, occurring in rabbits, that appears to be directed against antigenic determinants of the autologous IgG immunoglobulin that are hidden in the intact IgG molecule but are exposed when it is digested with the enzyme papain.

homoreactivation. The reactivation of a poxvirus by another poxvirus of the same immunological subgroup. *See also* reactivation (2).

homoserine. The homologue of serine that contains one CH_2 group more than serine.

homosomal aberration. An intrachromosomal aberration. *See also* chromosomal aberration.

homospecific antibody. An antibody in which both of the antibody combining sites are specific for the same antigen. *See also* heterospecific antibody.

homosteroid. A modified steroid that contains one or several CH_2 groups more than the parent compound; a homologous steroid.

homothallic. Descriptive of organisms, such as certain fungi and algae, that exist in the form of one mating type so that gametes from a single strain can fuse to form zygotes.

homotope. A monomer that can take the place of another monomer in a definable segment of a polymer; thus serine and glycine which occupy position 9 of the A chain of ox and sheep insulin, respectively, are said to be homotopes.

homotropic. Pertaining to a regulatory enzyme in which the substrate of the enzyme functions both as a substrate and as an effector.

homotropic-heterotropic. Pertaining to a regulatory enzyme in which both the substrate and some other metabolite function as effectors.

homozygosity. The state of having identical alleles at one or more loci in the homologous chromosomes.

homozygote. A zygote that carries identical alleles at one or more loci in the homologous chromosomes and that breeds true; a cell or an organism that carries two identical alleles of the same gene.

homozygous. Of, or pertaining to, a homozygote.

Hoogsten-type hydrogen bonding. Hydrogen bonding between adenine and uracil in which two hydrogen bonds are formed as follows: one between the nitrogen at position 3 of uracil and the nitrogen at position 7 of adenine; and one between the oxygen attached to carbon 4 of uracil and the nitrogen of the amino group of adenine.

hook model. A model for the attachment of the prophage to the bacterial DNA. According to this model, the prophage is joined through a point to the uninterrupted bacterial DNA, somewhat in the manner of a hook. *See also* insertion model.

Hopkins-Cole reaction. A colorimetric reaction for tryptophan and other compounds containing the indole ring; based on the formation of a violet color upon treatment of the sample with glyoxylic acid and sulfuric acid.

Horecker cycle. HEXOSE MONOPHOSPHATE SHUNT.

horizontal evolution. The simultaneous, parallel evolution of many sequences in the gene complement of a single species. This contrasts with the more common evolution of a single gene in two or more diverging species.

horizontal strip electrophoresis. An electrophoretic technique in which paper strips are used that are supported in a horizontal frame, with each end of the paper dipping into a buffer compartment.

horizontal transmission. The transmission of viruses between individual hosts of the same generation. *Aka* horizontal infection.

hormone. A regulatory substance that is synthesized by specialized cells of an organism, that is active at low concentrations, and that exerts its effect either on all of the cells of the organism or only on certain cells in specific organs. Hormones have three major functions: (a) an integrative function that deals with the interrelationships between different hormones and with the interrelationships between hormones and the nervous system, the blood flow, the blood pressure, and other factors; (b) a morphogenetic function that deals with the control of the type and rate of growth of various tissues; and (c) a regulatory function that deals with the maintenance of a constant internal environment with respect to the intra- and extracellular fluids. An animal hormone is a substance such as a polypeptide, a protein, or a steroid, that is secreted principally by an endocrine gland and that is transported by way of the circulation to target organs or target tissues; there the hormone exerts its effect either directly or indirectly and helps to regulate such overall physiological processes as metabolism, growth, and reproduction. A plant hormone is an organic compound that controls growth or some other function at a site removed from its place of production in the plant. The major groups of plant hormones are auxins, gibberellins, and cytokinins.

horror autotoxicus. An early synonym for autoimmunity, so named because it was originally believed to be nonexistent; it is now known to be widespread.

host. 1. An organism upon or in which a parasite lives. 2. The recipient of a transplant.

host-cell reactivation. The restoration of the activity of an ultraviolet-irradiated DNA phage by means of the excision-repair mechanism of the host, subsequent to the infection of the host cells by the phage. *Abbr* HCR.

host-controlled modification. HOST-INDUCED MODIFICATION.

host-controlled restriction. A host-induced modification that results in the formation of a restricted virus.

host-induced modification. A change in the properties of a virus that is brought about by the propagation of the virus in the host cells. In phages, the change amounts to a chemical alteration such as that produced by glucosylation or methylation of the phage DNA; in animal viruses, the change amounts to an incorporation of host cell membrane components into the viral envelope.

host range. The spectrum of hosts that can be infected by a specific virus.

host-range mutant. A mutant virus that can adsorb to, and infect, cells that are resistant to the wild-type virus.

hot. Containing one or more radioactive isotopes.

hot spot. A site on the DNA molecule that readily undergoes mutation, particularly a site in the rII region of T4 phage DNA.

Houssay animal. A hypophysectomized and depancreatized animal that is used in endocrinological studies.

HPG. Human pituitary gonadotropin.

HPL. Human placental lactogen.

HPLC. High-pressure liquid chromatography.

HPr. A cytoplasmic, heat-stable protein that functions in the transfer of glucose across a bacterial membrane. *See also* enzyme I; enzyme II.

hr. Hour.

HSA. Human serum albumin.

HSK-pathway. Hatch-Slack-Kortschak pathway.

5-HT. 5-Hydroxytryptamine.

5-HTP. 5-Hydroxytryptophan.

HTST method. High-temperature-short-time method.

Hudson's rule. An empirical rule for assigning anomeric configurations to carbohydrates. The rule states that the α-anomer of a given carbohydrate is that anomer which has a more positive specific rotation in the D-series and a more negative specific rotation in the L-series of the carbohydrates than the other anomer of the same carbohydrate.

Huebl number. IODINE NUMBER.

Huebl's iodine solution. A solution of iodine and mercuric chloride that is used in the determination of iodine numbers.

human chorionic gonadotropin. A gonadotropic hormone, produced by the placenta, that has similar biological effects to luteinizing hormone. *Var sp* human chorionic gonadotrophin. *Abbr* HCG.

human menopausal gonadotropin. A preparation derived from the urine of menopausal women that has both follicle-stimulating hormone activity and luteinizing hormone activity. *Var sp* human menopausal gonadotrophin. *Abbr* HMG.

humectant. A hygroscopic substance that is added to other materials, such as a food, to ensure a desired level of moisture content.

humic acid. A complex mixture of acid-insoluble and alkali-soluble substances that are extracted from the organic matter of soil. *See also* fulvic acid; humin.

humin. The heterogeneous fraction of soil organic matter that cannot be extracted with base. *See also* humic acid; fulvic acid.

humor. 1. A chemical substance, such as acetylcholine, that is formed in the body and that acts locally. 2. A fluid or a semifluid substance of an animal or a plant, such as the vitreous humor of the eye.

humoral. Of, or pertaining to, a humor.

humoral immunity. Immunity that is due to circulating antibodies in the blood in contrast to cellular immunity.

humus. The major insoluble portion of the organic substances in soil that is produced by the decomposition of animal and vegetable matter.

Hunter's syndrome. A sex-linked form of gargoylism.

H₂Urd. Dihydrouridine.

Hurler's syndrome. An autosomally linked form of gargoylism.

Huxley's theory. SLIDING FILAMENT MODEL.

HVL. Half-value layer.

hyaline. VITREOUS.

hyaluronic acid. A mucopolysaccharide of connective tissue that is composed of D-glucuronic acid and N-acetyl-D-glucosamine and that aids in blocking the spread of invading microorganisms and toxic substances.

hyaluronidase. The enzyme, present in snake venom and in bacteria, that catalyzes the hydrolysis of hyaluronic acid and thereby decreases the effectiveness of hyaluronic acid for blocking the spread of invading microorganisms and toxic substances in the tissues.

hybrid antibody. An artificially produced antibody molecule that is composed of fragments, such as intact light and heavy chains, which are derived from two purified and different antibodies.

hybrid duplex. A double-stranded nucleic acid molecule that is produced by hybridization.

hybrid hemoglobin. An artificially produced hemoglobin molecule that contains either globin chains or subunits which are derived from different sources.

hybridization. 1. A technique for assessing the extent of sequence homology between single strands of nucleic acids. The technique is based on allowing the polynucleotide strands to form double helical segments through hydrogen bonding between complementary base pairs. The greater the extent of complementarity between the strands, the greater is the extent of formation of double helical segments. The polynucleotide strands may be those of single-stranded nucleic acids or they may be derived from denatured double-stranded nucleic acids. The hybrids formed can be of the DNA/DNA, RNA/RNA, or DNA/RNA type. 2. The reconstitution of an oligomeric protein, such as hemoglobin, or of a molecular aggregate, such as a ribosome, from separate and different monomers and/or component parts. The monomers and/or the component parts may be either chemically or mutationally altered ones, or they may be derived from different sources. 3. The fusion of two cells in tissue culture. 4. The rearrangement of the atomic orbitals in an atom due to the effects of either neighboring or bonding atoms.

hydratase. An enzyme that catalyzes a hydration reaction.

hydrate. A compound that contains one or more molecules of water in loose combination.

hydration. 1. The process whereby water molecules surround, and bind to, solute ions and solute molecules in solution. 2. The formation of a hydrate.

hydration factor. *See* Oncley equation.

hydration shell. The layer of water molecules that are bound to an ion or a molecule in solution. *Aka* hydration sphere.

hydride ion. The anion H^-; a proton with an associated pair of electrons.

hydrion. A hydrogen ion; a proton.

hydrocarbon. An organic compound consisting only of carbon and hydrogen.

hydrodynamic. Of, or pertaining to, the motion of fluids and the force on, as well as the motion of, particles that are immersed in these fluids.

hydrodynamic method. A physical method for studying molecules, particularly macromolecules, on the basis of their movement in solution; includes such methods as sedimentation, diffusion, electrophoresis, and viscosity.

hydrodynamic unit. The unit that moves in solution and that consists of the solute particle together with the solvent that is bound tightly to it.

Hydrogemonas. *See* Knallgas reaction.

hydrogen. An element that is essential to all plants and animals. Symbol H; atomic number 1; atomic weight 1.00797; oxidation states -1, $+1$; most abundant isotope H^1; the stable isotope H^2; the radioactive isotope H^3, half-life 12.26 years, radiation emitted—beta particles.

hydrogenase. An enzyme that catalyzes the reduction of a substrate by molecular hydrogen.

hydrogenation. The introduction of hydrogen into an organic compound.

hydrogen bond. The attractive interaction that occurs between a covalently linked hydrogen atom and a neighboring atom or group of atoms. The hydrogen atom is linked covalently to an electronegative atom referred to as the donor, and is attracted to an electronegative atom or group of atoms referred to as the acceptor; the acceptor is frequently an oxygen or a nitrogen atom. The hydrogen bond is weaker and has a smaller bond energy than a covalent bond; it has a major electrostatic component and a minor covalent component. The hydrogen bond occurs both intra- and intermolecularly and it is known as a bifurcated bond if the hydrogen atom is attracted simultaneously to two acceptor atoms.

hydrogen carrier. An electron carrier that undergoes oxidation-reduction reactions by either the loss or the gain of hydrogen atoms.

hydrogen electrode. An electrode at which hydrogen gas is in equilibrium with hydrogen ions in solution. The hydrogen electrode serves as the primary standard for determinations of pH and electrode potentials.

hydrogen ion concentration. *See* pH.

hydrogen ion equilibrium. The dissociation and association reactions of hydrogen ions, particularly those pertaining to the functional groups in a protein.

hydrogen ion titration curve. A titration curve, particularly that of a protein, in which the number of hydrogen ions that are either bound or dissociated per molecule is plotted as a function of pH.

hydrogen isotope exchange. DEUTERIUM EXCHANGE.

hydrogenosome. A subcellular organelle of trichomonads that contains pyruvate synthase and hydrogenase and that produces molecular hydrogen by using protons as terminal electron acceptors.

hydrogen transport system. ELECTRON TRANSPORT SYSTEM.

hydrolase. An enzyme that catalyzes a hydrolysis reaction. *See also* enzyme classification.

hydrolysate. The solution that contains the mixture of compounds obtained upon hydrolysis. *Var sp* hydrolyzate.

hydrolysis (*pl* hydrolyses). The reaction of a substance with water in which the elements of water (H, OH) are separated: (a) The breakage of a molecule into two or more smaller fragments by the cleavage of one or more covalent bonds of acid derivatives; the elements of water are incorporated at each cleavage point such that one of the products combines with the hydrogen of the water while the other product combines with the hydroxyl group of the water; (b) the formation of the undissociated form of a weak electrolyte through the reaction of the ion of that electrolyte with either protons or hydroxyl ions.

hydrolytic. Of, or pertaining to, hydrolysis.

hydrolytic deamination. Nonoxidative deamination.

hydrolytic enzyme. HYDROLASE.

hydrolyze. To carry out a hydrolysis reaction.

hydrometer. An instrument for measuring the specific gravity of liquids.

hydronium ion. The hydrated proton H_3O^+.

hydrophilic. 1. POLAR. 2. Descriptive of the tendency of a group of atoms or of a surface to become either wetted or solvated by water. *See also* lyophilic.

hydrophobic. 1. NONPOLAR. 2. Descriptive of the tendency of a group of atoms or of a surface to resist becoming either wetted or solvated by water. *See also* lyophobic.

hydrophobic bond. HYDROPHOBIC INTERACTION.

hydrophobic chromatography. A column chromatographic technique for the separation of proteins in which the chromatographic support consists of a nonpolar material, such as a hydrocarbon-coated agarose. Fractionation is achieved primarily on the basis of the interactions that take place between accessible hydrophobic pockets or regions on the proteins and the hydrocarbon chains of the support.

hydrophobic interaction. The attractive force between nonpolar groups which leads to the association of these groups in an aqueous environment. Hydrophobic interactions occur due to the gain in entropy which results when nonpolar groups associate and liberate the water molecules that were previously organized around the nonpolar groups. Hydrophobic interactions are unique in that their strength increases as the temperature is raised in the range of 0 to 60°C.

hydroponics. The growth of plants in aqueous solutions that contain chemical nutrients.

hydrosol. A sol in which the dispersion medium is water.

hydrosphere. The aqueous envelope of the earth; includes oceans, lakes, streams, underground water, and water vapor in the atmosphere.

hydroxide ion. The anion OH^-. *Aka* hydroxyl ion.

hydroxyallysine. A derivative of hydroxylysine in which the epsilon amino group has been converted to an aldehyde group; hydroxyallysine undergoes an aldol condensation with allysine during the crosslinking of collagen chains.

hydroxyapatite. A calcium phosphate-calcium hydroxide complex, $Ca_{10}(PO_4)_6(OH)_2$, that occurs in bone and that is used in the purification of proteins by adsorption chromatography. *Abbr* HA.

β-hydroxybutyric acid. A compound that can be formed from acetyl coenzyme A and that is one of the ketone bodies.

5-hydroxyindoleacetic acid. A catabolite of serotonin. *Abbr* 5-HIAA.

hydroxylamine. A chemical mutagen that leads to the deamination of cytosine which then forms a base pair with adenine.

hydroxylase. MONOOXYGENASE.

hydroxylation. The introduction of a hydroxyl group into an organic compound.

hydroxyl group. The radical $-OH$. *See also* alcoholic hydroxyl group; phenolic hydroxyl group.

hydroxyl value. The number of milligrams of potassium hydroxide that are chemically equivalent to the hydroxyl content of 1 gram of fat.

hydroxylysine. An alpha amino acid that is derived from lysine and occurs in collagen and gelatin. *Abbr* Hyl; Hylys.

hydroxymethyl group. The one-carbon fragment $-CH_2OH$.

hydroxyproline. An alpha imino acid that is derived from proline and occurs in collagen and gelatin. *Abbr* Hyp; Hypro.

5-hydroxytryptamine. SEROTONIN.

5-hydroxytryptophan. An intermediate in the biosynthesis of serotonin. *Abbr* 5-HTP.

Hyflow supercel. Trademark for a preparation of diatomaceous earth.

hygrometer. An instrument for measuring the moisture content of a gas, a liquid, or a solid.

hygroscopic. Descriptive of a substance that readily absorbs moisture from the atmosphere.

Hyl. Hydroxylysine; also abbreviated Hylys.

Hyp. 1. Hydroxyproline; also abbreviated Hypro. 2. Hypoxanthine.

hyper-. Prefix meaning excessive.

hyperacidity. Excessive acidity, particularly that due to increased concentrations of gastric hydrochloric acid which is associated with "heartburn," "indigestion," and peptic ulcers.

hyperammonemia. A genetically inherited metabolic defect in man that is due to a deficiency of the enzyme ornithine carbamoyl transferase, in the case of hyperammonemia I, and due to a deficiency of the enzyme carbamoyl phosphate synthase, in the case of hyperammonemia II.

hyperbilirubinemia. CRIGLER-NAJJAR SYNDROME.

hyperbolic inhibition. Inhibition that yields a curve that is convex upward when either slopes or intercepts from a primary plot are plotted as a function of inhibitor concentration.

hyperbolic kinetics. *See* Michaelis-Menten kinetics.

hypercalcemia. The presence of excessive amounts of calcium in the blood.

hypercalcemic factor. PARATHORMONE.

hypercapnia. The presence of excessive amounts of carbon dioxide in the blood.

hyperchloremia. The presence of excessive amounts of chloride in the blood.

hyperchlorhydria. The presence of excessive amounts of hydrochloric acid in the gastric juice.

hypercholesterolemia. CHOLESTEROLEMIA.

hyperchromic effect. The increase in the ultraviolet absorbance of a solution containing either DNA or RNA that occurs upon either the denaturation or the degradation of the nucleic acid. The effect is due to changes in the electronic interactions between the bases and is generally measured at 260 nm.

hyperchromicity. The increase in absorbance that is due to the hyperchromic effect and that is measured at a specified wavelength.

hyperchromism. 1. The increase in absorbance that is due to the hyperchromic effect and that is measured between two wavelengths, specifically between zero and infinity, or between two forms, or two states, of the absorbing substance. 2. HYPERCHROMICITY.

hyperfine splitting. The breakup of a spectral peak into a series of peaks. In electron paramagnetic resonance this is due to the interaction of electrons with other electrons or nuclei; in nuclear magnetic resonance this is due to spin-spin coupling.

hyperglycemia. The presence of excessive amounts of glucose in the blood.

hyperglycemic factor. GLUCAGON.

hyperglycemic-glycogenolytic factor. GLUCAGON.

hyperimmunity. Immunity that is characterized by the presence of large amounts of antibodies in the blood, and that may be produced by repeated injections of antigens into an animal organism.

hyperinsulinism. The presence of excessive amounts of insulin in the body, resulting in hypoglycemia.

hyperkalemia. HYPERPOTASSEMIA.

hyperlipemia. LIPEMIA.

hypermagnesemia. The presence of excessive amounts of magnesium in the blood.

hypermorph. A mutant gene that has a similar, but stronger, effect than its wild-type gene.

hypernatremia. The presence of excessive amounts of sodium in the blood.

hyperon. A nuclear particle that has a mass greater than that of a nucleon.

hyperphosphatemia. The presence of excessive amounts of phosphate in the blood.

hyperplasia. The increase in the size of a tissue or an organ, excluding tumor formation, that results from an increase in the number of the component cells.

hyperploid state. The aneuploid state in which the chromosome number is greater than that of the characteristic euploid number. *Aka* hyperploidy.

hyperpotassemia. The presence of excessive amounts of potassium in the blood.

hyperprolinemia. A genetically inherited metabolic defect in man that is due to a deficiency of the enzyme proline oxidase, in the case of hyperprolinemia I, and due to a deficiency of the enzyme pyrroline-5-carboxylate reductase, in the case of hyperprolinemia II.

hypersensitive response. ALLERGIC RESPONSE.

hypersensitivity. 1. The altered immunological state that is produced in man and animals by their previous exposure to an antigen and that is characterized by the occurrence of different and pathological reactions upon their subsequent exposure to either the same antigen or a structurally related substance. *Aka* allergy. 2. The above average response of man, or an animal, to a drug.

hypertensin. ANGIOTENSIN.

hypertensinogen. ANGIOTENSINOGEN.

hypertension. High blood pressure.

hypertensive agent. An agent that brings about an increase in the blood pressure.

hyperthermia. A higher than normal temperature.

hyperthyroidism. Excessive activity of the thyroid gland that leads to an increase in oxygen consumption and to an increase in the overall metabolic rate.

hypertonic contraction. An alteration in the water and electrolyte balance in the body in which there is a decrease in the volume of the extracellular fluid but no equivalent loss of electrolytes, so that the osmotic pressure of the extracellular fluid increases.

hypertonic expansion. An alteration in the water and electrolyte balance in the body in which there is an increase in the electrolyte concentration and a less than equivalent increase in the volume of the extracellular fluid, so that the osmotic pressure of the extracellular fluid increases.

hypertonic solution. A solution that has a higher osmotic pressure than another solution.

hypertrophy. The increase in the size of a tissue or an organ, excluding tumor formation, that is due to either an increase in the volume of the component cells, or an increase in the functional activity of the tissue or the organ.

hyperuricemia. The presence of excessive amounts of uric acid in the blood.

hypervalinemia. A genetically inherited metabolic defect in man that is due to a deficiency of the enzyme valine transaminase.

hypervariable codon. A codon that has undergone an excessive number of mutations when it is considered in the context of the biochemical evolution of proteins.

hyperventilation. Excessive aeration of the blood in the lungs.

hypervitaminosis. A pathological condition resulting from an overdose of a vitamin, such as that produced by an overdose of either vitamin A or vitamin D.

hypha (*pl* hyphae). The filamentous and branched tube that forms the network that contains the cytoplasm of the mycelium of a fungus.

hypo-. Prefix meaning less than.

hypobaria. A lower than normal atmospheric pressure.

hypocalcemia. A deficiency of calcium in the blood.

hypocalcemic factor. CALCITONIN.

hypocapnia. A deficiency of carbon dioxide in the blood.

hypochloremia. A deficiency of chloride in the blood.

hypochlorhydria. A deficiency of hydrochloric acid in the gastric juice.

hypocholesterolemia. A deficiency of cholesterol in the blood.

hypochromic anemia. An anemia that is characterized by a decrease in the hemoglobin concentration of the red blood cells.

hypochromic effect. The decrease in the ultraviolet absorbance of a solution containing either DNA or RNA compared to the absorbance of the same solution but containing either the denatured or the degraded nucleic acid. The effect is due to changes in the electronic interactions between the bases and is generally measured at 260 nm.

hypochromicity. The decrease in absorbance that is due to the hypochromic effect and that is measured at a specified wavelength.

hypochromism. 1. The decrease in absorbance that is due to the hypochromic effect and that is measured between two wavelengths, specifically between zero and infinity, or between two forms, or two states, of the absorbing substance. 2. HYPOCHROMICITY.

hypofibrinogenemia. A pathological condition characterized by low levels of fibrinogen in the blood.

hypogammaglobulinemia. A genetically inherited metabolic defect in man that is characterized by an almost complete absence of immunoglobulins. *See also* agammaglobulinemia.

hypoglycemia. A deficiency of glucose in the blood.

hypoglycemic factor. INSULIN.

hypokalemia. HYPOPOTASSEMIA.

hypomagnesemia. A deficiency of magnesium in the blood.

hypomorph. A mutant gene that has a similar, but weaker, effect than its wild-type gene.

hyponatremia. A deficiency of sodium in the blood.

hypophosphatasia. A genetically inherited metabolic defect in man that is characterized by deficient bone formation and that is due to a deficiency of the enzyme serum alkaline phosphatase.

hypophosphatemia. A deficiency of phosphate in the blood.

hypophyseal. Of, or pertaining to, the pituitary gland.

hypophysectomy. The surgical removal of the pituitary gland.

hypophysin. An extract made of the posterior lobe of the pituitary gland.

hypophysis. PITUITARY GLAND.

hypoplasia. 1. The arrested development of a tissue or an organ. 2. The decrease in the size of a tissue or an organ that results from a decrease in the number of the component cells.

hypoploid state. The aneuploid state in which the chromosome number is less than that of the characteristic euploid number. *Aka* hypoploidy.

hypopotassemia. A deficiency of potassium in the blood.

hyposensitivity. The below average response of man, or an animal, to a drug.

hypostatic gene. A gene, the expression of which is suppressed by another gene in epistasis.

hypotension. Low blood pressure.

hypotensive agent. An agent that brings about a lowering of the blood pressure.

hypothalamic hormone. A hormone that is produced by the hypothalamus and that either stimulates or inhibits the release of another hormone. Hypothalamic stimulatory hormones stimulate the release of corticotropic hormone, luteinizing hormone, follicle-stimulating hormone, thyrotropic hormone, and growth hormone; hypothalamic inhibitory hormones inhibit the release of prolactin and melanocyte-stimulating hormone. *Aka* hypothalamic factors. *See also* chalone.

hypothalamus. A basal part of the brain.

hypothermia. A lower than normal temperature.

hypothesis (*pl* hypotheses). A statement describing a relationship that may be either true or untrue; the statement appears to be confirmed by experimental facts but is not as certain as a theory. A hypothesis is a tentative explanation or description of observed phenomena that remains to be critically tested by further experiments.

hypothyroidism. Deficient activity of the thyroid gland that leads to a decrease in the oxygen consumption and to a decrease in the overall metabolic rate.

hypotonic contraction. An alteration in the water and electrolyte balance in the body in which there is a decrease in the volume of the extracellular fluid and a more than equivalent decrease of electrolytes, so that the osmotic pressure of the extracellular fluid decreases.

hypotonic expansion. An alteration in the water and electrolyte balance in the body in which there is an increase in the volume of the extracellular fluid but no equivalent increase in electrolytes, so that the osmotic pressure of the extracellular fluid decreases.

hypotonic solution. A solution that has a lower osmotic pressure than another solution.

hypotrophy. 1. Subnormal growth. 2. The degeneration and loss of function of a tissue or an organ.

hypoventilation. Deficient aeration of the blood in the lungs.

hypovitaminosis. The deficiency disease that results from the inadequate dietary intake of one or more vitamins.

hypoxanthine. The purine 6-hydroxypurine that occurs in transfer RNA and that is formed in catabolism by the deamination of adenine. *Sym* I.

hypoxia. A condition in which the oxygen concentration of the blood and the tissues is below normal either as a result of environmental deficiency or as a result of impaired respiration and circulation.

Hypro. Hydroxyproline.

hypsochromic group. A group of atoms that, when attached to a compound, shifts the absorption of light by the compound to shorter wavelengths.

hypsochromic shift. A shift in the absorption spectrum of a compound toward shorter wavelengths.

hysteresis. The phenomenon in which changes of a property in one direction can be reversed back to the initial value of the property, but in so doing take on intermediate values that differ from those taken on in the forward direction. A plot of such a system results in two curves, coming together at both ends and forming a hysteresis loop in between.

hysteretic. Of, or pertaining to, hysteresis.

hysteretic enzyme. An enzyme that appears to undergo very slow changes in either kinetic or molecular properties after the addition or the removal of a ligand; believed to be due to conformational changes of the enzyme as a result of the change in ligand concentration.

Hz. Hertz.

H zone. The central, less dense portion of the A band of the myofibrils of striated muscle. *Aka* H band.

I

I. 1. Inosine. 2. Hypoxanthine. 3. Isoleucine. 4. Iodine. 5. Ionic strength. 6. Luminous intensity. 7. Electric current.

IAA. 1. Indoleacetic acid. 2. Monoiodoacetamide.

i-assay. The construction of a thermal denaturation profile by measurements of the system after it has been rapidly returned to standard conditions; the assay measures the reversibility of the transition. *See also* d-assay.

I band. The transverse light band that is seen in electron microscope preparations of the myofibrils from striated muscle and that is formed by the thin filaments.

ic. 1. *adj* Intracutaneous. 2. *adv* Intracutaneously.

ICD. Isocitrate dehydrogenase; an enzyme of the citric acid cycle.

iceberg. A cluster of water molecules that have become stabilized around a nonpolar group; the process is accompanied by a loss of entropy, and the water molecules possess greater crystallinity than that of the molecules in ordinary water.

ICH. Interstitial cell hormone.

Ichthyotocin. A peptide hormone of nine amino acids that is secreted by the posterior lobe of the pituitary gland of teleosts and that is related to oxytocin in its structure and in its function.

icosadeltahedron. The solid obtained by subdividing the surfaces of an icosahedron into smaller equilateral triangles.

icosahedral. Of, or pertaining to, an icosahedron.

icosahedral virion. A virus, such as poliovirus or adenovirus, in which the capsid has the shape of an icosahedron.

icosahedron. A symmetrical polyhedron having 12 vertices and 20 faces, with each face being an equilateral triangle; frequently descriptive of the structure of viruses.

ICSH. Interstitial cell stimulating hormone.

icterus index. A liver function test that measures the approximate level of bilirubin in either serum or plasma; based on a comparison of the color of either diluted serum or diluted plasma with a standard solution of potassium dichromate.

icy group. A group of compounds that have neither very high nor very low melting points and that are believed to have occurred in the original gas-dust of the solar nebula. *See also* earthy group; gaseous group.

i.d. Inside diameter.

ID$_{50}$. Median infectious dose.

ideal gas. A gas, the behavior of which is accurately described by the various gas laws.

ideal solution. A dilute solution in which all of the solutes follow Raoult's law; a solution such that the chemical potential μ_i of each component is given by $\mu_i = \mu_i^0 + RTlnX_i$, where μ_i^0 is the standard chemical potential (the potential of the substance in its pure state), R is the gas constant, T is the absolute temperature, and X_i is the mole fraction of component i.

identical twins. MONOZYGOTIC TWINS.

idiogram. A diagrammatic representation of the chromosome complement of an individual according to the size and/or a numbering system of the chromosomes.

idiopathic. 1. Denoting a disease of unknown cause. 2. Peculiar to an individual.

idiopathic pentosuria. PENTOSURIA.

idiotype. One of two or more antibodies that have different origins but are directed against the same antigenic determinant.

IDP. 1. Inosine diphosphate. 2. Inosine-5′-diphosphate.

IE. Immunoelectrophoresis.

IEF. Isoelectric focusing.

IF. 1. Initiation factor. 2. Intrinsic factor. 3. Isoelectric focusing.

I form. Independent form.

Ig. Immunoglobulin.

IgA. An immunoglobulin that is present in serum in moderate concentrations, has a variable sedimentation coefficient of 10 to 17S, is a small multimer of the 7S immunoglobulin, and is incapable of passing the placenta. *Aka* γA.

IgD. An immunoglobulin that is present in serum in low concentrations and that has a sedimentation coefficient of 7S. *Aka* γD.

IgE. An immunoglobulin that has a sedimentation coefficient of 8S, is present in serum in very low concentrations, and constitutes the reaginic antibody. *Aka* γE.

IgG. An immunoglobulin that is present in serum in high concentrations, has a sedimentation coefficient of 7S, has a molecular weight of 150,000, and is capable of passing the placenta. *Aka* γG.

IgM. An immunoglobulin that is present in serum in moderate concentrations, has a sedimentation coefficient of 19S, is a high-molecular weight multimer of the 7S immunoglobulin, and is incapable of passing the placenta. *Aka* γM.

ILA. Insulin-like activity.

Ile. 1. Isoleucine. 2. Isoleucyl.

ileum. The third and lowest portion of the small intestine.

Ilkovic equation. A basic equation of polarography.

illicit transport. The entry of a substance into a cell by means of a transport system that is designed for another substance; the transport of the derivative of a compound, where the compound itself cannot be transported, is an example.

im. 1. *adj* Intramuscular. 2. *adv* Intramuscularly.

imbalance theory. A theory of cancer according to which all organisms must evolve systems of growth regulation; a tumor arises when there is a failure in this regulation, and the imbalance in favor of growth passes a certain threshold.

imbibition. The uptake of fluid by a substance, as the uptake of water by a colloidal system.

ImD$_{50}$. Median immunizing dose.

imidazole group. The heterocyclic ring system of the amino acid histidine. *Aka* imidazolium group.

imine. An organic compound that contains the grouping $-N{=}C{-}CH{-}$.

imine-enamine tautomerism. The tautomerism that is due to a shift of a hydrogen atom so that one of the isomers is an imine and the other is an enamine.

imino acid. An acid derived from an imine; proline and hydroxyproline are alpha imino acids in which the nitrogen of the imino group and the carboxyl group are attached to the same carbon atom.

imino group. The grouping $-NH-$.

immediate early RNA. PREEARLY RNA.

immediate-type hypersensitivity. An allergic response that occurs soon, generally within a few minutes, after the administration of an antigen to an animal organism; the response is mediated by circulating antibodies.

immiscible. Incapable of being mixed.

immobile phase. STATIONARY PHASE (2).

immobilized enzyme. An enzyme that is physically confined while it carries out its catalytic function. This may occur naturally, as in the case of particulate enzymes, or it may be produced artificially by chemical or by physical methods. In the chemical methods, the enzyme is linked covalently to a support. These methods include attachment of the enzyme to a water-insoluble support, incorporation of the enzyme into a growing polymer chain, or cross-linking of the enzyme with a multifunctional low-molecular weight reagent. In the physical methods, the enzyme is not linked covalently to a support. These methods include adsorption of the enzyme to a water-insoluble matrix, entrapment of the enzyme within either a water-insoluble gel or a microcapsule, or containment of the enzyme within special devices equipped with semipermeable membranes.

immune. Of, or pertaining to, an organism that has been immunized.

immune adherence. The attachment of a complex—composed of particulate antigens, antibodies, and complement—to the surfaces of nonsensitized particles such as erythrocytes, platelets, yeast, or starch granules.

immune adsorbent. *See* immunoadsorbent.

immune antibody. ACQUIRED ANTIBODY.

immune clearance. IMMUNE ELIMINATION.

immune competent cell. *See* immunocompetent cell.

immune complex. ANTIGEN-ANTIBODY COMPLEX.

immune conglutination. A conglutination reaction caused by an immunoconglutinin.

immune conglutinin. *See* immunoconglutinin.

immune cytolysis. The lysis of cells by antibodies in the presence of complement.

immune elimination. The stage in an immune response during which the antigen is rapidly removed from the blood as a result of its combination with the antibody.

immune globulin. *See* immunoglobulin.

immune hemolysis. Hemolysis that results from complement fixation.

immune opsonin. *See* opsonin.

immune response. The formation of antibodies in an animal organism in response to an immunization and the reactions of these antibodies with the antigens used in the immunization.

immune serum (*pl* immune sera). ANTISERUM.

immunity. 1. The resistance of an individual or an animal to a specific disease, infecting agent, or toxic antigen. 2. The capacity of lysogenic bacteria to withstand infection by phage particles that are of the same kind as the prophage of the bacteria.

immunity substance. A cytoplasmic factor that is formed under the control of a phage gene and that confers immunity on a lysogenic bacterium; it functions as a repressor of the vegetative replication of the prophage in that bacterium.

immunization. 1. The administration of an antigen to an animal organism to stimulate the production of antibodies by that organism. 2. The administration of antigens, antibodies, or lymphocytes to an animal organism to produce the corresponding active, passive, or adoptive immunity.

immuno-. Combining form meaning immunology.

immunoadsorbent. An insoluble material that is used for the purification of antibodies by adsorbing them from a serum; a gel for trapping antibodies, or an inert solid to which either antigens or haptens have been covalently linked are two examples.

immunoassay. An assay that utilizes antigen-antibody reactions for the determination of biochemical substances.

immunoblast. A blast cell that is a forerunner of an immunocyte.

immunochemistry. The science that deals with the chemical aspects of immunology and combines the techniques of biochemistry and immunology.

immunochromatography. IMMUNO-GEL FILTRATION.

immunocompetent cell. A cell that has the capacity to recognize antigens and/or to synthesize antibodies.

immunoconglutinin. An antibody that is specific for antigenic determinants which are exposed in fixed complement, but which are unavailable for reaction in free complement. *See also* conglutinin.

immunocore electrophoresis. A separation technique that is based on the combined use of electrophoresis and immunodiffusion. Antigens or antibodies are first separated by disc gel electrophoresis, and the gel column is then extruded. A core of the gel column is removed and replaced with either an antiserum or a solution containing antigens. Following this, the gel is incubated to allow for the formation of precipitin bands of antigen-antibody complexes.

immunocyte. 1. IMMUNOCOMPETENT CELL. 2. An immunocompetent lymphocyte.

immunocyte adherence. A technique for detecting cells that carry antibodies on their surfaces either because they produce the antibodies or because the antibodies have become bound to the cells. The cells are reacted with the corresponding antigens or with other cells that are coated with soluble antigens. An antibody-bearing cell binds the antigens and forms a rosette-type structure, and the number of these rosettes is then determined microscopically. *Aka* rosette technique.

immunodiffusion. A method for carrying out the precipitin reaction in a gel that is based on the diffusion of antibody and/or antigen molecules through the gel. *See also* double diffusion; single diffusion.

immunoelectroadsorption. A method for measuring antibody concentrations in serum. A layer of the appropriate antigens is adsorbed onto a glass slide with the aid of an electrical current, and the antibodies in the given serum are then adsorbed onto the antigens. The thickness of the antibody layer is determined and it provides a measure of the antibody concentration in the serum.

immunoelectrofocusing. One of several separation techniques that are based on the combined use of gel electrofocusing and either immunodiffusion or immunoelectrophoresis; performed by incorporating either all or fractions of the gel from a gel electrofocusing experiment into a gel to be used for either immunodiffusion or immunoelectrophoresis.

immunoelectronmicroscopy. The use of electron microscopy in conjunction with immunochemical methods, as in the staining of electron microscope specimens with ferritin-labeled antibodies.

immunoelectrophoresis. A technique for identifying antigens in complex mixtures by first separating the antigens in one dimension by means of gel electrophoresis, and then allowing them to react with antibodies by means of two dimensional double diffusion through the gel; a pattern of precipitin arcs is thereby produced. *Abbr* IE.

immunoferritin. FERRITIN-LABELED ANTIBODY.

immunofiltration. The purification of an immunological solution by passing it through an immunoadsorbent.

immunofluorescence. *See* direct fluorescent antibody technique; indirect fluorescent antibody technique; anticomplement fluorescent antibody technique.

immuno-gel filtration. A separation technique that is based on the combined use of immunodiffusion and thin-layer gel filtration.

immunogen. 1. ANTIGEN. 2. A substance capable of producing an immune response that leads to the synthesis of antibodies.

immunogenetics. The branch of immunology that deals with the interrelationships of immunological reactions and the genetic makeup of an organism.

immunogenicity. ANTIGENICITY.

immunoglobulin. 1. A protein of animal origin that has a known antibody activity. 2. A protein that is closely related to an antibody by its chemical structure and by its antigenic specificity. 3. A protein of immunological significance, such as myeloma or Bence-Jones protein, for which antibody activity has not yet been demonstrated. *See also* IgA, IgD, etc.

immunoisoelectric focusing. IMMUNO-ELECTROFOCUSING.

immunologic. IMMUNOLOGICAL.

immunological. Of, or pertaining to, immunology.

immunological competence. *See* competence (2).

immunological enhancement. *See* enhancement.

immunological equivalence. The amounts of antigen and antibody that precipiate each other in the equivalence zone of the precipitin curve.

immunological inhibition. The competitive inhibition of antibody formation that is produced either by the administration of antibodies against the stimulating antigen, or by the administration of unconjugated hapten together with active hapten-protein conjugates.

immunologically competent cell. *See* immunocompetent cell.

immunological memory. The enhanced capacity of an animal to respond to a second dose of antigen that is characteristic of the secondary immune response.

immunological paralysis. 1. IMMUNOLOGICAL TOLERANCE. 2. Immunological tolerance produced by large doses of polysaccharide antigens.

immunological rejection. The destruction of foreign cells and tissues, which are either inoculated or transplanted into a recipient from a donor, by a specific immune reaction.

immunological tolerance. The decrease in, or the loss of, the ability of an animal to produce an immune response upon the administration of a particular antigen. Tolerance is induced by prior exposure of the animal to the same antigen and does not affect the ability of the animal to respond to other antigens.

immunological unresponsiveness. 1. IMMUNOLOGICAL TOLERANCE. 2. Immunological tolerance produced by large doses of protein antigens (overloading).

immunological zoo. The set of reagents required in the complement fixation test.

immunology. The science that deals with resistance to disease.

immunoprecipitation. The precipitation of either antigens or antibodies as a result of a precipitin reaction.

immunoreactive insulin. That fraction of serum insulin that is not protein-bound and that is readily neutralized by antiinsulin serum; the bulk of this insulin is called little insulin and a smaller fraction is called big insulin. *Abbr* IRI.

immunosorbent. An immunological sorbent. *See also* absorption (4); adsorption (2).

immunosuppressant. *See* immunosuppressive agent.

immunosuppression. The prevention of an immune response by physical, chemical, or biological means.

immunosuppressive agent. A physical, chemical, or biological agent that prevents the production of an immune response by an antigen.

immunotherapy. The treatment of a disease by immunization.

IMP. 1. Inosine monophosphate (inosinic acid). 2. Inosine-5′-monophosphate (5′-inosinic acid).

impedance. The resistance to the flow of an alternating electric current; equal to the ratio of the complex potential difference applied across a circuit element to the complex current flowing through the element. *Sym* Z.

impermeable. Not permeable.

impervious. Impenetrable; not permeable.

inactivation. The destruction of biological activity.

inactivation cross-section. A measure of the sensitivity of a target to inactivation by irradiation that is equal to the product of the quantum yield and the absorption cross-section.

inactivation probability. The ratio of the target molecular weight, calculated from radiation inactivation experiments, to the actual molecular weight.

inanition. The exhaustion that results from either the lack of food or the inability to assimilate it.

inapparent infection. A transient viral infection that does not produce overt disease symptoms. *See also* latent infection (2).

inborn error of metabolism. A genetically inherited metabolic defect that results in the synthesis of a modified enzyme or other protein, or in the complete lack of synthesis of an enzyme or other protein. *See also* genetic disease.

inbred strain. A strain of experimental animals produced by sequential brother-sister matings over many generations so that all the individuals are genetically identical.

inbreeding. The crossing of plants or animals that are closely related genetically.

incident. Falling on or striking.

incision. A break in a single strand of a nucleic acid; a nick.

incision enzyme. An enzyme that catalyzes the cleavage of a polynucleotide chain; an endonuclease.

inclusion. A discrete mass within the cell of either foreign or metabolically passive material.

inclusion body. A mass of virus particles within the cell of an animal that is infected with the virus.

incompatibility. *See* histoincompatibility.

incomplete antibody. An antibody that does not give the serologic reactions of precipitation and agglutination, and that can only be demonstrated either indirectly by means of special techniques or by means of its in vivo biological effect.

incomplete antigen. HAPTEN.

incomplete oxidation. The oxidation of organic compounds such that partially oxidized organic compounds are the final products; the term may refer either to a group of reactions or to a single reaction.

incomplete protein. A protein that is deficient in one or more of the amino acids commonly found in proteins.

incubation. The maintaining of organisms, reaction mixtures, and the like in an incubator or in some other constant temperature environment.

incubation mixture. A reaction mixture that is maintained at a specified temperature.

incubation period. The time interval between the invasion of an organism by a virus or by a bacterium and the appearance of overt disease symptoms.

incubator. A constant temperature chamber used principally to provide a controlled environment for the growth of cells and organisms.

independent assortment. The random distribution of genes, located on nonhomologous chromosomes, during gametogenesis.

independent form. The dephosphorylated form of the enzyme glycogen synthetase that is active in the absence of glucose-6-phosphate.

independent variable. A quantity that can assume any arbitrarily chosen value independent of the values of other related variables.

indeterminacy principle. UNCERTAINTY PRINCIPLE.

indeterminate error. An error in a measurement which is due to the fact that all physical measurements require a degree of estimation in their evaluation; such errors can be decreased in magnitude but cannot be eliminated entirely.

index fossil. A fossil that is of widespread occurrence in one or in a few contiguous geological layers and that can be used to correlate the ages of geological deposits at various locations.

index of discrimination. The ratio of the activities of a hormone that is assayed by two different methods.

index of hydrogen deficiency. A measure of the degree of unsaturation in a molecule; equal to the number of pairs of hydrogen atoms that must be removed from the formula of a saturated hydrocarbon, which has the same number of carbon atoms as a given compound, to yield the molecular formula of the given compound.

index of precision. 1. The standard deviation of the responses to a hormone divided by the slope of the dose-response curve. 2. The reciprocal of the slope of the dose-response curve of a hormone assay.

index of refraction. *See* refractive index.

indicator. A weak acid or a weak base in which the proton donor and the proton acceptor species have different colors; used for indicating end points of acid-base titrations, since the color of the indicator is determined by the relative concentrations of the two species which, in turn, are determined by the pH of the solution.

indicator enzyme. MARKER ENZYME.

indicator strain. SENSITIVE STRAIN.

indicator virus. A virus that has been heated so that its neuraminidase activity is destroyed but its hemagglutinating capacity is retained.

indirect-acting bilirubin. The water-insoluble and unconjugated form of bilirubin that does not give a color reaction with diazotized sulfanilic acid unless alcohol is added first to solubilize the bilirubin. *See also* direct-acting bilirubin.

indirect calorimetry. A method for determining the basal metabolic rate of an animal from measurements of the amount of nitrogen excreted in the urine, the volume of oxygen inhaled, the volume of carbon dioxide exhaled, and the respiratory quotient. *See also* direct calorimetry.

indirect complement fixation test. A complement fixation test for determining antibodies that fail to fix guinea pig complement; entails the addition of an antiserum of known ability to fix guinea pig complement to an antigen-antibody-guinea pig complement system, followed by the addition of a hemolytic system.

indirect Coombs' test. A Coombs' test in which the red blood cells are coated with antibody in vitro. *See also* direct Coombs' test.

indirect effect. The change brought about in a molecule as a result of its interaction with molecules, radicals, atoms, or electrons that are produced by a direct interaction of atoms or molecules with radiation. *See also* direct effect.

indirect fluorescent antibody technique. A fluorescent antibody technique in which the antigen of interest is first reacted with its immunoglobulin and then with a fluorescent antibody against the immunoglobulin. This technique is more sensitive than the direct one, since it multiplies the number of fluorochromes per antigen that is being stained. The technique can also be applied to the staining of antibodies. *Aka* sandwich technique; antiglobulin method. *See also* direct fluorescent antibody technique.

indirect induction. CROSS-INDUCTION.

indirect photoreactivation. The recovery of cells from the damage caused by their irradiation with ultraviolet light that is brought about by exposure of the damaged cells to light in the wavelength range of 310 to 370 nm; may be due to an inhibition of growth which allows more time for the repair of the damaged DNA.

indispensable enzyme. ESSENTIAL ENZYME.

indispensable gene. ESSENTIAL GENE.

indole. The aromatic and heterocyclic ring system of the amino acid tryptophan.

indoleacetic acid. An auxin that also has vaso-constrictor activity. *Abbr* IAA.

induced dipole moment. A dipole moment produced in a substance by the application of an external electric or magnetic field.

induced enzyme. *See* inducible enzyme.

induced-fit model. SEQUENTIAL MODEL.

induced-fit theory. A modification of the lock and key theory for the binding of a substrate to an enzyme. According to this theory, the substrate induces a conformational change in the enzyme such that the catalytic and the binding groups of the enzyme achieve the required orientation that permits their precise alignment and interaction with the substrate.

induced hypersensitivity. The hypersensitivity produced in an animal by contact with an antigen.

induced mutation. A mutation produced by the intentional exposure of an organism to a mutagen, in contrast to a spontaneous mutation.

induced radioactivity. Radioactivity that is artificially produced by bombardment of nuclei with high-velocity particles.

induced tumor. A tumor that arises subsequent to, and as a result of, the exposure of an organism to a carcinogen.

inducer. 1. The substance that brings about the synthesis of an inducible enzyme in the process of enzyme induction and that is generally either a substrate of the enzyme or a compound which is structurally similar to the substrate. 2. The substance used to induce an allergic state in an animal.

inducible enzyme. An enzyme that is normally either absent from a cell or present in very small amounts, but that is synthesized in appreciable amounts in response to an inducer in the process of enzyme induction. *See also* constitutive enzyme.

inducible system. The regulatory system consisting of the components that function in enzyme induction. *See also* enzyme induction.

inductance. The property of a conductor by which an electromotive force is induced in it by variations in an inducing electric current.

induction. 1. ENZYME INDUCTION. 2. The stimulation of a lysogenic bacterium that causes it to shift to a lytic cycle, to produce infective phage particles, and ultimately to lyse in a burst. *See also* detachment. 3. The reasoning from particulars to generals. 4. MORPHO-GENIC INDUCTION. 5. INDUCTIVE EFFECT.

induction effect. DIPOLE-INDUCED DIPOLE INTERACTION.

induction period. LAG PERIOD.

induction profile. The pattern of enzyme induction that is produced by a given inducer.

induction ratio. The ratio of the concentration of the induced enzyme to that of the basal enzyme. *Abbr* IR.

inductive effect. The partial charge induced in an atom, or in a group of atoms, of a molecule as a consequence of the electron-withdrawing and electron-donating properties of neighboring atoms and groups of atoms.

inductive phase. The time interval between the administration of an antigen to an animal and the appearance of antibodies in the serum.

inductor. 1. A substance that increases the rate of a chemical reaction and that is used up during the reaction. 2. A substance that acts like an organizer in affecting the development of embryonic and other undifferentiated tissue.

inert. Chemically and/or physiologically inactive.

infantile myxedema. CRETINISM.

infection. 1. The introduction of an infective agent, such as a virus or a bacterium, into a host cell or a host organism. 2. The condition produced by the introduction of an infective agent into a cell or an organism.

infectious hepatitis. A form of hepatitis that is caused by a viral infection that is transmitted through the mouth and the intestine.

infectious nucleic acid. A purified viral nucleic acid that can infect a host cell and lead to the production of infective viral particles.

infectious RNA. VIROID.

infectious titer. The number of infective units in a viral sample. *Aka* infective titer.

infective center. The phage particle or the phage-infected bacterium that forms a single plaque in the plaque assay.

infective unit. A virus particle in a viral sample that leads to the infection of a host cell. *Aka* infectious unit.

infectivity. The capacity of bacteria and viruses for interacting with, and altering, host cells.

infinite dilution. A solute concentration of zero to which the values of physical parameters are commonly extrapolated.

infinite thickness. SATURATION THICKNESS.

infinite thinness. A layer of solid radioactive material that is so thin that self-absorption is negligible and that the sample can be counted as if it were infinitely thin.

influent. ELUENT.

influenza virus. The flu virus that causes respiratory infections and that belongs to the group of myxoviruses.

influx. Inward flow, as that into a cell.

informational molecule. A molecule that carries genetic information in the form of specific sequences of building blocks. The term is generally restricted to DNA, messenger RNA, viral RNA, and to the anticodon segment of transfer RNA.

information theory. The branch of science that deals with the measurement, processing, and transmission of information; an extension of thermodynamics and probability theory that attempts to resolve information into a series of binary (yes/no) decisions.

informofer. A globular protein particle with which HnRNA is complexed.

informosome. A cytoplasmic messenger RNA-protein complex that is thought to function in the protection of messenger RNA against nuclease activity and in the transport of messenger RNA from the nucleus to the cytoplasm.

infrared dichroism. The dichroism of polarized infrared light that is used in the study of polypeptides.

infrared spectrum. That part of the electromagnetic spectrum that covers the wavelength range of about 7.5×10^{-5} to 4.2×10^{-2} cm and that includes photons which are emitted or absorbed during vibrational and rotational transitions. *Abbr* IR spectrum.

infusion. An extract made by soaking a substance, such as meat or a plant, in water.

ingest. To take material, such as food, into either a cell or a body.

inheritance of acquired characteristics. The hereditary transmission of structural changes in organisms that is postulated by the Lamarckian theory.

inherited immunity. NATURAL IMMUNITY.

inhibition. A decrease in the extent and/or the rate of an activity, or the complete abolition of an activity.

inhibition analysis. The study of secondary agents that can bring about the reversal of the inhibition of an enzyme; based on measurements of the precursors and products of the inhibited enzymatic reaction, and on measurements of substances that increase the enzyme concentration or influence the rates at which metabolites or inhibitors are destroyed.

inhibition coefficient. The lowest concentration of a bacteristatic agent that inhibits bacterial growth under defined conditions.

inhibition index. A measure of the potency of an antimetabolite; equal to the ratio of the concentration of antimetabolite that is required to inhibit the effect of an essential metabolite, to the concentration of the metabolite.

inhibition ratio. The ratio of the amount of an antivitamin that is required to inhibit the effect of a given amount of a vitamin.

inhibition zone. The antigen excess zone in a precipitin curve.

inhibitor. An agent that produces inhibition.

inhibitor constant. The equilibrium dissociation constant of the reaction $EI = E + I$, where E is the enzyme, and I is the inhibitor. *Sym* K_i.

inhibitor source material. A plasma glyceride that inactivates factor VIII of blood clotting.

inhibitory autacoid. CHALONE (2).

initial heat. The heat produced by a muscle when it is stimulated by either an electric shock or a nerve impulse.

initial velocity. The reaction velocity at the early stages of an enzymatic reaction; measured before the substrate concentration has decreased significantly (usually meaning that less than 5% of the initial substrate has been utilized) and while the concentrations of the products are low, so that the reverse reaction can be neglected. The initial velocity is given by the tangent, at the origin, to the curve which is obtained by plotting reaction velocity as a function of time. *Sym* v. *Aka* initial rate; initial steady-state rate; instantaneous velocity.

initiation. 1. The process of chain initiation during protein synthesis, specifically the formation of the ribosome-messenger RNA-initiator tRNA complex. 2. The first stage in a two-stage or multistage mechanism of carcinogenesis during which a normal cell is converted to a precancerous cell by the action of a carcinogen. 3. The second stage in the germination of a spore. *See also* germination. 4. The first step in a chain reaction.

initiation codon. The codon *AUG* that codes for the binding of N-formylmethionyl-tRNA which initiates protein synthesis in bacterial systems. *Aka* initiator codon.

initiation complex. The initial complex formed during protein synthesis which, in bacterial systems, consists of N-formylmethionyl-tRNA, messenger RNA, and a 30S ribosomal subunit. *Aka* initiator complex.

initiation factor. One of three protein factors that function in the initiation of protein synthesis and that are generally designated either as F_1, F_2, and F_3 or as IF_1, IF_2, and IF_3.

initiation point. The site on the DNA molecule at which DNA replication begins.

initiator. 1. A structural gene that forms part of the replicon and that forms a product, believed to be a protein, which interacts with the replicator and initiates the replication of the DNA that is attached to the replicator. 2. The product that is formed by the initiator structural gene and that interacts with the replicator to initiate DNA replication. 3. PRIMER. 4. A carcinogenic agent that brings about the first stage in a two-stage or multistage mechanism of carcinogenesis.

initiator transfer RNA. The transfer RNA molecule that is responsible for the initiation of protein synthesis; in bacterial systems, this is a molecule of methionine-tRNA.

innate immunity. NATURAL IMMUNITY.

inner compartment. MATRIX.

inner filter effect. The decrease in the intensity of light passing through a sample that is due to the absorption of light by the sample. The effect constitutes a source of error in emission spectroscopy, as in fluorescence, since under these conditions only some of the molecules of the sample will be excited by the incident beam.

inner membrane. 1. The internal mitochondrial membrane which is the site of the electron transport system. 2. The membrane forming the thylakoid disks in chloroplasts.

inner-membrane sphere. The elementary particle, or supermolecule, that is observed when inner mitochondrial membranes are examined with the electron microscope.

inner orbital. An orbital that functions in the bonding of a low-spin complex.

inner volume. The volume of solvent that is trapped within the gel particles of the bed in gel filtration chromatography.

Ino. Inosine.

inoculation. 1. The introduction of an inoculum into a culture or a culture medium. 2. The introduction of a substance into a cell or an organism, particularly the introduction of immunological substances for the purpose of producing immunity.

inoculum (*pl* inocula). A mass or a suspension of either cells or viruses that is used to initiate the growth of a new culture or to infect another culture.

inorganic. 1. Pertaining to compounds other than those of carbon. 2. Pertaining to substances other than those derived from plant, animal, or microbial sources.

inorganic phosphate. An anion, or a mixture of anions, derived from orthophosphoric acid (H_3PO_4). *Sym* P_i;P.

inorganic pyrophosphate. An anion, or a mixture of anions, derived from pyrophosphoric acid ($H_4P_2O_7$). *Sym* PP_i; PP.

inosine. The ribonucleoside of hypoxanthine. Inosine mono-, di-, and triphosphate are abbreviated, respectively, as IMP, IDP, and ITP. The abbreviations refer to the 5′-nucleoside phosphates unless otherwise indicated. *Abbr* Ino; I.

inosinic acid. The ribonucleotide of hypoxanthine.

inositol. An optically and biologically active, cyclic sugar alcohol; frequently classified with the B vitamins, since it is a growth factor for some organisms. *Aka i*-inositol; *meso*-inositol; *myo*-inositol; inosite.

inotropic effect. An effect on the contractility of muscular tissue.

insecticide. An agent that kills insects.

insertion. A mutation in either DNA or RNA in which one or more extra nucleotides are inserted into a polynucleotide chain.

insertion model. A model, proposed by Campbell, for the attachment of the prophage to the bacterial DNA. According to this model, the prophage first becomes attached in cyclic form to the bacterial DNA. The bacterial DNA is then interrupted by opening the ring at the point of contact, and the prophage is inserted linearly into the bacterial DNA which is attached to it at both ends. *See also* hook model.

in situ. In the normal, natural location or position.

in situ hybridization. A hybridization technique in which chromosomes, devoid of RNA and protein, are incubated with tritium-labeled nucleic acid and the hybridized segments are then visualized by radioautography.

insoluble enzyme. An enzyme that is linked covalently to a water-insoluble support, such as agarose or polyacrylamide, without destruction of its activity. Applications of such enzymes occur in batch type reactions, adsorption chromatography, and gel filtration. *See also* immobilized enzyme.

insoluble fibrin. HARD CLOT.

inspire. To inhale.

instability factor. One of a set of arbitrary values assigned to substituents, interactions, and the like, that affect the stability of monosaccharide conformations. The values obtained by summing these factors permit an estimation of the relative stabilities of different conformations of a given compound.

instructive theory. A theory of antibody formation according to which the information for antibody synthesis results from an instructive effect by the antigen, rather than from a genetic determination; the antigen instructs the biosynthetic machinery to synthesize specific antibodies which, in the absence of the antigen, would either not be formed at all or be formed only very rarely as a result of a chance event. *See also* selective theory.

insulin. A protein hormone that lowers the level of blood sugar and stimulates the utilization of glucose, presumably by affecting the rate of transport of glucose across the cell membrane. Insulin is secreted by the islets of Langerhans in the pancreas and also acts on protein and lipid metabolism.

insulin-like activity. A group of substances in serum that have some of the biological properties of insulin, but do not react with insulin antibodies. *Abbr* ILA.

insulinoma. A benign tumor of the insulin-producing cells of the pancreas.

insulin resistance index. The sum of the values

for the blood glucose concentration at 60, 90, and 120 min as determined in an insulin tolerance test. *Abbr* IRI.

insulin shock. A condition characterized by anxiety, delirium, and convulsions, that is occasionally produced upon the administration of insulin to an animal; it results from a lowering of the blood sugar by insulin to a level below that which is required for normal functioning of the brain.

insulin tolerance test. A test for evaluating insulin resistance and certain endocrine disorders. The test is performed by placing an individual on a carbohydrate diet, followed by an injection of insulin and a determination of the blood sugar level as a function of time.

integral counting. The counting, in a scintillation counter, of pulses that are above a certain level of intensity; the pulses are selected by means of a discriminator that rejects pulses of lower intensity.

integral discrimination. The selection of pulses that takes place in integral counting.

integral dose. The total, cumulative dose of radiation received by an individual.

integrase. An enzyme that functions in the incorporation of the prophage into the host bacterial DNA.

integrated state. The state of an episome in which it is incorporated into and replicates with the chromosome. *See also* integration.

integrating circuit. An electronic circuit for measuring the total number of ionizations and the resulting electrical currents that are produced in an ionization chamber in a given time interval. *See also* differentiating circuit.

integration. The incorporation of one DNA segment into another as in: (a) the incorporation of donor DNA into recipient DNA in genetic recombination; (b) the incorporation of episomal DNA into chromosomal DNA; and (c) the insertion of prophage DNA into the host bacterial DNA.

intensive property. A property of a system, such as density or viscosity, that can be defined by specifying the relative amounts of the substances involved. *See also* extensive property.

inter-. Prefix meaning between.

interacting flows. *See* interaction of flows.

interaction of flows. The interdependence of the diffusion of different solutes in a solution that contains two or more solutes; due to the effect of the concentration gradient of one solute on the diffusion of another.

interaction of heme groups. *See* heme-heme interaction.

interactive graphics. The interaction of an individual with a computer, the output of which is shown in graphical form; the interaction occurs

while the computer is in operation, and with the computer being coupled to oscilloscopes or other display devices.

interallelic complementation. INTRAGENIC COMPLEMENTATION.

inter-alpha-globulin. THYROXINE-BINDING GLOBULIN.

interband. A region between bands in a polytene chromosome.

intercalary deletion. A deletion in which genetic material is lost from some part of the chromosome other than its end.

intercalation. The process whereby a flat molecule, such as an acridine dye, becomes inserted between two adjacent, stacked bases in a double helical nucleic acid. Intercalation results in a frame shift mutation, since the subsequent replication of the nucleic acid leads to either a deletion or an insertion of a nucleotide.

intercept replot. *See* secondary plot.

interesterification. The formation of a new ester by the reaction of an ester with an acid, an alcohol, or another ester; these three reactions are known as acidolysis, alcoholysis, and ester interchange.

interface. 1. The boundary between two phases. 2. COMPUTER INTERFACE.

interface centrifugation. A centrifugal technique in which there are two immiscible phases in the centrifuge tube, and transfer of solute particles occurs from one phase to the other.

interfacial tension. The surface tension at the interface between two liquid phases.

interfacial test. RING TEST.

interference. 1. The mutual effect on meeting of two sets of waves, such as light waves, that results in neutralization (destructive interference) at some points, and in reinforcement (constructive interference) at other points. 2. The effect of a crossover at one locus on a chromosome on the probability of a crossover at another locus; the interference is said to be positive if the probability of the second crossover is decreased and it is said to be negative if that probability is increased. 3. VIRAL INTERFERENCE.

interference filter. A filter that allows the passage of a narrow range of wavelengths, about 10 to 20 nm wide. It is made by depositing semitransparent silver films on both sides of a dielectric so that constructive and destructive interference take place when light passes through the films.

interference fringe. One of either the light or the dark bands that are produced by the interference or the diffraction of light.

interference microscope. A microscope that permits the observation of transparent objects and the measurement of refractive indices.

interference optics. INTERFEROMETRIC OPTICAL SYSTEM.

interfering virus. A virus that interferes with the multiplication of another virus.

interferogram. The photographic record of an interference pattern.

interferometer. An instrument for the precise determination of wavelengths or distances; based on the separation of light in the instrument into two parts which travel unequal paths and which, when reunited, interact with each other to produce an interference pattern.

interferometric optical system. An optical system that focuses ultraviolet light passing through a solution in such a fashion that a photograph of interference fringes is obtained. A boundary in the solution appears as a break in the interference fringes and measurements are made on the photographic plate by counting the interference fringes. An optical system of this type, incorporating a Rayleigh interferometer, is used in the analytical ultracentrifuge.

interferon. A low-molecular weight, species-specific antiviral protein that is formed in many vertebrates in response to a variety of agents including viruses, microorganisms, and endotoxins. Interferon has an unusual stability to low pH, a moderate stability to heat, and functions by eliciting the formation of another protein that inhibits the translation of viral messenger RNA.

intergenic complementation. Complementation that is produced by two mutant chromosomes which carry a mutation in different genes. The unmutated gene of one chromosome makes up for, or complements, the mutated gene of the other chromosome so that the necessary gene products of both genes are formed, and the wild-type allele of each gene is expressed.

intergenic suppression. The restoration of a genetic function, which was lost by mutation, through a second mutation in a gene other than the one that sustained the primary mutation.

intergenic suppressor mutation. *See* suppressor mutation.

interkinesis. The interphase between the first and the second division in meiosis.

intermediary metabolism. 1. The enzyme-catalyzed reactions in cells whereby nutrients are transformed and energy is extracted from them for the growth and maintenance of the cells. 2. The sum total of all the chemical transformations of the nutrients in an animal subsequent to the absorption of the nutrients into the blood.

intermediate. A compound that participates in a reaction and that occurs between the starting materials and the final products of the reaction; in metabolism, an intermediate occurs between the nutrients on the one hand, and the cellular components and waste products on the other hand.

intermedin. MELANOCYTE-STIMULATING HORMONE.

internal conversion. 1. A mode of radioactive decay in which gamma rays that emanate from a nucleus collide with an orbital electron, transfer all their energy to it, and then eject the electron from the atom. 2. A mode of vibrational deexcitation in which the energy of an excited electronic state of a molecule is dissipated by conversion to vibrational energy of a lower electronic state that has the same multiplicity (i.e., singlet to singlet or triplet to triplet).

internal conversion electron. CONVERSION ELECTRON.

internal energy. The energy within a system. In chemistry, this usually refers only to those types of energy that can be modified by chemical processes; these include translational, vibrational, and rotational energies of molecules, energy involved in chemical bonding, and energy involved in noncovalent interactions between molecules.

internal gas counter. A radiation counter in which a radioactive gaseous sample is counted by being incorporated into the detector-filling gas mixture.

internal indicator. An indicator that is added to the titration vessel in which a liquid is being titrated.

internal monooxygenase. A monooxygenase in which the cosubstrate that incorporates the second oxygen atom is itself a product of the reaction.

internal protein. A protein that is complexed with DNA and occurs in the head of T-even phages.

internal quenching. The quenching of an ionization detector that occurs when a specific gas, such as butane or chlorine, is added to the mixture used for gas amplification. *See also* organic quenching.

internal radiation. The radiation emitted by radioactive substances that are deposited in the tissues.

internal respiration. *See* respiration (3).

internal-sample scintillation counter. A scintillation counter in which the sample and the fluor are in intimate contact, as in a liquid scintillation counter.

internal standard. A standard that is added to, and treated with, the sample.

internal standardization. A method for determining the counting efficiency of radioactive samples that is based on counting the sample both by itself and together with a known amount of added isotope.

International Union of Biochemistry. The organization that standardizes biochemical nomenclature, symbols, etc. *Abbr* IUB.

International Union of Pure and Applied Chemistry. The organization that standardizes chemical nomenclature, symbols, atomic weights, etc. *Abbr* IUPAC.

interphase. The period between two successive mitotic divisions.

interphase cycle. CELL CYCLE.

interpolation. The estimation of a function between two known values.

interrupted mating experiment. An experiment designed to study the transfer of genetic information during bacterial conjugation; performed by withdrawing samples at various times from a culture of mating bacteria and separating the mating bacteria by intense agitation in a blender. The bacterial fractions thus obtained contain recipient bacteria which have received varying amounts of the chromosome from the donor bacteria.

interrupted trough technique. An immunoelectrophoretic technique for comparing the antigens in two mixtures.

intersome. A collective term for the intermediate-size particles that either are naturally occurring precursors of ribosomes or are produced from ribosomes by the stepwise removal of proteins.

interstice. A small space between two structures such as cells, organs, or tissues.

interstitial. Of, or pertaining to, interstices.

interstitial cell hormone. LUTEINIZING HORMONE.

interstitial cell-stimulating hormone. LUTEINIZING HORMONE.

intersystem crossing. A nonradiative transition of a molecule from one energy state to another which has a different multiplicity; the transition from an excited singlet state to a triplet state by a decrease in the energy of the molecule as a result of collisions is an example.

intervent. A solute that intervenes in, and reduces, the interactions between macromolecules.

intervent dilution chromatography. A chromatographic technique for the separation of strongly interacting macromolecules that is based on forcing the macromolecules repeatedly across a boundary in an ion-exchange gel. On one side of the boundary the intervent concentration is high, so that the macromolecules are relatively independent of each other and macromolecular complexes dissociate; on the other side of the boundary the intervent concentration is low, so that the macromolecules are in an aggregated form and can be separated by adsorptive and ion-exchange processes.

intestinal juice. The digestive juice that consists of the secretion of the intestinal mucosa and that is discharged into the small intestine; contains lipase, enterokinase, peptidases, carbohydrases, nucleases, and phosphatases. *Aka* intestinal fluid.

intima. The inner lining of a blood vessel.

in toto. Totally; altogether.

intoxication. The abnormal state of an animal produced by relatively large amounts of a chemical agent such as a poison or a drug.

intra-. Prefix meaning within.

intracellular accumulation period. The period during viral infection that extends from the end of the eclipse to the first appearance of extracellular virus particles.

intracellular titer. The total titer of phage particles minus the extracellular titer.

intracellular transport. The transport across membranes of subcellular organelles, such as the transport across mitochondrial or chloroplast membranes. *See also* homocellular transport; transcellular transport.

intracellular virus. CELL-ASSOCIATED VIRUS.

intracistronic complementation. INTRAGENIC COMPLEMENTATION.

intracistronic suppression. INTRAGENIC SUPPRESSION.

intracristael space. The intermembrane space between the inner and the outer mitochondrial membranes.

intradermal injection. An injection into the skin.

intragenic complementation. Complementation that is produced by two mutant chromosomes which carry a mutation in the same gene but at different sites. Such complementation may arise when the product of the genes is a multimeric protein and when the individual gene products, or monomers, are nonfunctional but combine to produce an aggregate, or multimer, that is functional and nearly normal in its properties.

intragenic recombination. The genetic recombination between mutons of the same cistron.

intragenic suppression. The restoration of a genetic function, which was lost by mutation, through a second mutation in the same gene that sustained the primary mutation but at a different site in that gene.

intragenic suppressor mutation. *See* suppressor mutation.

intramuscular injection. An injection into the muscle. *Abbr* im.

intraperitoneal injection. An injection into the peritoneal cavity. *Abbr* ip.

intrapleural injection. An injection into the chest fluid.

intrathecal injection. An injection into the spinal cord.

intravenous injection. An injection into a vein. *Abbr* iv.

intravital staining. The staining of living cells without killing them.

intrinsic association constant. The association constant that describes the binding of a ligand to a particular site on a protein molecule, provided that all sites of this type, and present on the same molecule, are identical and noninteracting. *Aka* intrinsic binding constant.

intrinsic blood coagulation. INTRINSIC SYSTEM.

intrinsic Cotton effect. A Cotton effect that is caused by the protein itself and not by a small molecule that is bound to the protein.

intrinsic factor. A mucoprotein in the gastric juice that combines with free vitamin B_{12} and aids in its absorption from the intestine. *Aka* intrinsic factor of Castle.

intrinsic fluorescence. Fluorescence that is caused by the aromatic amino acids of the protein itself and not by a small molecule that is bound to the protein. *See also* autofluorescence.

intrinsic heterogeneity. The heterogeneity of antibodies that results from intrinsic factors, such as the different genes that are responsible for the synthesis of antibodies.

intrinsic system. The series of reactions in blood clotting that are initiated by the Hageman factor of plasma when the factor is activated by the surface of the blood capillaries.

intrinsic viscosity. The limiting value, at infinite dilution, of the reduced viscosity; it is equal to the product of the partial specific volume of the solute and its viscosity increment. *Sym* $[\eta]$.

intussusception. A mode of growth in which new material is incorporated within an existing matrix. The deposition of material in a plant cell wall or in a cell membrane are two examples.

inulin. A homopolysaccharide of D-fructose that occurs in some plants and that is used for measurements of renal clearance.

in utero. Within the uterus.

in vacuo. In a vacuum.

invagination. 1. The process of infolding and pocket formation. 2. The structure formed by infolding, as that produced by an infolded membrane.

invasin. HYALURONIDASE.

invasiveness. The spreading of bacteria, viruses, or cancer cells from the site of infection, or the site of the original tumor, to other sites and other tissues in the same organism.

inverse isotope dilution analysis. A technique for determining the amount of a labeled compound of known specific activity by the addition of a known amount of the same, but unlabeled, compound. The mixture of labeled and unlabeled compounds is then isolated and its specific activity is measured.

inverse square law. The law stating that a property, such as radiation intensity or Coulomb's force, varies inversely as the square of the distance from a given point.

inversion. 1. The hydrolysis of sucrose to an equimolar mixture of glucose and fructose; so called because the optical rotation changes sign as the hydrolysis proceeds. 2. A chromosomal aberration in which a block of genes is rotated by 180° so that the sequence of genes in that block is inverted.

inversion of configuration. The change from one enantiomeric configuration about an asymmetric carbon atom to the other in the course of a chemical reaction; requires the breaking of covalent bonds in a compound and the remaking of these bonds in the reverse sense.

invertase. The enzyme sucrase that catalyzes the hydrolysis of sucrose to glucose and fructose. *See also* inversion (1).

invertebrate. 1. *n* An animal that lacks a backbone. 2. *adj* Of, or pertaining to, an animal that lacks a backbone.

invert sugar. The equimolar mixture of glucose and fructose that is formed by the hydrolysis of sucrose. *See also* inversion (1).

Inv group. A group of allotypic antigenic sites in the constant region of the kappa chains of human immunoglobulins.

in vitro. Outside a living organism; pertaining to conditions of or to experiments with a perfused organ, a tissue slice, cells in tissue culture, a homogenate, a crude extract, or a subcellular fraction. *See also* in vivo.

in vitro complementation. The in vitro demonstration of intragenic complementation by the mixing of nonfunctional monomeric proteins and the formation of a functional multimeric protein.

in vitro marker. An induced mutation of mammalian cells that permits their phenotypic detection in tissue culture.

in vitro protein synthesis. CELL-FREE AMINO ACID INCORPORATING SYSTEM.

in vivo. Within a living organism; pertaining to conditions of or to experiments with a whole animal, an intact plant, an intact organ, or a population of microbial cells. *See also* in vitro.

in vivo marker. A naturally occurring mutation of mammalian cells that permits their phenotypic detection in tissue culture.

iodide pump. The active transport mechanism that concentrates iodide in the thyroid gland. *Aka* iodide trapping mechanism.

iodine. An element that is essential to several classes of animals and plants. Symbol I; atomic number 53; atomic weight 126.9044; oxidation states -1, $+1$, $+5$, $+7$; most abundant isotope I^{127}; a radioactive isotope I^{131}, half-life 8.1 days, radiation emitted—beta particles and gamma rays.

iodine number. A measure of the extent of unsaturation in a fat that is equal to the number of grams of iodine taken up by 100 grams of fat. *Aka* iodine value.

iodopsin. The visual pigment, consisting of cone opsin plus retinal, that occurs in mammals and other vertebrates and that has an absorption maximum at 562 nm.

ion. An atom, a group of atoms, or a molecule that carries an electrical charge. *See also* ionization.

ion antagonism. The phenomenon in which one or more ions lead to an activation of a reaction or a set of reactions, while one or more similar ions lead to an inhibition of the same reaction or set of reactions.

ion atmosphere. The region surrounding a charged atom, a charged group of atoms, or a charged molecule in which there is a statistical preference for ions of opposite charge.

ion chamber. IONIZATION CHAMBER.

ion cloud. ION ATMOSPHERE.

ion cluster. An aggregate of ion pairs that is formed in the immediate vicinity of a primary ionizing event, and that is produced by the secondary electrons which are generated by the incident ionizing particle or by the incident photon. *See also* ion pair (1).

ion-dipole interaction. The attractive or repulsive electrical force between an ion and a dipole; the energy of such interactions is proportional to r^{-4}, where r is the distance between the ion and the dipole.

ion etching. The exposure of a biological specimen to a beam of inert ions, such as those of argon, prior to an examination of the specimen with the scanning electron microscope.

ion-exchange chromatography. A chromatographic technique in which molecules are separated on the basis of their charge. The stationary phase is an ion-exchange resin, and the mobile phase is an aqueous solution. Ion-exchange chromatography is usually performed in columns, and the charged molecules are retarded in their movement through the column depending on the sign and the magnitude of their charge. *Aka* ion-exchange.

ion-exchange resin. A high-molecular weight, insoluble, branched, ionized polymer that is used as a support in ion-exchange chromatography; may be natural or synthetic.

ion exclusion. A process whereby strong electrolytes are separated from weak electrolytes or from nonelectrolytes by passage of the mixture through an ion-exchange resin; the strong electrolytes are excluded from the ionized resin and

elute more readily than the other substances. *See also* ion retardation.

ion-filtration chromatography. A chromatographic technique that is based on the combined use of gel filtration and ion-exchange chromatography; performed by eluting macromolecules in the sieving range from a cross-linked ion-exchange gel under precisely controlled conditions of pH and ionic strength.

ionic. Of, or pertaining to, ions.

ionic bond. The attractive force, described by Coulomb's law, between either a cation and an anion or between a cationic and an anionic group of atoms; the cationic and anionic groups may be on the same molecule or on different molecules.

ionic contact distance. The distance between two oppositely charged and interacting ions at which the attractive force due to the opposite charges is just balanced by the repulsive force due to the orbital electrons of the two ions.

ionic detergent. *See* anionic detergent; cationic detergent.

ionic double layer. The region encompassing the charges on an atom, a group of atoms, or a molecule together with the layer of counterions that surrounds these charges.

ionic migration. ISOTACHOPHORESIS.

ionic mobility analysis. ISOTACHOPHORESIS.

ionic orbital. An orbital that functions in the bonding of a high-spin complex.

ionic sieving. Gel filtration chromatography of ions.

ionic strength. A measure of the ionic concentration of a solution that is equal to $\frac{1}{2} \sum c_i Z_i^2$, where c_i is the concentration of the ith ion and Z_i is its charge. *Sym* $\Gamma/2; \mu; I$.

ion-induced dipole interaction. The attractive electrical force between an ion and a dipole which is induced by the ion; the energy of such interactions is proportional to r^{-4}, where r is the distance between the ion and the induced dipole.

ion-ion interaction. The attractive or repulsive electrical force between two ions that is described by Coulomb's law. *See also* electrostatic interaction.

ionizable. Capable of undergoing ionization.

ionization. 1. The breakup of a molecule into two or more ions when it is dissolved in water. 2. The formation of a charged group in a molecule by either the association or the dissociation of a group, most commonly a proton. 3. The formation of an ion pair by either an ionizing particle or a strong electrostatic field.

ionization chamber. A chamber for the production and collection of ions and for the measurement of the electric currents produced by these ions. The ions are commonly produced by bombardment of a gas with an ionizing radiation. Simple ionization chambers are used especially for measurements of x- or gamma rays; ionization chambers with gas amplification are used for measurements of alpha and beta rays, and are incorporated in proportional and Geiger-Mueller counters.

ionization constant. The dissociation constant for a reaction in which ions are produced, as in the dissociation of an acid to form a proton and an anion.

ionization detector. An instrument for the detection of ionizing events, such as those occurring in an ionization chamber.

ionization interference. The interference that occurs in atomic absorption spectrophotometry when the atoms, in addition to being dissociated from the molecule, are also excited by the flame; as a result, the excited atoms emit radiation which is of the same wavelength as that which is being absorbed.

ionization potential. The energy, in electron volts, that is required to remove an orbital electron from an atom.

ionization track. The trace of ion pairs that are produced by either an ionizing particle or a photon as it passes through matter.

ionized. In the form of an ion or ions.

ionizing energy. IONIZATION POTENTIAL.

ionizing event. Any process that produces an ion or a group of ions as a result of the interaction of matter with radiation; the formation of ions in an ionization chamber or in an irradiated cell are examples.

ionizing particle. A charged particle that has sufficient energy to dislodge an orbital electron from an atom and to produce an ion pair.

ionizing radiation. The electromagnetic or corpuscular radiation that produces ion pairs in the matter through which it passes; types of ionizing radiation include x-rays, protons, neutrons, alpha particles, and high-speed electrons.

ionogenic. Capable of forming ions; an amino group, for example, is ionogenic.

ionography. ZONE ELECTROPHORESIS.

ionophore. A permeability-inducing agent, such as an ionophorous antibiotic.

ionophoresis. 1. The electrophoresis of small ions. 2. ELECTROPHORESIS.

ionophoric agent. IONOPHORE.

ionophorous antibiotic. An antibiotic, such as gramicidin or valinomycin, that induces permeability to alkali metal cations in biological membranes. Ionophorous antibiotics are nonpolar compounds that are soluble in the lipid portion of the membrane and that are believed to function by chelating the metal ion and transporting it across the membrane in this form.

ionotropy. The ionization of a tautomeric compound in which a charged atom or a charged radical separates from an unsaturated molecule, thereby forming an oppositely charged fragment. The process is referred to as anionotropy or cationotropy depending on whether the separated atom or radical has a negative or a positive charge. *Aka* ion shifting.

ion pair. 1. The positive ion formed by the action of ionizing radiation together with the orbital electron that was ejected from the uncharged atom when the ion was formed. 2. The two ions composed of either a cation that has remained in the secondary hydration shell of an anion, or an anion that has remained in the secondary hydration shell of a cation.

ion pairing. SPECIFIC ADSORPTION.

ion product of water. The product of the hydrogen and hydroxyl ion concentrations, in moles per liter, in pure water at 25°; approximately equal to 10^{-14}. *Sym* K_w.

ion retardation. A process whereby nonelectrolytes are separated from electrolytes by passage of the mixture through a specially prepared ion-exchange resin; the resin absorbs electrolytes preferentially so that the nonelectrolytes elute more readily. *See also* ion exclusion.

ion-selective electrode. An electrode that responds in a reproducible manner to changes in the activity of a specific ion and that is almost insensitive to changes in the activities of other ions in the solution. The selectivity of the electrode results from the incorporation of an ion-exchanger, a specially formulated glass, or a specific crystal into the design of the electrode. *Aka* ion-specific electrode.

iontophoresis. An electrophoretic technique in which drugs, in ionic form, move through the skin under the influence of an electric field applied across the skin.

ion-transporting antibiotic. IONOPHOROUS ANTIBIOTIC.

ip. 1. *adj* Intraperitoneal. 2. *adv* Intraperitoneally.

IPTG. Isopropyl thiogalactoside.

IR. 1. Infrared. 2. Induction ratio.

IRI. 1. Immunoreactive insulin. 2. Insulin resistance index.

iron. An element that is essential to all animals and plants. Symbol Fe; atomic number 26; atomic weight 55.847; oxidation states +2, +3; most abundant isotope Fe^{56}; a radioactive isotope Fe^{59}, half-life 45 days, radiation emitted—beta particles and gamma rays.

iron-binding globulin. TRANSFERRIN.

iron protein. A subunit of the enzyme nitrogenase, present in *Clostridium pasteurianum*, that contains four iron atoms and four labile sulfur atoms and that has a molecular weight of about 50,000.

iron-sulfur protein. A protein in which iron is bound by means of sulfur-containing ligands of the protein. Three known types of iron-sulfur proteins are those in which: (a) a single iron atom is linked to four cysteine sulfur atoms as in rubredoxin; (b) two iron atoms are linked to four cysteine sulfur atoms and to two bridging, acid-labile sulfur atoms as in plant-type ferredoxin; and (c) four iron atoms are linked to four cysteine sulfur atoms and to four bridging, acid-labile sulfur atoms as in bacterial-type ferredoxin and in high-potential iron-sulfur proteins.

irradiation. The exposure to radiation.

irradiation chimera. *See* radiation chimera.

irreversible inhibitor. An inhibitor that binds to an enzyme (usually covalently) in an irreversible reaction so that the inhibition cannot be reversed by attempts to remove the inhibitor from the enzyme by such processes as dialysis or ultrafiltration. It may, however, be possible to restore the enzymatic activity by removing the inhibitor through a chemical reaction. *See also* active site-directed irreversible inhibitor; reactivation; reversible inhibitor.

irreversible process. A process in equilibrium thermodynamics in which a system goes from an initial equilibrium state to a final equilibrium state through stages that are not equilibrium states. In such a process, the net entropy change for the system plus its surroundings is greater than zero.

irreversible reaction. A chemical reaction that (a) proceeds to completion and cannot be reversed, or (b) proceeds essentially to completion because of an equilibrium constant that is much larger than 1.0, or (c) is made to proceed overwhelmingly in one direction because of other factors.

irreversible shock. A shock that has progressed so far that it cannot be reversed even by the proper therapy.

irreversible thermodynamics. A branch of thermodynamics that deals with changes between nonequilibrium states of open systems.

irritability. EXCITABILITY.

islets of Langerhans. Clusters of hormone-secreting cells in the pancreas some of which, the alpha cells, secrete glucagon while others, the beta cells, secrete insulin.

iso-. 1. Combining form meaning equal or like. 2. Prefix in the Cleland nomenclature of enzyme kinetics denoting a mechanism in which a stable enzyme form must isomerize to a different form before further reaction can occur.

3. Combining form meaning genetically different individuals of the same species. 4. Combining form meaning genetically identical individuals.

isoacceptor transfer RNA. One of two or more transfer RNA molecules that differ in their structure but are specific for the same amino acid. *Abbr* iso-tRNA. *Aka* isoaccepting transfer RNA.

isoagglutination. An agglutination reaction of isoagglutinins.

isoagglutinin. An antibody that is formed in one individual and that causes the agglutination of cells derived from another individual of the same species.

isoalloxazine. The heterocyclic ring structure that occurs in riboflavin and in the flavin nucleotides.

isoantibody. An antibody that is formed in one individual in response to antigens derived from another individual of the same species.

isoantigen. An antigen that is derived from one individual and that is immunogenic in other individuals of the same species.

isobar. 1. A graph or an equation that describes changes of temperature or volume at constant pressure. 2. One of two or more nuclides that have the same mass but have a different number of protons, and hence possess different atomic numbers.

isochore. A graph or an equation that describes changes of temperature or pressure at constant volume.

isocitric acid. An isomer of citric acid that is formed from *cis*-aconitic acid in the reactions of the citric acid cycle.

isocratic elution. The chromatographic elution with one solvent of constant composition as opposed to a gradient elution.

isodense. Of equal density.

isodensity centrifugation. A centrifugal technique in which both sedimentation and flotation take place in a single tube; performed by using a density gradient that straddles the buoyant densities of the solute molecules of interest.

isodensity equilibrium centrifugation. DENSITY GRADIENT SEDIMENTATION EQUILIBRIUM.

isodiametric. Having equal diameters, as a cell that has essentially the same length and width.

isodynamic enzyme. One of two or more different enzymes or enzyme forms that catalyze the same reaction but differ in their properties. *See also* heteroenzyme; isozyme; multiple forms of an enzyme.

isodynamic law. The generalization, enunciated by Rubner in 1878 and disproven since, that foodstuffs can replace each other in proportion to their caloric value; thus one food, containing a certain number of calories, is equivalent to any other food that contains the same number of calories, regardless of the composition of the foods.

isoelectric analysis. ISOELECTRIC FOCUSING.

isoelectric condensation. ISOELECTRIC FOCUSING.

isoelectric equilibrium electrophoresis. ISOELECTRIC FOCUSING.

isoelectric focusing. An electrophoretic technique for fractionating amphoteric molecules, particularly proteins, that is based on their distribution in a pH gradient under the influence of an electric field that is applied across the gradient. The molecules distribute themselves in the gradient according to their isoelectric pH values. Positively charged proteins are repelled by the anode and negatively charged proteins are repelled by the cathode; consequently, a given protein moves in the pH gradient and bands at a point where the pH of the gradient equals the isoelectric pH of the protein. The pH gradient is produced in a chromatographic column by the electrolysis of amphoteric compounds and is stabilized by either a density gradient or a gel. *Abbr* IEF.

isoelectric fractionation. ISOELECTRIC FOCUSING.

isoelectric pH. The pH at which a molecule has a net zero charge; the pH at which the molecule has an equal number of positive and negative charges, which includes those due to any ions bound by the molecule. The isoelectric pH is operationally defined as that pH at which the molecule does not move in an electric field. *Sym* pI. *Aka* isoelectric point. *See also* isoelectrophoretic pH; isoionic pH.

isoelectric precipitation. A fractional precipitation of proteins; a mixture of proteins is adjusted to the isoelectric pH of one of them so that all or most of that protein is precipitated.

isoelectric protein. A protein at its isoelectric pH; a protein that has a net charge of zero.

isoelectric separation. ISOELECTRIC FOCUSING.

isoelectric spectrum. The distribution pattern of the proteins that are separated by isoelectric focusing.

isoelectronic. Describing two atoms, two groups of atoms, two ions, or two molecules that have the same number and arrangement of valence electrons.

isoelectrophoretic pH. The pH at which the electrophoretic mobility of a protein is zero; this pH may coincide with the theoretical isoelectric pH of the protein, depending on the surface structure of the protein, the ionic strength, and the nature of the ionic double layer around the protein. *Aka* isoelectrophoretic point. *See also* isoelectric pH; isoionic pH.

isoenzyme. ISOZYME.

iso fatty acid. A fatty acid that is branched at the penultimate carbon atom at the hydrocarbon end of the molecule.

isogeneic. SYNGENEIC.

isogenic. Referring to genetically identical individuals.

isogenous. Having the same origin.

isograft. SYNGRAFT.

isohemagglutinin. A hemagglutinin that has the capacity to react with isoantigens on the surface of red blood cells.

isohormone. One of two or more different forms of the same hormone.

isohydric shift. A set of two reactions, occurring in the red blood cell, whereby the oxygenation and deoxygenation of hemoglobin during respiration are linked to the reversible ionization of carbonic acid. As a result, the intracellular pH of the red blood cell remains essentially constant despite the fact that protons participate in both reactions. *Aka* isohydric transport.

isoimmunity. The immunity acquired through reactions of antigens and antibodies that are derived from different individuals of the same species.

isoionic dilution. The dilution of a polyelectrolyte solution with a salt solution that has the same ionic strength as the polyelectrolyte solution, so that the total concentration of mobile counterions remains constant.

isoionic pH. The pH at which the number of positive and negative charges of a protein that arise exclusively from proton exchange are equal to each other. The isoionic pH is operationally defined as either (a) the pH of a solution of isoionic protein in water, or (b) the pH of a protein solution that does not change with increasing concentrations of the protein. *Aka* isoionic point. *See also* isoelectric pH; isoelectrophoretic pH.

isoionic protein. A protein that has an equal number of protonated basic groups and deprotonated acidic groups; operationally defined as a protein from which all bound ions have been removed by electrolysis or by mixed-bed ion-exchange chromatography. Such a protein contains no ions other than those arising from the dissociation of the solvent.

isokinetic gradient. A density gradient in which all the particles that have the same density sediment at a constant rate.

isokinetic sedimentation. Sedimentation in an isokinetic gradient.

isolable. Capable of being isolated. *Aka* isolatable.

isolate. To separate and purify a substance.

isolated rat uterus assay. A bioassay for neurohypophyseal hormones that is based on measurements of the contractility of rat uterus.

isolated system. A thermodynamic system that exchanges neither matter nor energy with its surroundings.

isoleucine. An aliphatic, nonpolar alpha amino acid. *Abbr* Ile; I.

isologous. 1. HOMOLOGOUS (1,6). 2. SYNGENEIC.

isomer. One of two or more compounds that have the same molecular composition but have different molecular structures and hence possess different properties.

isomerase. An enzyme that catalyzes the interconversion of one isomer into another. *See also* enzyme classification.

isomeric. Of, or pertaining to, isomers.

isomeric transition. The transition of one nuclear isomer to another that is accompanied only by the emission of gamma rays from the nucleus, and during which there is no change in either the atomic mass or the atomic number of the nuclide. *Abbr* IT.

isomerism. The phenomenon in which two or more compounds have the same molecular composition but have different molecular structures and hence possess different properties.

isomerization. The interconversion of one isomer into another; the formation of an isomer.

isomer number. The total number of possible stereoisomers that have the same structural formula.

isometric. Having the same dimensions.

isometric contraction. The exertion of force by a muscle without a shortening in the length of the muscle; may be achieved by stimulating a muscle while maintaining it mechanically at constant length.

isometric virus. A virus, the structure of which can be described by a symmetrical polyhedron of the isometric crystal system; this system is characterized by three identical axes at right angles to each other and includes the cube and the regular octagon.

isomorphic. Morphologically alike; having the same shape and structure. *Aka* isomorphous.

isomorphous replacement. A method for solving the phase problem in an x-ray diffraction study of a protein by using a heavy atom derivative of the protein. The introduction of the heavy atom or atoms into the protein must be isomorphous, so that there is no change in the conformation of the protein nor any change in the size or in the symmetry of the unit cell. A comparison of the diffraction patterns of the original protein and of its heavy metal derivative permits the localization of the heavy atom in the unit cell and hence permits a determination of phase angles. The determination of all the phase angles requires the use of several heavy atom derivatives.

isoosmolar solution. ISOTONIC SOLUTION.

isophane insulin. NPH INSULIN.

isophile antibody. An antibody that reacts only with antigens which are derived from the same species.

isopiestic technique. A technique for measuring the binding of solute and/or solvent molecules to a macromolecule in a solution in which only one of the solvent components, generally water, is volatile. A number of solutions of known initial composition, and all containing the same volatile solvent component, are allowed to equilibrate with each other with respect to the activity of the volatile solvent component. Equilibration may be carried out in a vacuum desiccator and the equilibrium compositions of the solutions are determined by weighing; any change in composition from an initial to an equilibrated solution is due to the transfer of the volatile solvent component.

isoplith. One of a group of oligonucleotides that have the same chain length but differ in their base composition and/or base sequence.

isoprene. The five-carbon compound, 2-methyl-1,3-butadiene, that occurs in the structure of several biochemically important compounds, including coenzyme Q, vitamin A, and vitamin K.

isoprenoid. 1. *n* A compound containing two or more isoprene units or derivatives of isoprene units. 2. *adj* Of, or pertaining to, isoprene.

isopropyl thiogalactoside. A gratuitous inducer of the enzyme β-galactosidase. *Abbr* IPTG.

isoprotic pH. The pH of a solution of an ampholyte that does not change upon increasing the concentration of ampholyte. *Aka* isoprotic point. *See also* isoionic pH; isoelectrophoretic pH.

isopycnic. Having the same density, specifically buoyant density.

isopycnic centrifugation. ISODENSITY CENTRIFUGATION.

isopycnic gradient centrifugation. DENSITY GRADIENT SEDIMENTATION EQUILIBRIUM.

isopycnotic. ISOPYCNIC.

isorotation rule. The rule, proposed by Hudson in 1902, which states that the optical rotation of a carbohydrate can be approximated by a sum of two factors; one factor represents a contribution due to the anomeric carbon, and the other factor represents a contribution due to the rest of the molecule. The rule further states that these two contributions are approximately independent of each other and have similar values for similar molecules.

isosbestic point. The wavelength at which the

absorptivity of two or more compounds is identical; the wavelength at which the absorption spectra of two or more compounds intersect when the absorbance is measured for solutions of equimolar concentrations.

isosemantic substitution. The incorporation of an amino acid into a polypeptide chain that results from a mutation in which the normal codon of an amino acid has been mutated to a synonym codon; consequently, the amino acid that is incorporated into the polypeptide chain is identical to the normal one, but is incorporated in response to one of its synonym codons.

isosmotic. Having the same osmotic pressure.

isosmotic solution. ISOTONIC SOLUTION.

isostere. 1. A compound that has apparently different chemical characteristics from those of a naturally occurring essential metabolite, but that can substitute for the metabolite because it fits a specific binding site for that metabolite; the binding site may be the active site of an enzyme, the receptor site of a hormone, or a similar site. 2. One of two or more compounds that are similar in their physical properties and that have the same number and arrangement of valence electrons; compounds such as CO_2 and N_2O, or N_2 and CO are examples.

isosteric. Of, or pertaining to, isosteres.

isostich. A cluster of either purine or pyrimidine oligonucleotides, as one of the clusters obtained by the hydrolysis of apurinic or apyrimidinic acid.

isotachophoresis. An electrophoretic technique in which the sample is introduced into a capillary tube between a leading electrolyte and a terminator electrolyte that have mobilities which are, respectively, greater and smaller than those of any of the sample ions; a displacement analysis in which the separation and the sharpness of the fronts between two substances are a function of the properties of the substances, and not of their concentration in the original sample.

isotactic polymer. A polymer in which the monomers have been polymerized in a stereospecific manner so that all the R groups of the monomers are on one side of the plane which contains the main chain.

isotherm. A graph or an equation that describes changes of volume or pressure at constant temperature.

isothermal process. A process in which the temperature is maintained constant.

isotocin. ICHTHYOTOCIN.

isotone. One of two or more nuclides that have the same number of neutrons but have a different number of protons and hence possess different atomic numbers.

isotonic contraction. 1. An alteration in the water and electrolyte balance in the body in which there is an increase in the volume of the extracellular fluid and an equivalent increase in electrolyte, so that the osmotic pressure of the extracellular fluid remains unchanged. 2. The exertion of force by a muscle that is accompanied by a shortening in the length of the muscle.

isotonic expansion. An alteration in the water and electrolyte balance in the body in which there is a decrease in the volume of the extracellular fluid and an equivalent decrease in electrolyte, so that the osmotic pressure of the extracellular fluid remains unchanged.

isotonic solution. A solution that has the same osmotic pressure as another solution.

isotope. One of two or more nuclides of the same element that have the same number of protons and the same atomic number, but that have a different number of neutrons and different atomic masses.

isotope derivative method. A quantitative analytical method that may be used as an enzyme assay; based on adding an isotopically labeled compound to a sample, reacting it with the sample, and determining the labeled derivative that is formed.

isotope dilution analysis. The quantitative determination of a substance by means of isotopes. *See also* direct isotope dilution analysis; double isotope dilution analysis; inverse isotope dilution analysis; modified inverse isotope dilution analysis.

isotope effect. An effect on the rate of a reaction due to the differences in mass of the isotopes in question. *See also* primary isotope effect; secondary isotope effect.

isotope exchange. The replacement of one isotope in a compound by another. *See also* deuterium exchange.

isotope incorporation. The introduction of an isotope into a compound by synthesizing the compound in the presence of one or more labeled precursors.

isotopically enriched. Descriptive of a substance in which the relative amounts of one or more isotopes has been increased.

isotopic competition. A method for demonstrating that compound X is a precursor of metabolite Y. A labeled form of X is added to the in vivo, or the in vitro, system in the presence of a general, unlabeled carbon source, and the extent of label in Y is then determined. The appearance of label in Y indicates that X is a direct precursor of Y.

isotopic dilution analysis. *See* isotope dilution analysis.

isotopic enrichment. *See* isotopically enriched.

isotopic label. The isotope that serves to label a compound.

isotopic tracer. TRACER (1).

isotopic trapping. A method for demonstrating that compound X is an intermediate in a metabolic pathway. A labeled precursor of the pathway is added to the in vivo or the in vitro system, followed by the addition of large amounts of unlabeled X. The appearance of label in X indicates that it is an intermediate of the pathway at a position subsequent to that of the labeled precursor.

iso-tRNA. Isoacceptor transfer RNA.

isotropic. Of, or pertaining to, isotropy.

isotropic band. I BAND.

isotropy. The constancy in the physical properties of a substance, regardless of the direction in which these properties are measured. *Aka* isotropism.

isotype. One of two or more antibodies produced by the same species; the IgA, IgD, IgE, IgG, and IgM immunoglobulins of a given species are examples.

isotypic variation. The antigenic differences of isotypic antibodies.

isovalent resonance. Resonance in which the various resonance structures have the same number of bonds.

isovaleric acidemia. A genetically inherited metabolic defect in man that is characterized by elevated levels of α-ketoisovaleric acid in blood and urine, and that is due to a deficiency of the enzyme isovaleryl CoA dehydrogenase.

isozyme. 1. One of two or more isodynamic enzymes derived from a single homogeneous biological source; isozymes may occur within a single species, a single organism, or a single cell. 2. One of two or more multiple forms of an enzyme that arise from genetically determined differences in primary structure; excludes enzymes derived by modification of the same primary structure. *See also* heteroenzyme; isodynamic enzyme; multiple forms of an enzyme.

IT. Isomeric transition.

ITP. 1. Inosine triphosphate. 2. Inosine-5′-triphosphate. 3. Isotachophoresis.

IU. International unit.

IUB. International Union of Biochemistry.

IUPAC. International Union of Pure and Applied Chemistry.

iv. 1. *adj* Intravenous. 2. *adv* Intravenously.

J

J. Joule.

Jablonski diagram. The energy diagram of a molecule in which discrete molecular energy states are indicated by a series of horizontal and parallel lines.

Jacob and Monod hypothesis. *See* operon hypothesis.

Jerne plaque technique. *See* plaque technique.

Jimpy mutation. An X-linked recessive mutation in mice that produces a myelin-deficient animal.

Johnston-Ogston effect. The changes in the values of the sedimentation coefficients and of the apparent concentrations that are obtained for two or more components when they are present in a mixture as compared to the values obtained when each component is present alone. The effect is due to the dependence of the sedimentation coefficient on concentration, and it leads to a decrease in the observed sedimentation coefficients for both the slow and the fast components in a mixture. The apparent concentration, which is proportional to the area under the ultracentrifuge peak, will be increased for the slow component and will be decreased for the fast component.

joinase. LIGASE.

joining enzyme. 1. LIGASE. 2. DNA LIGASE.

joint transduction. LINKED TRANSDUCTION.

joint transformation. LINKED TRANSFORMATION.

Jones-Mote reaction. A delayed-type hypersensitivity reaction of the skin that is of low intensity and that is produced only by several daily injections of antigen.

joule. A unit of work equal to 10^7 erg.

junction potential. The potential that arises either across the junction between two half cells or across the boundary between two solutions of different concentrations; it is due to differences in diffusion rates of the ions on the two sides.

juvenile hormone. One of a group of insect hormones, composed of modified linear isoprenoid units, that promote larval development; produced by neuroendocrine structures known as corpora allata and hence also referred to as allatum hormone.

K

k. 1. Kilo. 2. Boltzmann constant. 3. Rate constant.

K. 1. Equilibrium constant. 2. Degree Kelvin. 3. Potassium. 4. Lysine.

K_0. Average intrinsic association constant.

K'. Apparent equilibrium constant; an equilibrium constant that is based on molar concentrations, as opposed to the thermodynamic equilibrium constant K that is based on activities.

K_a. Dissociation constant of an acid.

kallidin I. LYSYL-BRADYKININ.

kallidin II. BRADYKININ.

kallikrein. One of a group of proteolytic enzymes in plasma that catalyzes the formation of kinins from a plasma globulin.

kanamycin. An aminoglycoside antibiotic.

kaolin. A clay that consists principally of hydrated aluminum silicate.

kappa. 1. A cytoplasmic, DNA-containing, self-replicating particle that is present in killer strains of *Paramecium aurelia*; this particle is transformed to a *P* particle which releases a toxin, called paramecin, that kills other, sensitive, paramecia. 2. RECIPROCAL ION-AT-MOSPHERE RADIUS.

kappa chain. One of the two types of light chains of the immunoglobulins.

karyokinesis. The division of the cell nucleus in either mitosis or meiosis.

karyolymph. NUCLEOPLASM.

karyon. NUCLEUS (1).

karyoplasm. NUCLEOPLASM.

karyotype. 1. The sum of the characteristics (size, number, shape, etc.) of the chromosomes of an individual or a species. 2. The sum of the characteristics of the somatic metaphase chromosomes of an individual or a species, frequently described by photomicrographs arranged as in an idiogram.

kat. Katal.

μ**kat.** Microkatal.

katabolism. Variant spelling of catabolism.

katal. The amount of enzymatic activity that converts 1 mole of substrate per second; the katal (*kat*) is related to an enzyme unit (*U*) by the relationship $1kat = 6 \times 10^7 U$. In some cases "conversion of 1 mole of substrate" is equivalent to the number of reaction cycles which equals the number of carbon atoms in 0.012 kg of the nuclide C^{12}.

Kat F unit. A unit of catalase activity no longer in use and equal to the unimolecular rate constant divided by the number of grams of enzyme under specified conditions.

katharometer. A thermal conductivity detector that is similar to a thermistor and that is used for determining the composition of gas mixtures in gas chromatography and in studies of basal metabolism.

K_b. Dissociation constant of a base.

kb. Kilo base.

KB cells. An established cell line of cells derived from a human carcinoma in 1954 and maintained in tissue culture since then.

K-capture. 1. The capture of a K-shell electron by an atomic nucleus. 2. ELECTRON CAPTURE.

Keilin-Hartree particles. A particulate submitochondrial preparation from heart muscle that can carry out electron transport, but cannot carry out oxidative phosphorylation.

Kelvin temperature scale. ABSOLUTE TEMPERATURE SCALE.

Kendall's compound. A designation for some steroids; Kendall's compounds B, E, and F refer, respectively, to corticosterone, cortisone, and cortisol.

K_{eq}. Equilibrium constant.

keratin. A fibrous scleroprotein occurring in hair, wool, nails, and other epidermal structures.

keratinization. The formation of keratin-rich horny sections and skin appendages.

kernicterus. A pathological condition in newborn humans that is characterized by discoloration and degeneration of brain tissue and by elevated serum levels of unconjugated bilirubin.

Kerr effect. ELECTRIC BIREFRINGENCE.

ketal. A compound derived from a ketone and two alcohol molecules by splitting out a molecule of water.

ketimine. An organic compound that has the general formula $R-\overset{\overset{\displaystyle R'}{|}}{C}=NH$.

keto-. 1. Combining form meaning ketose. 2. Combining form meaning ketone.

keto acid. An organic acid that also carries a ketone group.

β**-ketoadipate pathway.** A branched pathway in bacteria that leads to succinyl-coenzyme A, by way of β-ketoadipate, from either *cis, cis*-muconate or from β-carboxy-*cis, cis*-muconate.

keto-enol tautomerism. The tautomerism that is

due to a shift of a hydrogen atom so that one of the isomers is a ketone and the other is an enol.

ketofuranose.   A ketose in furanose form.

ketogenesis.   The formation of ketone bodies.

ketogenic amino acid.   An amino acid that can serve as a precursor of acetyl coenzyme A and ketone bodies in metabolism.

α-ketoglutarate dehydrogenase.   A multienzyme system that is similar in its structure and in its properties to the pyruvate dehydrogenase system and that catalyzes the reduction of α-ketoglutaric acid to succinic acid in the citric acid cycle.

α-ketoglutarate pathway.   The catabolic pathway whereby arginine, proline, histidine, glutamine, and glutamic acid enter the citric acid cycle by way of α-ketoglutaric acid.

α-ketoglutaric acid.   A dicarboxylic acid that is an intermediate in the citric acid cycle where it is formed from isocitric acid.

ketone.   An organic compound that contains a ketone group.

ketone body.   One of the three compounds—acetoacetic acid, acetone, and β-hydroxybutyric acid—that arise from acetyl coenzyme A and that may accumulate in excessive amounts as a result of starvation, diabetes mellitus, or other defects in carbohydrate metabolism.

ketone group.   The carbonyl group attached to

$$\overset{O}{\underset{}{\overset{\|}{}}}$$

two carbons; the grouping $-\overset{\overset{O}{\|}}{C}-$.

ketonemia.   The presence of excessive amounts of ketone bodies in the blood.

ketonuria.   The presence of excessive amounts of ketone bodies in the urine.

ketopyranose.   A ketose in pyranose form.

ketose.   A monosaccharide, or its derivative, that has a ketone group.

ketosis.   The presence of excessive amounts of ketone bodies in the body.

ketosteroid.   One of a group of steroids that are known as either 17-ketosteroids or neutral-17-oxosteroids and that represent degradation products of steroids; they are excreted in the urine and provide an index of androgen production in the body.

ketostix test.   A rapid, semiquantitative test for the determination of ketone bodies in urine or serum by means of impregnated paper strips.

keto sugar.   A carbohydrate that carries a ketone group.

ketotic.   Of, or pertaining to, ketosis.

keV.   Kilo electron volt; 1000 eV.

key enzyme.   An enzyme that is unique to a metabolic pathway which has several enzymes in common with other pathways.

kg.   Kilogram; 1000 grams.

K$_i$.   Inhibitor constant.

kidney stone.   *See* calculus.

kieselguhr.   A fine-grain diatomaceous earth.

Kiliani synthesis.   The reaction whereby a one-carbon fragment is added to a carbonyl group by means of cyanide, as in the extension of a monosaccharide from a pentose to a hexose.

killed vaccine.   A vaccine that consists of originally infectious bacteria or viruses that have been rendered noninfectious; a vaccine of either killed bacteria or of inactivated viruses.

killer strain.   A strain of cells that kills sensitive cells of the same species; such strains have been isolated from both yeast and *Paramecium*.

killing efficiency.   INACTIVATION PROBABILITY.

killing titer.   The titer of a phage suspension that is determined from the number of bacterial cells before infection and the number of cells surviving after infection.

kilo-.   Combining form meaning one thousand and used with metric units of measurements. *Sym* k.

kilo base.   A unit of length equal to either 1000 bases in a single-stranded nucleic acid or 1000 base pairs in a double-stranded nucleic acid.

kilocalorie.   A large calorie; 1000 small calories. *Abbr* kcal; Cal.

kinase.   An enzyme that catalyzes the transfer of a phosphoryl group from ATP, and occasionally from other nucleoside triphosphates, to another compound.

kinematic viscosity.   The dynamic viscosity of a liquid divided by the density of the liquid.

kinetic.   Of, or pertaining to, kinetics.

kinetic coefficient.   A rate constant that depends on the concentration of either a reactant or a product.

kinetic constant.   RATE CONSTANT.

kinetic energy correction.   A correction term that is applied to viscosity measurements made in a capillary viscometer and that takes into account the kinetic energy that is acquired by the liquid during its flow through the capillary.

kinetic equation.   RATE EQUATION.

kinetic isolation.   The state of a chemical reaction in which it is not affected by changes in another reaction that ties in with it; may be produced, for example, by the preceding reaction either discharging a product at zero concentration or accepting a substrate that is already present at a saturation level.

kinetic law.   RATE EQUATION.

kinetics.   1. The science that deals with the rate behavior of physical and chemical systems. 2. The rate behavior of a physical or a chemical system.

kinetin.   6-Furfurylaminopurine; a cytokinin.

kinetochore.   CENTROMERE.

kinetogene.   A plasmagene that is attached to a kinetosome.

kinetosome. The cytoplasmic structure to which a cilium or a flagellum is attached.

King-Altman method. A mathematical method for solving simultaneous equations of steady-state enzyme kinetics by means of determinants.

kinin. A basic peptide that is a pharmacologically active mediator of the allergic response. Kinins are formed from a plasma globulin by enzymes known as kallikreins and cause vasodilation, increased capillary permeability, and contraction of smooth muscle.

kininogenase. KALLIKREIN.

Kirkwood-Shumaker interactions. The attractive charge-fluctuation interactions between ionized macromolecules that result from fluctuations in their dipole moments caused by the movement of protons over the surfaces of the macromolecules. Since at any instant only a random fraction of the ionizable groups on the macromolecule are fully ionized, the protons move in random fashion from one group to another; for example from such groups as $—NH_3^+$ to such groups as $—CO_2^-$ in the case of a protein. The fluctuations in the dipole moment of one molecule then induce fluctuations in the dipole moment of a neighboring molecule which results in an attractive interaction between the two molecules.

Kirschner value. The number of milligrams of butyric acid in the fraction of volatile water-soluble fatty acids obtained from 5 grams of fat.

Kjeldahl method. A procedure for determining protein nitrogen by digesting the sample with concentrated sulfuric acid in the presence of a catalyst. The protein nitrogen is converted to ammonia that is distilled over into standard acid, and the excess acid is then titrated. The Kjeldahl method is used especially for the standardization of other protein determinations. *Aka* Kjeldahl digestion.

Kjeldahl nitrogen. The protein nitrogen determined by means of the Kjeldahl method.

Kleinschmidt technique. A technique for preparing monomolecular films of DNA for electron microscopy in which the DNA is stabilized by complexation with a basic protein, such as cytochrome c.

K_m. Michaelis constant.

Knallgas bacteria. Bacteria of the genus *Hydrogemonas* which utilize the Knallgas reaction.

Knallgas reaction. The reaction between hydrogen and oxygen, forming water, that is utilized as an energy-yielding reaction by bacteria of the genus *Hydrogemonas*.

knife and fork model. KORNBERG MECHANISM.

Knoop's hypothesis. The hypothesis, proposed by Knoop in 1905, that fatty acids are oxidized in metabolism by means of successive removals of two-carbon fragments in the form of acetic acid. *See also* beta oxidation.

Knop's solution. A solution that contains the major inorganic constituents which are required for the growth of plant cells: calcium nitrate, potassium chloride, magnesium sulfate, and potassium dihydrogen phosphate. The solution will support plant growth if it is supplemented with a carbon source, trace elements, vitamins, etc.

Koagulations Vitamin. VITAMIN K.

Koch phenomenon. A delayed-type hypersensitivity reaction in which skin inflammation to injected tubercle bacilli is more intense in a previously infected individual than in a noninfected one.

Koettstorfer number. SAPONIFICATION NUMBER.

Kok effect. The phenomenon in which the quantum efficiency above the compensation point differs from that below it. *See also* compensation point.

Kornberg enzyme. The DNA-dependent DNA polymerase, first isolated by Kornberg in 1958 from *Escherichia coli*, that is believed to function in the repair-synthesis of DNA.

Kornberg mechanism. A mechanism for the replication of DNA that requires the action of three enzymes—endonuclease, DNA polymerase, and DNA ligase—and allows for the essentially simultaneous replication of both strands of an antiparallel duplex DNA.

Kostoff genetic tumor. A plant tumor that develops spontaneously in certain interspecific hybrids of the genus *Nicotiana*.

Kovats retention index system. A system for characterizing retention volumes in gas chromatography in which the retention volume of a compound is compared with the retention volumes of a series of saturated aliphatic hydrocarbons, chromatographed on the same column as that used for the compound.

Krabbe's disease. A genetically inherited metabolic defect in man that is associated with mental retardation and that is characterized by an accumulation of galactocerebrosides and by the almost complete absence of myelin; due to a deficiency of the enzyme galactosyl ceramide β-galactosyl hydrolase.

Krafft temperature. The temperature at which the solubility of a surfactant is sufficient to achieve the critical micelle concentration. Since unassociated surfactant molecules have limited solubility while micelles are highly soluble, the solubility of a surfactant increases sharply above the Krafft temperature.

Krebs cycle. CITRIC ACID CYCLE.

Krebs-Henseleit cycle. UREA CYCLE.

Krebs-Ringer solution. A modified Ringer's solution that contains magnesium sulfate and phosphate buffer in addition to the other components of Ringer's solution.

K region. A bond in an aromatic hydrocarbon that is active in addition reactions; the presence of such bonds has been correlated with the carcinogenic activity of some hydrocarbons. *See also* L region.

Kronig-Kramer transformation. An equation that permits the interconversion of circular dichroism data and optical rotatory dispersion data.

K_s. Substrate constant.

K system. A system composed of a regulatory enzyme and an effector, such that the effector alters the substrate concentration at which one half of the maximum velocity of the reaction is obtained but does not alter the value of the maximum velocity.

Kuhn statistical length. The length of a polymer that is equal to twice its persistence length.

Kunitz inhibitor. A soybean trypsin inhibitor.

K_w. Ion product of water.

kwashiorkor. A disease caused by nutritional deficiency, primarily protein deficiency, that occurs in children between the ages of one and three. The disease generally occurs in children of underdeveloped areas of the tropical belt when the children are weaned from the breast and are placed on a low-protein diet.

kymograph. An apparatus that consists of a smoked drum and that is used for recording physiological responses.

kynurenine. An intermediate in the metabolism of tryptophan that is excreted in the urine in cases of vitamin B_6 deficiency.

kynureninase. An enzyme that functions in tryptophan metabolism and requires pyridoxal phosphate as a coenzyme.

L

l. 1. Levorotatory. 2. Liter.

L. 1. L-configuration. 2. Leucine.

label. 1. A radioactive or a stable isotope that is introduced into a molecule. 2. A group of atoms or a molecule that is linked covalently to another molecule for purposes of identification.

labeled compound. A compound containing a label.

label triangulation method. A method for determining the quaternary structure of protein complexes. The subunits of the complex are separated, labeled with heavy atoms, and allowed to aggregate and reform the complex. The distances between the subunits are then measured by some technique such as x-ray diffraction, neutron diffraction, or spectroscopy.

labile. Unstable; readily undergoing change.

labile factor. PROACCELERIN.

labile methyl group. A methyl group that can be transferred from one compound to another by transmethylation, such as the methyl group attached to either a nitrogen or a sulfur atom of certain organic compounds.

labile phosphate group. A phosphate group that is readily liberated from a compound by hydrolysis at 100°C in 1 N hydrochloric acid for 7 to 10 min. The terminal phosphate group of nucleoside triphosphates and the phosphate groups of pyrophosphate fall into this category; the phosphate group of ordinary esters generally requires longer times for hydrolysis. *Aka* 7-min phosphate; 10-min phosphate.

lac operon. The operon in *Escherichia coli*, isolated in pure form in 1969, that consists of the three genes that code for the enzymes that function in the hydrolysis and transport of β-galactosides.

lactalbumin. The fraction of whey proteins that consists principally of beta lactoglobulin and alpha lactalbumin.

lactam. The keto form of a cyclic amide that contains the grouping $-NH-\overset{|}{C}=O$.

lactam-lactim tautomerism. The tautomerism that is due to a shift of a hydrogen atom so that one of the isomers is a lactam and the other isomer is a lactim.

lactate dehydrogenase. The pyridine-linked dehydrogenase that catalyzes the oxidation of lactic acid to pyruvic acid and that occurs in the form of five isozymes. *Abbr* LDH.

lactation. 1. The formation of milk by the mammary gland. 2. The period following childbirth during which milk is formed by the mammary gland.

lactic acid. The hydroxy acid that is formed from pyruvic acid when glycolysis proceeds under anaerobic conditions.

lactic acid bacteria. Bacteria of different genera that are characterized by the production of lactic acid as their main metabolic product. *See also* homofermentative lactic acid bacteria; heterofermentative lactic acid bacteria.

lactic acid fermentation. *See* homofermentative lactic acid bacteria; heterofermentative lactic acid bacteria.

lactic acidosis. An acidosis that is caused by the accumulation of lactic acid and that may arise from oxygen deprivation of the tissues.

lactim. The enol form of a cyclic amide that contains the grouping $-N=\overset{|}{C}-OH$.

Lactobacillus bulgaricus factor. PANTE-THEINE.

Lactobacillus casei factor. FOLIC ACID.

lactochrome. An impure preparation of riboflavin.

lactoflavin. An impure preparation of riboflavin from milk.

lactogenic. Stimulating lactation; having prolactin activity.

lactogenic hormone. PROLACTIN.

lactoglobulin. *See* beta lactoglobulin.

lactonase. An enzyme that catalyzes the hydrolysis of a lactone.

lactone. An intramolecular ester that is formed by elimination of a molecule of water between a hydroxyl group and a carboxyl group.

lactone rule. A rule, proposed by Hudson in 1910, which states that the specific rotation of the γ or δ-lactone of a D-aldonic acid is positive when the hydroxyl group on carbon atom 4 or 5 of the acid is to the right in the Fischer formula, and that the specific rotation is negative when the hydroxyl group is to the left.

lactopoiesis. LACTATION.

lactose. A disaccharide of galactose and glucose that is present in milk.

lactose synthetase. The enzyme system composed of two proteins, denoted A and B, that catalyzes the synthesis of lactose from glucose and UDP-galactose in the mammary gland.

lactose tolerance test. A test used for evaluating intestinal lactase activity by the measurement of the level of blood glucose as a function of time following the administration of a dose of lactose to an individual.

lactosuria. The presence of excessive amounts of lactose in the urine.

lag period. 1. The delay in cell-free protein synthesis that occurs when synthetic polyribonucleotides are used as messenger RNA and that is thought to be caused by the absence of an initiating codon in the polyribonucleotide. 2. INDUCTIVE PHASE.

lag phase. That phase of growth of a bacterial culture that precedes the exponential phase and during which there is only little or no growth.

lake. To lyse erythrocytes by suspending them in a hypotonic medium.

Laki-Lorand factor. FIBRINASE.

Lamarckian theory. A theory of evolution, proposed by Lamarck in 1809, according to which the environment leads to structural changes in organisms, especially through new or increased uses of organs and through disuse and atrophy of other organs; these acquired characteristics are then transmitted to the offspring.

lambda. 1. A temperate phage that infects *Escherichia coli* and that contains double-stranded DNA. 2. A microliter. *Sym* λ.

lambda chain. One of the two types of light chains of the immunoglobulins.

lambda pipet. A pipet for the transfer of volumes in the microliter range.

Lambert-Beer Law. BEER'S LAW.

lamella (*pl* lamellae). LAMINA.

lamina (*pl* laminae). A thin layer; a thin plate.

laminar. Arranged in the form of layers or plates.

laminar flow. The undisturbed flow of a liquid along a tube; flow without obstacles. *See also* turbulent flow.

laminar-flow burner. PREMIX BURNER.

Lamm equation. CONTINUITY EQUATION.

Lamm scale displacement method. A method for measuring diffusion coefficients from the distortion of a transparent scale that is placed in front of a diffusion cell and that is photographed with light which has passed through the cell. The displacements of the scale lines are proportional to the refractive index gradient in the cell.

Lande G-factor. G-VALUE.

Landsteiner's rule. The rule that a blood group antigen and its antibody do not coexist in one individual.

Langmuir adsorption isotherm. An equation that describes the adsorption of a gas onto a solid and that has the same mathematical form as the Michaelis-Menten equation.

Langmuir trough. A surface balance.

lanolin. The wax of sheep wool.

lanosterol. The first biosynthetic sterol formed from squalene; the immediate precursor of cholesterol.

LAP. Leucine aminopeptidase.

Laplacian distribution. NORMAL DISTRIBUTION.

lard factor. VITAMIN A.

large-angle x-ray diffraction. A method of x-ray diffraction in which the scattering of the x-rays is measured at large angles; used for the analysis of small molecular spacings, as those between individual atoms.

large calorie. A kilocalorie; 1000 small calories. *Abbr* kcal; Cal.

laser. Acronym for "light amplification by stimulated emission of radiation"; a device capable of producing intense beams of monochromatic light.

late enzyme. A virus-specific enzyme that is transcribed from a late gene.

late gene. A viral gene that is transcribed late after the infection of a host cell by the virus. *Aka* late function gene.

latence. The mean time, generally measured in days, between the exposure of animals to a carcinogen and the appearance of tumors.

latency. The phenomenon of an inactive particulate enzyme that either becomes active after it is exposed, while still attached to an insoluble matrix, or becomes active after it is detached from the insoluble matrix. *See also* crypticity.

latent enzyme. An enzyme, especially a particulate enzyme, that exhibits activity only when the conditions are changed. *See also* latency.

latent image. The invisible image that is produced on a photographic film when it is exposed to radiation and that is rendered visible by subsequent photographic development of the film.

latent infection. 1. A bacterial infection that does not produce overt disease symptoms and in which the bacteria cannot be detected by currently available techniques. 2. A persistent viral infection that does not produce overt disease symptoms. *See also* inapparent infection.

latent iron-binding capacity. UNSATURATED IRON-BINDING CAPACITY.

latent period. 1. The time interval between the infection of a bacterium by a phage and the first appearance of extracellular phage progeny. 2. The time interval between the injection of a sensitizing antiserum into an animal and the time at which a second injection will elicit an anaphylactic response. 3. LATENCE.

latent virus. A tumor virus that has pathogenic activity and that has infected a host, but that does not lead to manifestations of the disease for a period of time.

late protein. A virus-specific protein that is transcribed from a late gene.

Latin square. An array of elements that are distributed over an equal number of columns and rows, such that any element occurs only once in each column and once in each row; used for the design of experiments in which the effects of combinations of variables are to be studied. A 4 × 4 latin square consists of four columns, four rows, and four elements, spread over a total of 16 positions.

LATS. Long-acting thyroid stimulator.

lattice. A three-dimensional network of elements that are arranged in geometric patterns, such as the atoms in a crystal lattice or the antigens and antibodies in an antigen-antibody lattice.

lattice theory. The theory according to which the reaction between antigens and antibodies leads to the formation of an insoluble antigen-antibody network. Each antibody has at least two binding sites for antigens, while each antigen has may binding sites for antibodies. Upon mixing antigens and antibodies an insoluble network, or lattice, is formed in which each antibody is bound to at least two antigens and each antigen is bound to several antibodies.

Laue pattern. An x-ray diffraction pattern that is produced by nonmonochromatic x-rays. *Aka* Laue photograph.

lauric acid. A saturated fatty acid that contains 12 carbon atoms.

law. A statement that describes a general truth or a general relationship.

lawn. The layer of bacterial cells growing on a solid medium.

law of mass action. The law that, at a given temperature, the rate of a chemical reaction is directly proportional to the active masses of the reactants. The rate of the reaction is proportional to the product of the molar concentrations of the reactants (more precisely, the activities), with each concentration raised to a power equal to the number of reactant molecules of the corresponding type which participate in the reaction.

layered metabolic pathway. TIERED METABOLIC PATHWAY.

layer line. One of the parallel lines of spots that are obtained by the rotating crystal method of x-ray diffraction.

LBF. *Lactobacillus bulgaricus* factor.

LC. Liquid chromatography.

L cell. A cell belonging to a strain of normal mouse fibroblast cells that have been maintained in tissue culture for 25 years. *See also* L-form.

L chain. Light chain.

L-configuration. The relative configuration of a molecule that is based upon its stereochemical relationship to L-glyceraldehyde.

LD. 1. Low-density lipoprotein. 2. Lethal dose.

LD$_{50}$. Median lethal dose.

LDH. Lactate dehydrogenase.

LDL. Low-density lipoprotein.

leading. The chromatographic and electrophoretic phenomenon in which a peak appears lopsided and a band, or a spot, appears ill-defined; due to the fact that in the support the front edge of the region which contains the component is diffuse while the back edge is sharp. *See also* tailing; trailing.

leading reactant. The first substrate that is bound to an enzyme in an ordered mechanism.

leakage. The loss of material across a cell membrane.

leak current. The transfer of charge that occurs when an electrolyte moves through a tube and that is due to back conduction by ion diffusion and to electroosmosis.

leaky gene. HYPOMORPH.

leaky mutant. A mutant that has a leaky gene.

leaky patch model. A model of the conformational changes produced in a cell membrane by the action of complement. According to this model, the phospholipid bilayer of the cell membrane is temporarily disrupted either by direct enzymatic action of complement or through the production of a lytic substance by complement. The disrupted membrane represents a leaky patch that allows the passage of water and ions across the membrane. *See also* doughnut model.

leaky protein. A protein, formed by a mutant gene, that has a fraction of the activity of the normal protein.

least squares. *See* method of least squares.

leaving group. A group of atoms that is displaced from a carbon atom by the attack of a nucleophile in a nucleophilic substitution reaction. *See also* S_N1 mechanism; S_N2 mechanism.

LeChatelier's principle. The principle that a system in equilibrium reacts to any change in its conditions in a manner that would tend to abolish this change.

lecithin. PHOSPHATIDYL CHOLINE.

lectin. A naturally occurring protein that has antibody-like activity in that it causes red blood cell agglutination. Some lectins are mitogenic and stimulate lymphocyte transformations; some agglutinate malignant cells preferentially. Lectins can be extracted from plants, especially from the seeds of legumes, and from several other sources such as snails and fish.

lente insulin. A slightly soluble form of insulin that is produced by crystallizing insulin from an acetate buffer in the presence of zinc ions. *See also* NPH insulin.

lepton. A subatomic particle that has a mass equal to, or less than, that of an electron; a negatron, a positron, or a neutrino.

lesion. 1. A pathological change in a tissue. 2. A

genetic defect; a mutation. *See also* biochemical lesion.

-less mutant. Combining form indicating an auxotrophic mutant, as in "thymineless mutant."

LET. Linear energy transfer.

lethal. Fatal; causing death.

lethal dose. The dose of a toxic compound, virus, etc., that kills 100% of the animals in a test group within a specified time. *Abbr* LD.

lethal gene. A gene, the expression of which leads to the death of the organism that carries the gene. *Aka* lethal factor.

lethal hit. The hit by a photon or by an ionizing particle that kills a cell or inactivates a virus.

lethal mutation. A mutation that leads to the premature death of the organism that possesses the mutation.

lethal mutation model. A model for the evolution of the genetic code according to which the code evolved so as to minimize the possibilities of deleterious mutations. *See also* frozen accident theory.

lethal synthesis. The process whereby an enzyme catalyzes a reaction with a compound, other than its normal substrate, and leads to the formation of a product which then acts as an inhibitor of a different enzyme.

letter. A nucleotide in a codon.

Leu. 1. Leucine. 2. Leucyl.

leucine. An aliphatic, nonpolar alpha amino acid. *Abbr* Leu; L.

leucine aminopeptidase. An aminopeptidase that acts on most amino acids and that catalyzes the sequential hydrolysis of a polypeptide chain from the N-terminal. *Abbr* LAP.

leuco-. Combining form meaning white or colorless.

leucocyte. Variant spelling of leukocyte.

leuco dye. The colorless form of a dye.

leuco methylene blue. The colorless, reduced form of methylene blue.

leucoplast. A plastid that does not contain pigments.

leucovirus. Variant spelling of leukovirus.

leucovorin. FOLINIC ACID.

leukemia. One of a group of generally fatal, cancerous diseases of the blood. The disease affects the white blood cells, which are usually overproduced, and the blood-forming organs.

leukocyte. A white blood cell that protects the organism against infection by eliminating invading bacteria through phagocytosis.

leukopenia. A condition characterized by a decrease in the number of white blood cells.

leukopoiesis. The formation of white blood cells.

leukosis. An abnormal proliferation of one or more of the leukocyte-forming tissues.

leukovirus. An enveloped, complex, animal virus that contains single-stranded RNA. Leukoviruses mature by budding from cytoplasmic membranes and some leukoviruses are oncogenic.

lev. Levorotatory.

levan. A homopolysaccharide of fructose.

levorotatory. Having the property of rotating the plane of plane-polarized light to the left, or counterclockwise, as one looks toward the light source. *Abbr* lev; l.

levulose. D-Fructose; a levorotatory monosaccharide.

Lewis acid. An atom, an ion, or a molecule that acts as an electron pair acceptor.

Lewis acid-base catalysis. The catalysis in solution in which the catalysts are Lewis acids and/or Lewis bases.

Lewis base. An atom, an ion, or a molecule that acts as an electron pair donor.

Lewis factor. One of two antigens that are closely related to those of the ABO blood group system.

L form. One of a number of bacterial variants of different genera that lack a rigid mucopeptide cell wall and that are capable of growth and multiplication.

LFT. Low-frequency transduction.

LFT lysate. Low-frequency transduction lysate; a lysate prepared by induction of a prophage that possesses normal, low transducing power.

LH. Luteinizing hormone.

LHR. Liquid holding recovery.

LHRF. Luteinizing hormone releasing factor; *see* luteinizing hormone releasing hormone.

LHRH. Luteinizing hormone releasing hormone.

LIBC. Latent iron-binding capacity.

librational motion. A rotational oscillation about an equilibrium position.

Liebermann-Burchard reaction. A colorimetric reaction for cholesterol and related sterols that is based on the successive production of a red, blue, and blue-green color upon treatment of the sample with acetic anhydride and concentrated sulfuric acid.

life. The sum of the properties that distinguish animals, plants, and microorganisms from nonliving matter, such as metabolism, reproduction, growth, excitability, movement, function, and complexity; the state of existence of a functioning cell, a group of cells, or an organism.

life cycle. 1. The sequence of the developmental stages of an organism from its formation to its death, or from any specified stage to the recurrence of that stage. 2. CELL CYCLE.

ligand. 1. An atom, a group of atoms, or a molecule that binds to a metal ion. 2. An atom, a group of atoms, or a molecule that binds to a macromolecule.

ligand chromatography. A column

chromatographic method in which metal ion complexes, as those formed with amino acids, are chromatographed on an ion-exchange resin.

ligand exchange chromatography. A column chromatographic technique in which a chelating ion-exchange resin is saturated with metal ions which are then removed by chelation with ligands in the sample solution that is passed through the column.

ligand field theory. A theory that describes the way in which the electrons of a metal ion in a metal ion-ligand complex reduce the repulsion of ligand electrons through angular polarization.

ligase. An enzyme that catalyzes the joining together of two different molecules or of two ends of the same molecule in a reaction that is coupled to the breakdown of ATP, or to the breakdown of a similar nucleoside triphosphate. *See also* enzyme classification.

ligate. To bind a ligand.

light. 1. *n* A form of electromagnetic radiation that has both wave and particle aspects. 2. *adj* Unlabeled, as opposed to being labeled with a heavy isotope. 3. *adj* Labeled with a light isotope, as opposed to containing either the natural or a heavy isotope.

light chain. One of two polypeptide chains that are linked to two heavy chains to form the immunoglobulin molecule. The molecular weight of a light chain is about 25,000 and that of a heavy chain is about 50,000. The light chains of type K and type L immunoglobulins are known, respectively, as κ and λ chains. *Abbr* L-chain.

light chopper. *See* chopper.

light isotope. An isotope that contains a smaller number of neutrons in the nucleus than the more frequently occurring isotope.

light label. A light isotope that is generally introduced into a molecule to facilitate its separation from identical molecules containing the more frequently occurring isotope.

light meromyosin. The terminal tail fragment of the myosin molecule. *Abbr* L-meromyosin; LMM. *Aka* F_3 fragment.

light microscope. An ordinary microscope that consists of an optical system for use with visible light.

light path. The thickness of sample through which light passes for optical measurements.

light reaction. A photosynthetic reaction or reaction sequence that depends directly on light energy and that serves to convert light energy into chemical energy.

light respiration. PHOTORESPIRATION.

light scattering. The dispersion of light rays by matter in directions other than that of the incident beam; commonly refers to the light scattered by solutions of macromolecules and to the

use of this scattering for determining molecular weights of the solute macromolecules. *See also* Mie scattering; Raman effect; Rayleigh scattering.

light strand. 1. A polynucleotide chain that is either not labeled with a heavy isotope or that is labeled with a light isotope. 2. The naturally occurring polynucleotide chain of a duplex that has a lower density than the complementary chain.

lignification. The formation of lignin.

lignin. The complex phenylpropanoid polymer that strengthens the cellulose framework of wood fibers and vascular plants.

limit dextrin. The branched core of amylopectin that remains after its digestion with either alpha or beta amylase.

limited proportional region. That portion of the characteristic curve of an ionization chamber that is above the proportional region and similar to it except for the fact that the amplification that can be achieved has a limiting value.

limiting current. The maximum current obtainable in a polarographic system which, in simple cases, may be equal to the diffusion current.

limiting velocity. MAXIMUM VELOCITY.

limiting viscosity number. INTRINSIC VISCOSITY.

lincomycin. An antibiotic, produced by *Streptomyces lincolnensis*, that inhibits protein synthesis by acting on the ribosome.

linear absorption coefficient. The fractional decrease in the intensity of a beam of radiation per unit thickness of the absorber.

linear accelerator. An instrument for imparting high kinetic energy to subatomic particles that are made to move in a long and straight path.

linear chain. OPEN CHAIN.

linear correlation. A relationship between two variables so that, as one increases, the other either increases or decreases, and a plot of one variable against the other yields a straight line.

linear density gradient. A density gradient in which the density increases in such a fashion that a plot of density versus distance in the tube yields a straight line.

linear dichroism. The dichroism that occurs when linearly polarized light is absorbed by partially or completely oriented molecules.

linear electric field effect. The change in the electron paramagnetic resonance properties of a sample by the application of an electric field across the sample.

linear energy transfer. The energy dissipation of a radiation as it passes through a tissue or other matter; generally expressed either in terms of kiloelectron volts per micron, or in terms of megaelectron volts per centimeter divided by the density of the substance in grams per cubic centimeter. *Abbr* LET.

linear growth. The growth of a culture such that a plot of the number of cells (or the cell mass) as a function of time yields a straight line.

linear inhibition. Inhibition that yields a straight line when either slopes or intercepts from a primary plot are plotted as a function of inhibitor concentration.

linear metabolic pathway. A metabolic pathway of the type $A \rightarrow B \rightarrow C$.

linear polymer. A polymer composed of open, unbranched chains.

linear polyphosphate. One of a group of compounds that contain a polymetaphosphate grouping, H_2PO_3-$(HPO_3)_n$-H_2PO_4, and that have been used to drive the polymerization of amino acids in studies on the origin of life. *Aka* polymetaphosphate.

linear-sweep polarography. Oscillographic polarography in which the entire potential scan is applied to the dropping mercury electrode either once as a single sweep, or several times as a multisweep, during the life of a single mercury drop.

linear velocity. The straight line distance moved per unit time.

line emission. The emission of light of either one or several specific wavelengths that is produced in flame photometry by a given ion.

line of best fit. *See* method of least squares.

line of stability. The line drawn through the band that represents the stable nuclides in a plot of the number of protons versus the number of neutrons in the nuclide.

line spectrum. A spectrum in which either the absorption or the emission of radiation is limited to only a few wavelengths.

Lineweaver-Burk plot. The double reciprocal plot of the Michaelis-Menten equation; a plot of $1/v$ versus $1/S$, where v is the velocity of the reaction and S is the substrate concentration.

linkage. Any association of genes in inheritance that exceeds that to be expected from independent assortment and that is due to their being located on the same chromosome; linkage is assessed by the tendency of two markers to remain together during recombination.

linkage group. A group of linked genes; all the genes located on the same chromosome.

linkage map. A scale representation of a chromosome that shows the relative positions of all its known genes.

linked gene. A gene showing linkage; a gene on the same chromosome as another gene.

linked reactions. COUPLED REACTIONS.

linked transduction. A bacterial transduction in which there is a simultaneous transfer of two or more genes that lie close together on a chromosome.

linked transformation. A bacterial transformation in which there is a simultaneous transfer to the bacterium of two or more genes that lie close together on a chromosome.

linoleic acid. An unsaturated fatty acid that contains 18 carbon atoms and two double bonds.

linolenic acid. An unsaturated fatty acid that contains 18 carbon atoms and three double bonds.

lipamino acid. Variant spelling of lipoamino acid.

lipase. An enzyme that catalyzes the hydrolysis of fats.

lipectomy. The surgical removal of adipose tissue.

lipemia. The presence of excessive amounts of lipid in the blood.

lipid. One of a heterogeneous group of compounds that are synthesized by living cells and that are sparingly soluble in water but are soluble in nonpolar solvents; they can be extracted from tissues by nonpolar solvents, and they have as a major part of their structure long hydrocarbon chains that may be branched or unbranched, straight or cyclic, saturated or unsaturated. Various classifications of lipids are in use; these include (a) simple lipids (neutral fats and waxes); complex lipids (phospholipids, sphingolipids, glycolipids, etc.); derived lipids (steroids, vitamins, carotenoids, etc.); and (b) neutral lipids (neutral fats, waxes, carotenoids, etc.); amphipathic lipids (glycerolipids, sphingolipids, etc.); redox lipids (quinones, etc.).

lipid bilayer. A layer of amphipathic lipid molecules that is two molecules thick and that is believed to form most or all of the central portion of biological membranes. In a lipid bilayer which is surrounded by a polar environment, the nonpolar parts of the lipid molecules are directed inward and the polar parts are on the outside.

lipide. LIPID.

lipidemia. LIPEMIA.

lipid-globular protein mosaic model. A model of biological membranes in which amphipathic lipids and globular proteins are arranged in an alternating mosaic pattern throughout the membrane; the nonpolar portions of the lipids and the bulk of the nonpolar amino acid residues of the proteins are in the interior of the membrane, while the polar portions of the lipids and the charged amino acid residues of the proteins are both at the surface of the membrane. The lipids are arranged primarily in a bilayer form, and they, as well as some of the proteins, possess a degree of fluidity that allows some lateral movement within the lipid matrix of the membrane.

lipid imbibition theory. A theory of atherosclerosis according to which the formation of atheromatous plaques is caused by the uptake of lipids, such as cholesterol, from the blood stream by the walls of the arteries.

lipid intermediate. The compound, undecaprenyl phosphate, that functions in peptidoglycan synthesis.

lipid mobilization. *See* mobilization.

lipidosis (*pl* lipidoses). 1. One of a number of genetically inherited or acquired diseases that are characterized by the deposition of one or more types of lipids in specific tissues or organs. 2. One of a number of genetically inherited metabolic defects in man that are characterized by lipid accumulation in specific tissues or organs and that result from a defect in the metabolism of glycosphingolipids. *See also* Fabry's disease; Gaucher's disease; Krabbe's disease; Niemann-Pick disease; Tay-Sachs disease.

lipid-soluble vitamin. FAT-SOLUBLE VITAMIN.

lipid solvent. A nonpolar solvent—such as chloroform, acetone, or methanol—that will extract lipids from tissues.

lipid storage disease. LIPIDOSIS.

lipin. LIPID.

lipoamide. The dipeptide-like structure, formed by linking a molecule of lipoic acid through its carboxyl group to the ϵ-amino group of a lysine residue, that is part of the enzyme for which lipoic acid serves as a coenzyme. *See also* lipoylprotein.

lipoamino acid. 1. A compound formed by joining a fatty acid or a long chain alcohol to an amino acid by means of either an ester or an amide bond. 2. An amino acid ester of phosphatidylglycerol.

lipocaic. A substance, secreted by the pancreas, that prevents the fatty infiltration of liver and stimulates the oxidation of fatty acids.

lipochrome. A naturally occurring, fat-soluble pigment.

lipogenesis. The biosynthesis of fatty acids from acetyl coenzyme A.

lipoic acid. A compound that is generally classified with the B vitamins, since it is a growth factor for some microorganisms; it functions as a coenzyme in the multienzyme systems that catalyze the oxidative decarboxylation of pyruvic acid to α-ketoglutaric acid and of α-ketoglutaric acid to succinic acid.

lipoid. 1. LIPID. 2. Resembling a fat or an oil.

lipolysis. The hydrolysis of lipids.

lipolytic. Of, or pertaining to, lipolysis.

lipolytic hydrolysis. LIPOLYSIS.

lipoma. A benign tumor of adipose tissue.

lipopeptide. 1. A compound formed by joining a fatty acid or a long chain alcohol to a peptide by means of either an ester or an amide bond. 2. A peptide ester of phosphatidylglycerol.

lipophilic. NONPOLAR.

lipophilic sephadex. A sephadex preparation that can be used with organic solvents.

lipophilic stain. A stain for lipids.

lipopolysaccharide. A water-soluble, lipid-polysaccharide complex. *Abbr* LPS.

lipoprotein. A conjugated, water-soluble protein in which the nonprotein portion is a lipid; the lipid is usually a glyceride, a phospholipid, cholesterol, or a combination of these. *See also* proteolipid.

lipoprotein lipase. The lipase that catalyzes primarily the hydrolysis of triglycerides which are associated with proteins, as those present in serum lipoproteins and in chylomicrons. *Abbr* LPL.

liposarcoma. A malignant tumor of adipose tissue.

liposome. An artificially prepared, cell-like structure in which a bimolecular layer or layers of lipid enclose an aqueous compartment; a membrane-bound vesicle, frequently formed by dispersion of phospholipid in aqueous salt solutions.

lipotrophic hormone. LIPOTROPIC HORMONE.

lipotrophin. LIPOTROPIC HORMONE.

lipotropic. Descriptive of a compound that can contribute methyl groups for the synthesis of choline and that can prevent or alleviate a fatty liver condition which results from a deficiency of choline.

lipotropic agent. A compound, such as choline or methionine, that aids in the transport of fat and thereby prevents or alleviates the condition of fatty infiltration of the liver.

lipotropic hormone. One of three peptides that are secreted by the anterior lobe of the pituitary gland and that stimulate the mobilization of lipids, especially fatty acids, from lipid deposits. *Var sp* lipotrophic hormone.

lipotropin. LIPOTROPIC HORMONE.

lipovitellenin. 1. A degraded form of a low-density lipoprotein component that is present in hens' egg yolk. 2. LOW-DENSITY FRACTION.

lipovitellin. 1. A high-density lipoprotein in hens' egg yolk. Two such proteins, denoted α and β, have been isolated; they are similar in their composition except for their content of protein-bound phosphorus. 2. A high-density yolk lipoprotein from any species.

lipoyllysine. LIPOAMIDE.

lipoylprotein. A conjugated protein in which lipoic acid is covalently bound to the protein by way of an amide link between its carboxyl group and the ϵ-amino group of a lysine residue in the protein. *See also* lipoamide.

lipuria. The presence of lipid in the urine.

liquefying amylase. ALPHA AMYLASE.

liquid chromatography. A collective term for liquid-liquid, liquid-solid, paper, thin-layer, ion-exchange, and molecular sieve chromatography. *Abbr* LC.

liquid crystal. A phase that has a mobility like that of a liquid and a high degree of order like that in a crystal. *Aka* mesophase; mesomorphic phase.

liquid holding recovery. The phenomenon that bacterial cells, when allowed to stand in buffer following their irradiation with ultraviolet light, show an increased viability as compared to cells that are plated out immediately after irradiation. *Abbr* LHR.

liquid junction potential. *See* junction potential.

liquid-liquid chromatography. Partition chromatography in which the mobile phase is a liquid and the stationary phase is an inert support, coated with a liquid. *Abbr* LLC.

liquid medium. A solution of nutrients.

liquid scintillation. The emission of light flashes by a solution containing a fluorescent chemical when the chemical is struck by either an ionizing particle or a photon; used as a method for measuring the radioactivity of a sample dissolved in the solution. When used in this fashion, the light flashes are transformed into electrical pulses by means of a photomultiplier tube and are then counted.

liquid scintillation counter. A radiation counter in which incident ionizing particles or incident photons are counted by the scintillations which they induce in a liquid fluor; the sample and the fluor are either dissolved in a common solvent or one is suspended in a solution of the other. *Abbr* LSC.

liquid-solid chromatography. Adsorption chromatography in which the mobile phase is a liquid and the stationary phase is a solid. *Abbr* LSC.

liquid surfactant membrane. A water-immiscible phase that contains emulsion size droplets and that consists of surfactants, a hydrocarbon solvent, and other compounds; the droplets may be used to hold a reagent or to encapsulate an enzyme.

liter. A metric unit of volume equal to 1 dm^3.

lithiasis. The formation of calculi, particularly biliary and urinary ones.

lithogenic. Leading to the formation of calculi; stone-producing.

lithosphere. The solid, mineral part of the earth.

lithotroph. A cell or an organism that uses inorganic compounds as electron donors for its energy-yielding, oxidation-reduction reactions.

little insulin. That fraction of free serum insulin that is indistinguishable from pancreatic insulin.

live. Alive, not dead; viable.

liver. The principal metabolic organ of animals that is capable of carrying out all the major metabolic reactions and that has a variety of functions in both anabolism and catabolism. It is a large gland, located in the abdominal cavity.

liver filtrate factor. PANTOTHENIC ACID.

liver function test. A quantitative determination of either a metabolite or an enzyme that is used in evaluating the functional capacity of the liver. Liver function tests include measurements of serum bilirubin concentration, alkaline phosphatase activity, and galactose tolerance.

liver profile. The composite results obtained from a battery of liver function tests.

live-timing. A method of timing, used in scintillation counters, in which the timing device is turned off during the interval that is required for the electronic processing of a pulse. *See also* clock-timing.

live vaccine. A vaccine that consists of infectious bacteria or of viruses, the virulence of which has been attenuated.

living. Possessing the properties of life; alive, not dead.

LLC. Liquid-liquid chromatography.

L.L.D. factor. VITAMIN B_{12}.

LLF. Laki-Lorand factor.

L-L factor. Laki-Lorand factor.

L-meromyosin. Light meromyosin.

LMM. Light meromyosin.

ln. Natural logarithm.

LNPF. Lymph node permeability factor.

LnRNA. Low-molecular weight nuclear RNA.

loading. The in vitro process whereby cellular structures, such as erythrocyte ghosts or mitochondria, are made to take up or accumulate specific substances.

load test. A tolerance test in which an individual is given a dose of a specific metabolite and the urinary concentration of this metabolite, or of a related compound, is determined as a function of time. A phenylalanine load test is used as a diagnostic tool for phenylketonuria, and a tryptophan load test is used as a diagnostic tool for schizophrenia. *See also* glucose tolerance test; galactose tolerance test.

Lobry De Bruyn-Alberta Van Eckenstein transformation. The interconversion of monosaccharides that occurs in alkaline solution as a result of the formation of enediol intermediates.

localized bond. A chemical bond involving two atoms.

localized orbital. A molecular orbital that is spread over two bonding atoms.

localized infection. A viral or a bacterial infection in which the infective agents remain primarily at the site of entry into the host.

lock and key theory. A theory of the mechanism

of an enzymatic reaction according to which the substrate binds to the enzyme to form an enzyme-substrate complex. The binding site on the enzyme is called the active site and it is structurally complementary to the substrate, so that the substrate fits onto the enzyme much as a key fits into a lock. *See also* flexible active site; induced fit theory.

Locke's solution. A solution that is similar in composition to Ringer's solution; it contains (in w/v) 0.9% sodium chloride, 0.024% calcium chloride, 0.042% potassium chloride, 0.02% sodium bicarbonate, and 0.1% glucose.

locoweed. One of a number of plants that contain selenium or other poisons and that cause alkali disease when eaten by animals.

locus (*pl* loci). A place or a position, particularly that occupied by a gene on a chromosome.

log. Logarithm; also denoted log_{10}.

logarithm. The exponent that indicates the power to which a fixed number has to be raised to produce a given number; the fixed number is known as the base. In ordinary, or Briggsian, logarithms the base is 10; in natural, or Naperian, logarithms, the base is the constant *e*. Ordinary logarithms are denoted log or log_{10}; natural logarithms are denoted ln or log_e.

logarithmic growth. EXPONENTIAL GROWTH.

logarithmic paper. Graph paper in which both scales have been distorted to allow the plotting of the logarithm of one variable versus the logarithm of a second variable.

logarithmic phase. EXPONENTIAL PHASE.

log_e. Natural logarithm.

log phase. Logarithmic phase.

Lohmann reaction. The reversible reaction, catalyzed by the enzyme creatine kinase, in which ATP and creatine are formed from ADP and phosphocreatine.

London dispersion forces. DISPERSION FORCES.

long-acting thyroid stimulator. A thyroid-stimulating substance that appears to be an immunoglobulin against a thyroid component; it is present in the blood of individuals suffering from hyperthyroidism and exerts its effect over a long period of time. *Abbr* LATS.

Long cat. A pancreatectomized and adrenalectomized cat that is used in endocrinological studies.

long-chain fatty acid thiokinase. A fatty acid thiokinase that catalyzes the activation of fatty acids having more than 12 carbon atoms to fatty acyl coenzyme A.

long-range hydration. The hydration by water molecules that are located in the secondary hydration shell.

long-range interactions. The attractive and repulsive forces between atoms and molecules

that do not decrease rapidly with distance. *See also* strong interactions.

long spacing fibrils. *See* fibrils long spacing.

long spacing segments. *See* segments long spacing.

Longsworth scanning method. An optical method for obtaining a photograph that depicts the refractive index gradient of a boundary; produced by photographing the boundary while a knife edge moves upward in front of the camera lens and while the photographic plate is driven horizontally at a speed which is proportional to that of the knife edge.

loop. One of three or four nonhydrogen-bonded regions in the cloverleaf model of transfer RNA.

loose coupling. The state of cellular respiration in which the mitochondria are characterized by having a low acceptor control ratio.

low-angle x-ray diffraction. SMALL-ANGLE X-RAY DIFFRACTION.

low background counter. A specially shielded radiation counter in which the level of background counts has been greatly reduced.

low-density fraction. 1. A low-density fraction of hens' egg yolk that contains about 89% lipid and two lipoproteins which differ primarily in their size. 2. LIPOVITELLENIN.

low-density lipoprotein. A lipoprotein that contains approximately 60% neutral lipid, 20% phospholipid, and 20% protein; such lipoproteins have molecular weights of the order of 2 to 10×10^6, a density of about 1.02 to 1.06 g/l, and a flotation coefficient of about 0 to 20S. *Abbr* LDL; LD.

low-energy compound. A compound that, upon hydrolysis under standard conditions, yields a small amount of free energy; the standard free energy change for the hydrolysis reaction is less than 7 kcal/mole.

low-energy phosphate acceptor. A low-energy compound that can function as an acceptor for the phosphoryl group transferred from high energy phosphate donors by way of the ADP-ATP phosphoryl group carrier system. *Aka* low-energy phosphate compound.

low-frequency transduction. Transduction in which the phages that are capable of transducing constitute a small proportion of the total phage population. *Abbr* LFT.

low gate. The cut-off level in integral discrimination.

low-lipid lipoprotein. HIGH-DENSITY LIPOPROTEIN.

low-order antibody. INCOMPLETE ANTIBODY.

Lowry method. A modification of the Folin-Ciocalteau reaction that is used as a colorimetric reaction for the quantitative determination of proteins.

low-speed sedimentation equilibrium. SEDI-MENTATION EQUILIBRIUM.

low spin. The state of a complex in which there is a maximum of paired electrons; referred to as a state of essentially covalent bonding and ascribed to certain hemoproteins.

low-temperature heat method. The pasteurization of material by heating it at 61.6°C for 30 min. *Abbr* LTH method.

LPL. Lipoprotein lipase.

LPS. Lipopolysaccharide.

L-region. Two reactive para positions in an aromatic hydrocarbon; the presence of such a region has been correlated with the lack of carcinogenic activity of some hydrocarbons. *See also* K-region.

LRF. Luteinizing hormone releasing factor; *see* luteinizing hormone releasing hormone.

LRH. Luteinizing hormone releasing hormone.

LS antigen. An antigen that can be dissociated from a poxvirus and that consists of a heat-labile and a heat-stable component.

LSC. 1. Liquid-solid chromatography. 2. Liquid scintillation counter.

LSD. Lysergic acid diethylamide.

l-strand. WATSON STRAND.

LTH. 1. Lactogenic hormone. 2. Luteotropic hormone.

LTH method. Low-temperature heat method.

Lubrol. A nonionic detergent.

luciferase. The enzyme that functions in the bioluminescence reactions of the firefly; it catalyzes the oxidation of luciferin with the concomitant release of visible light.

luciferin. The substrate of the enzyme luciferase. *See also* luciferase.

lucite. PLEXIGLASS.

Lugol's solution. A solution containing (in w/v) 5% iodine and 10% potassium iodide.

lumen. 1. A passageway in a small tube or duct. 2. A unit of luminous flux; equal to the quantity of visible light falling on 1 cm² at a distance of 1 cm from a light source of 1 Icd.

lumichrome. A blue fluorescent compound formed by photolysis of riboflavin.

lumiflavin. A yellow-green fluorescent compound formed by photolysis of riboflavin.

luminescence. The emission of light that results from chemical reactions within, or a flow of energy into, an emitter, rather than from an increase in temperature. *See also* fluorescence; phosphorescence; luciferase.

luminescent. Of, or pertaining to, luminescence.

lumirhodopsin. A structurally altered form of rhodopsin that is produced after the exposure of rhodopsin to light and prior to its dissociation into opsin and retinal$_1$.

lumisterol. A compound produced from ergosterol by irradiation with ultraviolet light.

Luria-Delbrueck fluctuation test. *See* fluctuation test.

Luria-Latarjet experiment. An experiment for determining the sensitivity of intracellular vegetative phage to irradiation. The phage-infected bacteria are irradiated at different stages of the phage multiplication cycle with varying doses of ultraviolet light or other radiation. The cells are then analyzed for the fraction of surviving infective phage particles.

luteal. Of, or pertaining to, the corpus luteum.

luteinizing hormone. A gonadotropic protein hormone, secreted by the anterior lobe of the pituitary gland, that stimulates the final ripening and rupture of the ovarian follicles in the female and that stimulates the production of testosterone in the male. *Abbr* LH.

luteinizing hormone releasing hormone. The hypothalamic hormone that controls the secretion of luteinizing hormone. *Abbr* LRH; LHRH. *Aka* luteinizing hormone releasing factor (LRF; LHRF).

luteotrophic hormone. Variant spelling of luteotropic hormone.

luteotrophin. Variant spelling of luteotropin.

luteotropic hormone. PROLACTIN.

luteotropin. PROLACTIN.

LVP. Lysine vasopressin.

Lyase. An enzyme that catalyzes a nonhydrolytic reaction whereby a group is either removed from a substrate, thereby forming a double bond, or added to a double bond, thereby forming a single bond. *See also* enzyme classification.

lyate ion. A solvent molecule minus a proton; in the case of water, the lyate ion is the hydroxyl ion.

lycopene. A red pigment in tomatoes that is the parent compound of the carotenoids.

lymph. The fluid, derived from the interstitial fluid, that bathes the tissues, circulates through the lymphatic vessels, and is ultimately discharged into the blood.

lymphatic. 1. *n* A small lymph vessel. 2. *adj* Of, or pertaining to, lymph.

lymph node permeability factor. A factor that occurs in extracts of lymph node cells and which, when injected into the skin, increases vascular permeability. *Abbr* LNPF.

lymphocyte. A cell that occurs in lymphatic tissue, spleen, and blood and that is characterized by having a large round nucleus; a lymph cell.

lymphocyte transformation. The formation of immunoblasts from lymphocytes, a process that is accompanied by rapid synthesis of DNA and RNA.

lymphoid. 1. Resembling lymph or lymphatic tissue. 2. LYMPHATIC.

lymphoid cell. LYMPHOCYTE.

lymphokine. A soluble, humoral mediator that is released by primed lymphocytes when they come in contact with specific antigens. *See also* migration enhancement factor; migration inhibition factor.

lympholytic agent. An immunosuppressive agent that causes the destruction of lymphocytes.

lymphoma. A tumor of lymphatic tissue.

lymphosarcoma. A malignant lymphoma composed of abnormal and immature lymphocytes.

lyochrom. An impure preparation of riboflavin.

lyoenzyme. SOLUBLE ENZYME.

lyolysis. Solvolysis in which the solvent either donates or accepts protons.

lyonium ion. A solvent molecule plus a proton; in the case of water, the lyonium ion is the hydronium ion.

lyophilic. Descriptive of the tendency of a group of atoms or of a surface to become either wetted or solvated by the solvent. *See also* hydrophilic.

lyophilization. The removal of water under vacuum from a frozen sample; a relatively gentle process for the removal of water in which the water sublimes from the solid to the gaseous state. *See also* cryosublimation.

lyophobic. Descriptive of the tendency of a group of atoms or of a surface to resist becoming either wetted or solvated by the solvent. *See also* hydrophobic.

lyophobic bond. LYOPHOBIC INTERACTION.

lyophobic interaction. The association of nonpolar groups with each other in nonaqueous, but polar, solvents. These interactions are weaker than hydrophobic interactions, and their strength decreases with an increase in temperature.

lyotropic mesophase. A liquid crystal, the structure of which is loosened as a result of solvent action.

lyotropic series. An arrangement of cations and anions in a series according to their effect on the solubility of proteins. *See also* salting out.

Lys. 1. Lysine. 2. Lysyl.

lysate. The suspension of ruptured cells obtained upon lysis.

lysergic acid diethylamide. A hallucinogenic drug that occurs naturally in a fungus and that is believed to produce chromosomal aberrations. *Abbr* LSD.

lysin. An antibody that can lead to cell lysis.

lysine. An aliphatic, basic, and polar alpha amino acid. *Abbr* Lys; K.

lysine intolerance. A genetically inherited metabolic defect in man that is due to a deficiency of the enzyme L-lysine: NAD-oxidoreductase.

lysine vasopressin. A vasopressin, occurring in hogs, in which the eighth amino acid residue has been replaced by a lysine residue. *Abbr* LVP.

lysis. The rupture and dissolution of cells.

lysis from within. The lysis of bacterial cells that occurs as a result of the intracellular multiplication of phage particles.

lysis from without. The lysis of bacterial cells that occurs without intracellular phage multiplication and that is due to the large number of holes produced in the cell wall by lytic enzymes of the phage; the enzymes either are released into the medium or are contained within the adsorbed phage particles.

lysis inhibition. The delay in the lysis of bacterial cells that occurs when a cell culture is inoculated with a phage and is then heavily reinoculated with the same phage a few minutes later.

lysochrome. A substance that dissolves in lipids and thereby colors them.

lysogenic. Of, or pertaining to, lysogeny.

lysogenic bacterium. 1. A bacterium that has survived the infection by a temperate phage and that has incorporated the prophage into the bacterial DNA; a lysogenized bacterium. 2. A bacterium that can be infected by a temperate phage.

lysogenic conversion. The phage-mediated, phenotypic changes of a bacterium, as those relating to growth and morphology, that may accompany the infection of the bacterium by a temperate phage.

lysogenic cycle. The sequence of reactions whereby a bacterial cell becomes infected with a temperate phage, incorporates the prophage into the bacterial DNA, and then divides.

lysogenic response. The incorporation of phage DNA into the bacterial DNA that follows the infection of the bacterium by a temperate phage.

lysogenic virus. A virus that can become a prophage; a temperate phage.

lysogenization. The production of lysogenic bacteria by infection of a sensitive bacterial strain with a temperate phage.

lysogenized bacterium. LYSOGENIC BACTERIUM (1).

lysogeny. The phenomenon of bacterial infection by temperate phages.

lysolecithin. A lysophosphoglyceride of lecithin that causes lysis of erythrocytes.

lysophosphatide. LYSOPHOSPHOGLYCERIDE.

lysophosphoglyceride. A phosphoglyceride from which the fatty acid, attached to the second carbon atom of glycerol, has been removed by hydrolysis.

lysophospholipid. LYSOPHOSPHOGLYCERIDE.

lysosomal disease. A genetically inherited metabolic defect in man that is due to a deficiency of an enzyme located in the lysosomes; examples of lysosomal diseases are Fabry's, Gaucher's, Niemann-Pick's, Pompe's, and Wolman's disease.

lysosome. A membrane-enclosed cytoplasmic structure that is rich in hydrolytic enzymes.

lysozyme. The enzyme that catalyzes the hydrolysis of polysaccharides which occur in the glycopeptide layer of bacterial cell walls; a bacteriolytic enzyme that is present in such biological fluids as egg white and saliva.

lysyl-bradykinin. A decapeptide kinin.

lytic. Of, or pertaining to, lysis.

lytic cycle. 1. The sequence of reactions whereby a virulent phage infects a bacterial cell, multiplies inside the cell, and ultimately leads to the lysis of the cell. 2. The sequence of reactions whereby a prophage is induced in an infected bacterial cell, multiplies inside the cell, and ultimately leads to the lysis of the cell.

lytic enzyme. An enzyme that catalyzes a hydrolysis reaction; a hydrolytic enzyme.

lytic response. The intracellular multiplication of a virulent virus that leads to the lysis of the infected cell.

lytic virus. A virus that can cause cell lysis; a virulent virus.

M

m. 1. Meta. 2. Milli. 3. Meter.

M. 1. Molar concentration. 2. Molecular weight. 3. Metal ion. 4. Methionine. 5. Mitosis.

macroevolution. Evolution that extends over long time periods, involves large steps, and leads to marked changes in the genetic makeup of an organism.

macroglobulin. A globulin, such as the IgM immunoglobulin, that has a molecular weight above 400,000.

macroglobulinemia. The presence of excessive amounts of macroglobulins, specifically IgM immunoglobulins, in the blood.

macroion. A charged macromolecule.

macrolide antibiotic. One of a number of antibiotics, such as erythromycin and oleandomycin, that are similar in their structure, action, and antimicrobial spectrum, and that are characterized by having a large lactone ring that contains anywhere from 14 to 20 carbon atoms.

macromolecule. A high-molecular weight molecule; a polymer.

macromutation. A mutation in which a large segment of a chromosome is altered, as distinct from a point mutation.

macronutrient. An essential nutrient that is needed by an organism in appreciable amounts.

macrophage. A cell, derived from the reticuloendothelial system, that functions in phagocytosis and that is also believed to function in the formation of antibodies.

magic number. 1. The number of the different amino acids that occur in proteins. 2. The number of either the neutrons or the protons which, when they occur in an atomic nucleus, contribute to the stability of the nucleus.

magic spot. One of two unusual nucleotides, originally discovered as spots on a radioautogram and designated I and II, that occur in microorganisms and are believed to serve in a regulatory capacity; the nucleotides are: guanosine-5′-diphosphate, 3′-diphosphate ($_{pp}G_{pp}$) and guanosine-5′-triphosphate, 3′-diphosphate ($_{ppp}G_{pp}$). *Abbr* MS.

magnesium. An element that is essential to all plants and animals. Symbol Mg; atomic number 12; atomic weight 24.312; oxidation state $+2$; most abundant isotope Mg^{24}; a radioactive isotope Mg^{28}, half-life 21.3 hours, radiation emitted—beta particles and gamma rays.

magnetic circular dichroism. Circular dichroism that is measured in the presence of a magnetic field. Magnetic circular dichroism depends on the nature of, and the couplings between, the ground and the excited states of the chromophore; it reflects the intrinsic geometry of the chromophore and is not sensitive to the surroundings of the chromophore. *Abbr* MCD. *See also* circular dichroism.

magnetic dipole. A substance that has two magnetic poles; the separation between the poles is measured by the magnetic dipole moment.

magnetic dipole moment. A measure of the tendency of a substance to become oriented in a magnetic field; equal to the product of the strength of the magnetic pole and the length of the magnet. The magnetic dipole moment depends on the presence or absence of paired electrons, since a spinning electron behaves as a magnet. Substances with paired electrons have no net magnetic dipole moment and are diamagnetic; substances with unpaired electrons have a permanent magnetic dipole moment and are paramagnetic.

magnetic field. The space surrounding magnetic poles in which a mechanical force will be exerted on a magnet introduced into it.

magnetic moment. MAGNETIC DIPOLE MOMENT.

magnetic resonance. *See* electron paramagnetic resonance; nuclear magnetic resonance.

magnetic stirrer. A plastic- or glass-coated magnet that is used as a stirring bar.

magnetic susceptibility. A measure of the tendency of a substance to become oriented in a magnetic field; the proportionality constant between the total magnetic moment per unit volume in the direction of an applied magnetic field and the magnetic field strength.

magnetic transition moment. The magnetic moment that is induced in a molecule during a transition, such as that associated with the absorption of a photon, as a result of a rotation of charge.

magnification. The apparent enlargement of an object when viewed through a microscope; expressed as the number of times that the diameter of the object appears to have been enlarged.

Maillard reaction. One of a group of nonenzymatic reactions in which aldehydes, ketones, or reducing sugars react with amino acids, peptides, or proteins; believed to occur as part of the browning reactions of food.

main diffusion coefficient. The diffusion coefficient that a component has when no other diffusing components are present; used in the treatment of diffusion data in a system showing interaction of flows. *See also* cross-term diffusion coefficient.

major gene. A gene that has a marked phenotypic effect as opposed to a modifying gene.

major groove. The groove in Watson-Crick type DNA that is approximately 22 Å across.

major-minor code. An early version of the genetic code according to which the central nucleotide of a codon is the major factor in positioning the correct amino acid on the template, while the two adjoining nucleotides are minor factors in selecting the amino acid.

MAK. Methylated albumin-kieselguhr.

malaria. An acute, and sometimes chronic, disease of man and other vertebrates that is caused by protozoans of the genus *Plasmodium* which are transmitted by mosquitoes of the genus *Anopheles*.

malate shuttle. A shuttle, the components of which are malic acid, oxaloacetic acid, and the enzyme malate dehydrogenase.

male hormone. TESTOSTERONE.

maleic acid. The cis isomer of fumaric acid that is not an intermediate in the citric acid cycle.

male sex hormone. An androgen that affects the development of secondary sex characteristics in the male. The principal male sex hormones are testosterone and dihydrotestosterone which are produced by the testes; other compounds that have male sex hormone activity, such as adrenosterone, are produced by the adrenal gland.

malic acid. A dicarboxylic acid that is formed from fumaric acid in the reactions of the citric acid cycle.

malic enzyme. The enzyme that catalyzes the anaplerotic reaction whereby pyruvic acid is carboxylated to malic acid.

malignant. Descriptive of a tumor that metastasizes and endangers the life of the organism.

malnutrition. A condition that is caused by inadequate quantity, quality, digestion, absorption, or utilization of ingested nutrients.

malonic acid. A competitive inhibitor of the enzyme succinate dehydrogenase.

maltose. A disaccharide that is composed of two glucose residues linked by means of an $\alpha(1\rightarrow4)$ glycosidic bond and that constitutes the repeating unit of starch.

malt sugar. MALTOSE.

mammal. A vertebrate of the class Mammalia that is characterized by possession of hair and mammary glands.

mammalian. Of, or pertaining to, mammals.

mammary tumor agent. MOUSE MAMMARY TUMOR VIRUS.

mammotropin. PROLACTIN.

Man. Mannose.

mandelate pathway. A degradative pathway of mandelic acid, catechol, and related aromatic compounds that occurs in *Pseudomonas fluorescens*.

manganese. An element that is essential to all animals and plants. Symbol Mn; atomic number 25; atomic weight 54.9380; oxidation states $+2$, $+3$, $+4$, $+7$; most abundant isotope Mn^{55}; a radioactive isotope Mn^{54}, half-life 313 days, radiation emitted—gamma rays.

mannan. A homopolysaccharide of mannose that occurs in bacteria, molds, and plants.

mannitol. A sugar alcohol derived from mannose.

mannose. A six-carbon aldose. *Abbr* Man.

manometer. An instrument for measuring the pressure of a liquid or a gas.

mantissa. The fractional part of a logarithm.

Mantoux test. An intradermal tuberculin test for delayed-type hypersensitivity in man.

MAO. Monoamine oxidase.

MAOI. Monoamine oxidase inhibitor.

map. 1. *n* GENETIC MAP. 2. *n* CYTOGENETIC MAP. 3. *v* To establish the structure or structural details for a portion of a macromolecule, as in the mapping of an active site. 4. *v* To establish the location of either a mutable site on a genetic map or a gene on a cytogenetic map. *See also* peptide map; nucleotide map.

map distance. The distance between any two markers on a genetic map. *See also* map unit.

maple syrup urine disease. A genetically inherited metabolic defect in man that is associated with mental retardation and that is characterized by the presence of urinary keto-acids derived from valine, leucine, and isoleucine; due to a deficiency of the enzyme, branched-chain keto-acid decarboxylase.

map unit. A measure of distances along a linkage map that is equal to a recombination frequency of 1%.

Mariotte flask. A device used in column chromatography for maintaining a constant pressure head of eluent.

marijuana. A naturally occurring, hallucinogenic drug.

marker. 1. A mutable site on a chromosome that is useful for cell identification and for genetic studies; the site of a gene of known function and known location on the chromosome. *See also* biochemical marker; genetic marker. 2. A group or a molecule that is linked chemically to another molecule for purposes of identification. 3. A reference substance that is used in a physical technique such as chromatography, electrophoresis, or density gradient centrifugation.

marker enzyme. An enzyme, the intracellular location of which is known, so that an assay of the enzyme can be used as an aid in following the isolation and purification of subcellular fractions.

marker rescue. The incorporation of a genetic marker from a mutated virus into the DNA of an active progeny virus during cross-reactivation.

maser. Acronym for "microwave amplification by stimulated emission of radiation"; a device capable of producing intense beams of microwave radiation.

masked residue. An amino acid residue in a protein that is not accessible to, and cannot undergo a reaction with, specific reagents. The lack of activity may be due to the occurrence of the residue in the internal portion of the protein or to stereochemical, electrostatic, and other properties of its immediate environment.

masked virus. A tumor virus that lacks pathogenic activity.

Mason's theory. A theory that describes the first step in chemical carcinogenesis in terms of the electronic interactions between the carcinogen and either a protein or a nucleic acid molecule of the affected cell or organism.

mass absorption coefficient. The linear absorption coefficient divided by the density of the absorber.

mass action. *See* law of mass action.

mass balance equation. CONSERVATION EQUATION.

mass chromatogram. A paper chromatogram of preparative-type separations in which the sample is streaked as a band on a large sheet of filter paper.

mass fragmentogram. The photographic record obtained from mass fragmentography.

mass fragmentography. The combined use of gas chromatography and mass spectrometry in which the effluent from a gas chromatographic column is fed into a mass spectrometer which serves as a detector for the recording of from one to three fragments at preselected mass-to-charge ratios; the mass spectrometer is preset to these mass-to-charge ratios and serves as a sensitive detector for small quantities of specific ions. *Aka* single-ion monitoring.

mass number. The sum of the number of protons and neutrons in the nucleus of an atom. *Sym* A.

mass spectrogram. The photographic record of a mass spectrum as obtained with a mass spectrograph.

mass spectrograph. A mass spectrometer in which a photographic record of the mass spectrum is obtained.

mass spectrometer. An instrument for the separation of charged particles according to their mass-to-charge ratios. Molecules and ions are fragmented by bombardment with electrons, and the ions thus formed are focused by means of electrostatic and magnetic fields and ultimately strike a photographic plate or some other detector. Both the mass-to-charge ratios and the relative amounts of the charged fragments can be determined.

mass spectrum. A plot of the number of fragments as a function of their mass-to-charge ratio as measured with a mass spectrometer or a mass spectrograph.

mass stopping power. LINEAR ENERGY TRANSFER.

mass unit. *See* atomic mass unit.

mast cell. A basophilic cell of connective tissue that contains heparin and histamine and that functions in immediate-type hypersensitivity.

master plate. The mounted piece of sterile velvet that is used in replica plating and that is covered with the original bacterial culture.

master strand. WATSON STRAND.

mate killer. A *Paramecium aurelia* cell that carries *mu* particles and that kills or injures sensitive paramecia with which it conjugates; the *mu* particles protect the organism against other mate killers. *See also* kappa particle.

mathematical model. An equation that describes the behavior of an actual physical system and that is derived on the basis of theoretical considerations and pertinent numerical parameters.

mating pool. *See* Visconte-Delbrueck hypothesis.

matrix (*pl* matrices). A gel-like substance that fills the space between the cristae of mitochondria and that is the site of many of the enzymes of the citric acid cycle.

matrix interference. The interference that occurs in atomic absorption spectrophotometry when light is absorbed either by the organic solvent in which the sample is dissolved, or by the solids that are formed from this solvent by its evaporation in the flame.

matrix method. A mathematical method for treating a steady-state enzyme system in which the time derivatives of the differential equations describing this state are set equal to zero, and the resultant algebraic equations are then solved by matrix inversion.

maturation. 1. The development of a red blood cell from its formation to its final form as an erythrocyte. 2. The assembly of the different components of a virus that results in a complete and infectious virion. 3. The development of a spore.

maturation-defective mutant. A phage mutant that can synthesize DNA but not the viral structural proteins.

maturation protein. A phage-specific protein of small RNA phages—such as MS2, f2, and

R17—that is required for the production of complete and infectious phage particles. *Aka* maturation factor.

mature virion. A fully assembled, infectious virus particle.

max. Maximum.

maximal medium. A rich medium that contains all the necessary metabolites for the growth of cells and that frequently consists of a protein hydrolysate, a yeast extract, and inorganic salts.

maximum height-area method. A method for calculating translational diffusion coefficients from the height of the peak of the gradient curve and from the area under the peak.

maximum permissible body burden. The greatest amount of total, cumulative exposure of an individual to radioactive radiation that is permitted by federal safety standards.

maximum permissible concentration. The greatest concentration of radioactive isotopes in air and water that is permitted by federal safety standards. *Abbr* MPC.

maximum permissible dose. The greatest amount of radioactive radiation that an individual may receive over a given period of time according to federal safety standards. *Abbr* MPD.

maximum stationary phase. STATIONARY PHASE (1).

maximum velocity. The greatest velocity of an enzymatic reaction that is attainable with a fixed amount of enzyme under defined conditions; the velocity that is obtained when the enzyme is saturated with substrates and cosubstrates. *Sym* V_{max}; V.

Maxwell effect. FLOW BIREFRINGENCE.

Mb. Myoglobin; related compounds are abbreviated as MbO_2-oxymyoglobin and MbCo-carbon monoxide myoglobin.

MBSA. Methylated bovine serum albumin.

mc. Millicurie.

McArdle-Schmid-Pearson disease. GLYCOGEN STORAGE DISEASE TYPE V.

McArdle's disease. GLYCOGEN STORAGE DISEASE TYPE V.

MCD. Magnetic circular dichroism.

M chain. One of the two types of polypeptide chains of lactate dehydrogenase isozymes; denoted M, since the tetramer of M chains is found predominantly in muscle tissue.

mCi. Millicurie.

McLeod gauge. A laboratory pressure gauge for measuring gas pressures of vacuum systems to as low as 10^{-6} mm Hg.

MDH. Malate dehydrogenase.

Me. Methyl group.

mean. The value obtained by summing the values of a set of measurements and dividing the sum by the number of individual measurements in the set; arithmetic average.

mean activity coefficient. The average activity coefficient for the anion and cation of an electrolyte; specifically, $log f_{\pm} = -0.5Z_+Z_-\sqrt{\Gamma/2}$, where $f_{\pm}$ is the mean activity coefficient, Z_+ and Z_- are the charges of the cation and the anion, and $\Gamma/2$ is the ionic strength.

mean free path. The average distance traveled by a molecule, an ionizing particle, or some other particle between collisions.

mean generation time. DOUBLING TIME.

mean ion activity coefficient. MEAN ACTIVITY COEFFICIENT.

mean life. AVERAGE LIFE.

mean range. The distance between the source of a radiation, particularly a source of alpha particles, and the point at which the intensity of the beam is reduced to one half.

mean residue rotation. The specific rotation of a polymer that is calculated on the basis of the concentration of the monomers, or residues, of the polymer rather than on the basis of the concentration of the intact polymer molecules. Specifically, $[m] = M_0 [\alpha]/100$, where $[m]$ is the mean residue rotation, M_0 is the mean residue weight, and $[\alpha]$ is the specific rotation of the polymer. If the mean residue rotation is corrected to that in a medium having a refractive index of one, it is given by the expression $[m] = [3/(n^2 + 2)] \times M_0[\alpha]/100$, where n is the refractive index of the medium.

mean residue weight. The average molecular weight of a monomer, or residue, in a polymer.

mean square. VARIANCE.

mean square end-to-end distance. ROOT MEAN SQUARE END-TO-END-DISTANCE.

measuring pipet. A pipet consisting of a tube of uniform diameter that is drawn out to a tip and is graduated uniformly along its length.

mechanism. 1. A step-by-step description of a chemical or a physical reaction or reaction sequence; for a chemical reaction this includes the electron shifts and the bond-making and bond-breaking aspects of the reaction. 2. MECHANISTIC PHILOSOPHY.

mechanistic philosophy. The doctrine that life and its phenomena are explicable entirely in terms of the laws and processes of chemistry and physics. *Aka* mechanism. *See also* vitalism.

mechanistic process. A deterministic process in which each step is a necessary and direct result of a preceding step. *See also* stochastic process.

mechanistic theory. 1. A theory according to which the evolution of the genetic code is based on a necessary physical-chemical relationship between an amino acid and its codons; consequently, the code could have evolved in one, or at most in a few, possible ways. *See also* selective theory. 2. A theory according to which the formation of atheromatous plaques in atherosclerosis is due to the precipitation and/or the coagulation of one or more of the components of the blood.

mechanochemical coupling hypothesis. A hypothesis of the coupling of ATP synthesis to the electron transport system in oxidative phosphorylation. According to this hypothesis, the transport of electrons leads to the formation of energy-rich conformations of mitochondrial proteins, and the energy associated with the relaxation of these energized proteins then drives the phosphorylation of ADP to ATP.

median. The value of a set of measurements around which the measurements are equally distributed; one half of the measurements are numerically greater and one half are numerically smaller than this value.

median effective dose. The dose of a drug that produces therapeutic effects in 50% of the animals in a test group within a specified time. *Sym* ED_{50}.

median immunizing dose. The dose of a vaccine or an antigen that produces immunity in 50% of the animals in a test group within a specified time.

median infectious dose. The dose of bacteria or viruses that produces demonstrable infection in 50% of the animals in a test group within a specified time. *Sym* ID_{50}. *Aka* median infective dose.

median lethal dose. The dose of bacteria, viruses, or a toxic compound that produces deaths in 50% of the animals in a test group within a specified time. *Sym* LD_{50}. *See also* half-value dose.

median paralysis dose. The dose of virus that produces paralysis in 50% of the animals in a test group within a specified time. *Sym* PD_{50}.

median tissue culture dose. The dose of virus that produces tissue culture degeneration in 50% of the test units within a specified time. *Sym* TC_{50}.

median toxic dose. The dose of a toxic agent that produces toxic effects in 50% of the animals in a test group within a specified time. *Sym* TD_{50}.

mediated transport. The movement of a solute across a biological membrane that requires the participation of one or more transport agents.

medicine. 1. The science and art of diagnosing, treating, and preventing disease. 2. A substance used in either the treatment or the prevention of a disease.

medium (*pl* media). A liquid or a solid preparation of nutrients that is used for the maintenance and growth of cells.

medium-chain fatty acid thiokinase. A fatty acid thiokinase that catalyzes the activation of fatty acids having 4 to 12 carbon atoms to fatty acyl coenzyme A.

medulla. *See* adrenal medulla.

MEF. Migration enhancement factor.

mega-. Combining form meaning one million

(10^6) and used with metric units of measurement. *Sym* M.

megamitochondrion. An enlarged mouse liver mitochondrion that is produced artificially by exposure of the animal to the drug cuprizone (biscyclohexanone oxaldihydrazone).

Mehler reaction. A photosynthetic reaction whereby hydrogen peroxide is formed from water and molecular oxygen.

meiosis. The nuclear division of the germ cells of sexually reproducing organisms in which the chromosome number is halved; it occurs during gametogenesis in animals and during sporogenesis in plants.

meiospore. A spore produced by meiosis.

meiotic. Of, or pertaining to, meiosis.

meiotic drive. An irregularity in the segregation of the chromosomes during meiosis that leads to alterations in the allele frequencies of a population.

meiotic effect. The phenomenon in which the mutation rate during meiosis differs from that during mitosis.

melanic. 1. Of, or pertaining to, melanism. 2. Having a dark pigmentation.

melanin pigment. One of a group of dark coloring substances that are responsible for the pigmentation of the skin and that are formed in melanocytes by the oxidation of phenylalanine, tyrosine, and other aromatic compounds.

melanism. The abnormal coloration of the skin or other tissues that is caused by the accumulation of melanin pigments.

melanocyte. A cell that synthesizes melanin pigments in its cytoplasm.

melanocyte-stimulating hormone. One of two peptide hormones, denoted α and β, that are produced by the posterior lobe of the pituitary gland and that have a darkening effect by causing the dispersion of melanin pigments in the melanocytes. *Abbr* MSH.

melanocyte-stimulating hormone regulatory hormone. One of two hypothalamic hormones (or factors) that, respectively, stimulate or inhibit the release of melanocyte-stimulating hormone from the pituitary gland. The melanocyte-stimulating hormone releasing hormone (or factor) is abbreviated variously as MRH (MRF), or MSHRH (MSHRF). The melanocyte-stimulating hormone release-inhibiting hormone (or factor) is abbreviated variously as MIH (MIF), MRIH (MRIF), MSHIH (MSHIF), or MSHRIH (MSHRIF).

melanocyte-stimulating hormone release-inhibiting hormone. *See* melanocyte-stimulating hormone regulatory hormone.

melanocyte-stimulating hormone releasing hormone. *See* melanocyte-stimulating hormone regulatory hormone.

melanoma. A malignant tumor derived from melanocytes.

melanosome. A tyrosinase-containing intracellular organelle of melanocytes.

melanotrophin. Variant spelling of melanotropin.

melanotropin. MELANOCYTE-STIMULATING HORMONE.

melatonin. A tryptophan-related compound that is formed in the pineal gland and that reverses the darkening effect of the melanocyte-stimulating hormone by causing aggregation of the melanin granules in the melanocytes.

melituria. GLYCOSURIA.

melphalan. A mutagenic, alkylating agent.

melting. 1. The transition of double-helical nucleic acid segments to random coil conformations that is produced by an increase in the temperature of the solution containing the nucleic acid and that is due to the breaking of the hydrogen bonds of the paired bases. *Aka* thermal-denaturation; melting out. 2. The transition of a solid to a liquid that is produced by an increase in temperature.

melting out profile. THERMAL DENATURATION PROFILE.

melting out temperature. The temperature of a thermal denaturation profile at which one half of the maximum change in absorbance (or other property) is obtained. *Sym* T_m; $T_{1/2}$. *Aka* melting temperature.

melting point. 1. MELTING OUT TEMPERATURE. 2. The temperature at which a solid changes to a liquid as a result of the application of heat; at this temperature the solid and the liquid are in equilibrium.

melting profile. THERMAL DENATURATION PROFILE.

membrane carrier. *See* carrier (3).

membrane electrode. An electrode, such as the glass electrode, that has a membrane incorporated into its design.

membrane equilibrium. GIBBS-DONNAN EQUILIBRIUM.

membrane osmometer. An osmometer that has a semipermeable membrane incorporated into its design and that is used for measurements of osmotic pressure and for measurements of number average molecular weights of macromolecules.

membrane potential. The electrical potential across a membrane, particularly a biological membrane, that arises from the charges in the membrane and from the charges present on either side of the membrane. *See also* diffusion potential; dipolar potential; distribution potential; equilibrium potential.

membrane structure. *See* Benson model; Davson-Danielli model; lipid-globular protein mosaic model; supermolecule.

membrane transport. The movement of materials across a biological membrane.

membron. A functioning, regulatable, translating complex of a polysome and a specific surface area of a biological membrane.

membron theory of cancer. The theory according to which tumor cells differ from normal cells in the stability of their membrons.

memory. 1. A device for storing information. 2. IMMUNOLOGICAL MEMORY.

memory cell. A cell that is responsible for immunological memory.

memory response. SECONDARY IMMUNE RESPONSE.

menadione. Vitamin K_3.

Mendelian. Of, or pertaining to, Mendel or to his laws of heredity.

Mendelian character. A character that follows Mendel's laws in inheritance.

Mendel's laws. The laws of inheritance proposed by Mendel in 1866 and known as the law of segregation and the law of independent assortment.

meniscus. The flat or crescent-shaped interface between a liquid in a tube and air.

meniscus depletion sedimentation equilibrium. A variation of the sedimentation equilibrium method in which the ultracentrifuge is operated at sufficiently high speeds so that all the macromolecules are sedimented out of the region near the meniscus. The method obviates a separate run to determine the initial concentration of the solution, is especially useful for monodisperse systems, and is generally used in conjunction with an interferometric optical system.

meq. Milliequivalent; also denoted mEq.

meractinomycin. ACTINOMYCIN D.

mercaptan. THIOL.

2-mercaptoethanol. A sulfhydryl group containing compound that is used to protect the sulfhydryl groups of enzymes and other proteins against oxidation.

6-mercaptopurine. A purine analogue used in cancer chemotherapy. *Abbr* MP.

mercerization. The treatment of cellulose with 20% sodium hydroxide which produces a cellulose that has an increased capacity for dyes and a greater tensile strength.

mercurial. An organic compound, such as a drug, that contains mercury.

meridian. The line in an x-ray diffraction pattern that passes through the zero point and that is perpendicular to the layer lines.

meridional. At right angles to the equator of an x-ray diffraction pattern.

meridional reflection. An x-ray diffraction spot that lies on the meridian.

merocrine gland. A gland that produces a secretion without significantly damaging its secreting cells.

meromyosin. One of two fragments produced from myosin by treating it with either trypsin or chymotrypsin. *See also* heavy meromyosin; light meromyosin.

merozygote. A partially diploid zygote that contains one complete and one partial genome; produced as a result of a partial genetic exchange, as that which may occur during bacterial transformation or transduction.

Merrifield method. SOLID PHASE TECHNIQUE.

mersalyl. A drug that inhibits the exchange of phosphate and hydroxyl ions across the inner mitochondrial membrane.

MES. 2-(*N*-Morpholino)-ethanesulfonic acid; used for the preparation of buffers in the pH range of 5.5 to 7.0.

mescaline. A hallucinogenic drug that occurs naturally in a cactus and that has the structure of a phenylethylamine.

Meselson-Stahl experiment. An experiment that provides support for the semiconservative mode of DNA replication. The experiment consists of labeling the DNA in a growing culture of *Escherichia coli* with a heavy isotope, isolating the DNA at various stages of the growth curve, and analyzing the DNA by means of density gradient sedimentation equilibrium.

mesh size. A standard screen for designating the particle size of ion-exchange resins, gels, and other chromatographic supports; a larger mesh size indicates a smaller particle diameter.

meso carbon. A carbon atom to which are attached two identical and two different substituents. The two identical substituents bear a mirror image relationship to each other and react differently with a given enzyme.

meso compound. An optical isomer that possesses asymmetric elements, such as asymmetric carbons, but has overall molecular symmetry and is, therefore, optically inactive.

mesomorphic. Of, or pertaining to, a liquid crystal.

meson. A subatomic particle that has a mass greater than that of a lepton but smaller than that of a nucleon.

mesophase. LIQUID CRYSTAL.

mesophile. An organism that grows at moderate temperatures in the range of 20 to 45°C, and that has an optimum growth temperature in the range of 30 to 37°C.

mesophilic. Of, or pertaining to, mesophiles.

mesosome. An infolding of the bacterial cell membrane.

mesotocin. A peptide hormone that is related to oxytocin in its structure and in its function; it is secreted by the posterior lobe of the pituitary gland and occurs in reptiles and amphibia.

mesotrophic lake. A lake that has intermediate properties between those of an oligotrophic lake and those of a eutrophic lake.

Mesozoic era. The geologic time period that extends from about 63 to 225 million years ago and that is characterized by the development of the reptiles.

message. 1. A messenger RNA molecule. 2. The segment of a polycistronic messenger RNA molecule that codes for one polypeptide chain.

messenger RNA. A single-stranded RNA molecule that is synthesized during transcription, is complementary to one of the strands of double-stranded DNA, and serves to transmit the genetic information contained in DNA to the ribosomes for protein synthesis. *Abbr* mRNA. *Aka* messenger.

messenger RNA hypothesis. The hypothesis, proposed by Jacob and Monod, that an RNA molecule serves as the template for the synthesis of proteins; this RNA molecule, the messenger RNA, is transcribed from DNA, has a base sequence which is complementary to that of one of the strands of duplex DNA, and carries the genetic information from the DNA to the ribosomes where the proteins are synthesized.

Met. 1. Methionine. 2. Methionyl.

meta-. Prefix indicating two substituents on alternate carbon atoms of the ring in an aromatic compound. *Sym* m.

metabolic. Of, or pertaining to, metabolism.

metabolic acidosis. A primary acidosis that results from changes in the concentrations of acids and bases other than carbon dioxide and carbonic acid.

metabolic alkalosis. A primary alkalosis that results from changes in the concentrations of acids and bases other than carbon dioxide and carbonic acid.

metabolic antagonist. ANTIMETABOLITE.

metabolic balance study. *See* balance study.

metabolic block. A block in a biochemical reaction, generally due to a mutation, that results in the lack of synthesis of an enzyme or in the synthesis of a defective enzyme.

metabolic disease. A pathological abnormality of metabolism such as acidosis, alkalosis, or an inborn error of metabolism.

metabolic pathway. A sequence of consecutive enzymatic reactions that brings about the synthesis, breakdown, or transformation of a metabolite from a key intermediate to some terminal compound. A metabolic pathway may be linear, cyclic, branched, tiered, directly reversible, or indirectly reversible. *See also* linear metabolic pathway; cyclic metabolic pathway; etc.

metabolic poison. A substance that inhibits a metabolic reaction.

metabolic pool. *See* pool.

metabolic quotient. A measure of the rate of uptake or discharge of a metabolite by a tissue or an organism. The uptake of oxygen, denoted Q_{O_2}, and the evolution of carbon dioxide, denoted Q_{CO_2}, are frequently expressed in terms of microliters taken up or evolved per hour per milligram dry weight of tissue. *Sym* Q.

metabolic transformation. BIOTRANS-FORMATION.

metabolic turnover. The rate at which cellular components are replaced by degradation and synthesis under steady-state conditions; turnover time.

metabolism. 1. The sum total of all the chemical and physical changes that occur in a living system, which may be a cell, a tissue, an organ, or an organism. The reactions of metabolism are almost all enzyme-catalyzed and include transformation of nutrients, excretion of waste products, energy transformations, synthetic and degradative processes, and all the other functions of a living organism. Metabolism is broadly divided into anabolism, which encompasses the synthetic reactions, and catabolism, which encompasses the degradative reactions. 2. The sum total of all the chemical and physical changes in a living system with respect to one class of compounds, as in "amino acid metabolism."

metabolite. Any reactant, intermediate, or product in the reactions of metabolism.

metachromatic dye. A dye that stains a tissue with two or more different colors depending on the extent to which the dye molecules are stacked on the chromotrope.

metachromatic granule. VOLUTIN GRAN-ULE.

metachromatic leukodystrophy. A genetically inherited metabolic defect in man that is associated with mental retardation and that is characterized by an accumulation of sulfatides; it is due to a deficiency of the enzyme sulfatidase which catalyzes the conversion of sulfatides to cerebrosides.

metagon. An RNA particle of the protozoan *Paramecium aurelia* which acts like a messenger RNA molecule in that organism, but replicates like an RNA virus when ingested by the protozoan *Didnium*.

metakentrin. LUTEINIZING HORMONE.

metal chelate. *See* chelate.

metalloenzyme. A conjugated enzyme that contains one or more metal ions as prosthetic groups.

metalloflavoprotein. A complex flavoprotein that contains a metal ion in addition to either FMN or FAD.

metalloporphyrin. A complex composed of a metal ion that is chelated by a porphyrin.

metalloprotein. A conjugated protein that contains one or more metal ions as prosthetic groups.

metal shadowing. SHADOWCASTING.

metamorphosis. A transformation in the form of an animal, such as that from larval to adult form in insects, or that from tadpole to adult form in amphibians.

metaphase. The second stage in mitosis during which the chromosomes arrange themselves in an equatorial region.

metaphosphate. The anionic radical PO_3^- of metaphosphoric acid.

metaprotein. A denatured protein, formed by the action of acids or bases, that is soluble in weak acids and bases but is insoluble in neutral solutions.

metarhodopsin. A structurally altered form of rhodopsin that is produced after the exposure of rhodopsin to light and prior to its dissociation into opsin and retinal₁.

metastable. Describing an unstable condition or substance that changes readily either to one which is more stable, or to one which is less stable.

metastable nuclide. An excited nuclear isomer that emits a gamma ray upon its return to the ground state.

metastable state. 1. The excited state of an atom or a molecule that is characterized by a delayed emission of the excitation energy as the atom or the molecule returns to the ground state. 2. The excited state of a nuclear isomer that is characterized by the emission of a gamma ray as the nucleus returns to the ground state.

metastasis (*pl* metastases). 1. The detachment of cells from a tumor and their transport, by way of the blood and the lymphatic system, to distant sites in the organism where they grow to form additional tumors. 2. The tumor formed at a site distant from that of the original tumor, and produced by detachment and transport of cells from the original tumor.

metastasize. To invade by means of metastasis.

metastatic tumor. A tumor that is undergoing metastasis; a metastasizing tumor.

metathesis. A chemical reaction in which there is a double exchange of elements or groups, as in the reaction $AB + CD = AD + BC$.

methanolysis. An alcoholysis reaction in which the alcohol is methanol; frequently used for the formation of fatty acid methyl esters which are then analyzed by gas chromatography.

MetHb. Methemoglobin.

methemoglobin. A hemoglobin molecule in which the iron has been oxidized to the trivalent state. *Abbr* MetHb.

methemoglobinemia. A genetically inherited metabolic defect in man that is characterized by high concentrations of methemoglobin in the blood and that is due to a deficiency of the enzyme methemoglobin reductase.

methionine. A sulfur-containing, nonpolar alpha amino acid. *Abbr* Met; M.

methionyl transfer RNA. A transfer RNA molecule that exists in two forms, designated tRNA^met and tRNA^fmet. The former is ordinary methionyl-tRNA, and the latter allows for the formylation of methionine after it has become attached to the tRNA; N-formylmethionyl-tRNA^fmet serves as the initiator aminoacyl-tRNA in bacterial protein synthesis.

method of continuous variation. A method for studying the interaction between two components in solution, as in the formation of a duplex from two single strands. The method requires the plotting of a property that is characteristic of one component versus the mole fraction of that component in a two-component system. In the absence of interaction between the components, a straight line is obtained which connects the points corresponding to a mole fraction of one for each of the components. In the presence of interaction between the components, various deviations from a straight line are obtained.

method of least squares. A method for fitting a straight line to a set of experimental points such that the sum of the squares of the deviations of the experimental points from the line has a minimum value; the line is referred to as a line of best fit.

method of optimal proportions. A method for determining the equivalence zone of a precipitin reaction by measuring the proportion of antigen to antibody at which precipitation occurs most rapidly. The zone of optimal proportions measured in this fashion is generally near the equivalence zone of the precipitin curve. *See also* Dean and Webb method; Ramon method.

method of ultimate precision. A method for measuring the absorbance of a solution in a single beam photometer with maximum sensitivity and with minimum error; measurement is made by adjusting the instrument to 0 and 100% transmission with two solutions of known concentration in the compound of interest.

methotrexate. AMETHOPTERIN.

methylated albumin-kieselguhr. An adsorbent used for the chromatographic fractionation of nucleic acids; prepared by converting the carboxyl groups of the glutamic and aspartic acid residues of serum albumin to methyl esters, and then precipitating the methylated albumin onto kieselguhr particles. *Abbr* MAK.

methylating agent. S-ADENOSYLMETHIONINE.

methylation. The introduction of a methyl group into an organic compound.

methylcholanthrene. A carcinogenic hydrocarbon.

O-methyl derivative. A carbohydrate derivative in which one or more hydroxyl groups have been methylated.

methylene blue. A dye used as an oxidation-reduction indicator; the oxidized form is blue, and the reduced form is colorless.

methylene blue technique. THUNBERG TECHNIQUE.

methylene group. The grouping $-CH_2-$.

methyl green. A basic dye used in cytochemistry for the staining of DNA.

methyl group. The radical $-CH_3$. *Sym* Me.

ϵ-N-methyllysine. A rare amino acid that is present in the protein actin.

methylmalonicaciduria. A genetically inherited metabolic defect in man that is caused by a deficiency of the enzyme methylmalonyl-CoA carboxymutase.

methylneogenesis. The biosynthesis of methyl groups from one-carbon fragments.

methylpherase. Methyltransferase.

methyl-poor transfer RNA. A transfer RNA molecule that contains less than the usual amount of methylated bases.

N^5-methyl tetrahydrofolic acid. SERUM FOLATE.

MetMb. Metmyoglobin.

metmyoglobin. A myoglobin molecule in which the iron has been oxidized to the trivalent state. *Abbr* MetMb.

metopon. A semisynthetic drug, made by converting morphine to methyldihydromorphinone.

metric combining forms. *See* tera; giga; mega; kilo; hecto; deca; deci; centi; milli; micro; nano; pico; femto; atto.

metyrapone test. A clinical test for assessing the pituitary reserve of adrenocorticotropin by administering the drug metyrapone; the drug inhibits the synthesis of cortisol, which controls the release of adrenocorticotropin from the pituitary gland by a feedback mechanism.

MeV. Mega electron volt; 10^6 eV.

mevalonic acid. An intermediate in the biosynthesis of cholesterol and a precursor of squalene and other isoprenoid compounds.

Meyerhof oxidation quotient. A measure of the Pasteur effect that is equal to the difference between the rates of anaerobic and aerobic glycolysis, divided by the rate of oxygen uptake.

mg. Milligram.

Mg. Magnesium.

mg percent. Referring to a solution, the concentration of which is expressed in terms of the number of milligrams of solute per 100 ml of solution.

MGT. Mean generation time.

MHD. 1. Minimum hemagglutinating dose. 2. Minimum hemolytic dose.

MHO. The reciprocal of ohm.

micellar. Of, or pertaining to, micelles.

micelle. An organized colloidal structure that consists of a large number of oriented surface-active molecules; micelles are generally charged and are typically formed by soaps and by phospholipids.

micellisation. The formation of micelles.

Michaelis constant. A kinetic constant for a given substrate of an enzymatic reaction; it is numerically equal to that substrate concentration which yields one half of the maximum velocity of the reaction at saturating concentrations of all cosubstrates. The Michaelis constant K_m is composed of the rate constants for the individual steps in the reaction; for the sequence

$$E + S \underset{k_{-1}}{\overset{k_{+1}}{\rightleftharpoons}} ES \overset{k_{+2}}{\longrightarrow} E + P \text{ it is given by}$$

$K_m = (k_{-1} + k_{+2})/k_{+1}$; for the sequence $E +$

$$S \underset{k_{-1}}{\overset{k_{+1}}{\rightleftharpoons}} ES \underset{k_{-2}}{\overset{k_{+2}}{\rightleftharpoons}} EP \overset{k_{+3}}{\longrightarrow} E + P \text{ it is given by}$$

$K_m = (k_{-1}k_{-2} + k_{-1}k_{+3} + k_{+2}k_{+3})/[k_{+1}(k_{+2} + k_{-2} + k_{+3})]$, where the k's are the rate constants, E is the enzyme, S is the substrate, and P is the product. *Sym* K$_m$.

Michaelis-Menten-Briggs-Haldane equation. BRIGGS-HALDANE EQUATION.

Michaelis-Menten equation. 1. The rate equation for an enzymatic reaction which is derived on the assumptions that (a) a rapid equilibrium is established between the enzyme, the substrate, and the enzyme substrate complex; and (b) the velocity of the reaction is an initial velocity, proportional to the concentration of enzyme-substrate complex, so that the reverse reaction from products to enzyme-substrate complex can be neglected. Specifically, $v = V[S]/(K_s + [S])$, where v is the initial velocity of the reaction, V is the maximum velocity, $[S]$ is the substrate concentration, and K_s is the substrate constant. 2. BRIGGS-HALDANE EQUATION.

Michaelis-Menten kinetics. The kinetics of an enzymatic reaction that can be described by the Michaelis-Menten equation; such a reaction yields a typical hyperbolic curve when the velocity of the reaction is plotted as a function of the substrate concentration.

Michaelis pH function. An expression for the concentration of either the undissociated or the dissociated form of a substance which can undergo one or more ionizations by the loss of protons; the concentration of the substance is expressed in terms of the hydrogen ion concentration and the appropriate ionization constants.

micro-. 1. Combining form meaning one millionth (10^{-6}) and used with metric units of measurements. *Sym* μ. 2. Combining form meaning microscopic or minute.

microaerophilic. Descriptive of bacteria that grow best at partial pressures of oxygen which are considerably lower than those in air.

microbial genetics. The genetics of microorganisms.

microbiological assay. The assay of a biochemical compound, such as a vitamin or an amino acid, that is based on measuring the growth of microorganisms for which the compound is an essential growth factor.

microbiology. The study of microorganisms.

microbioscope. A microscope that permits the observation of living tissues.

microbody. PEROXISOME.

microcinematography. The study of cells by means of motion pictures taken through a phase contrast microscope. In time-lapse microcinematography, cells are photographed at selected time intervals and the film is then projected at a faster speed to provide a better understanding of the time-relationships of cellular processes.

microcomparator. An optical instrument for making measurements on photographic plates as those used with schlieren and interferometric optical systems. The instrument resembles a toolmaker's microscope that is equipped with a screen upon which is reflected an enlarged image from a photographic plate, and measurements are made on the screen by means of cross-hairs and micrometers.

microcrystalline wax. A wax derived from petroleum and consisting of large numbers of small crystals.

microcurie. One millionth of a curie.

microdrop technique. A technique for studying antibody formation in individual lymphocytes by suspending a single lymphocyte in a microdrop and determining whether a bacterium or a bacteriophage becomes immobilized in the drop.

microencapsulated enzyme. An enzyme that is immobilized in a microcapsule, equipped with a semipermeable membrane, such that substrates and products, but not the enzyme, can diffuse across the membrane.

microenvironment. A very small environment, as that surrounding a molecule or a functional group of a molecule.

microevolution. Evolution that extends over short time periods, involves small steps, and leads to small changes in the genetic makeup of an organism.

microfossil. A fossil of microorganisms.

microglobulin. A globulin that has a molecular weight of less than 40,000.

micrograph. *See* electron micrograph.

microheterogeneity. The state of a given preparation of macromolecules of one kind, especially proteins, in which the macromolecules exhibit slight differences with respect to their charge, state of aggregation, extent of denaturation, or other properties.

microincineration. The combustion of the organic material of a thin tissue slice or of a cell suspension that is placed on a glass slide. The remaining ash provides information about the quality, quantity, and distribution of inorganic compounds in the sample.

micromanipulator. An instrument for positioning and handling needles, electrodes, pipets, and the like for experimentation with microscopic specimens, including single cells.

micrometer. 1. A device for measuring minute distances that is used in conjunction with optical instruments. 2. MICRON.

micromethod. A method of chemical analysis that requires very small amounts of sample and reagents.

micron. A unit of length equal to 10^{-6} meter and useful for describing cellular dimensions; a micrometer. *Sym* μ.

micronutrient. An essential nutrient that is needed by an organism in minute amounts.

microorganism. An organism that is too small to be seen with the naked eye.

microperoxisome. A small cytoplasmic organelle, bound by a single membrane and devoid of a nucleoid, that is found in varying amounts in different mammalian cells and that can be visualized by the use of reagents which are specific for the enzyme catalase.

microscope. An instrument for magnifying and visualizing objects that are too small to be seen with the naked eye.

microscope electrophoresis. PARTICLE ELECTROPHORESIS.

microsome. A subcellular fraction that consists of ribosomes and endoplasmic reticulum.

microspectrophotometer. A cytophotometer that consists of a microscope and a spectrophotometer.

microsphere. A spherical cell-like structure that is formed spontaneously from proteinoids under suitable conditions and that is believed by some to have been a forerunner of primitive cells.

microtome. An instrument for cutting thin sections of tissues, about 1 to 10 μ in thickness, for staining and microscopic examination.

microtomy. The methodology connected with the use of a microtome.

microtubule. A small tubule occurring in the mitotic spindle and also in flagella and cilia of unicellular eucaryotic organisms.

micrurgy. Microsurgery performed on a specimen that is viewed through a microscope; frequently entails operations on single cells with the aid of a micromanipulator.

middle mesophase. A mesophase consisting of cylindrical micelles, arranged in a hexagonal array.

midpoint potential. The electrode potential, measured at 25°C and 1 atm, at the midpoint of an oxidation-reduction titration curve; the electrode potential at which the oxidant and the reductant are present at equal concentrations.

Mie scattering. The scattering of light by spherical particles that are neither very large nor very small in comparison to the wavelength of the incident light.

MIF. 1. Melanocyte-stimulating hormone release-inhibiting factor; *see* melanocyte-stimulating hormone regulatory hormone. 2. Migration inhibition factor.

migration. 1. The movement of a molecule in either electrophoresis or chromatography. 2. An intramolecular rearrangement of atoms, groups of atoms, or bonds.

migration enhancement factor. A factor extractable from lymphocytes that enhances the migration of macrophages out of tissue slices and out of capillary tubes. *Abbr* MEF. *See also* lymphokine.

migration inhibition factor. A factor extractable from lymphocytes that inhibits the migration of macrophages out of tissue slices and out of capillary tubes. *Abbr* MIF. *See also* lymphokine.

MIH. *See* melanocyte-stimulating hormone regulatory hormone.

milieu intérieur. The internal environment that consists of the extracellular fluid that surrounds the tissue cells of multicellular organisms.

milk agent. MOUSE MAMMARY TUMOR VIRUS.

milk sugar. LACTOSE.

Miller experiment. An experiment in which organic compounds are synthesized under conditions believed to simulate those that have existed during the early stages of chemical evolution on the earth. The experiment demonstrates the formation of amino acids and other compounds from a mixture of reducing gases that have been subjected to an electric discharge.

Miller index (*pl* Miller indices). One of three numbers that designate a plane in which the atoms of a crystal lie.

milli-. Combining form meaning one thousandth (10^{-3}) and used with metric units of measurements. *Sym* m.

millicurie. One thousandth of a curie. *Abbr* mc; mCi.

milliequivalent. One thousandth of a gram equivalent weight. *Abbr* meq.

millimicron. A unit of length equal to 10^{-9} meter; a nanometer. *Abbr* nm; mμ.

Millipore filter. Trademark for a group of

synthetic filters having pores of specified diameter.

Millon reaction. A colorimetric reaction for tyrosine and other phenolic compounds that is based on the production of a red color upon treatment of the sample with a solution of mercurous and mercuric nitrates in concentrated nitric acid.

min. 1. Minute. 2. Minimum.

mineralization. The conversion of organic matter to inorganic matter, and the infiltration of organic matter by inorganic matter.

mineralocorticoid. A 21-carbon steroid, such as deoxycorticosterone or aldosterone, that is secreted by the adrenal cortex and that acts primarily on water and electrolyte metabolism by stimulating the retention of sodium and the excretion of potassium.

mineralocorticosteroid. MINERALOCORTICOID.

minicell. A spherical, anucleate bacterial body that results from an abnormal fission near the polar extremity of a parent cell. Such cells do not contain a bacterial chromosome and they lack a number of DNA-associated enzymes, such as DNA-dependent RNA polymerase, but they may contain plasmid DNA.

minimal deviation hepatoma. A hepatoma that is cancerous but resembles a normal liver cell so closely in its enzyme content and in its histological and other properties that it can be compared with normal cells.

minimal medium. A synthetic medium that contains only those compounds essential for the growth of the wild-type organism and that is incapable of supporting the growth of auxotrophs.

minimal stable length. The minimum size of the segment of nucleotide pairs that must be formed during renaturation with the base-pairing in perfect register to lead to the formation of duplex DNA or RNA. *See also* snapback.

minimum hemagglutinating dose. The smallest dose of a virus that produces complete agglutination of a standard volume of red blood cells within a specified time. *Abbr* MHD.

minimum hemolytic dose. The smallest dose of complement that produces complete lysis of a standard volume of sensitized red blood cells within a specified time. *Abbr* MHD.

minimum lethal dose. The smallest dose of bacteria, virus, or a toxic compound that produces deaths in 100% of the animals in a test group within a specified time. *Abbr* MLD.

minimum molecular weight. The molecular weight of a substance that is determined by an assay for some structural element on the assumption that there must be at least one such structural element per molecule of the substance. The structural element may be a metal ion, a functional group, a ligand, a monomer, etc.

minisome. The smallest ribonucleoprotein particle that is either a naturally occurring precursor of ribosomes, or is produced from ribosomes by the stepwise removal of proteins.

minor base. One of a group of purines and pyrimidines that generally occur only in small amounts in most nucleic acids but that are found in relatively large amounts in transfer RNA; many of the minor bases are methylated derivatives of the commonly occurring purines and pyrimidines.

minor groove. The groove in Watson-Crick type DNA that is approximately 12 Å across.

minority codon. MODULATING CODON.

minus strand. 1. The nucleic acid strand that is formed during the replication of a phage which contains single-stranded nucleic acid, and that is complementary to the original, or plus, strand of the phage. 2. WATSON STRAND.

minute phage. One of a group of small phages that contain single-stranded DNA and that are either spherical phages, such as ϕX-174 and S-13, or filamentous phages, such as M-13 and fd.

7-minute phosphate. *See* labile phosphate.

10-minute phosphate. *See* labile phosphate.

minute plaque mutant. A plaque-type mutant that produces very small plaques.

mirror image. ENANTIOMER.

mischarging. The covalent linking of an amino acid to a transfer RNA molecule that is specific for a different amino acid.

miscible. Capable of being mixed.

miscoding. MISTRANSLATION.

miscopying. The occurrence of an error during transcription.

misincorporation. The incorporation of either a wrong monomer or an analogue into a polymer; used specifically with regard to the synthesis of DNA.

mismatching. MISPAIRING.

mispairing. The occurrence of a base in one strand of a double-stranded nucleic acid molecule that is not complementary to the base in the corresponding position in the second strand, resulting in incomplete base-pairing.

misreading. The occurrence of an error during translation.

misreplication. The occurrence of an error during replication.

missense codon. A codon that has been altered from its normal, sense form in which it codes for one amino acid to a codon that codes for a different amino acid.

missense mutation. A mutation in which a normal, sense codon is altered so that it becomes a missense codon.

missense suppression. The suppression of a missense mutation.

mistranslation. The incorporation of an amino acid into a polypeptide chain in response to a codon for a different amino acid.

MIT. MONOIODOTYROSINE.

mitochondrion (*pl* mitochondria). A subcellular organelle in aerobic eucaryotic cells that is the site of cellular respiration and that carries out the reactions of the citric acid cycle, electron transport, and oxidative phosphorylation. Mitochondria have a high degree of biochemical autonomy; they contain DNA and ribosomes, carry out protein synthesis, and are capable of self-replication.

mitogen. An agent that stimulates mitosis.

mitomycin C. An antibiotic, produced by *Streptomyces caespitosus*, that prevents DNA replication by cross-linking the complementary strands of DNA.

mitoribosome. A ribosome occurring in mitochondria.

mitosis (*pl* mitoses). The division of the nucleus of eucaryotic cells which occurs in four stages designated prophase, metaphase, anaphase, and telophase. *Abbr* M.

mitotic. Of, or pertaining to, mitosis.

mitotic cycle. 1. MITOSIS. 2. CELL CYCLE.

mitotic recombination. The crossing-over between homologous chromosomes during mitosis which leads to segregation of heterozygous alleles.

mixed acid fermentation. The fermentation of glucose that is characteristic of *Escherichia coli* and related bacteria and that yields formic, acetic, lactic, and succinic acids, as well as a number of other products.

mixed amino acid fermentation. The fermentation of amino acids, as that occurring in putrefaction.

mixed anhydride. *See* acid anhydride.

mixed bed demineralizer. A mixture of cation and anion exchange resins that is used for the removal of ions in the preparation of deionized water. *Aka* mixed bed ion exchanger; mixed bed resin.

mixed complex. A metal ion complex that contains two or more different ligands.

mixed function oxidase. MONOOXYGENASE.

mixed indicator strain. A mixture of two related bacterial strains, such as a wild-type and a mutant, that is used in determining the relative amounts of two corresponding viral genotypes in a mixed population of virions.

mixed inhibition. The inhibition of an enzyme that cannot be fully described in terms of one of the basic types of inhibition. In the Cleland nomenclature of enzyme kinetics most mixed-type inhibitions are considered to be varieties of noncompetitive inhibition.

mixed-order reaction. A chemical reaction, the observed rate of which cannot be fully described by a simple first-, second-, or third-order rate equation.

mixed surface film. A monolayer composed of two or more different components.

mixed triglyceride. A triglyceride that contains two or three different fatty acids.

mixed-type inhibitor. An enzyme inhibitor that alters both the maximum velocity of the reaction and the Michaelis constant of the enzyme.

mixed vaccine. A vaccine that contains antigens derived from different infectious agents.

mixotropic series. A series of solvents arranged in the order of their relative polarity, and hence their miscibility with water.

ml. Milliliter.

MLD. Minimum lethal dose.

M line. The dark line that bisects the H zone of the myofibrils of striated muscle.

mM. Millimolar concentration.

M macroglobulin. An abnormal immunoglobulin of the IgM type that is produced by individuals suffering from Waldenstroem's macroglobulinemia.

Mn. Manganese.

$\bar{M}_n$. Number average molecular weight.

Mo. Molybdenum.

mobile phase. The liquid or gas phase that is the bulk moving phase in chromatography.

mobility. *See* electrophoretic mobility.

mobility spectrum. The distribution pattern of compounds which are separated by electrophoresis.

mobilization. The release of lipids, particularly fatty acids, from adipose tissue and their conversion to lipids that are transported by the blood.

mobilizing lipase. A lipase that functions in the mobilization of fatty acids from adipose tissue.

modal class. The category in a statistical distribution that contains a larger number of observations or measurements than any other category.

mode. The value of the variable that has the maximum frequency in a statistical distribution.

model. A three-dimensional representation of a molecule.

model system. A system that is studied and that is considered to either simulate or be representative of other systems in which the same or similar reactions take place.

moderate virus. An animal virus that resembles a temperate phage in its properties and that establishes a stable complex with the host cell.

moderator. A substance that alters the rate of an enzymatic reaction; an activator or an inhibitor.

modification allele. *See* modification gene.

modification enzyme. An enzyme that catalyzes the introduction of minor bases into RNA or DNA and that generally functions by modifying a normal base subsequent to its insertion into the polynucleotide strand.

modification gene. A gene the product of which is a modification enzyme.

modified air storage. GAS STORAGE.

modified inverse isotope dilution analysis. A variation of the inverse isotope dilution analysis in which the amount of labeled material is determined by means of a second radioactive substance.

modifier. 1. MODIFYING GENE. 2. EFFECTOR.

modifying gene. A gene that effects the expression of another, nonallelic gene.

modulating codon. A codon that codes for a rare transfer RNA molecule and that does not lead to the insertion of an amino acid into the growing polypeptide chain during translation. Instead, such a codon acts as a regulatory agent, leading either to release of the ribosome and interruption of translation, or to a slowing down of the rate of translation. *Aka* modulator codon.

modulation. 1. The regulation of the frequency with which a specific gene is transcribed. 2. The decrease in the rate of translation of a messenger RNA brought about by a modulating codon. 3. The control of a regulatory enzyme by means of an effector.

modulator. EFFECTOR.

modulator transfer RNA. A rare transfer RNA molecule that is coded for by a modulating codon.

Moffit plot. A plot based on the Moffit-Yang equation in which $[m'] (\lambda^2 - \lambda_0^2)$ is plotted as a function of $(\lambda^2 - \lambda_0^2)^{-1}$ so that a straight line is obtained for a chosen value of λ_0.

Moffit-Yang equation. An equation that describes the variation of optical rotation with wavelength; specifically, $[m'] = a_0 \lambda_0^2/(\lambda^2 - \lambda_0^2) + b_0 \lambda_0^4/(\lambda^2 - \lambda_0^2)^2$, where $[m']$ is the reduced mean residue rotation, λ is the wavelength, and $a_0, b_0,$ and λ_0 are constants. *Aka* Moffit equation.

Mohr pipet. A measuring pipet in which the calibration marks are contained between two marks on the stem of the pipet and do not extend to the tip of the pipet.

MOI. Multiplicity of infection.

moiety. 1. One of two, approximately equal parts. 2. One of two parts.

mol. Variant spelling of mole.

molality. The concentration of a solution expressed in terms of the number of moles of solute per 1000 grams of solvent.

molal solution. A solution that contains one mole of solute dissolved in 1000 grams of solvent.

molar absorbancy index. MOLAR ABSORPTIVITY.

molar absorptivity. The absorbance of a one molar solution when the light path through the solution is 1 cm; frequently denoted by the symbol ϵ. The molar absorptivity is related to the absorption cross section s by the equation $s = 3.8 \times 10^{-21} \epsilon$.

molar activity. A measure of enzymatic activity that is equal to the number of katals per mole of enzyme. *See also* molecular activity.

molar ellipticity. A measure of circular dichroism that is equal to 3300 times the difference between the molar extinction coefficients for the left-and right-circularly polarized light beams.

molar extinction coefficient. MOLAR ABSORPTIVITY.

molar growth yield. The dry weight of bacteria, in grams, that is obtained per mole of substrate utilized by the bacteria during their growth.

molarity. The concentration of a solution expressed in terms of the number of moles of solute per liter of solution.

molar rotation. The optical rotation of a solute that is calculated on the basis of its molar concentration and that is corrected to that in a medium having a refractive index of one. Specifically, $[m'] = [3/(n^2 + 2)] \times M[\alpha]/100$, where $[m']$ is the molar rotation, M is the molecular weight of the solute, $[\alpha]$ is the specific rotation of the solute, and n is the refractive index of the medium. Also used in its uncorrected form as $[m'] = M[\alpha]/100$.

molar solution. A solution that contains 1 mole of solute in 1 liter of solution.

mold. 1. A fungus characterized by having long mycelia. 2. TEMPLATE.

mole. 1. The molecular weight expressed in grams; the weight of a compound in grams that is numerically equal to its molecular weight. 2. The amount of substance that contains as many elementary entities as there are carbon atoms in 0.012 kg of the nuclide C^{12}.

molecular. 1. Of, or pertaining to, molecules. 2. Indicating the molecularity of a reaction when used with the prefixes mono-, bi-, or ter-.

molecular activity. A measure of enzyme activity equal to the number of moles of substrate (or the number of equivalents of the group concerned) that are transformed into products per minute per mole of enzyme at optimal substrate concentration. *See also* molar activity; catalytic center activity.

molecular biology. 1. The science that deals with the study of biological processes at the molecular level, particularly with respect to the physical-chemical properties and changes of cellular components and the relationship of these properties and changes to biological phenomena. Nerve impulse conduction, vision, membrane transport, and molecular genetics are some of the topics of molecular biology. 2. The science that deals with molecular genetics; the replication and transcription of both DNA and RNA, and the translation of RNA.

molecular disease. A disease that can be traced to a change in a single type of molecule; sickle cell anemia, which is caused by an abnormal hemoglobin, is an example.

molecular evolution. CHEMICAL EVOLUTION.

molecular-exclusion chromatography. GEL FILTRATION CHROMATOGRAPHY.

molecular fossil. CHEMICAL FOSSIL.

molecular genetics. The study of genetics at the molecular level; the replication and transcription of both DNA and RNA, and the translation of RNA.

molecular hybrid. *See* hybridization (1,2).

molecular ion. A molecule that has lost one electron, as that produced in a mass spectrometer.

molecularity. The number of reactant molecules that participate in a chemical reaction.

molecular mass. MOLECULAR WEIGHT.

molecular orbital. A composite orbital in a molecule that is derived from the overlapping, hybridized or unhybridized, atomic orbitals of the component atoms. *Abbr* MO.

molecular orbital theory. The theory of chemical bonding that is developed by considering the bonding atomic nuclei to occupy their equilibrium positions in the molecule, and then feeding orbital electrons into the resultant force field. *Abbr* MO theory.

molecular photosensitization. *See* photosensitization.

molecular radioautography. High-resolution radioautography, as that applied to isolated DNA strands.

molecular rotation. The rotation of an entire molecule about an axis.

molecular sieve. 1. A substance used for fractionating molecules according to their size by means of either gel filtration or gel permeation chromatography. 2. ZEOLITE.

molecular sieve chromatography. 1. GEL FILTRATION CHROMATOGRAPHY. 2. GEL PERMEATION CHROMATOGRAPHY.

molecular sieve coefficient. The partition coefficient of a solute in gel filtration chromatography; equal to the ratio of the equilibrium concentration of the solute within the gel to its concentration in the mobile phase.

molecular taxonomy. CHEMOTAXONOMY.

molecular vibration. The stretching or bending of the bonds between atoms that results in a displacement of the atomic nuclei but does not affect their equilibrium positions.

molecular weight. 1. The sum of the atomic weights of all the atoms in a molecule. 2. The sum of the atomic weights of all the atoms in a molecular aggregate such as an oligomeric protein, a ribosome, or a virus. *Abbr* M; MW; mol. wt. *Aka* particle weight.

molecule. The smallest unit of a compound; a structural unit of matter that retains all its properties, has an independent existence, and is composed of like or unlike atoms.

molecule microscope. A microscope for the study of surface materials by molecular beam techniques. The microscope reveals spatial variations in the evaporation of neutral molecules from the surface of a sample that is exposed to reduced pressure. The evaporating molecules may be part of the sample, may have been applied previously to the surface, or may be passed through the sample during the microscopic observation.

mole fraction. A measure of concentration expressed in terms of the number of moles of a substance divided by the total number of moles of all the substances in a solution or in a mixture.

mole percent. A measure of concentration expressed in terms of the number of moles of a substance per 100 moles of all related substances.

Molisch test. A test for carbohydrates that is based on the production of a purple color upon treatment of the sample with concentrated sulfuric acid and α-naphthol.

Moloney leukemia virus. A mouse leukemia virus that belongs to the leukovirus group.

molting hormone. ECDYSONE.

mol. wt. Molecular weight.

molybdenum. An element that is essential to several classes of animals and plants. Symbol Mo; atomic number 42; atomic weight 95.94; oxidation state +6; most abundant isotope Mo^{98}; a radioactive isotope Mo^{99}, half-life 66.7 hours, radiation emitted—beta particles and gamma rays.

molybdoferredoxin. MOLYBDOIRON PROTEIN.

molybdoiron protein. A subunit of the enzyme nitrogenase, present in *Clostridium pasteurianum*, that contains both iron and molybdenum and that has a molecular weight of about 200,000.

molybdoprotein. A conjugated protein containing molybdenum as a prosthetic group.

monestrous. Having one estrous cycle per year.

mongolism. A congenital abnormality characterized by imbecility and due to the presence of one of the autosomes in the triploid rather than in the diploid state.

monitor. A detector for determining a physical or a chemical variable either periodically or continuously; frequently refers either to a radiation detector for measuring the amount of ionizing radiation or of radioactive contamination, or to a spectrophotometer for measuring the absorbance of visible or ultraviolet light.

mono-. Prefix meaning one.

monoamine oxidase. A flavoprotein enzyme that catalyzes the oxidative deamination of monoamines such as epinephrine and norepinephrine. *Abbr* MAO.

monochromatic radiation. Electromagnetic radiation of a single wavelength; electromagnetic radiation in which all of the photons have the same energy.

monochromator. An instrument for the isolation of narrow band widths of radiation by means of filters, prisms, or diffraction gratings.

monocistronic messenger RNA. A messenger RNA molecule that carries the information for the synthesis of only one polypeptide chain.

monodentate. Designating a ligand that is chelated to a metal ion through one donor atom.

monodisperse. Consisting of macromolecules that are all alike in size.

Monod, Wyman, and Changeaux model. *See* symmetry model.

monoenergetic radiation. Radiation that consists of either photons or particles and in which all of the photons or all of the particles have the same energy.

monoenoic fatty acid. A fatty acid that has one double bond.

monoesterase. An enzyme that catalyzes the hydrolysis of an ester which is formed by the esterification of one of the hydroxyl groups of phosphoric acid.

monoglyceride. A glyceride formed by the esterification of one glycerol molecule with one fatty acid molecule.

monolayer. 1. A monomolecular layer formed either at a surface or at an interface. 2. A single layer of cells that are growing on a surface.

monomer. 1. The repeating unit in a polymer. 2. The basic unit in a molecular aggregate, regardless of the number of polypeptide chains or the number of subunits of which it is composed; thus the 70S ribosome is a monomer while the 30 and 50S ribosomes are subunits and the 100S ribosome is a dimer. 3. The individual polypeptide chain in an oligomeric protein. 4. A protein that is composed of a single polypeptide chain. 5. STRUCTURAL UNIT. *See also* protomer; subunit.

monomolecular layer. A layer, one molecule thick.

monomolecular reaction. A chemical reaction in which one molecule of a single reactant is converted into products.

monomorphism. The occurrence of only one form or one shape. *See also* doctrine of monomorphism.

mononuclear complex. A metal ion-ligand complex that is formed from a single metal ion. *See also* polynuclear complex.

mononucleotide. 1. A single nucleotide 2. A compound that is structurally related to a nucleotide, such as flavin mononucleotide or nicotinamide mononucleotide.

monooxygenase. An enzyme that catalyzes a reaction with molecular oxygen in which only one of the oxygen atoms is introduced into a compound.

monoploid state. The chromosome state in which the number of chromosomes is the basic one in a polyploid series; the haploid state. *Aka* monoploidy.

monoprotic acid. An acid that has one dissociable proton.

monosaccharide. A polyhydroxy alcohol containing either an aldehyde or a ketone group; a simple sugar.

monosome. 1. The complex that consists of a single ribosome attached to a strand of messenger RNA. 2. A chromosome that lacks a homologue.

monospecific antiserum. A purified antiserum that reacts with only one type of antigen or one type of antigenic determinant.

monovalent. 1. Having a valence of one. 2. Descriptive of a regulatory enzyme that responds to only one effector.

monozygotic twins. Twins that are genetically identical and that are derived from one fertilized egg; they are formed by a division of the embryo into two halves at some stage of its development.

Monte Carlo method. A method for calculating the time course of a reaction from the probability that the reaction will occur during a given time interval.

MOPS. 2-(*N*-Morpholino)-propanesulfonic acid; used for the preparation of buffers in the pH range of 6.5 to 7.9.

Morawitz theory. An early formulation of the blood clotting mechanism in terms of a two-stage process, consisting of the activation of prothrombin to thrombin and the conversion of fibrinogen to fibrin.

Morgan unit. A measure of the distance between genes on a chromosome that is equal to a crossover value of 100%.

Morner's test. 1. A test for tyrosine that is based on the production of a green color upon treatment of the sample with sulfuric acid and formaldehyde. 2. A test for cysteine that is based on the production of a purple color upon treatment of the sample with sodium nitroprusside.

morphine. An alkaloid narcotic drug that occurs in opium.

morphogenesis. The developmental processes that lead to the mature size, form, and structure of organelles, cells, tissues, organs, or whole organisms.

morphogenetic. Of, or pertaining to, morphogenesis. *Aka* morphogenic.

morphogenetic gene. A gene that plays a role in morphogenesis through some function other than that of specifying the synthesis of a structural protein.

morphogenic induction. The determination of the differentiation of one cell mass of an embryo by its interaction with another cell mass of the same embryo.

morphology. The science that deals with the structures and forms of organisms.

morphopoiesis. MORPHOGENESIS.

morphopoietic gene. MORPHOGENETIC GENE.

mosaic. An organism that contains portions of genetically different tissues.

mosaic theory. The theory of hypertension that describes the regulation of blood pressure in terms of the interactions between eight variables placed diagrammatically at the corners of an octagon. The eight variables are chemical factors, neural factors, elasticity, cardiac output, viscosity, vascular caliber, volume, and reactivity.

Mossbauer effect. The inelastic collision between a nucleus and a gamma ray in which the gamma ray is absorbed by and excites the nucleus; the nucleus remains in the excited state for a brief period of time (10^{-6} to 10^{-10} sec) and subsequently returns to its ground state with the emission of the gamma ray.

Mossbauer spectrometer. An instrument for detecting small changes in the interaction between a nucleus and its environment that are produced by variations in temperature, pressure, or chemical state.

MO theory. Molecular orbital theory.

motility model. FLUCTUATION THEORY (2).

motor end plate. NEUROMUSCULAR JUNCTION.

motor neuron. A neuron that conveys impulses to a muscle, resulting in muscle contraction.

mottled enamel. A pitted and corroded form of enamel that can be produced by the drinking of water that contains excessive amounts of fluoride ions.

mottled plaque. A plaque produced by the joint growth of two related phages, such as two mutants, in the same infectious center.

mouse antialopecia factor. INOSITOL.

mouse leukemia virus. An oncogenic virus that contains single-stranded RNA and causes leukemia in mice; the Friend, Graffi, Gross, Moloney, and Rauscher leukemia viruses, which belong to the leukovirus group, are examples.

mouse mammary tumor virus. An oncogenic virus that is transmitted through the milk and that causes mammary cancer in mice; it contains single-stranded RNA and belongs to the group of leukoviruses.

mouse satellite DNA. A satellite DNA that has been isolated from a variety of mouse tissues

and that constitutes about 10% of the total mouse DNA; it consists of highly repetitious DNA that contains about one million copies of a segment some 400 nucleotide pairs in length.

moving boundary analysis. ISOTACHOPHORESIS.

moving boundary centrifugation. Centrifugation in which an initially uniform solution is centrifuged so that boundaries are formed in the solution and move across the centrifuge cell; generally performed in an analytical-type ultracentrifuge.

moving boundary electrophoresis. Electrophoresis, performed in a Tiselius apparatus, in which an initially uniform solution is partially separated so that boundaries are formed that move toward or away from an electrode.

moving zone centrifugation. DENSITY GRADIENT CENTRIFUGATION.

m.p. Melting point; also abbreviated m.pt.

6-MP. 6-Mercaptopurine.

MPC. Maximum permissible concentration.

MPD. Maximum permissible dose.

M-protein. 1. A galactoside carrier protein in the permease system of *Escherichia coli*. 2. A structural protein present in the M line of the myofibrils of striated muscle.

MR. 1. Multiplicity reactivation. 2. Metabolic rate.

MRF. *See* melanocyte-stimulating hormone regulatory hormone.

MRH. *See* melanocyte-stimulating hormone regulatory hormone.

MRIF. *See* melanocyte-stimulating hormone regulatory hormone.

MRIH. *See* melanocyte-stimulating hormone regulatory hormone.

mRNA. Messenger RNA.

mRNA coding triplet. CODON.

MS. 1. Magic spot. 2. Mass spectrometry.

MS2. A phage that infects *Escherichia coli* and that contains single-stranded RNA.

MSG. Monosodium glutamate.

MSH. Melanocyte-stimulating hormone.

MSHIF. *See* melanocyte-stimulating hormone regulatory hormone.

MSHIH. *See* melanocyte-stimulating hormone regulatory hormone.

MSHRF. *See* melanocyte-stimulating hormone regulatory hormone.

MSHRH. *See* melanocyte-stimulating hormone regulatory hormone.

MSHRIF. *See* melanocyte-stimulating hormone regulatory hormone.

MSHRIH. *See* melanocyte-stimulating hormone regulatory hormone.

MTA. Mammary tumor agent.

mtDNA. Mitochondrial DNA.

mu chain. The heavy chain of the IgM immunoglobulins. *Sym µ*.

mucilage. A complex, colloidal, carbohydrate material that is derived from plants; it can form gels and has adhesive properties.

mucin. A mucoprotein secreted by mucous glands and mucous cells.

mucin clot. The clot, composed of hyaluronic acid and small amounts of protein, that is formed upon acidification of some biological fluids such as the vitreous humor of the eye and the synovial fluid.

muco-. Combining from meaning amino sugar.

mucoid. MUCOPROTEIN.

mucopeptide. A peptide that is covalently linked to an amino sugar.

mucopolysaccharide. A polysaccharide that contains either an amino sugar or a derivative of an amino sugar; hyaluronic acid, chondroitin sulfate, dermatan sulfate, and heparin are some examples.

mucoprotein. A conjugated protein in which the nonprotein portion is a mucopolysaccharide that is covalently linked to the protein and that contains more than 4% of amino sugars.

mucosa. MUCOUS MEMBRANE.

mucosal block. The permeability barrier of the intestinal mucosa to the absorption of iron from the intestine.

mucous. Of, or pertaining to, mucus.

mucous gland. A gland that secretes mucus.

mucous membrane. An epithelial membrane, the surface of which is bathed by mucus.

mucus. The viscous secretion of mucous glands that consists largely of mucin and water and that serves to bathe mucous membranes.

multi-. Combining form meaning many.

multichannel analyzer. A scintillation spectrometer that can record pulses in a number of different channels.

multicomponent survival curve. MULTI-TARGET SURVIVAL CURVE.

multienzyme complex. MULTIENZYME SYSTEM.

multienzyme system. The structural and functional entity that is formed by the association of several different enzymes which catalyze a sequence of closely related reactions; the aggregate may contain one or more molecules of a given enzyme.

multifactorial. Of, or pertaining to, a polygene; polygenic.

multihit survival curve. 1. A survival curve that describes a radiation phenomenon in which two or more photons must be absorbed by one target before the viability of the active unit is lost 2. MULTITARGET SURVIVAL CURVE.

multimer. OLIGOMER.

multiphasic zone electrophoresis. DISC GEL ELECTROPHORESIS.

multiple alleles. A group of three or more al-ternative alleles, any one of which may occur at the same locus on a chromosome.

multiple binding. MULTIPLE EQUILIBRIA.

multiple codon recognition. The binding of a given molecule of transfer RNA to more than one codon, as postulated by the Wobble hypothesis.

multiple development. A chromatographic technique, used particularly with paper or thin-layer chromatography, in which the sample is developed repeatedly with either the same or different solvents.

multiple equilibria. The interactions that occur between a macromolecule that has several binding sites and the ligands that bind to these sites.

multiple factor hypothesis. The hypothesis that quantitative traits, such as size and weight, result from the cumulative effect of a group of genes. See also polygene.

multiple forms of an enzyme. A collective term for all the proteins that possess the same enzyme activity and that occur naturally in a single species; includes genetically independent proteins, heteropolymers, genetic (allelic) variants, proteins conjugated with other groups, proteins derived from one polypeptide chain, polymers of a single subunit, and forms differing in conformation.

multiple gene. POLYGENE.

multiple-hit survival curve. See multihit survival curve.

multiple inhibition analysis. A kinetic analysis of the interactions of two or more inhibitors of an enzymatic reaction. The analysis indicates whether the inhibitors are mutually exclusive or whether they can bind simultaneously to the enzyme and, if so, whether the binding of one inhibitor to the enzyme facilitates or hinders the binding of another.

multiple myeloma. A malignant disease of antibody-producing plasma cells that is characterized by the formation of large amounts of Bence-Jones protein.

multiple sclerosis. A disease that is characterized by partial paralysis, changes in speech, and inability to walk; caused by demyelination, by sclerosis of the brain, and by sclerosis of the spinal cord. Aka demyelination disease.

multiplet. A multiple peak, as that obtained in nuclear magnetic resonance.

multiplication cycle. The sequence of steps from the infection of a cell by a virus to the formation of new virus particles and their release from the cell.

multiplicity. See enzyme multiplicity; transfer RNA multiplicity.

multiplicity of infection. 1. The number of virus particles that have either adsorbed to, or infected, cells in a culture, divided by the total number of cells in the culture. 2. The number of virus particles added to a culture divided by the total number of cells in the culture. *Abbr* MOI.

multiplicity reactivation. The restoration of the activity of a virus that carries a lethal mutation by the simultaneous infection of a host cell with this and one or more other mutant viruses. The process involves a genetic exchange whereby a viable genome is produced from the undamaged sections of the mutant, and from otherwise nonviable, genomes. *Abbr* MR. *See also* cross-reactivation.

multitarget survival curve. A survival curve that describes a radiation phenomenon in which two or more photons must be absorbed by two or more targets before the viability of the active unit is lost.

multivalent. POLYVALENT.

multivalent allosteric inhibition. The inhibition of a regulatory enzyme by two or more negative effectors; the inhibition of a multivalent regulatory enzyme.

multivalent feedback inhibition. CONCERTED FEEDBACK INHIBITION.

multivalent regulatory enzyme. A regulatory enzyme, the activity of which can be altered by more than one effector.

multivalent vaccine. *See* polyvalent vaccine.

mu particle. *See* mate killer.

muramic acid. A compound derived from glucosamine and lactic acid, the acetylated form of which is a major building block of the bacterial cell wall.

muramidase. LYSOZYME.

Murayama hypothesis. The hypothesis that the replacement of glutamic acid by valine at position 6 in the beta chains of sickle cell hemoglobin permits the formation of intermolecular hydrophobic bonds which lead to a head-to-tail stacking of hemoglobin molecules; the filaments thus formed distort the red blood cell and convert it to a sickle cell.

murein. PEPTIDOGLYCAN.

murexide test. A test for purines that is based on the production of a red color upon treatment of the sample with concentrated nitric acid and then with ammonium hydroxide.

murine. Of, or pertaining to, mice and rats.

murine leukemia. Leukemia in mice that is produced by a mouse leukemia virus.

murine leukemia virus. MOUSE LEUKEMIA VIRUS.

muropeptide. The repeating unit in peptidoglycan that consists of *N*-acetylglucosamine, *N*-acetylmuramic acid, and a tetrapeptide side chain.

muscle. A contractile organ of the body.

muscle contraction. *See* contraction; sliding filament model.

muscle fiber. A long, multinucleated cell of striated muscle.

muscle hemoglobin. MYOGLOBIN.

muscle phosphorylase. *See* phosphorylase a; phosphorylase b.

muscle sugar. INOSITOL.

muscular dystrophy. One of a number of degenerative diseases of muscle, some of which are genetically inherited.

mustard gas. SULFUR MUSTARD.

mustard oil glycoside. A plant toxin that is a sulfur-containing glycoside.

mutability. The capacity to undergo mutation.

mutability spectrum. MUTATIONAL SPECTRUM.

mutable. Capable of undergoing mutation.

mutable gene. An unstable gene that is characterized by a high rate of spontaneous mutation.

mutable site. Any site along the chromosome at which a mutation can occur.

mutagen. A physical or chemical agent that is capable of inducing mutations; a mutagen raises the frequency of mutation above that due to spontaneous mutations.

mutagenesis. The production of mutations.

mutagenic. Capable of inducing mutations.

mutagenic agent. MUTAGEN.

mutagenize. To expose nucleic acids, viruses, cells, or organisms to a mutagen.

mutant. 1. A cell, a virus, or an organism that carries a gene that has undergone mutation. 2. A gene that has undergone mutation.

mutant allele. *See* mutant gene.

mutant gene. A gene that has undergone a mutation; the modified nucleotide sequence of a wild-type gene.

mutant protein. A protein formed from a mutant gene.

mutarotation. The change in optical rotation that occurs when an optical isomer, such as a carbohydrate, is dissolved in water and is converted to an equilibrium mixture of several different optical isomers.

mutase. An enzyme that catalyzes the intramolecular transfer of a group, specifically a phosphate group.

mutation. 1. The process whereby a gene undergoes a structural change leading to a sudden and stable change in the genotype of a cell, a virus, or an organism; any heritable change in the genome of a cell, a virus, or an organism other than that due to the incorporation of genetic material from other sources. 2. MUTANT GENE. 3. The cell, the virus, or the organism that carries a mutant gene.

mutational load. The genetic inadequacy of a population as a result of the mutational accumulation of deleterious genes.

mutational spectrum. The genetic map of the point mutations that either arise spontaneously or are produced by exposure to a mutagen.

mutation frequency. 1. The proportion of mutants in a population of organisms. 2. MUTATION RATE.

mutation frequency decline. The phenomenon that ultraviolet-induced reversions from auxotroph to prototroph decrease as a function of time if they are determined under conditions where postirradiation protein synthesis is inhibited. *See also* mutation stabilization.

mutation index. An estimate of the mutation frequency; it is equal to the proportion of mutants in a population of cells that have been grown from an inoculum that contained such a small number of organisms that it is unlikely that any mutants were among them.

mutation pressure. The continued production of specific mutants.

mutation rate. The total number of mutations, or the number of mutations of a specified kind, that are produced in a population of cells or organisms per cell division, or per replication, over a given period of time.

mutation stabilization. The phenomenon that ultraviolet-induced reversions from auxotroph to prototroph achieve a constant value as a function of time if they are determined under conditions allowing for postirradiation protein synthesis. *See also* mutation frequency decline.

mutation theory. *See* somatic mutation theory.

mutator gene. A gene that increases the mutation rate of other genes, as in a system where the mutator gene produces a DNA polymerase that makes errors during replication.

mutator mutant. A mutant carrying a mutator gene.

mutator strain. A strain carrying a mutator gene.

mutein. A mutant protein such as the cross-reacting material.

muton. The unit of genetic mutation; the smallest section of a chromosome, which may be as small as a single nucleotide, a change in which can result in a mutation.

$\overline{M}_w$. Weight average molecular weight.

MW. Molecular weight.

mycelium (*pl* mycelia). The vegetative structure of a fungus that consists of a multinucleate mass of cytoplasm, enclosed within a branched network of filamentous tubes known as hyphae.

mycolic acid. One of a group of long-chain hydroxy fatty acids that contain from 60 to 90 carbon atoms and that have varying degrees of branching and unsaturation.

mycology. The branch of botany that deals with fungi.

mycoplasma. A genus of bacteria that are the smallest known, independently living organisms, and that differ from other bacteria in not having a cell wall.

mycoside. A lipid composed of a long-chain, highly branched, and hydroxylated hydrocarbon that is terminated by a phenol to which a trisaccharide is linked by means of a glycosidic bond.

mycosis. A disease caused by a fungus.

mycosterol. A sterol of fungi.

myelination. The formation and the deposition of the myelin sheath around an axon.

myelin sheath. The lipid-rich, insulating covering of an axon that is formed by wrapping the plasma membrane of a Schwann cell around the axon. Myelin contains lipids, especially sphingolipids, proteins, polysaccharides, salts, and water.

myeloma. A tumor of cells that are derived from the hematopoietic tissue of bone marrow.

myeloma protein. The Bence-Jones protein produced by individuals suffering from multiple myeloma.

myo-. Combining form meaning muscle.

myocardial. Of, or pertaining to, the heart muscle.

myofibril. A small, contractile, threadlike structure of striated muscle; the myofibrils are arranged in parallel bundles within the cytoplasm of a muscle fiber.

myofibrillar ghost. FIBRIL GHOST.

myofilament. A minute, contractile, threadlike structure of striated muscle; the myofilaments are arranged in parallel bundles within a myofibril and consist of thin and thick filaments.

myogen. An aqueous extract of the sarcoplasm of striated muscle that consists largely of glycolytic enzymes.

myogenesis. The formation of muscle tissue.

myoglobin. The oxygen-storing protein of muscle that consists of a single polypeptide chain surrounding a heme group, and that is closely related to the monomeric unit of hemoglobin. *Abbr* Mb.

myograph. An instrument for recording the forces of muscular contractions.

myohematin. CYTOCHROME.

myokinase. The enzyme that catalyzes the interconversion between two molecules of ADP and one molecule each of ATP and AMP.

myoneural. Of, or pertaining to, muscles and nerves.

myoneural junction. NEUROMUSCULAR JUNCTION.

myosin. The most abundant protein of the myofilaments of striated muscle and the protein that forms the thick filaments. Myosin is an asymmetric molecule composed of two polypeptide chains which form a head, or globular portion, and a long, fibrous tail. Both the ATPase activity and the actin-binding capacity of myosin reside in the globular head portion of the molecule.

myosin B. ACTOMYOSIN.

myosin filament. A thick filament of striated

muscle from which cross-bridges protrude that link the thick filament to the thin filaments; a myofilament.

myotropic activity. The anabolic effect of androgens on nitrogen metabolism that leads to nitrogen retention by the body and to a limited increase in muscle strength and development.

myristic acid. A saturated fatty acid that contains 14 carbon atoms and that occurs in animal fat.

myxedema. HYPOTHYROIDISM.

myxovirus. A large, enveloped animal virus that contains single-stranded RNA. Myxoviruses are divided into two subgroups: the orthomyxoviruses, which include the influenza virus, and the paramyxoviruses, which include the mumps virus.

$\overline{M}_z$. Z-average molecular weight.

MZE. Multiphasic zone electrophoresis.

N

n. 1. Refractive index. 2. Nano. 3. Neutron.

N. 1. Nitrogen. 2. Normal concentration. 3. Nucleoside. 4. Asparagine. 5. Haploid number. 6. Neutron number. 7. Newton.

Na. Sodium.

NA. Noradrenaline.

NAcneu. *N*-Acetylneuraminic acid.

NAD⁺. Nicotinamide adenine dinucleotide. *Aka* DPN⁺.

NADH. Reduced nicotinamide adenine dinucleotide. *Aka* DPNH.

Nadi reagent. A reagent that consists of a mixture of α-naphthol and *p*-phenylenediamine and that is used for the detection of certain oxidases.

NADP⁺. Nicotinamide adenine dinucleotide phosphate. *Aka* TPN⁺.

NADPH. Reduced nicotinamide adenine dinucleotide phosphate. *Aka* TPNH.

NAGA. *N*-Acetyl galactosamine.

Nagarse. Tradename for subtilisin.

Na,K-ATPase. The adenosine triphosphatase that is located in the cell membrane and that is implicated in the membrane transport of sodium and potassium ions. The enzyme is vectorial in its action, is greatly stimulated by the addition of both sodium and potassium ions, but is only slightly stimulated by the addition of either sodium or potassium ions alone.

naked virion. A virion that consists of a nucleocapsid that is not surrounded by a membrane.

NANA. *N*-Acetylneuraminic acid.

nano-. Combining form meaning one billionth (10^{-9}) and used with metric units of measurements. *Sym* n.

Naperian logarithm. NATURAL LOGARITHM.

narcosis. A state of stupor produced by a narcotic drug.

narcotic. 1. Of, or pertaining to, narcosis. 2. NARCOTIC DRUG.

narcotic drug. A substance that, if taken in appropriate doses, produces sedation or sleep, as well as relief of pain.

nascent. Being formed; in the process of being synthesized, particularly in reference to the synthesis of macromolecules.

nascent polypeptide chain. A polypeptide chain that is in the process of being formed and that is attached to a transfer RNA molecule which, in turn, is bound to a ribosome.

Natelson microgasometer. An instrument for the manometric measurement of the oxygen, carbon monoxide, and carbon dioxide content of blood.

National Formulary. A pharmacopeia published in the United States. *Abbr* N.F.

native. 1. Descriptive of a protein or a nucleic acid molecule in its natural, in vivo state as opposed to its denatured state. 2. Descriptive of a protein or a nucleic acid molecule which has been isolated by mild procedures so that it is undenatured, or only slightly denatured, and is taken to represent the in vivo state of the molecule.

native conformation. 1. The in vivo conformation of a macromolecule. 2. The normal conformation of a macromolecule that has been isolated by mild procedures, shows no apparent structural alterations, and is investigated under suitable conditions of pH, temperature, and ionic strength.

native immunity. NATURAL IMMUNITY.

native plasma. Plasma obtained without the addition of an anticoagulant.

natriuretic hormone. A polypeptide hormone that increases the excretion of sodium ions, apparently by inhibiting the reabsorption of sodium ions by the kidney.

natural abundance. The relative proportion of an isotope in nature, based on the concentration of all the other isotopes of that particular element.

natural antibody. An antibody that is present in the blood and that is capable of reacting with specific antigens even though the organism had no known exposure to those antigens.

natural auxin. INDOLEACETIC ACID.

natural chain elongation. The elongation of a chain by tailward growth.

natural immunity. The immunity that is characteristic of an organism and that is a result of the genetic makeup of the organism.

natural immunization. An immunization brought about by natural exposure, as by inhalation, ingestion, skin contact, or infection.

natural logarithm. A logarithm to the base *e*. *Abbr* $\log_e$; ln. *See also* logarithm.

natural pH gradient. The pH gradient that is used in isoelectric focusing and that is produced during the experiment by the electrolysis of carrier ampholytes. The carrier ampholytes band at positions where the isoelectric pH of the ampholyte equals the pH of the solution; a pH gradient is thereby established which, once formed, is stable for prolonged times.

natural product. 1. A secondary metabolite that has no known function. 2. Any organic compound produced by a living organism.

natural selection. The principle, proposed by Darwin in 1859, that natural processes favor those members of a species that are better adapted to their environment and tend to eliminate those that are unfitted to their environment. Thus the "fittest" organisms survive and through successive generations changes become established that lead to the production of new types and new species.

n_D. Refractive index, measured with the light of a sodium D line.

NDP. 1. Nucleoside diphosphate. 2. Nucleoside-5′-diphosphate.

nearest-neighbor base frequency analysis. A method for assessing base sequences in nucleic acids by a comparison of the frequencies with which any pair of adjacent bases occurs in these nucleic acids. The method is used especially for DNA and, in that case, requires the synthesis of DNA by DNA polymerase in the presence of a DNA template and the four 5′-deoxyribonucleoside triphosphates, one of which is labeled with P^{32} in its alpha phosphorus. The synthesized product is digested to the 3′-deoxyribonucleoside monophosphates which results in a shift of the P^{32} label from the incorporated nucleotide to its nearest neighbor in the synthesized polynucleotide strand. The experiment is performed four times, using a different, labeled deoxyribonucleoside triphosphate at each time. The extent of label in the isolated 3′-deoxyribonucleoside monophosphates can then be used to calculate the frequencies of pairs of adjacent bases in the newly synthesized DNA.

nebulizer. ATOMIZER.

necrosis. The pathological death of cells, tissues, or organs.

NEFA. Nonesterified fatty acids.

negative catalysis. Catalysis that leads to a decrease in the rate of a chemical reaction. This may occur, for example, through the tight binding of an intermediate to an enzyme; as a result, the intermediate is prevented from reacting with a "wrong" compound by having the energy of activation for that reaction increased due to its binding to the enzyme.

negative contrast staining. NEGATIVE STAINING.

negative control. The prevention of a biological activity by the presence of a specific molecule; the prevention by a repressor of either inducible enzyme synthesis or initiation of messenger RNA synthesis are two examples.

negative cooperativity. Cooperative binding in which the binding of one ligand to one site on the molecule decreases the affinity for the binding of subsequent ligands to other sites on the same molecule.

negative electron. ELECTRON (1).

negative feedback. A feedback mechanism, as in many biological systems, in which a large output of the system leads to a decrease in the subsequent output while a small output of the system leads to an increase in the subsequent output.

negative hydration. Hydration in which the water molecules in the primary hydration shell of an ion have greater mobility than those in pure water.

negative labeling. The process whereby a specific residue in a protein is either masked or protected so that it will not undergo a reaction with a given reagent, and all the remaining, available, and unprotected residues are then labeled by reaction with the reagent.

negative phase. The stage that follows the administration of a second dose of antigen to an animal during which there is a temporary decrease in the concentration of free antibodies in the circulation due to the fact that the added antigens combine with preexisting circulating antibodies.

negative staining. A staining technique, used in electron microscopy, in which the material to be examined is mixed with an electron-dense substance, such as phosphotungstic acid, and appears transparent against an opaque background.

negatron. A negatively charged electron.

negentropy. The informational content of a molecule.

neighboring group effect. The effect of a group in a molecule on a nucleophilic displacement reaction in which the molecule participates; the effect is due to the group functioning as an internal nucleophile for an intramolecular displacement reaction.

neighbor restoration. The reactivation of cells, damaged by irradiation with ultraviolet light, that is brought about when the cells are incubated in a liquid medium at high concentrations of cells from either the same or other strains.

nematic liquid crystal. A liquid crystal in which the constituent molecules are parallel, can rotate about their axes, and can move both up and down and from side to side.

nematosome. A cytoplasmic inclusion occurring in certain neurons.

Nembutal. Trademark for sodium pentobarbital, an anesthetic.

neobiogenesis. The repeated formation of life from nonliving, inorganic matter.

neomorph. A mutant gene that produces a qualitatively new effect which is not produced by the wild-type gene.

neomycin. An aminoglycoside antibiotic produced by *Streptomyces fradiae.*

neonatal. 1. Of, or pertaining to, a neonate. 2. Of, or pertaining to, the period immediately following birth.

neonate. A newborn.

neoplasia. The pathological condition characterized by tumor formation and tumor growth.

neoplasm. A new and abnormal growth; a proliferation of cells that is not subject to the usual limitations of growth; a tumor. *See also* benign neoplasm; malignant neoplasm.

neoplastic. Of, or pertaining to, a neoplasm.

neosome. A collective term to describe both intersomes and minisomes.

nephelometry. The quantitative determination of a substance in suspension that is based on measurements of the light scattered by the suspended particles at right angles to the incident beam. *See also* turbidimetry.

nephrectomy. The surgical removal of a kidney.

nephron. The structural and functional unit of the kidney; consists of the glomerulus, Bowman's capsule, Henle's loop, and the proximal and distal tubules.

Nernst equation. An expression that relates the actual electrode potential E of a given redox couple to its standard electrode potential E_0 and to the concentrations of the oxidant and the reductant. For a half reaction, the expression is $E = E_0 + \dfrac{0.06}{n} \log_{10} ([Ox]/[Red])$ where n is the number of electrons participating in the half reaction, $[Ox]$ is the concentration of oxidant, and $[Red]$ is the concentration of reductant.

nerve. An elongated structure of nervous tissue that consists of nerve fibers enclosed within a sheath, and that serves to connect the nervous system with other organs and tissues of the body.

nerve-end particle. SYNAPTOSOME.

nerve fiber. The process of a neuron.

nerve gas. A mixture of compounds, including diisopropylfluorophosphate, that react with the serine group of the enzyme acetylcholinesterase and thereby inhibit the transmission of nerve impulses.

nerve growth factor. One of a family of closely related proteins that produce hypertrophy and hyperplasia of nerve cells, growth of nerve cell processes, and an increase in the metabolism of various nerve cells. *Abbr* NGF.

nerve impulse. An electrical stimulus that passes along a nerve and that leads to excitation of the nerve along the way.

nerve impulse conduction. The passage of a nerve impulse along a nerve cell.

nerve impulse transmission. The passage of a nerve impulse from one nerve cell to another.

nervous system. The nervous tissue which, in vertebrates, consists of the central and the peripheral nervous systems.

Nesslerization. The treatment of a sample with Nessler's reagent.

Nessler's reagent. A solution of mercuric iodide and potassium iodide in potassium hydroxide that is used for the colorimetric determination of nitrogen in biological materials.

Neuberg ester. FRUCTOSE-6-PHOSPHATE.

neural. Of, or pertaining to, nerves.

neuraminic acid. A compound, derived from mannosamine and pyruvic acid, the acetylated form of which is a major building block of animal cell coats.

neuraminidase. The enzyme that catalyzes the cleavage of N-acetylneuraminic acid from mucopolysaccharides; the enzyme is present on the surface of certain viruses and destroys the receptor activity of many cells for these viruses.

neurochemistry. The science that deals with the biochemistry of nervous tissue.

neuroendocrine. Of, or pertaining to, both the nervous and the endocrine systems.

neurofibril. A small, threadlike structure in a neuron.

neurohormone. A hormone, such as a hypothalamic hormone, that is released into the circulation at nerve endings and that acts upon cells located at some distance from its point of release. *See also* neurohumor.

neurohumor. A substance, such as acetylcholine, that is released at nerve endings and that acts upon adjacent nerve cells. *See also* neurohormone.

neurohypophyseal. Of, or pertaining to, the posterior lobe of the pituitary gland.

neurohypophysis. The posterior lobe of the pituitary gland that produces vasopressin, oxytocin, and melanocyte-stimulating hormone.

neurological mutant. A mutant that leads either to pronounced malformations of the central nervous system or to pronounced abnormalities in locomotion.

neurology. The science that deals with the structure and function of the nervous system.

neuromuscular. Of, or pertaining to, both nerves and muscles.

neuromuscular junction. The junction between the axon of a motor neuron and a striated muscle fiber.

neuron. A nerve cell; the structural and functional unit of the nervous system that consists of a cell body and its processes, the axon and the dendrites.

neurophysin. A protein, derived from the posterior lobe of the pituitary gland, that binds both vasopressin and oxytocin and serves as a biological carrier of these two hormones.

neuroplasm. The cytoplasm of a neuron.

neurosecretion. 1. The secretion of chemical substances, such as neurohormones and neurohumors, by nerve cells. 2. The chemical substances secreted by nerve cells.

neurosecretory granule. A particle, derived from the posterior lobe of the pituitary gland, that contains oxytocin, vasopressin, and neurophysin.

neurosecretosome. A pinched-off nerve ending.

Neurospora crassa. The red bread mold; a fungus used for biochemical and genetic studies.

neurotoxin. A toxin that acts specifically on nervous tissue.

neurotransmitter. A substance, such as acetylcholine, that functions in the transmission of nerve impulses.

neurotropic virus. A virus, the target organ of which is the nervous system.

neutral. 1. Being neither acidic nor basic; a neutral solution has a pH of 7.0. 2. Being neither positively nor negatively charged; a neutral atom has an equal number of protons and orbital electrons.

neutral amino acid. An amino acid that has one amino group and one carboxyl group.

neutral fat. An ester formed from a molecule of glycerol and one to three molecules of fatty acids; a mono-, di-, or triglyceride.

neutral glycolipid. CEREBROSIDE.

neutralization. 1. The reaction between an acid and a base, forming water. 2. The inactivation of a soluble antigen, such as a toxin, or a particulate antigen, such as a virus, by reaction with the appropriate antibodies.

neutral lipid. A lipid, such as a glyceride or a steroid, that is devoid of pronounced polar groups. *Abbr* NL.

neutrino. A subatomic particle that has zero charge and essentially zero mass; it accounts for that part of the energy of beta decay that is not associated with the emitted beta particle.

neutron. A neutral, subatomic particle of the nucleus that has a mass of 1.009 amu; it is equivalent to a combined proton and electron. *Sym* n.

neutron-activated phosphorus-bakelite plaque. A radioactive disk that contains P^{32} atoms and that is used as a source of beta particles.

neutron activation analysis. *See* activation analysis.

neutron capture. The capture of a neutron by an atomic nucleus that frequently occurs during the production of artificial radioactive isotopes.

neutron number. The number of neutrons in the nucleus of an atom. *Sym* N.

Newcastle disease virus. A virus that infects the respiratory tract of birds and that belongs to the group of paramyxoviruses.

Newman projection. The representation of a molecule in which the arrangement of the atoms is such as would be seen by an observer viewing the molecule from one end along the carbon-to-carbon bond closest to the observer. When viewed in this fashion, some atoms appear to be fully or partially hidden by other atoms and are called eclipsed, while other atoms are clearly visible and are called staggered.

Newtonian fluid. A fluid, the viscosity of which is independent of the rate of shear.

N.F. National Formulary; used to denote a chemical that meets the specifications set out in the National Formulary.

n-fold symmetry. Having an *n*-fold axis of rotational symmetry. *See also* axis of rotational symmetry.

NGF. Nerve growth factor.

NHI. Nonheme iron.

niacin. 1. NICOTINIC ACID. 2. A generic descriptor for pyridine-3-carboxylic acid and its derivatives that exhibit qualitatively the biological activity of nicotinic acid.

niacinamide. NICOTINAMIDE.

nick. A break in a single strand of a nucleic acid, particularly a break in a single strand of a double-stranded nucleic acid.

nickase. ENDONUCLEASE.

nicked DNA. A DNA molecule having one or more breaks in either one or both of its strands.

Nicol prism. One of two prisms used in a polarimeter. *See also* analyzer; polarizer.

nicotinamide. The amide of nicotinic acid.

nicotinamide adenine dinucleotide. A coenzyme form of the vitamin nicotinic acid; a coenzyme for pyridine-linked dehydrogenases. *Abbr* NAD^+; DPN^+.

nicotinamide adenine dinucleotide phosphate. A coenzyme form of the vitamin nicotinic acid; a coenzyme for pyridine-linked dehydrogenases. *Abbr* $NADP^+$; TPN^+.

nicotinamide mononucleotide. A precursor of nicotinamide adenine dinucleotide in which nicotinamide is linked to ribose-5-phosphate. *Abbr* NMN^+.

nicotine. An alkaloid derived from tobacco and considered to be carcinogenic.

nicotinic acid. A B vitamin, the deficiency of which causes the disease pellagra and the coenzyme forms of which are NAD^+ and $NADP^+$. Nicotinic acid is unique among the B vitamins in that it can be synthesized in animal tissues (from tryptophan).

nicotinic acid amide. NICOTINAMIDE.

Niemann-Pick disease. A genetically inherited metabolic defect in man that is associated with mental retardation and that is characterized by an accumulation of sphingomyelin in the tissues; it is due to a deficiency of the enzyme sphingomyelinase.

night blindness. An early manifestation of vitamin A deficiency in which the retinal rods have an elevated visual threshold and do not respond normally to faint light.

night vision. The capacity to see in dim light that is due to the rods in the retina.

NIH. National Institutes of Health; an agency of the U.S. Public Health Service.

NIH shift. The reaction, discovered at the National Institutes of Health, in which a proton is shifted from the para position in phenylalanine to the meta position in tyrosine during the hydroxylation of phenylalanine to tyrosine.

ninhydrin reaction. The reaction of ninhydrin (triketohydrindene hydrate) with the free alpha amino groups of amino acids, peptides, or proteins; the reaction yields colored compounds that are useful for the chromatographic detection and for the quantitative determination of amino acids and peptides.

nisin. A peptide antibiotic, produced by *Streptococcus lactis,* that consists of 34 amino acids.

Nissl substance. The granular bodies that are largely aggregates of ribosomes and that are associated with the endoplasmic reticulum of neurons.

nitrate. A salt of nitric acid.

nitrate assimilation. The biological conversion of nitrate to ammonia, amino acids, or related compounds.

nitrate reductase. The enzyme that catalyzes the reduction of nitrate to nitrite in nitrate assimilation by plants and fungi; it is a molybdenum-containing flavoprotein.

nitrate reduction. The reduction of nitrate to nitrite, ammonia, or molecular nitrogen, that is carried out in nature by bacteria and fungi.

Nitrazine paper. Trademark for a dye-impregnated paper that is used for estimation of pH values.

nitrification. The oxidation of ammonia to nitrite or nitrate that is carried out in nature by nitrifying bacteria.

nitrifying bacteria. Bacteria that oxidize ammonia to nitrite and that oxidize nitrite to nitrate.

nitrite. A salt of nitrous acid.

nitrite reductase. The enzyme that catalyzes the reduction of nitrite to ammonia in nitrate assimilation by plants.

nitrogen. An element that is essential to all plants and animals. Symbol N; atomic number 7; atomic weight 14.0067; oxidation states -1, $+1$, $+2$, $+3$, $+4$, $+5$; most abundant isotope N^{14}; a stable isotope N^{15}.

nitrogenase. The nitrogen-fixing enzyme system that catalyzes the reduction of atmospheric nitrogen to ammonia and that is a complex of two proteins, called "iron protein" and "molybdoiron protein."

nitrogen balance. The difference between the nitrogen intake and the nitrogen excretion of an animal. The nitrogen balance is denoted as zero, positive, or negative depending on whether the intake is equal to, greater than, or smaller than the excretion.

nitrogen cycle. The cyclic set of reactions involving plants, animals, and bacteria, whereby (a) atmospheric nitrogen is converted to inorganic compounds and these are then converted to complex organic compounds; and (b) the organic compounds are broken down to inorganic compounds which ultimately yield atmospheric nitrogen.

nitrogen equilibrium. The state of an animal in which the nitrogen balance is equal to zero.

nitrogen fixation. The conversion of atmospheric nitrogen to ammonia by means of either a biological or a synthetic reaction.

nitrogen mustard. Di-(2-chloroethyl)methylamine; a chemical mutagen and alkylating agent. *See also* alkylating agent.

nitrogenous. Nitrogen-containing.

nitrogenous base. 1. A purine or a pyrimidine. *Aka* nitrogen base. 2. A nitrogen-containing basic compound.

p-**nitrophenol.** *See p*-nitrophenylphosphate.

p-**nitrophenylphosphate.** A synthetic substrate for assaying both acid and alkaline phosphatase activity; these enzymes catalyze the removal of a phosphate group from *p*-nitrophenylphosphate and the *p*-nitrophenol which is formed is then determined spectrophotometrically. *Abbr* PNPP.

nitroprusside reaction. A colorimetric reaction for cysteine and free sulfhydryl groups in a protein that is based on the production of a red color upon treatment of the sample with sodium nitroprusside and ammonia.

nitrous acid. A chemical mutagen that leads to the deamination of purines and pyrimidines and that converts adenine to hypoxanthine and cytosine to uracil. Since hypoxanthine base pairs with cytosine, the effect of nitrous acid treatment is that an original adenine is read as guanine and an original cytosine is read as uracil.

nitrous acid mutant. A mutant produced by treatment of a nucleic acid with nitrous acid; such mutants have been produced, for example, by treating the RNA from tobacco mosaic virus with nitrous acid, infecting tobacco leaves with the mutated RNA, and isolating the mutant viral particles formed in the infected leaves. *See also* nitrous acid.

nkat. Nanokatal.

NL. Neutral lipid.

nm. Nanometer; also indicated as mμ (millimicron).

NMN⁺. Nicotinamide mononucleotide.

NMP. 1. Nucleoside monophosphate. 2. Nucleoside-5′-monophosphate.

NMR. Nuclear magnetic resonance.

nodal compound. The compound that is common both to a linear metabolic pathway and to its branch.

node. NODAL COMPOUND.

nodoc. ANTICODON.

noise. The background interference in an instrument that may be caused, for example, by electronic, optical, or chemical disturbances.

Nomarski differential interference microscope. A microscope that provides a three-dimensional view of an object and that permits the observation of transparent structures in the living cell.

nomenclature. A system of names, designations, and symbols that are used in a given discipline.

nominally labeled. Designating a compound in which some, and usually a significant amount, of the label is at a given position or positions in the molecule, but for which no further information is available as to the extent of label, if any, at other positions in the molecule. *Sym* N.

nomogram. An alignment chart for the rapid determination of a variable from the given values of two or more variables. A typical nomogram consists of a minimum of three scales such that when known values of two scales are connected by a straight line, the line intersects the third scale at the sought value. The hemoglobin concentration in blood, for example, can be determined from a scale of the specific gravity of plasma and a scale of the specific gravity of whole blood.

nona-. Combining form meaning nine.

nonagglutinating antibody. INCOMPLETE ANTIBODY.

non-AIS-suppressible insulin. INSULIN-LIKE ACTIVITY.

nonamer. An oligomer that consists of nine monomers.

nonbonding interaction. NONCOVALENT INTERACTION.

nonbonding orbital. A molecular orbital that contains electrons that take little part in the actual bonding of the atomic nuclei. Such orbitals usually have energies intermediate between those of bonding orbitals (sigma, pi) and those of antibonding orbitals (sigma star, pi star).

noncollisional energy transfer. The energy transfer from an excited molecule to another molecule that occurs when the two molecules remain farther apart than the contact distance that they attain during molecular collisions.

noncompetitive inhibition. The inhibition of the activity of an enzyme that is characterized by an increase in the slope of a double reciprocal plot (1/velocity versus 1/substrate concentration) and by a decrease in the maximum velocity compared to those of the uninhibited reaction.

noncompetitive inhibitor. An inhibitor that produces noncompetitive inhibition and that in general is structurally unrelated to the substrate.

noncoupled pump. A pump for the transport of one solute across a membrane that also drives the transport of a second solute across the same membrane in the opposite direction and in such a fashion that the transport of the second solute is physically independent of the pump.

noncovalent bond. A bond between atoms and/or molecules that does not involve shared pairs of electrons and that is due to other types of interactions. Examples of such bonds are electrostatic bonds, hydrogen bonds, and hydrophobic bonds.

noncovalent interaction. An interaction between atoms and/or molecules that does not involve the formation of chemical bonds and that is based on the formation of noncovalent bonds.

noncyclic electron flow. The light-induced, photosynthetic electron flow in which the electrons flow from water, or some other electron donor, to NADP⁺, or some other electron acceptor; in chloroplasts, it is the electron flow from water through photosystems II and I to NADP⁺.

noncyclic photophosphorylation. The synthesis of ATP that is coupled to the noncyclic electron flow of photosynthesis.

nonelectrolyte. A substance that does not dissociate into ions in water; solutions of nonelectrolytes do not conduct an electric current.

nonenergized conformation. CONDENSED CONFORMATION.

nonequilibrium thermodynamics. IRREVERSIBLE THERMODYNAMICS.

nonessential enzyme. An enzyme that is not required for either the growth or the survival of a cell or an organism.

nonessential gene. A gene, the product of which is a nonessential enzyme.

nonexclusive binding. The binding to a regulatory enzyme that takes place when both the relaxed and the constrained forms of the enzyme are present in significant amounts. *See also* symmetry model.

nonexclusive binding coefficient. The ratio of the ligand dissociation constant for the relaxed form of a regulatory enzyme to that for the constrained form. *See also* symmetry model.

nonheme iron. Iron that occurs in biological systems but that is not in the form of a heme group. *Abbr* NHI.

nonheme-iron chromophore. A pair of iron atoms, as in plant-type ferredoxin, that are located close enough to each other in a molecule so that they can engage in antiferromagnetic coupling; such atoms possess a characteristic absorption spectrum and a distinct electron paramagnetic resonance spectrum.

nonheme-iron protein. A conjugated protein that contains iron but not in the form of a heme group.

noninducible enzyme. CONSTITUTIVE ENZYME.

nonionic detergent. A surface-active agent that has polar and nonpolar groups but carries no charges. *Aka* nonionic surface-active agent.

nonionized acids. COMBINED ACIDITY.

non-Newtonian fluid. A fluid, the viscosity of which depends on the rate of shear; solute molecules of a non-Newtonian fluid, especially asymmetric molecules, tend to become oriented as the rate of shear is increased.

nonnucleic acid base. A base that rarely, if ever, occurs in a nucleic acid under normal circumstances.

nonose. A monosaccharide that has nine carbon atoms.

nonpalindromic helix. A helix that has no end-to-end symmetry so that both ends are different either in composition and/or by virtue of the helix having a sense of direction. The rotation of such a helix by 180° about an axis perpendicular to the longitudinal axis of the helix produces a structure that is not identical to that before rotation.

nonparametric statistics. Statistical calculations that are not based on any prior assumptions with respect to the variable and the probability distribution of the data.

nonpermissible substitution. RADICAL SUBSTITUTION.

nonpermissive cell. 1. A cell in which a conditional lethal mutant cannot grow. 2. A cell that does not support the lytic infection by a virus.

nonpermissive conditions. Conditions that do not permit the growth of a conditional lethal mutant.

non-plasma-specific enzyme. An enzyme that is present in blood plasma but has no known specific function in the plasma.

nonpolar. Lacking polarity; lacking a permanent dipole moment.

nonpolar amino acid. An amino acid that has a nonpolar side chain.

nonpolar bond. 1. A covalent bond in which the electron pair or pairs of the bond are held with equal strength by the two bonded atoms. 2. HYDROPHOBIC BOND.

nonpolar solvent. A solvent that is devoid of significant concentrations of charged groups and/or dipoles.

nonprecipitating antibody. COPRECIPITATING ANTIBODY.

nonproductive complex. An enzyme-substrate complex in which the substrate is bound to the enzyme in such a fashion that catalysis is impossible and that products cannot be formed.

nonprotein amino acid. An amino acid that rarely, if ever, occurs in a protein under normal circumstances.

nonprotein nitrogen. The nitrogen in serum that is not present in the form of serum proteins. *Abbr* NPN.

nonprotein respiratory quotient. The respiratory quotient that is calculated on the basis of a volume of oxygen equal to the total volume utilized minus that utilized for protein catabolism, and on the basis of a volume of carbon dioxide equal to the total volume produced minus that produced by protein catabolism.

nonsaponifiable lipid. A lipid that cannot be hydrolyzed with alkali to yield soap as one of the products; steroids and terpenes are two major nonsaponifiable lipids.

nonsecretor. An individual who does not secrete water-soluble forms of the blood group substances into his body fluids.

nonsense codon. A codon that does not code for an amino acid; a termination codon.

nonsense mutation. A mutation in which a normal codon that specifies an amino acid is changed to one of the three termination codons.

nonsense suppression. The suppression of a nonsense mutation.

nonsequential mechanism. PING PONG MECHANISM.

nonspecific immunity. The immunity that is produced by nonimmunological mechanisms such as lysozyme action, or phagocytosis, or interferon action.

nonsymbiotic nitrogen fixation. The conversion of atmospheric nitrogen to ammonia by organisms, such as photosynthetic bacteria or blue-green algae, without the participation of plants.

nonviable. Descriptive of a cell or an organism that is dead and incapable of reproduction.

noradrenaline. NOREPINEPHRINE.

n orbital. Nonbonding orbital.

norepinephrine. A catecholamine hormone that is secreted by the adrenal medulla and that has a biological activity that is similar to that of epinephrine but less pronounced. *Var sp* norepinephrin.

Norit. Trademark for a purified charcoal made from birch; used for the decolorization of solutions and for the adsorption of compounds in adsorption chromatography. *Var sp* Norite.

Norit eluate factor. FOLIC ACID.

norleucine. An amino acid analogue that can be incorporated into protein during protein synthesis.

normal configuration. The configuration of steroids in which substituents at positions 5 and 10 are cis with respect to the plane of rings A and B.

normal distribution. A frequency distribution characterized by a bell-shaped curve and described by the equation $Y = \dfrac{1}{\sigma\sqrt{2\pi}} e^{(X-m)^2/2\sigma^2}$, where m is the mean, σ is the standard deviation, e is the base of natural logarithms, π is a constant equal to 3.1416 . . . , and Y is the height of the ordinate for a given value of X on the abscissa. Different values of m shift the curve along the abscissa without changing its shape. Different values of σ change the shape of the curve without changing the position of the center.

normal electrode potential. STANDARD ELECTRODE POTENTIAL.

normal enzyme. An enzyme, the substrates of which are metabolites normally occurring within the organism, as distinct from a drug-metabolizing enzyme, the substrates of which are compounds foreign to the organism.

normal error curve. NORMAL DISTRIBUTION.

normal frequency distribution. NORMAL DISTRIBUTION.

normality. The concentration of a solution expressed in terms of the number of gram-equivalent weights of solute in 1 liter of solution.

normalizing. The adjustment of data to an arbitrary standard; the normalizing of a spectrum, for example, is done by multiplying the observed absorbance values at all measured wavelengths by a factor that is equal to the ratio of the desired absorbance to the observed absorbance at one particular wavelength.

normal saline. PHYSIOLOGICAL SALINE.

normal solution. A solution that contains 1 gram-equivalent weight of solute per liter of solution.

normal temperature and pressure. STANDARD TEMPERATURE AND PRESSURE.

normal value. The amount of a chemical constituent in, or the value of a physical property of, a body fluid or an excretion that is found in 95% of a population of clinically normal and apparently healthy individuals.

norsteroid. A modified steroid that lacks one or more carbon atoms compared to the number of carbon atoms in the parent compound.

notatin. Glucose oxidase; a flavoprotein enzyme that catalyzes the oxidation of glucose to the delta lactone and that can be isolated in a highly active form from the mold *Penicillium notatum.*

nothing dehydrogenase effect. An abnormality in the electrophoretic determination of lactate dehydrogenase isozymes in which blank preparations, from which substrate has been omitted, exhibit faint replicas of the normal isozyme pattern. The effect is believed to be due to the presence of alcohol dehydrogenase in the enzyme preparation.

NP antigen. A nucleoprotein core antigen of poxviruses.

NPH insulin. A neutralized zinc salt of protamine insulin developed by Hagedorn. The salt is insoluble and, when injected into an animal, provides a slowly adsorbed insulin depot so that fewer injections of insulin are required in clinical treatments of diabetes.

n-pi star transition. The excitation of an electron fron an n orbital to a pi star orbital.

NPN. Nonprotein nitrogen.

nRNA. Nuclear RNA.

NSF. National Science Foundation.

N-terminal. The end of a peptide or of a polypeptide chain that carries the amino acid that has a free alpha amino group; in representing amino acid sequences, the N-terminal is conventionally placed on the left-hand side. *Aka* N-terminus.

NTP. 1. Nucleoside triphosphate. 2. Nucleoside-5′-triphosphate. 3. Normal temperature and pressure.

Nuc. Nucleoside.

nuclear. Of, or pertaining to, the nucleus of either an atom or a cell.

nuclear column. A column of cell nuclei that are immobilized with small pieces of membrane filters and through which a solvent is passed.

nuclear cycle. CELL CYCLE.

nuclear division. KARYOKINESIS.

nuclear emulsion. A photographic emulsion that has been specially sensitized for the detection of alpha or beta particles; it is generally thicker and more concentrated in silver halide than ordinary photographic emulsions.

nuclear envelope. The envelope that surrounds the nucleus of eucaryotic cells and that consists of two membranes that enclose perinuclear cisternae.

nuclear equivalent. NUCLEOID.

nuclear isomer. One of two or more nuclides that have the same atomic number and the same atomic mass but that have nuclei that are at different energy levels.

nuclear magnetic resonance. A method for studying the interaction of an atomic nucleus, having an odd mass number and an odd number of protons, with the environment of the nucleus. A nucleus of this kind has a spin and a magnetic moment and may exist in one of several allowed energy levels. When placed in an applied magnetic field of suitable magnitude, the nucleus will undergo a transition from one energy level to another, accompanied by the absorption of electromagnetic radiation. The relative magnitudes of the applied magnetic field and of the absorbed radiation are interpreted in terms of the interaction of the nucleus with its environment. The method is used particularly for protons and, when so used, is also referred to as proton magnetic resonance. *Abbr* NMR.

nuclear membrane. NUCLEAR ENVELOPE.

nuclear reaction. 1. A reaction taking place in the nucleus of a cell. 2. A physical reaction in which there are changes in the nuclei of reacting atoms as distinct from a chemical reaction in which there are changes in the orbital electrons.

nuclear resonance scattering. MOSSBAUER EFFECT.

nuclear zone. NUCLEOID.

nuclease. An enzyme that catalyzes the hydrolysis of phosphodiester bonds in nucleotides and nucleic acids.

nucleated. Possessing a nucleus.

nucleation. 1. The formation of regions of three-dimensional structure in separate portions of a protein molecule prior to attainment of the complete tertiary structure of the molecule. 2. The formation of a crystal or an aggregate by the condensation of matter on minute particles that serve as nuclei.

nucleic acid. A polynucleotide of high molecular weight that is synthesized by living cells. Nucleic acids occur as either DNA or RNA and may be either single-stranded or double-stranded. DNA functions in the transfer of genetic information, and RNA functions in the biosynthesis of proteins.

nuclein. The nucleoprotein discovered by Miescher in 1868.

nucleocapsid. The protein coat of a virus together with the nucleic acid which it encloses.

nucleohistone. A conjugated protein consisting of histone and nucleic acid.

nucleoid. The DNA-containing region of procaryotes and viruses that is analogous to the nucleus of eucaryotes.

nucleolar organizer. NUCLEOLUS ORGANIZER.

nucleolus (*pl* nucleoli). An RNA-rich region in the cell nucleus.

nucleolus organizer. A portion of the chromosome that is associated with the nucleolus and that functions in the synthesis of ribosomal RNA.

nucleon. A constituent particle occurring within the atomic nucleus; a proton or a neutron.

nucleonics. The application of nuclear phenomena to other fields.

nucleon number. MASS NUMBER.

nucleophile. An atom or a group of atoms that is electron pair donating.

nucleophilic. Of, or pertaining to, either a nucleophile or a reaction in which a nucleophile participates.

nucleophilic catalysis. Catalysis in which the catalyst donates a pair of electrons to a reactant.

nucleophilic displacement. A chemical reaction in which a nucleophilic group attacks and displaces a susceptible group in a compound and then binds covalently to the compound at that site. *See also* S_N1 mechanism; S_N2 mechanism.

nucleophilic substitution. NUCLEOPHILIC DISPLACEMENT.

nucleoplasm. The protoplasm of the cell nucleus.

nucleoprotein. A conjugated protein in which the nonprotein portion is a nucleic acid, and the protein portion is frequently either a histone or a protamine.

nucleosidase. An enzyme that catalyzes the hydrolysis of a nucleoside to the pentose and the base.

nucleoside. A glycoside composed of D-ribose or 2-deoxy-D-ribose and either a purine or a pyrimidine. *Abbr* Nuc; N.

nucleoside cyclic monophosphate. A nucleotide in which the phosphoric acid residue is esterified to two hydroxyl groups on the sugar.

nucleoside diphosphate. A high-energy derivative of a nucleoside in which a pyrophosphate group is esterified to a hydroxyl group of the sugar. *Abbr* NDP.

nucleoside diphosphate kinase. An enzyme that catalyzes the transfer of a phosphate group from a nucleoside-5′-triphosphate to a nucleoside-5′-diphosphate.

nucleoside diphosphate sugar. One of a group of compounds that consist of a sugar linked to a nucleoside diphosphate and that serve as glycosyl group donors in the biosynthesis of starch, glycogen, and other oligo- and polysaccharides. *Abbr* NuDP-sugar.

nucleoside monophosphate. NUCLEOTIDE.

nucleoside monophosphate kinase. An enzyme that catalyzes the transfer of a phosphate group from a nucleoside-5′-triphosphate to a nucleoside-5′-monophosphate.

nucleoside triphosphate. A high-energy derivative of a nucleoside in which three phosphate groups are linked in succession to one hydroxyl group of the sugar. *Abbr* NTP.

nucleotidase. An enzyme that catalyzes the hydrolysis of a nucleotide to the nucleoside and orthophosphate.

nucleotide. The building block of the nucleic acids that consists of a nucleoside plus a phosphoric acid residue which is esterified to one of the hydroxyl groups of the sugar.

nucleotide anhydride. NUCLEOTIDE CO-ENZYME.

nucleotide coenzyme. One of a group of compounds, derived from uracil, cytosine, or thymine, that function as either coenzymes or substrates in carbohydrate and lipid metabolism; UDP-glucose, CDP-ethanolamine, and TDP-ribose are some examples.

nucleotide exchange reaction. The reaction, catalyzed by the enzyme nucleoside diphosphate kinase, in which a nucleoside diphosphate and a nucleoside triphosphate are converted to the opposite pair of nucleoside di- and triphosphates by the transfer of a phosphate group.

nucleotide map. A fingerprint of nucleotides.

nucleus (*pl* nuclei). 1. The structure in eucaryotic cells that contains the chromosomes. 2. The central core of an atom that consists of protons and neutrons. 3. The ring structure of an organic compound.

nuclide. An atom that is characterized by the composition of its nucleus, having a specified atomic number and mass number; an atom that has a specific number of protons, a specific number of neutrons, and a specific energy content in its nucleus.

nuclidic mass. ATOMIC MASS.

NuDP-sugar. Nucleoside diphosphate sugar.

null hypothesis. A hypothesis stating that there is no difference between two values, such as between the means of two populations. The hypothesis is advanced to evaluate data; it is subsequently tested statistically and either accepted or rejected.

number average molecular weight. An average molecular weight that is weighted toward those molecules present in largest number; specifically, $\bar{M}_n = \sum n_i M_i / \sum n_i$, where n_i is the number of moles of component i, and M_i is the molecular weight of component i. *Sym* $\bar{M}_n$. *See also* average molecular weight.

nutrient. A substance that promotes the growth, maintenance, function, and reproduction of a cell or an organism.

nutrient agar. A solid medium that contains a meat extract and that is used in bacteriology.

nutrient broth. A liquid medium that contains a meat extract and that is used in bacteriology.

nutrilite. A compound that fits the definition of a vitamin with the exception that it is required by a microorganism and not by an animal; a vitamin of microorganisms.

nutriment. Nourishment; food.

nutrition. The supplying to, and the intake and utilization by, an organism of all the necessary elements required by it for normal growth, maintenance, function, and reproduction.

nutritional mutant. AUXOTROPH.

nyctalopia. NIGHT BLINDNESS.

Nylander's reagent. A reagent that contains bismuth subnitrate, potassium-sodium tartrate, and potassium hydroxide, and that is used for the detection of reducing sugars; the sugars yield a black precipitate of metallic bismuth when treated with the reagent.

nylon. Polyhexamethylene adipamide; a synthetic polymer.

O

o. Ortho.

O. 1. Oxygen. 2. Orotic acid. 3. Orotidine.

OA. Ovalbumin.

OAA. Oxaloacetic acid.

O-antigen. A cell wall antigen of bacteria.

obesity. The state of being overweight, frequently due to overnutrition; usually interpreted as being 10% or more above the average standard weight for the individual.

objective. The microscope lens closest to the specimen.

oblate ellipsoid of revolution. An ellipsoid of revolution formed by rotation of an ellipse about its minor axis.

obligate. Limited to living under a specific set of conditions.

obligate aerobe. An organism or a cell that can grow only in the presence of molecular oxygen.

obligate anaerobe. An organism or a cell that can grow only in the absence of molecular oxygen.

obligatory reactant. LEADING REACTANT.

occlude. To produce an occlusion.

occlusion. 1. The blocking of a passageway, such as the blocking of an artery. 2. The trapping of a substance by adsorption and adhesion, such as the trapping of soluble substances in a precipitate.

occult blood. Small amounts of blood, as those in urine and feces, that cannot be detected visually.

occult virus. A virus that has infected a host but cannot be detected.

occupation theory. A theory according to which the action of drugs is due to their interaction with specific receptor sites in the organism that is being treated.

ochre codon. The codon UAA; one of the three termination codons.

ochre mutation. A mutation in which a codon is mutated to the ochre codon, thereby causing the premature termination of the synthesis of a polypeptide chain.

ochre suppression. The suppression of an ochre codon.

octa-. Combining form meaning eight.

octamer. An oligomer that consists of eight monomers.

octose. A monosaccharide that has eight carbon atoms.

ocular. The eyepiece of an optical instrument, such as that of a microscope.

o.d. Outside diameter.

OD. Optical density.

odd base. MINOR BASE.

odd-carbon fatty acid. ODD-NUMBERED FATTY ACID.

odd electron. UNPAIRED ELECTRON.

odd-numbered fatty acid. A fatty acid molecule that has an odd number of carbon atoms.

ODP. 1. Orotidine diphosphate. 2. Orotidine-5'-diphosphate.

-OEt. Ethoxy group.

-ogen. Suffix meaning inactive precursor of an enzyme.

ohm. A unit of electrical resistance; equal to the resistance of a conductor that carries a current of 1 A when a potential difference of 1 V is applied across its terminals.

Ohm's law. The law stating that the strength of a direct electric current is proportional to the potential difference and inversely proportional to the resistance.

oil. A fat that is liquid at room temperature.

oil-immersion objective. An objective lens used to increase the resolution attainable with the light microscope; based on filling the space between the coverslide of the specimen and the objective with an oil that has the same refractive index as the coverslide.

Okazaki fragments. A group of short DNA fragments that are produced during the initial stages of DNA replication and that are believed to be subsequently joined by means of a ligase to form longer fragments. Okazaki fragments have been demonstrated, in the case of *Escherichia coli*, by exposing the cells to tritiated thymine for short periods during their growth. The finding of Okazaki fragments has been taken as supportive evidence for the Kornberg mechanism of DNA replication.

old cells. 1. Cells in the stationary phase of growth. 2. Cells that have been stored for a prolonged time.

old tuberculin. Tuberculin prepared by concentrating and filtering a culture of the tubercle bacillus, *Mycobacterium tuberculosis*. *Abbr* OT.

old yellow enzyme. A flavoprotein from yeast that catalyzes the oxidation of NADPH and that was isolated in 1932 by Warburg and Christian.

olefin. An unsaturated aliphatic hydrocarbon; an alkene.

oleic acid. An unsaturated fatty acid that contains 18 carbon atoms and one double bond.

oleophilic. HYDROPHOBIC.

oleophobic. HYDROPHILIC.

oligomer. A protein molecule that consists of two or more polypeptide chains, referred to as either monomers or protomers, linked together covalently or noncovalently. *See also* monomer; protomer.

oligomycin. A polyene antibiotic produced by *Streptomyces diastatochromogenes.*

oligonucleotide. A linear nucleic acid fragment that consists of from 2 to 10 nucleotides joined by means of phosphodiester bonds; oligoribonucleotides consist of ribonucleotides, and oligodeoxyribonucleotides consist of deoxyribonucleotides.

oligopeptide. A linear peptide that consists of from 2 to 10 amino acids joined by means of peptide bonds.

oligosaccharide. A linear or branched carbohydrate that consists of from 2 to 10 monosaccharide units joined by means of glycosidic bonds.

oligotrophic lake. A deep and clear-water lake, having a depth of 15 m or more, that has a plant population at various depths, and that has a low rate of nutrient supply in relation to its volume of water. In such a lake, both the biomass and the productivity are low. The bottom layers of the lake are saturated with dissolved oxygen throughout the year. *See also* eutrophic lake; mesotrophic lake.

O-locus. The locus of the operator.

-oma. Suffix meaning tumor.

-OMe. Methoxy group.

omega oxidation. A minor oxidative pathway of fatty acids.

ommochrome. One of a group of pigments that are derived from tryptophan and that occur in the eyes of insects.

OMP. 1. Orotidine monophosphate (orotidylic acid). 2. Orotidine-5'-monophosphate (5'-orotidylic acid).

Oncley equation. An equation expressing the frictional ratio of a macromolecule as a product of two factors, one of which is a measure of the hydration, and the other is a measure of the asymmetry of the molecule. Specifically, $f/f_0 = (f/f_h)(f_h/f_0)$, where f, f_0, and f_h are, respectively, the frictional coefficients for the macromolecule, an anhydrous sphere, and a hydrated sphere; f_h/f_0 is the hydration factor, and f/f_h is the shape, or asymmetry, factor.

onco-. Combining form meaning tumor.

oncogene. An abnormal gene derived from the RNA of an oncogenic RNA virus. *See also* oncogene theory.

oncogenesis. The origin and growth of a tumor.

oncogene theory. A theory of cancer according to which the RNA of oncogenic RNA viruses becomes part of the genome of animals in the form of abnormal genes. These genes, called oncogenes, may function in normal metabolic reactions, and they may become activated by carcinogens to become determinants of cancer, possibly through the synthesis of specific enzymes and/or the synthesis of complete, oncogenic virus particles. The information for the production of oncogenic viruses and malignant transformations of cells is, therefore, vertically transmitted through the germ line and is present in the DNA of all the cells of all the animals prone to cancer. *Aka* virogene theory; virogene-oncogene theory. *See also* protovirus theory.

oncogenic. Capable of inducing tumors.

oncogenic virus. A DNA- or RNA-containing virus that can transform infected cells so that they proliferate in an uncontrolled fashion and may form a tumor. *See also* oncornavirus.

oncology. The study of tumors.

oncolytic. Capable of destroying tumor cells.

oncornavirus. Acronym for oncogenic RNA virus. Oncornaviruses are characterized by (a) their content of a high-molecular weight RNA genome (60-70S RNA, 10^7 daltons), (b) their banding at a particular density level in density gradient centrifugation, and (c) their content of RNA-dependent DNA polymerase (reverse transcriptase). The oncornaviruses are divided into three classes, denoted A, B, and C. Type A viruses constitute a small group of protein-encapsulated viruses that have not been shown to be oncogenic; they occur either in the cytoplasm, believed to be immature forms of type B viruses, or in body fluids, believed to be immature forms of type C viruses. Type B viruses have a somewhat eccentric nucleoid and glycoprotein surface spikes; they are associated primarily with the formation of carcinomas. Type C viruses have a roughly spherical nucleoid surrounded by an electron-translucent lipid layer; they infect a large number of animal species and cause leukemias, lymphomas, and sarcomas.

oncotic pressure. The effective colloid osmotic pressure; it is equal to the difference between the osmotic pressure of the plasma proteins and that of the tissue fluid proteins.

one-carbon fragment. A group of atoms or a compound that contains one carbon atom.

one gene-one enzyme hypothesis. The hypothesis that there is a large group of genes among the genes of an organism in which each gene codes for a specific enzyme or other protein. Since it is now known that many enzymes, as well as other proteins, consist of several polypeptide chains coded for by different genes, this hypothesis has been replaced by the one gene-one polypeptide chain hypothesis.

one gene-one polypeptide chain hypothesis. The hypothesis that there is a large group of genes among the genes of an organism in which each gene codes for a specific polypeptide chain; the polypeptide chain may be part of a protein, or constitute an entire functional protein.

one-hit theory. A theory according to which the damage of one site on the erythrocyte membrane, resulting from a reaction with complement, is sufficient to bring about cell lysis.

one-sigma level. The confidence interval corresponding to the standard deviation; a confidence interval such that there is a 68.27% chance that a measurement will fall within it.

one-step conditions. A set of conditions that are used for infecting cells with viruses when it is desired to produce infected cells which contain only a single virus particle per cell. For bacterial cultures this may be achieved by incubating the cells briefly with the phage to allow attachment of the phage to the bacteria, and then diluting the phage-cell suspension drastically prior to additional incubation.

one-step growth experiment. An experiment in which virus growth is carried out under one-step conditions.

one-step multiplication curve. A curve that describes the production of progeny virus under one-step conditions.

Onsager's equations. PHENOMENOLOGICAL EQUATIONS.

ontogenetic. Of, or pertaining to, ontogeny.

ontogeny. The development of an individual. *Aka* ontogenesis.

ontogeny recapitulates phylogeny. RECAPITULATION THEORY.

oocyte. A cell that develops into a mature ovum upon meiosis.

oogenesis. The formation of a mature egg.

oolemma. The cell membrane of an ovum.

ooplasm. The cytoplasm of an ovum.

oosperm. A fertilized ovum; a zygote.

Oparin's hypothesis. The hypothesis that simple organic compounds were formed spontaneously during an early stage of the earth as a result of physical and chemical processes in the primitive atmosphere. These compounds are believed to have dissolved in the primitive oceans and to have led, by a large number of small spontaneous reaction steps, to the formation of macromolecules which ultimately gave rise to the first living cell. *See also* chemical evolution; biochemical evolution; biological evolution.

open chain. A chain of atoms in which the two ends are not linked together covalently.

open-circuit system. A system for measurements of indirect calorimetry in which both the oxygen consumption and the carbon dioxide production are determined.

open gene. A gene that is engaged in transcription.

open hemoprotein. A hemoprotein in which either the fifth and/or the sixth coordination positions of the heme are unoccupied.

opening transformation. The transformation of a micellar membrane from one having small spaces between the micelles ("closed") to one having large spaces betwen them ("open").

open system. A thermodynamic system that can exchange both matter and energy with its surroundings.

operating potential. The potential at which a Geiger-Mueller plateau is obtained and at which a Geiger-Mueller counter is normally operated.

operational definition. A definition that is based upon properties relevant to one or more specific experimental procedures or conditions, regardless of the possibility that different and more fundamental properties may apply to that which is defined.

operator. A gene that is adjacent to a structural gene or to a group of contiguous structural genes and that controls the transcription of this gene or group of genes. *Aka* operator gene.

operator-constitutive mutant. A mutant resulting from a constitutive mutation in which the operator gene has been mutated so that the repressor cannot combine with it; as a result, a previously inducible enzyme becomes a constitutive one.

operator gene. OPERATOR.

operon. The combination of operator and the adjacent structural gene or genes which are controlled by it.

operon hypothesis. The model, proposed by Jacob and Monod, according to which enzyme induction and enzyme repression result from the control of one or more structural genes by an adjacent operator gene. The operator is blocked ("turned off") during enzyme repression and is unblocked ("turned on") during enzyme induction. The blocking occurs through the binding to the operator of either a repressor or a repressor-corepressor complex. An inducer binds to the repressor and thus prevents the blocking of the operator; a corepressor binds to the repressor and leads to an even more effective blocking of the operator than that produced by the repressor alone. The synthesis of the repressor is controlled by a regulator gene which need not be adjacent to the operon, the term used to describe the combination of operator and structural gene or genes. Inducers are generally substrates of the enzyme or compounds similar to the substrates, while corepressors are generally products of the enzymatic reaction or compounds similar to the products.

operon network. A system of interacting operons and their associated regulator genes such that the product of a structural gene from one operon acts as a repressor or an inducer for another operon.

opiate. A narcotic that resembles opium in its action and that may be a natural, a semisynthetic, or a synthetic compound.

opium. The dried exudate derived from the seeds of the oriental poppy *Papaver somniferum* that contains the narcotic drugs, morphine and codeine.

opportunistic fungus. A fungus that generally does not cause infection in a healthy individual but may cause serious illness in an individual who suffers from one of a number of unrelated diseases.

opposing rolling circle model. A variation of the rolling circle model for the replication of duplex circular DNA. According to this model, replication begins with two nicks, not far from each other, and with one in each of the strands of the circular duplex. As a result, both strands grow in opposite directions and the molecule replicates in the form of two rolling circles that move in opposite directions.

opsin. A protein, occurring in both the rods and the cones of the retina, that combines with either retinal$_1$ or retinal$_2$ to form the major visual pigments of vertebrates.

opsonic effect. The enhancement of the phagocytosis of antigen-antibody complexes that is produced by antibodies.

opsonic index. The ratio of the phagocytic index of an immune serum to that of a normal serum.

opsonin. A substance that enhances phagocytosis by modifying the particles to be engulfed so that they are more readily taken up by the phagocytic cells. Some of these substances are antibodies, called immune opsonins, against surface antigens of the bacteria or the particles that are engulfed; others are nonantibody, heat-labile substances, believed to be related to components of complement.

opsonization. The enhancement of phagocytosis by the action of opsonins.

optical activity. 1. The capacity of a substance to interact with radiation in such processes as optical rotation and circular dichroism. 2. OPTICAL ROTATION.

optical antipode. ENANTIOMER.

optical density. ABSORBANCE.

optical isomer. One of two or more isomers that differ from each other in their symmetry as a result of the presence of either asymmetric carbon atoms or overall molecular asymmetry; many, but not all, optical isomers exhibit optical rotation.

optically active. Possessing optical activity.

optical path. LIGHT PATH.

optical quenching. The quenching that occurs in liquid scintillation counting as a result of changes in the light-transmitting properties of the sample, such as those produced by the partial freezing of the sample solution, or by the fogging of the outside of the sample vial.

optical rotation. The rotation of the plane of plane-polarized light by a substance when such light is passed through a solution containing the substance. Optical rotation is shown by a substance that can exist in the form of mirror images as a result of the presence of either asymmetric carbon atoms or overall molecular asymmetry. Because of this asymmetry, the substance has different refractive indices for left and right circularly polarized light and hence shows optical rotation.

optical rotatory dispersion. The variation of optical rotation as a function of the wavelength of the light that is used for measuring it; optical rotatory dispersion is useful for studying the secondary structure of macromolecules. *Abbr* ORD.

optical system. *See* absorption optical system; interferometric optical system; schlieren optical system.

optimal proportions. *See* method of optimal proportions.

optimum pH. The pH at which an enzyme exhibits maximal activity under specified conditions.

optimum temperature. The temperature at which an enzyme exhibits maximal activity under specified conditions.

oral insulin. A synthetic drug that lowers the level of blood sugar by stimulating the pancreas to produce insulin.

orbital. The probability distribution, or the wave function, for an electron that is in a particular energy level in either an atom or a molecule.

orbital electron capture. *See* electron capture.

orbital steering. The precise angular juxtaposition of reacting substrate atoms and the "steering" of the orbitals of these atoms in the course of an enzyme-catalyzed reaction.

orchiectomy. The surgical removal of a testis. *Aka* orchidectomy.

orcinol reaction. A colorimetric reaction for carbohydrates that is based on the production of a green color upon treatment of the sample with orcinol and with ferric chloride dissolved in concentrated hydrochloric acid. The reaction is used particularly for pentoses and for the determination of RNA.

Ord. Orotidine.

ORD. Optical rotatory dispersion.

ordered mechanism. The mechanism of an enzymatic reaction in which two or more substrates are added to the enzyme in an orderly fashion; thus in the case of a reaction with two substrates, the formation of a ternary complex, composed of the enzyme and the two substrates, is preceded by the formation of a complex between the enzyme and one of the substrates.

order of a reaction. *See* reaction order.

order with respect to concentration. The order of a chemical reaction that is determined on the basis of a number of experiments in which the initial reactant concentration is varied; the order of the reaction is then determined from a plot of the logarithm of the initial velocity of the reaction versus the logarithm of the initial reactant concentration.

order with respect to time. The order of a chemical reaction that is determined on the basis of a single experiment in which a single initial reactant concentration is used; the order of the reaction is then determined from the decrease in the reactant concentration as a function of time.

ordinate. The vertical axis, or Y-axis, in a plane rectangular coordinate system.

organ. A differentiated part of an organism that has a specific structure and performs specific functions.

organ culture. The in vitro maintenance of an organ or of parts of an organ so that the structure and/or the function are retained.

organelle. A specialized structure in the cell that has definite functions.

organic. 1. Pertaining to the compounds of carbon. 2. Pertaining to an organ. 3. Pertaining to a living organism.

organic quenching. The quenching of an ionization detector that occurs when organic compounds are added to the mixture used for gas amplification. *See also* internal quenching.

organism. A living plant, animal, or protist.

organizer. A portion of an embryo that, through a group of substances produced by it, affects the development (determination and differentiation) of another part of the embryo. *See also* inductor.

organogenesis. The development of an organ or organs.

organotroph. A cell or an organism that utilizes organic compounds as electron donors in its energy-yielding oxidation-reduction reactions.

organ-specific enzyme. TISSUE-SPECIFIC ENZYME.

orientation effect. 1. The contribution to the catalytic activity of an enzyme that is due to the stereochemical arrangement of the reactants on the surface of the enzyme in a manner that favors their undergoing a reaction. *See also* proximity effect. 2. DIPOLE-DIPOLE INTERACTION.

origin. 1. The point of sample application on an electrophoretic or a chromatographic support.

2. The point of intersection of the vertical and horizontal axes in a plane rectangular coordinate system.

original antigenic sin. *See* doctrine of original antigenic sin.

origin of life. The processes whereby biomolecules, subcellular structures, and, ultimately, living cells have come into existence. *See also* biochemical evolution: biological evolution; chemical evolution.

Orn. Ornithine.

ornithine. A nonprotein, alpha amino acid that is an intermediate in the urea cycle. *Abbr* Orn.

ornithine cycle. UREA CYCLE.

Oro. 1. Orotic acid. 2. Orotate.

orosomucoid. A conjugated plasma protein that contains a number of carbohydrates.

orotic acid. A nonnucleic acid pyrimidine that is a precursor in the biosynthesis of the pyrimidine nucleotides of nucleic acids. *Abbr* O.

orotic aciduria. A genetically inherited metabolic defect in man that is characterized by the excretion of orotic acid and that is due to a deficiency of the enzyme' orotidine-5'-phosphate pyrophosphorylase.

orotidine. The ribonucleoside of orotic acid. Orotidine mono-, di-, and triphosphate are abbreviated, respectively, as OMP, ODP, and OTP. The abbreviations refer to the 5'-nucleoside phosphates unless otherwise indicated. *Abbr* Ord; O.

orotidylic acid. The nucleotide of orotic acid.

orphan virus. A virus that has not been implicated as the cause of a specific disease.

ortho-. Prefix indicating two substituents of adjacent carbon atoms of the ring in an aromatic compound. *Sym* o.

orthochromatic dye. A dye that stains cells or tissues with a single color as opposed to a metachromatic dye.

orthodox conformation. A high-energy conformation of mitochondria that occurs in mitochondrial preparations containing little or no ADP and that is characterized by a mitochondrial matrix which is squeezed together tightly and which stains heavily. *See also* condensed conformation.

orthomolecular medicine. The maintaining of good health and the treatment of disease that is achieved by varying the concentrations in the human body of substances which are normally present in, and are required by, the body. The increased dietary intake of ascorbic acid for the control of respiratory infections is an example.

orthomyxovirus. *See* myxovirus.

orthophosphate. INORGANIC PHOSPHATE.

orthophospate cleavage. The hydrolytic removal of an orthophosphate group from a nucleoside di- or triphosphate.

oryzamin. VITAMIN B_1.

oscillographic polarography. A polarographic technique in which the polarographic wave is depicted on an oscilloscope.

Osm. Osmolar.

osmiophilic. Descriptive of a specimen, prepared for electron microscopy, that has a tendency to take up the electron-dense stain of osmium tetroxide.

osmium tetroxide. A compound used for fixing and staining specimens for electron microscopy.

osmolality. The concentration of a solution expressed in terms of the number of osmoles of solute per 1000 grams of solvent.

osmolal solution. A solution that contains one osmole of solute dissolved in 1000 grams of solvent.

osmolarity. The concentration of a solution expressed in terms of the number of osmoles in 1 liter of solution.

osmolar solution. A solution that contains one osmole of solute in 1 liter of solution.

osmole. A measure of the osmotically effective amount of solute; equal to the mole of the solute divided by the number of ions formed per molecule of the solute, on the assumption that electrolytes dissociate completely into ions in solution. Thus one mole of sodium chloride is equivalent to two osmoles and one mole of phosphoric acid is equivalent to four osmoles.

osmometer. An instrument for measuring osmotic concentration, osmotic pressure, and number average molecular weight.

osmophile. An organism that grows preferentially in solutions that have high osmotic pressures.

osmophilic. Of, or pertaining to, an osmophile.

osmoreceptor. A receptor in the central nervous system that responds to changes in the osmotic pressure of the blood.

osmoregulator. An organism that maintains a constant internal osmotic concentration irrespective of variations in the osmotic concentration of its external environment.

osmosis. The movement of water or another solvent across a semipermeable membrane from a region of low solute concentration to one of higher solute concentration.

osmotic. Of, or pertaining to, osmosis.

osmotic barrier. PERMEABILITY BARRIER.

osmotic coefficient. A factor that relates the observed osmotic pressure of a solution to that of an ideal solution.

osmotic concentration. 1. OSMOLARITY. 2. OSMOLALITY.

osmotic potential. The osmotic pressure that a solution is capable of producing if it were separated from the pure solute by a semipermeable membrane.

osmotic pressure. The pressure that causes water or another solvent to move in osmosis from a solution having a low solute concentration to one having a high solute concentration; it is equal to the hydrostatic pressure that has to be applied to the more concentrated solution to prevent the movement of water (solvent) into it.

osmotic shock. The sudden and drastic dilution of a suspension of cells that results in the movement of water into the cells and leads to cell rupture.

osmotic work. The work performed by cells or cellular organelles in transporting substances across biological membranes, with the result that some substances accumulate inside the cell or inside the organelle, while others are eliminated.

ossification. The formation of bone.

osteomalacia. The softening and bending of bones in an adult due to a deficiency of vitamin D. *Aka* adult rickets.

Ostwald-Folin pipet. A blow-out pipet that is similar to a volumetric pipet but has the bulb closer to the tip of the pipet; used for the delivery of viscous fluids such as blood or serum.

Ostwald viscometer. A simple capillary viscometer constructed in the shape of a U-tube with a small upper bulb in one arm, a larger lower bulb in the second arm, and a capillary connecting the two bulbs. The instrument is especially useful for measuring relative viscosities.

OT. 1. Oxytocin. 2. Old tuberculin.

OTP. 1. Orotidine triphosphate. 2. Orotidine-5′-triphosphate.

ouabain. A cardiac glycoside that is a specific inhibitor of the sodium and potassium ion transport across the cell membrane.

Ouchterlony method. A method for double immunodiffusion in which antigen and antibody solutions are placed in wells that are cut in agar and are spaced in one of a number of geometric arrangements. Both the antigens and the antibodies diffuse through the agar and interact to form precipitation arcs.

Oudin technique. A method for single immunodiffusion in which the antibodies are mixed with agar in a narrow vertical tube and a solution of antigens is layered above the agar. The antigens diffuse through the gel and interact with the antibodies to form precipitation zones.

outbreeding. The crossing of genetically unrelated plants or animals.

outer coat. CELL COAT.

outer membrane. 1. The external membrane of mitochondria. 2. The external membrane of chloroplasts.

outer orbital. An orbital that functions in the bonding of a high-spin complex.

outer sphere. SECONDARY HYDRATION SHELL.

outer sphere activated complex. An oxidation-reduction transition state that is formed by the transfer of electrons without a change in the inner coordination shells of the participating ions.

outer sphere complexing ligand. A ligand that is within the secondary hydration shell of a metal ion and that participates in ion-pairing.

outer volume. The volume of solvent that is contained in the spaces among the gel particles of the bed in gel filtration chromatography.

outgrowth. The third stage in the conversion of a spore to a vegetative cell during which the spore core grows and the wall of the vegetative cell develops.

ovalbumin. The principal protein of egg white. *Abbr* OA.

ovariectomy. The surgical removal of an ovary.

overlapping code. A genetic code in which one or more nucleotides of one codon also serve as nucleotides for an adjacent codon.

overloading. 1. The application of excessive amounts of material to an electrophoretic or a chromatographic support. 2. The addition of large amounts of a labeled or unlabeled intermediate to a system to study metabolic pathways. 3. The administration of large doses of protein antigens to an animal to produce immunological unresponsiveness.

overnutrition. Malnutrition that results from an excessive intake of one or more nutrients.

overspeeding technique. The initial spinning of an analytical ultracentrifuge rotor at high speeds to decrease the time required to establish sedimentation equilibrium.

oviduct. The duct that serves for the passage of eggs from the ovary to the uterus.

ovine. Of, or pertaining to, sheep.

ovoflavin. An impure preparation of riboflavin from egg white.

ovogenesis. OOGENESIS.

ovomucoid. A mucoprotein from egg white.

ovum (*pl* ova). The female reproductive cell; an egg cell; the female gamete.

oxaloacetate pathway. A metabolic pathway whereby aspartic acid and asparagine are catabolized through conversion to oxaloacetic acid, which then feeds into the citric acid cycle.

oxaloacetic acid. A dicarboxylic acid that initiates the reactions of the citric acid cycle by condensing with acetyl coenzyme A to form citric acid and coenzyme A. *Abbr* OAA.

oxamycin. CYCLOSERINE.

oxidant. The electron acceptor species of a given half reaction; the species that undergoes reduction in an oxidation-reduction reaction.

oxidase. An enzyme that catalyzes the oxidation of a substrate, with molecular oxygen serving as the electron acceptor.

oxidation. The change in an atom, a group of atoms, or a molecule that involves one or more of the following: (a) gain of oxygen; (b) loss of hydrogen; (c) loss of electrons.

oxidation number. A measure of the oxidized or the reduced state of an atom. For monoatomic ions, the oxidation number is equal to the charge of the ion. For complex ions and for molecules containing covalent bonds, the oxidation number is equal to the number of electrons gained or lost when (a) the electrons of a covalent bond are assigned completely to the more electronegative atom, and (b) the sum of the oxidation numbers of all the atoms is set equal to either zero in the case of a neutral molecule, or to the charge of the ion in the case of a complex ion. For elements in the free state, the oxidation number is taken as equal to zero.

oxidation potential. The electrode potential that measures the tendency of an oxidation-reduction reaction to occur through a loss of electrons. *Sym* E. *See also* electrode potential; standard electrode potential.

oxidation quotient. *See* Meyerhof oxidation quotient.

oxidation-reduction enzyme. OXIDOREDUCTASE.

oxidation-reduction potential. The electrode potential expressed either as a reduction potential or as an oxidation potential.

oxidation-reduction reaction. A reaction composed of two half-reactions, one of which is a reduction half-reaction and one of which is an oxidation half-reaction.

oxidation state. OXIDATION NUMBER.

oxidative assimilation. The metabolism of microorganisms in which a large fraction of the substrates that are being oxidized are converted to cellular components in order to permit growth; this contrasts with the metabolism in a fully grown animal or human where the bulk of the substrates are oxidized to carbon dioxide and water which are then excreted.

oxidative deamination. A deamination reaction with a concomitant oxidation, as in the conversion of an alpha amino acid to an alpha keto acid.

oxidative pathway of hexose metabolism. HEXOSE MONOPHOSPHATE SHUNT.

oxidative phosphorylation. The synthesis of ATP from ADP that is coupled to the operation of the mitochondrial electron transport system.

oxidizing agent. OXIDANT.

oxidizing atmosphere. An atmosphere that is rich in gases which are readily reduced; the present-day atmosphere of the earth, which is rich in oxygen, is an example.

oxidizing power. The capacity of a substance to function as an oxidizing agent.

oxidoreductase. An enzyme that catalyzes an oxidation-reduction reaction. *See also* enzyme classification.

oximeter. An instrument for determining the degree of oxygenation of blood by a direct measurement of a translucent part of the organism.

oxoacid. KETO ACID.

oxosteroid. KETOSTEROID.

oxybiontic. Capable of using molecular oxygen for growth. *Aka* oxybiotic. *See also* aerobic (2,3).

oxybiotin. A synthetic analogue of biotin in which the sulfur has been replaced by oxygen and which can generally substitute for biotin.

oxygen. An element that is essential to all plants and animals. Symbol O; atomic number 8; atomic weight 15.9994; oxidation number -2; most abundant isotope O^{16}; a stable isotope O^{18}.

oxygenase. An enzyme that catalyzes an oxidation reaction by molecular oxygen, in which both of the oxygen atoms are inserted into a compound. *See also* monooxygenase; dioxygenase.

oxygenation. 1. The saturation of a liquid with oxygen, as in the oxygenation of blood. 2. The introduction of oxygen into a compound, as in the oxygenation of hemoglobin.

oxygen capacity of blood. OXYGEN COMBINING POWER OF BLOOD.

oxygen carrier. A pigment, such as hemoglobin, myoglobin, or hemerythrin, that combines with and serves to transport oxygen.

oxygen combining power of blood. The volume of oxygen that is required for combining with all of the hemoglobin in 100 ml of blood.

oxygen content of blood. The total amount of oxygen, both physically dissolved and combined with hemoglobin, that is present in 100 ml of blood and that is determined at partial pressures of oxygen corresponding to those of the blood when it flows through either the arteries or the veins.

oxygen cycle. The group of reactions whereby oxygen is generated from water by photosynthetic organisms and is reconverted to water by heterotrophic organisms.

oxygen debt. The extra amount of oxygen that is required by a mammal after a period of strenuous exercise. The oxygen is used to oxidize a portion of the lactic acid that accumulated during the exercise, and the energy thus obtained drives the conversion of the remainder of the lactic acid to glycogen.

oxygen saturation. The percentage of oxygen actually bound to either hemoglobin or myoglobin at a given partial pressure of oxygen, compared to the maximum amount that can be bound at high partial pressures of oxygen.

oxygen saturation curve. A plot of oxygen saturation of either hemoglobin or myoglobin as a function of the partial pressure of oxygen.

oxygen transferase. OXYGENASE.

oxygen transport. The carrying of oxygen by the blood from the lungs to the tissues.

oxyhemoglobin. An oxygenated hemoglobin molecule. *Abbr* HbO_2; $HHbO_2$.

oxymyoglobin. An oxygenated myoglobin molecule. *Abbr* MbO_2.

oxysome. A macromolecular aggregate that functions as a unit in oxidative phosphorylation. It is referred to as either a lumped or a distributed oxysome depending on whether the components for oxidative phosphorylation are all present in one aggregated particle or are distributed over a number of different aggregated particles.

oxytocic hormone. OXYTOCIN.

oxytocin. A cyclic peptide hormone that consists of nine amino acids and that causes the contraction of smooth muscle; it is secreted by the posterior lobe of the pituitary gland. *Abbr* OT.

Oz group. A group of allotypic antigenic sites in the constant region of the lambda chain of human immunoglobulins.

ozone. The triatomic allotropic form of oxygen that is present in the stratosphere and that is generated by a silent electric discharge in oxygen or air.

ozonolysis. The oxidation of a compound by means of ozone; used for locating the double bonds in a fatty acid molecule, since ozonolysis leads to a cleavage of the molecule at those points.

P

p. 1. Para position. 2. Pico. 3. Proton. 4. Phospho-, as in p-creatine. 5. Phosphate, as in glucose-1-p.

P. 1. Inorganic phosphate. 2. Phosphate in an abbreviation, as in ATP. 3. Phosphorus. 4. Product. 5. Probability. 6. Proline. 7. Poise. 8. Pressure.

~P. A phosphate group in a high-energy compound which, when removed by hydrolysis, results in a reaction that has a large, negative free energy change.

P_{450}. CYTOCHROME P_{450}.

P_{700}. The pigment of photosystem I of chloroplasts that is thought to be a special chlorophyll molecule, the properties of which are influenced by its environment.

P^{32}. A radioactive isotope of phosphorus that is a strong beta emitter and has a half-life of 14.3 days.

PABA. *p*-Aminobenzoic acid; also abbreviated PAB.

pacemaker enzyme. An enzyme that catalyzes a reaction that is essentially irreversible in the chemical sense; such enzymes frequently catalyze either the initial, the final, or a branch point reaction of a metabolic pathway.

PAGE. Polyacrylamide gel electrophoresis.

PAH. *p*-Aminohippuric acid.

pair annihilation. The production of a photon from the energy dissipated when an electron collides with a positron.

paired. Designating bases in two nucleic acid strands, or in one strand folded back upon itself, that are linked by hydrogen bonding according to the base pairing rules.

paired electrons. The two electrons, of opposite spin, that are normally present in an atomic orbital.

paired sera. An acute serum and a convalescent serum.

paired source method. A method for determining the resolving time of a radiation counter by counting two radiation sources, first separately and then together.

pair production. A unique reaction of energy to mass transition in which a high-energy photon is converted to a positron and an electron in the strong magnetic field of an atomic nucleus.

PAL. Pyridoxal phosphate.

Palade granule. RIBOSOME.

paleobiochemistry. A branch of biochemistry that deals with the study of the organic constituents of fossils.

paleontology. The science that deals with life of earlier geologic eras and that is based on the study of plant and animal fossils.

Paleozoic era. The geologic time period that extends from about 225 to 600 million years ago and that is characterized by the development of land animals and plants.

palindromic helix. A helix that has end-to-end symmetry so that both ends are indistinguishable. A palindromic helix can be rotated by 180° about an axis perpendicular to the longitudinal axis of the helix and thereby produce a structure that is identical to that before rotation.

palmitic acid. A saturated fatty acid that contains 16 carbon atoms and is present in animal fat.

palmitoleic acid. An unsaturated fatty acid that contains 16 carbon atoms and one double bond.

PALP. Pyridoxal phosphate.

pancreas. An endocrine gland that secretes the hormones, insulin and glucagon, and that is located behind the stomach.

pancreatectomy. The surgical removal of the pancreas.

pancreatic. Of, or pertaining to, the pancreas.

pancreatic deoxyribonuclease. DEOXYRIBONUCLEASE I.

pancreatic diabetes. Diabetes that is caused by either a lesion or the removal of the pancreas.

pancreatic juice. The digestive juice that consists of the secretion of the pancreas and that is discharged into the small intestine; contains proteolytic enzymes, secreted as zymogens, nucleases, carbohydrases, and lipase. *Aka* pancreatic fluid.

pancreatic ribonuclease. An endonuclease that catalyzes the hydrolysis of RNA and leads to the production of mono- and oligonucleotides consisting of, or terminating in, a 3′-pyrimidine nucleotide.

pancreatropic. Exerting an effect on the pancreas.

pancreozymin. CHOLECYSTOKININ.

pandemic. Of, or pertaining to, an epidemic disease of unusually widespread occurrence.

panspermy hypothesis. The hypothesis that life originated elsewhere in the universe before the solar system was formed and that it was later transported in some fashion through interstellar space and deposited on the earth in the form of heat-resistant spores or in some other form. *Aka* panspermia hypothesis.

pantetheine. An intermediate in the biosynthesis of coenzyme A in mammalian liver and in some microorganisms, and a growth factor for a species of lactic acid bacteria, *Lactobacillus bulgaricus*; consists of pantothenic acid linked to β-mercaptoethylamine.

pantothenic acid. One of the B vitamins, the coenzyme form of which is coenzyme A.

pantropic virus. A virus that affects more than one tissue or more than one organ.

papain. A proteolytic enzyme that is derived from papaya and that has a broad specificity.

paper chromatogram. The developed strip or sheet of filter paper that is obtained in paper chromatography.

paper chromatography. Partition chromatography in which the stationary phase is a moistened strip or sheet of filter paper and the mobile phase is a solvent that either ascends or descends along the paper. *Abbr* PC.

paper electrophoresis. Electrophoresis in which a strip or sheet of filter paper is used as the supporting medium; the filter paper is moistened with buffer and dips into two buffer compartments that contain the electrodes.

papergram. PAPER CHROMATOGRAM.

papilloma. A benign tumor of the skin or of a mucous surface; a wart.

papilloma virus. A small, naked, icosahedral virus that contains double-stranded DNA and that produces papillomas in animals; belongs to the group of papovaviruses.

papovavirus. A small, naked, icosahedral virus that contains double-stranded DNA; most papovaviruses produce either benign or malignant tumors.

PAPS. 3'-Phosphoadenosine-5'-phosphosulfate.

para-. Prefix indicating two substituents on opposite carbon atoms of the ring in an aromatic compound. *Sym* p.

parabiosis. The natural or artificial joining of two organisms so that they are linked both anatomically and physiologically.

parabolic inhibition. Inhibition that yields a curve that is concave upward when either slopes or intercepts from a primary plot are plotted as a function of inhibitor concentration.

paracrystalline. Descriptive of a solid that has a somewhat lesser order than the regular three-dimensional arrangement of the atoms in a true crystal.

paraffin method. A method of preparing a tissue specimen for microscopic examination in which the tissue is embedded in paraffin and then cut with a microtome to produce thin paraffin sections.

paraffin section. *See* paraffin method.

paraffin wax. A macrocrystalline wax obtained from petroleum.

parahematin. FERRIHEMOCHROME.

parahemophilia. A genetically inherited metabolic defect in man that is caused by a deficiency of accelerator globulin, a factor in the blood clotting system.

parallel chains. Two polypeptide chains that run in the same direction; both progress from the C-terminal to the N-terminal, or vice versa.

parallel spin. The spin of two particles in the same direction.

parallel strands. Two polynucleotide strands that run in the same direction; both progress from the 3'-terminal to the 5'-terminal, or vice versa.

paralysis. *See* immunological paralysis.

paralysis time. COINCIDENCE TIME.

paramagnetic. Descriptive of a substance that has unpaired electrons and that has a permanent magnetic dipole moment as a result of the magnetic properties of its spinning electrons; when such a substance is placed in a magnetic field, it tends to become oriented in line with the applied field.

paramecin. A toxin that is present in the kappa particles of *Paramecium aurelia.*

Paramecium. A genus of cilia-possessing protozoa.

parameter. 1. An independent variable through functions of which other variables may be expressed. 2. A property, associated with a molecule or some other particle, that can be expressed quantitatively and that has specific values; molecular weight, sedimentation coefficient, and axial ratio are some examples.

parametric statistics. Statistical calculations that are based on prior assumptions with respect to the variable and the probability distributions of the data, and that are valid only if these assumptions hold.

paramylon. A linear homopolysaccharide that occurs in *Euglena* and that is composed of D-glucose units linked by means of β (1→3) glycosidic bonds.

paramyosin. TROPOMYOSIN A.

paramyxovirus. *See* myxovirus.

paranemic coiling. The coiling of two threads in opposite directions so that they can be separated without uncoiling the threads. *See also* plectonemic coiling.

paranuclein. Obsolete designation for a phosphorus-containing protein that is not a nucleoprotein.

paraprotein. An immunoglobulin that is present in the serum of individuals suffering from multiple myeloma or from Waldenstroem's macroglobulinemia.

paraproteinemia. The presence of paraproteins in the blood.

parasite. An organism that lives in or upon another organism from which it derives some or all of its nutrients.

parathormone. A polypeptide hormone, secreted by the parathyroid glands, that stimulates the release of calcium from bone and leads to an increase in the level of calcium in the blood.

parathyroidectomy. The surgical removal of a parathyroid gland.

parathyroid gland. One of two endocrine glands that secrete parathormone and that are adjacent to, or embedded within, the thyroid gland.

parathyroid hormone. PARATHORMONE.

paratope. ANTIBODY-COMBINING SITE.

parenchymal cell. A cell of the functional tissue of a gland or an organ.

parent. 1. A cell in relation to the cells formed from it by cell division. 2. A molecule of DNA or a chromosome in relation to the molecules or the chromosomes formed from it by replication. 3. A virus in relation to its progeny. 4. A radioactive nuclide in relation to the nuclides formed from it by radioactive decay.

parenteral. Referring to the introduction of a substance into an animal organism by ways other than that of the digestive tract, as in the case of an intradermal injection.

parent gelatin. The product of the thermal helix-coil transition of collagen in solution; ordinary gelatin is derived from parent gelatin by heat denaturation or by degradation with acid or alkali.

parietal cell. A hydrochloric acid-secreting cell of the gastric glands.

parthenogenesis. The development of an organism from an unfertilized egg.

partial hydrolysate. The solution that contains the mixture of substances obtained by partial hydrolysis. *Var sp* partial hydrolyzate.

partial hydrolysis. Incomplete hydrolysis, as the hydrolysis of a polymer by an enzyme that is specific for only some of the bonds between the monomers.

partial inhibition. The inhibition of the activity of an enzyme in which a saturating concentration of inhibitor cannot cause the reaction velocity to become zero.

partial molar quantity. The rate of change of an extensive property of a substance with the number of moles of the substance.

partial pressure. The pressure exerted by a gas when it is part of a mixture of gases; it is equal to the pressure that the gas would exert if it alone occupied the entire volume occupied by the mixture at the same temperature.

partial specific quantity. The rate of change of an extensive property of a substance with the number of grams of the substance.

partial specific volume. The rate of change of solution volume with the number of grams of solute; the volume increase of a very large volume of solution upon the addition of 1 gram of solute. *Sym v̄*.

particle diffusion. The diffusion of ions through the granules of an ion-exchange resin which proceeds at such a rate that the diffusion controls the rate of ion-exchange taking place.

particle electrophoresis. The electrophoretic movement of large colloidal particles in an electrophoresis cell that is mounted under a microscope; used for the direct measurement of the electrophoretic mobility of such particles.

particle scattering factor. A factor that allows for the angular dependence of light scattered by particles that have at least one dimension that is greater than one twentieth of the wavelength of the incident light; the factor depends on the radius of gyration of the particles.

particle weight. MOLECULAR WEIGHT (2).

particulate antigen. An antigen that is part of an insoluble structure of a microbial or other cell, a virus, or some other particle.

particulate enzyme. An enzyme that is bound to an insoluble structure of a cell, a cellular organelle, or some other particle.

particulate fraction. A fraction consisting of cellular and intracellular insoluble structures but devoid of the intracellular fluids and their dissolved solutes.

partition chromatography. Chromatography in which the distribution of compounds between a mobile and a stationary liquid phase is based on the solubilities of the compounds in the two phases; the stationary liquid phase is held in place by a porous solid such as a sheet of filter paper or a column of starch. Partition chromatography refers particularly to those cases in which the stationary phase is a polar liquid and the mobile phase is a nonpolar liquid. *Abbr* PC.

partition coefficient. The ratio of the concentrations of a substance in two immiscible phases at equilibrium; the phases may be two liquids, or a liquid and a gas.

partitioner phase. STATIONARY PHASE (2).

partition function. A thermodynamic function that describes the distribution of molecules over all possible energy levels.

partition isotherm. PARTITION COEFFICIENT.

partition law. The law that a solute, added to a system composed of two immiscible phases, will distribute itself between the two phases according to its solubility in each phase; the phases may be two liquids, or a liquid and a gas.

parturition. Childbirth; giving birth to young.

parvovirus. A small, naked, icosahedral animal virus that contains single-stranded DNA.

PAS. *p*-Aminosalicylic acid.

PA/S procedure. Periodic acid/Schiff procedure.

PAS reagent. SCHIFF'S REAGENT.

passage. SUBCULTURE.

passenger virus. A virus, the presence of which is not associated with the disease, if any, that is occurring in the tissue from which the virus is isolated.

passive agglutination. The agglutination reaction of soluble antigens attached covalently, or noncovalently, to cell surfaces.

passive anaphylaxis. The anaphylactic reaction produced in an animal by injecting it first with antibodies from another animal and then with the corresponding antigens.

passive hemagglutination. The passive agglutination of red blood cells.

passive hemolysis. The lysis of red blood cells in the presence of complement that is brought about by antibodies to antigens which have been artificially attached to the surface of the cells.

passive immunity. The immunity acquired by an animal organism as a result of the injection of antibodies into it.

passive transfer. 1. The transfer of immunity by the injection of serum, antibodies, or lymphocytes from an immune individual to a normal one. 2. The transfer of hypersensitivity by the injection of serum, antibodies, or lymphocytes from an allergic individual to a normal one.

passive transport. The movement of a solute across a biological membrane that is produced by diffusion, is directed downward in a concentration gradient, does not require carriers, and does not require the expenditure of energy.

Pasteur effect. The inhibition of glycolysis and the decrease of lactic acid accumulation that is produced by increasing concentrations of oxygen. *See also* Crabtree effect.

pasteurization. The brief heating of a food, such as milk or wine, that is designed to kill pathogenic microorganisms without actually sterilizing the food.

Pasteur-Liebig controversy. A controversy that raged during the second half of the 19th century between Louis Pasteur and Justus Liebig. Pasteur believed that fermentation and similar processes were carried out by the metabolic activities of living cells; Liebig believed that such processes were due to chemical substances and that the reactions resulted from self-perpetuating instabilities in the solution that were initiated by exposure of the solution to air.

Pasteur pipet. A drawn-out ungraduated pipet with a constriction in the wider part of the pipet for insertion of a cotton plug.

patch and cut repair. A repair mechanism of DNA in which repair replication begins after the first incision by a nuclease, and the damaged segment is fully excised only after the repair replication is complete. The final step requires the action of DNA ligase to join the newly synthesized segment to the existing strand. *See also* cut and patch repair.

patch test. A test for delayed-type hypersensitivity to the cutaneous application of tuberculin.

pathogen. A virus, a microorganism, or some other substance that can produce a specific disease.

pathogene. VIROID.

pathogenesis. The origin and development of a disease.

pathogenic. Disease-producing.

pathogenicity. The disease-producing capacity of a microorganism, a virus, or other substance.

pathogenic RNA. VIROID.

pathological. Of, or pertaining to, pathology.

pathological biochemistry. The biochemistry of pathological tissues and fluids.

pathology. The science that deals with the origin, nature, and course of diseases.

pathway. *See* metabolic pathway.

pattern method. A viral assay based on measuring the extent of hemagglutination by dilutions of the viral sample.

Patterson function. A mathematical function used in the construction of a Patterson map.

Patterson map. A graphical projection used in determining the positions of heavy atoms in x-ray diffraction patterns that have been obtained by the isomorphous replacement method.

paucidisperse. Consisting of macromolecules that fall into a small number of classes with respect to their size.

Pauli exclusion principle. The principle that no two electrons in an atom can be in the same detailed state described by the same four quantum numbers; as a result, a maximum of two electrons can occupy a single atomic orbital.

Pauly reaction. A colorimetric reaction for histidine and other imidazole compounds that is based on the production of a red color on treatment of the sample with an alkaline solution of diazotized sulfanilic acid.

Pauly's reagent. A sulfanilamide-containing reagent that is used for the detection of tyrosine and other phenolic compounds in chromatograms.

PBI. Protein-bound iodine.

PC. 1. Phosphatidylcholine. 2. Phosphocreatine. 3. Paper chromatography. 4. Partition chromatography.

PCA. 1. Passive cutaneous anaphylaxis. 2. Perchloric acid.

PCB. Polychlorinated biphenyl.

PCMB. *p*-Chloromercuribenzoic acid.

P$_{CO_2}$. 1. Carbon dioxide tension. 2. Partial pressure of carbon dioxide.

pD. A term that is equivalent to pH for a system that contains deuterons.

PD$_{50}$. Median paralysis dose.

PDC. Pyruvate dehydrogenase complex; *see* pyruvate dehydrogenase system.

P^{32} decay. *See* radiophosphorus decay.

PE. 1. Phosphatidylethanolamine. 2. Polyethylene.

pectate lyase. The enzyme that catalyzes the degradation of pectins whereby the glycosidic bonds between the monosaccharide residues are cleaved and water molecules are eliminated. *Aka* eliminase; transeliminase.

pectic acid. A polysaccharide of galacturonic acid that occurs in fruits.

pectic substance. One of a group of substances that occur in plants and that consist of pectin, pectic acid, and related compounds.

pectin. A polysaccharide that occurs in fruits and that consists of a form of pectic acid in which many of its carboxyl groups have been methylated.

PEI-cellulose. Polyethyleneimine-cellulose; an anion exchanger.

pellagra. The disease caused by a deficiency of the B vitamin nicotinic acid.

pellagra-preventative factor. NICOTINIC ACID.

pellet. The material collected by sedimentation of a solution in a centrifuge tube.

pellicular. Descriptive of a column chromatographic packing, used in high-pressure liquid chromatography, that consists of small, spherical, glass particles which are coated with a thin layer of a chromatographic support such as a gel, an adsorbent, or an ion-exchange resin.

penicillin. One of a group of antibiotics, produced by the mold *Penicillium notatum,* that function by inhibiting the synthesis of the bacterial cell wall; penicillin G, or benzylpenicillin, is the most widely used of the group.

penicillinase. The enzyme that catalyzes the hydrolysis of the lactam bond in penicillin; it is produced by many bacteria and is sometimes a constitutive and sometimes an inducible enzyme.

penicillin enrichment. A method for concentrating bacterial auxotrophs. Wild-type or mutagenized cells are grown on a minimal medium and are subjected to penicillin-induced lysis. This destroys the growing cells but not the nongrowing auxotrophs. The latter can then be supplied with an enriched medium to allow their growth subsequent to the destruction of the penicillin by means of penicillinase. *Aka* penicillin method.

penta-. Combining form meaning five.

pentagonal capsomer. *See* capsomer.

pentamer. 1. An oligomer that consists of five monomers. 2. PENTAGONAL CAPSOMER.

penton. The fibrous morphological subunit of adenoviruses that is surrounded by five capsomers in the intact viral particle.

penton antigen. An antigen of the penton of adenoviruses.

pentosan. A polysaccharide of pentoses.

pentose. A monosaccharide that contains five carbon atoms.

pentose cycle. HEXOSE MONOPHOSPHATE SHUNT.

pentose nucleic acid. RIBONUCLEIC ACID.

pentose oxidation cycle. HEXOSE MONOPHOSPHATE SHUNT.

pentose phosphate cycle. HEXOSE MONOPHOSPHATE SHUNT.

pentose phosphoketolase pathway. A pathway that is related to the hexose monophosphate shunt and that occurs in some bacteria.

pentosuria. A genetically inherited metabolic defect in man that is characterized by the presence of excessive amounts of L-xylulose in the urine and that is due to a deficiency of the enzyme L-xylulose dehydrogenase. *Aka* idiopathic pentosuria.

penultimate carbon. The carbon atom preceding the last one in a chain.

PEP. Phosphoenolpyruvic acid.

pepsin. A proteolytic enzyme in the stomach that is unique because of its very low optimum pH. The major sites of pepsin action are peptide bonds in which the amino function is contributed by either aromatic or acidic amino acids.

pepsin inhibitor. A polypeptide fragment that inhibits pepsin and that has a molecular weight of about 3000; it is removed, together with other peptides, from pepsinogen in the course of its activation to pepsin.

pepsinogen. The inactive precursor of pepsin.

peptic. 1. Of, or pertaining to, pepsin. 2. Of, or pertaining to, the stomach and gastric digestion.

peptic peptides. The peptides obtained by digestion of a protein with the enzyme pepsin.

peptidase. An enzyme that catalyzes the hydrolysis of peptide bonds in both peptides and proteins.

peptide. A linear compound that consists of two or more amino acids that are linked by means of peptide bonds.

peptide antibiotic. An antibiotic, such as gramicidin or actinomycin D, that consists largely or entirely of a peptide.

peptide bond. A covalent bond formed by splitting out a molecule of water between the carboxyl group of one amino acid and the amino group of a second amino acid. *Aka* peptide link.

peptide map. A fingerprint of peptides.

peptide synthetase. PEPTIDYL TRANSFERASE.

peptidoglycan. The rigid framework of bacterial cell walls that consists of a cross-linked network of mucopeptides; the mucopeptides are pentapeptides which are linked to the disaccharide *N*-acetylglucosamine-*N*-acetylmuramic acid.

peptidoglycolipid. A compound composed of a peptide, a carbohydrate, and a lipid.

peptidolipid. LIPOPEPTIDE.

peptidyl-puromycin. A peptide attached to puromycin; formed between a growing polypeptide chain and puromycin when protein synthesis is inhibited by puromycin.

peptidyl site. The site on the ribosome at which the peptidyl-transfer RNA is bound at a time when the next aminoacyl-transfer RNA becomes bound to the aminoacyl site.

peptidyl transferase. The enzyme in protein synthesis that catalyzes peptide bond formation between the growing polypeptide chain and the next amino acid to be added; in bacteria, the enzyme is one of the proteins of the 50S ribosomal subunit.

peptidyl-tRNA. A transfer RNA molecule with an attached peptide, as the tRNA with the attached growing polypeptide chain in protein synthesis.

peptidyl-tRNA site. PEPTIDYL SITE.

peptone. A partially hydrolyzed protein that is not precipitated by ammonium sulfate; a secondary protein derivative used as a component of microbiological culture media.

peptonization. The enzymatic conversion of a protein into a peptone.

percentage average deviation. The ratio of the average deviation to the mean, multiplied by 100; $100A/M$, where A is the average deviation, and M is the mean.

percentage error. PERCENTAGE AVERAGE DEVIATION.

percentage law. The law stating that the percentage of virus particles which are neutralized by a given antiserum is constant and does not depend on the virus titer, provided that the formation of antibody-virus complexes is reversible and that the antibody is present in excess.

percentile. The value of a statistical variable below which the indicated percentage of the measurements of the frequency distribution fall; thus the 10th percentile is that value below which 10% of the measurements fall.

percent saturation. The salt concentration of a solution relative to that of a saturated solution of the same salt.

percent solution. A measure of concentration that, unless otherwise indicated, refers to the number of grams of solute in 100 ml of solution (w/v). Other expressions for percent concentration are volume/volume (v/v) and weight/weight (w/w).

percent transmittance. The ratio of the intensity of the transmitted light to that of the incident light, multiplied by 100; $100I/I_0$, where I is the intensity of the transmitted light, and I_0 is the intensity of the incident light. *Aka* percent transmission.

percutaneous. Through the skin.

performance index. A rating of ultracentrifuge rotors that expresses their relative effectiveness in accomplishing the complete sedimentation of a given material under idealized conditions.

perfusate. The liquid leaving a perfused organ.

perfused organ. An organ that either has been or is being subjected to perfusion.

perfusion. The passage of blood, plasma, or other fluids through the blood vessels of an isolated organ or a tissue; used for metabolic studies and for keeping organs alive during organ transplantation.

perhydrocyclopentanophenanthrene. The system of four fused rings that is the parent structure of the steroids.

pericyclic reaction. A chemical reaction that is characterized by a concerted regrouping of bonding orbitals in the molecule and that proceeds by way of a cyclic transition state.

perikaryon. The cytoplasmic cell body that surrounds the nucleus of a neuron.

perinuclear. Surrounding the nucleus of a cell.

perinuclear cisternae. A group of fluid-filled compartments that are enclosed by the inner and outer nuclear membranes.

periodate oxidation. The oxidative cleavage by periodate of the bond between two carbon atoms, one of which carries a hydroxyl group while the other carries an amino, a carbonyl, a carboxyl, or a hydroxyl group. The reaction is used for elucidating the structure of an unknown carbohydrate.

periodic acid/Schiff procedure. A staining procedure for polysaccharides in which periodate oxidation of the polysaccharide is followed by staining with Schiff's reagent for aldehydes. *Abbr* PA/S procedure.

periodicity. An occurrence at regular intervals in either time or space; used to describe the spacing of diffraction spots in x-ray diffraction patterns of biopolymers.

periodic polymer. REPEATING POLYMER.

periodic table. The arrangement of the chemical elements as a function of their atomic number; elements with similar properties are placed one under the other to form groups of elements.

peripheral. At or near an outer surface or a boundary.

peripheral nervous system. That part of the nervous system of vertebrates that consists of the nerves and the ganglia but excludes the components of the central nervous system. *Abbr* PNS.

periplasmic enzyme. A bacterial enzyme that exists either in free or in bound form in a region between the cell wall and the cell membrane.

peristalsis. The progressive wave-like movements occurring in the intestine and in other hollow, muscular structures that serve to mix the contents present in the structure and to move it forward.

peristaltic. Of, or pertaining to, peristalsis.

peristaltic pump. A pump that moves contents along a flexible tube by intermittently pressing on the tube from the outside; this may be achieved by the rotation of a cylinder with spaced protuberances over the tube.

peritrichous. Descriptive of a bacterium that has flagella all over its surface.

permanent cell strain. CELL LINE.

permanent dipole moment. A dipole moment that is due to the structure of the substance and not due to the influence of an external electric or magnetic field.

permeability. The property of a membrane that is measured by the qualitative and quantitative aspects of the passage of ions and molecules across it.

permeability barrier. The limited ability, or the complete inability, of a substance to cross a biological membrane.

permeability coefficient. The net number of molecules of solute that cross each square centimeter of a membrane in unit time when the concentration difference across the membrane is 1 mole/cc.

permeable. Descriptive of a membrane that permits the passage of both solutes and solvent across it.

permeant. 1. *n* A substance that permeates. 2. *adj* Capable of permeating.

permease. 1. TRANSPORT AGENT. 2. The transport agent for galactosides in *Escherichia coli.*

permeate. To pass into or through a substance, such as a gel or a membrane.

P.E.R. method. Protein efficiency ratio method.

permissible dose. *See* maximum permissible dose.

permissible substitution. CONSERVATIVE SUBSTITUTION.

permissive cell. 1. A cell in which a conditional lethal mutant can grow. 2. A cell that supports the lytic infection by a virus.

permissive conditions. Conditions that permit the growth of a conditional lethal mutant.

permittivity. An electrical unit that is identical to the dielectric constant when the electric field is static and of moderate intensity.

permselective membrane. A membrane that is impermeable to water and that is selectively permeable to positive ions only or to negative ions only; such a membrane may be prepared synthetically by incorporating a polyelectrolyte into a suitable matrix.

permutation of the map. CIRCULAR PERMUTATION.

permutite. A synthetically produced, alkali metal- or alkaline earth-aluminum silicate that is used as an ion-exchange resin for water softening. *See also* zeolite.

pernicious anemia. A disease caused by the inadequate absorption of vitamin B_{12} from the intestine.

per os. By way of the mouth, as in the giving of food or medicine.

peroxidase. An enzyme that catalyzes the oxidation of a substrate by using hydrogen peroxide as the electron acceptor.

peroxisome. A membrane-enclosed, cytoplasmic organelle that contains a variety of oxidoreductases, such as urate oxidase, catalase, and D-amino acid oxidase. *Aka* peroxidosome.

Perrin equation. One of two equations that relate the axial ratio of either an oblate or a prolate ellipsoid of revolution to its frictional ratio.

persistence length. A parameter related to the conformational rigidity of a worm-like coil; it is equal to the length of the projection of the end-to-end distance of the coil onto an axis tangential to one end of the coil.

persistent fraction. 1. The fraction of stable virus-antibody complexes that is formed during the neutralization of a virus by antibodies and that contains nonneutralized virus particles. 2. The fraction of interferon-resistant virus particles.

persistent induction. Enzyme induction in which enzyme synthesis does not drop off rapidly when the inducer is removed.

perspective formula. A two-dimensional representation of a molecule in which bonds projecting forward with respect to the plane of the page are indicated by wedges, and bonds projecting backward are indicated by dotted lines.

perspiration. 1. The excretion of fluid by the sweat glands. The perspiration is referred to as sensible or insensible depending on whether the loss of water through the skin is accompanied or unaccompanied by visible sweat formation. 2. The fluid excreted by the sweat glands.

perturbation. *See* solvent perturbation.

pervaporation. The evaporation of a solvent through a membrane; used for concentrating solutions.

pesticide. A chemical, such as an insecticide or a herbicide, that kills forms of animal or plant life.

petite mutant. A mutant of the yeast *Saccharomyces cerevisiae* that grows in small colonies and that lacks several respiratory enzymes. A segregational petite carries a mutated nuclear gene; a vegetative, or neutral, petite carries mutated mitochondrial DNA.

petri dish. A flat, covered glass container used for growing bacteria on a nutrient gel.

petri plate. A petri dish containing agar mixed with cells and viruses for use in the plaque assay of viruses.

Petroff-Hausser counting chamber. A special hollowed-out microscope slide that holds a known volume of liquid in a ruled grid and that is used for the direct counting of bacteria under a microscope.

PFC. Plaque-forming cell.

PFU. Plaque-forming unit.

PG. 1. Prostaglandin. 2. Phosphatidyl glycerol. 3. Phosphoglycerate.

PGA. 1. A prostaglandin that has a double bond between carbon atoms 10 and 11. 2. Pteroylglutamic acid. 3. Phosphoglyceric acid.

PGAH$_4$. Tetrahydropteroylglutamic acid.

PGB. A prostaglandin that has one double bond between carbon atoms 8 and 12 and another between carbon atoms 13 and 14.

PGE. A prostaglandin that carries a hydroxyl group at carbon atom 11.

PGF. A prostaglandin that carries a hydroxyl group at carbon atom 9.

pH. A measure of the hydrogen ion concentration in solution. The pH was originally defined as the negative logarithm, to the base 10, of the hydrogen ion concentration expressed in terms of equivalents (or grams, or moles) per liter. The pH is now defined operationally by reference to standard solutions of assigned pH values. The difference in pH between an unknown and a standard solution is directly proportional to the difference in electromotive force between a cell containing the unknown solution and a cell containing the standard solution when these are measured with the same electrodes (a hydrogen or a glass electrode, and a reference electrode). *See also* pH scale; pH unit.

phage. BACTERIOPHAGE.

phage conversion. *See* conversion.

phage cross. The production of recombinant phage progeny, carrying genes of two or more parental phage types, that is brought about when a single bacterium is infected with two or more phages that differ in one or more of their genes.

phage induction. *See* induction (2).

phage lysozyme. The enzyme, present in many phages, that has a specificity similar to that of lysozyme and that functions in the injection of phage nucleic acid into, and the release of progeny phage from, the bacterial cell.

phagocyte. A cell that engulfs bacteria and other foreign particles by phagocytosis.

phagocytic. Of, or pertaining to, phagocytosis or phagocytes.

phagocytic cell. PHAGOCYTE.

phagocytic index. 1. The average number of bacteria or particles that are taken up by phagocytosis per phagocytic cell; frequently measured in vitro. 2. The rate at which inert particles are removed from the blood and are taken up by the phagocytic cells; a measure of the activity of the reticuloendothelial system that is usually determined by means of an injection of carbon particles into an animal.

phagocytin. A protein believed to function in phagocytosis.

phagocytosis. The engulfment and destruction of foreign cells and of foreign particulate matter by a cell.

phagolysosome. The fusion of a phagosome with a lysosome that results in the digestion of the phagosome by the hydrolytic enzymes of the lysosome. *Aka* phagolyosome.

phagosome. A cytoplasmic vesicle that is formed in endocytosis by invagination of the cell membrane and that contains particles taken up by the cell through phagocytosis.

phantom. A mass of material that approximates a tissue in its physical properties and that is used in determining the dose of radiation to be applied to the tissue.

pharmaceutical. 1. *n* A drug. 2. *adj* Of, or pertaining to, pharmacy.

pharmaceutical chemistry. The branch of chemistry that deals with the preparation, composition, and testing of drugs.

pharmacodynamics. The branch of pharmacology that deals with the reactions between drugs and living structures, specifically the action and the fate of drugs in animal organisms.

pharmacogenetics. The area of molecular genetics that deals with the genetic mechanisms that underlie individual differences in the response to drugs.

pharmacognosy. A branch of pharmacology that deals with the identification of drugs.

pharmacokinetics. The area of pharmacology that deals with the quantitative distribution of drugs in the body.

pharmacology. The science that deals with the origin, the composition, and the identification of drugs, and with the effect of drugs on living systems.

pharmacopeia. An official compilation of the names, the composition, and the medicinal doses of drugs, and of tests and procedures relating to these drugs. *Var sp* pharmacopoeia.

pharmacy. The branch of pharmacology that deals with the origin, the composition, the preparation, and the dispensing of drugs.

phase. A solid, liquid, or gaseous homogeneous substance that exists as a distinct and mechanically separable fraction in a heterogeneous system.

phase contrast microscope. A microscope that converts differences in refractive index into visible variations of light intensity and permits the observation of transparent structures in the living cell. *Aka* phase microscope.

phase partition. A technique for the isolation and purification of subcellular fractions in which the material is allowed to partition itself between two or more immiscible or partially miscible phases.

phase plate. A plate that serves as a schlieren diaphragm in the schlieren optical system; it has a coating over half of its area so that incident light will be retarded by half a wavelength when it strikes this area.

phase problem. A problem in the interpretation of x-ray diffraction patterns which is due to the fact that reflections from different sets of atomic planes can be evaluated with respect to their intensities but not with respect to their phase angles.

phase rule. A mathematical generalization of systems in equilibrium; expressed as $P + F = C + 2$, where P is the number of independent phases, F is the number of degrees of freedom, and C is the number of independently variable components. The expression can be used to assess the purity of a protein preparation on the basis of the solubility behavior of the protein.

phase shift mutation. FRAME SHIFT MUTATION.

phase test. A test for chlorophyll that is based on the change in color produced by treating chlorophyll with cold alcoholic potassium hydroxide.

Phe. 1. Phenylalanine. 2. Phenylalanyl.

pH electrode. An electrode that is sensitive to the hydrogen ion concentration of solutions.

phene. A phenotypic character that is controlled by genes.

phenocopy. 1. A nonhereditary change, resembling the change caused by a mutation, that occurs in the phenotype, but not in the genotype, of an organism. The change is brought about by nutritional or environmental factors and results in an organism that resembles another organism; the change leads to an effect that is

characteristic of that produced by a specific gene of the other organism. 2. The organism produced by a nonhereditary change in the phenotype of a parent organism.

phenol. 1. Hydroxybenzene. 2. An aryl hydroxide.

phenol coefficient. A measure of the sterilizing capacity of a compound; equal to the ratio, under standard conditions, of the minimal sterilizing concentration of phenol to the minimal sterilizing concentration of the compound.

phenolic group. *See* phenolic hydroxyl group.

phenolic hydroxyl group. A hydroxyl group attached to a benzene ring.

phenol reagent. The reagent used in the Folin-Ciocalteau reaction.

phenolsulfonphthalein test. A clinical test in which the renal blood flow and the secretory capacity of the renal tubules are assessed by means of the dye phenolsulfonphthalein.

phenome. A collective term for all of the components of a living cell other than the genome.

phenomenological coefficients. A set of coefficients in the phenomenological equations that are functions of temperature, pressure, and composition.

phenomenological equations. A set of equations of irreversible thermodynamics that represent the fluxes that take place within a system as a linear combination of driving forces and that describe the possible couplings which may give rise to new effects.

phenomic lag. PHENOTYPIC LAG.

phenon. A taxonomic grouping of organisms of similar phenotype.

phenotype. 1. The physical appearance and the observable properties of an organism that are produced by the interaction of the genotype with the environment. 2. A group of organisms that have the same physical appearance and the same observable properties. *See also* genotype.

phenotypic. Of, or pertaining to, phenotype.

phenotypic adaptation. The preferential growth of a phenotypically varied organism.

phenotypic curing. The restoration of biological activity, lost through mutation, that is produced by changes in the environmental conditions; the changes result in a temporary alteration of transcription and/or translation that leads to the production of a functional protein even though the protein is specified by a mutant gene.

phenotypic lag. The time between the exposure of an organism to a mutagen and the phenotypic expression of a mutation by that organism or by its progeny.

phenotypic mixing. The production of a virus in which the phenotype differs from the genotype;

may be achieved by infecting a bacterial cell with two related phages, such as two mutants of the same phage, so that components of one phage are incorporated into the structure of the second phage during phage assembly in the host cell.

phenotypic variation. A change that is within the range of the potential changes of a phenotype and that is shown by essentially all of the organisms in a population.

phenylaceturic acid. A compound formed by the conjugation of phenylacetic acid with glycine; the form in which phenylacetic acid is detoxified and excreted in the urine.

phenylalanine. An aromatic, nonpolar alpha amino acid. *Abbr* Phe; F.

phenylalanine hydroxylase. The enzyme that catalyzes the synthesis of tyrosine from phenylalanine. *See also* phenylketonuria.

phenylalanine load test. *See* load test.

phenylalanine tolerance index. The composite of the concentrations of phenylalanine in serum after 1, 3, and 4 hours following the administration of a dose of phenylalanine to an individual; used in characterizing individuals who are suffering from, or are carriers for, phenylketonuria.

p-**phenylenediamine.** A substrate used for assaying ceruloplasmin activity. *Abbr.* PPD.

phenylhydrazone. The compound formed by the reaction of a monosaccharide with equimolar amounts of phenylhydrazine.

phenyl isothiocyanate. *See* Edman degradation.

phenylketonuria. A genetically inherited metabolic defect in man that is characterized by mental retardation, if the defect is not corrected for in childhood, and that is due to a deficiency of the enzyme phenylalanine hydroxylase. *Abbr* PKU.

phenylosazone. The compound formed by the reaction of a monosaccharide with excess phenylhydrazine; phenylosazones are useful for the identification of unknown monosaccharides.

phenylpyruvic oligophrenia. PHENYLKE-TONURIA.

phenylthiocarbamyl amino acid. An intermediate in the formation of a phenylthiohydantoin amino acid. *Abbr* PTC-amino acid.

phenylthiohydantoin amino acid. An amino acid derivative formed by the reaction of phenylisothiocyanate with the free alpha amino groups of amino acids, peptides, or proteins. *Abbr* PTH-amino acid. *See also* Edman degradation.

pH 5 enzyme. AMINOACYL-tRNA SYNTHETASE.

pherogram. ELECTROPHEROGRAM.

pheromone. A chemical that is produced and discharged by an organism and that elicits a physiological response in another organism of the same species; the sex attractant of insects is an example.

pH 5 fraction. A subcellular fraction obtained by centrifuging a suspension of broken cells at $100,000 \times g$ and then precipitating the transfer RNA and the aminoacyl-tRNA synthetases by adjusting the pH of the solution to 5. *See also* S-100 fraction.

pH gradient electrophoresis. ISOELECTRIC FOCUSING.

phi X174. A phage that infects *Escherichia coli* and that contains single-stranded DNA. *Sym* ϕX174.

Philpot-Svensson optics. A schlieren optical system that incorporates a cylindrical lens in its design.

phloretin. A derivative of trihydroxyaceto-phenone that acts as a competitive inhibitor of glucose transport across the erythrocyte membrane. *Aka* dihydronaringenin.

phloridzin. A toxic glycoside that blocks the reabsorption of glucose by the kidney tubules.

pH meter. An instrument for measuring pH values of solutions, commonly by means of a glass electrode and a reference electrode or by means of a combination electrode.

phosphagen. A high-energy phosphate compound, such as phosphocreatine or phosphoarginine, that serves as a storage form of free energy.

phosphatase. An enzyme that catalyzes the hydrolysis of the esters of orthophosphoric acid.

phosphate. 1. An anionic radical of phosphoric acid, specifically one of orthophosphoric acid; inorganic phosphate. *Sym* P_i; P. 2. A salt of phosphoric acid.

phosphate bond energy. The free energy change of a reaction in which a phosphorylated compound is hydrolyzed to yield either inorganic phosphate or inorganic pyrophosphate as one of the products.

phosphate group. The molecule $HO-\overset{\displaystyle OH}{\underset{\displaystyle OH}{\overset{|}{\underset{|}{P}}}}=O$ or

the radical $-O-\overset{\displaystyle OH}{\underset{\displaystyle OH}{\overset{|}{\underset{|}{P}}}}=O$ that, under physiological conditions, is dissociated to give the corresponding anionic forms. *See also* phosphoryl group.

phosphate-group transfer. A reaction in which a phosphate group is transferred from one compound to another, as in the reactions involving high- and low-energy phosphate compounds; in actuality, a phosphoryl group, rather than a phosphate group, is transferred. *Aka* phosphoryl-group transfer.

phosphatemia. HYPERPHOSPHATEMIA.

phosphate potential. The concentration of ATP in a system divided by the product of the concentrations of ADP and inorganic phosphate; the term [ATP]/[ADP][P_i]. *See also* energy charge.

phosphate transfer potential. PHOSPHORYL TRANSFER POTENTIAL.

phosphatidal choline. A choline-containing plasmalogen.

phosphatidal ethanolamine. An ethanolamine-containing plasmalogen.

phosphatidal group. The parent structure of the plasmalogens; consists of a molecule of glycerol in which phosphoric acid is esterified to the first carbon, a fatty acid is esterified to the central carbon, and an α,β-unsaturated ether is linked to the third carbon of the glycerol.

phosphatidal serine. A serine-containing plasmalogen.

phosphatide. 1. PHOSPHOGLYCERIDE. 2. PHOSPHOLIPID.

phosphatidic acid. The parent compound of many phosphoglycerides; consists of a molecule of glycerol in which phosphoric acid is esterified to the first carbon and fatty acids are esterified to the remaining two carbons of the glycerol.

phosphatidyl choline. A major phosphoglyceride in higher plants and animals; consists of choline that is esterified to the phosphoric acid residue of phosphatidic acid. *Abbr* PC. *Aka* lecithin.

phosphatidyl ethanolamine. A major phosphoglyceride in higher plants and animals; consists of ethanolamine that is esterified to the phosphoric acid residue of phosphatidic acid. *Abbr* PE. *Aka* cephalin.

phosphatidyl glycerol. A condensation product of phosphatidic acid and glycerol. *Abbr* PG.

phosphatidyl group. The group derived from phosphatidic acid by removal of a hydrogen from the phosphate group.

phosphatidyl serine. A phosphoglyceride consisting of serine that is esterified to the phosphoric acid residue of phosphatidic acid. *Abbr* PS. *Aka* cephalin.

phosphatidyl sugar. GLYCOPHOSPHO-GLYCERIDE.

3′-phosphoadenosine-5′-phosphosulfate. *See* active sulfate (1).

phosphoarginine. A phosphagen that is present in many invertebrates.

phosphocozymase. NICOTINAMIDE ADENINE DINUCLEOTIDE PHOSPHATE.

phosphocreatine. A phosphagen occurring in the muscle of many vertebrates. *Abbr* PC.

phosphodiester. A compound consisting of two alcohols that are esterified to a molecule of phosphoric acid; a phosphoric acid molecule that is esterified twice.

phosphodiesterase. An enzyme that catalyzes the hydrolysis of a doubly esterified phosphoric acid molecule, as that occurring in oligo- and polynucleotides; phosphodiesterases may be of either the endo- or the exonuclease type.

phosphodiester bond. A linkage between two molecules by means of phosphoric acid to which each of the molecules is esterified once.

3′,5′-phosphodiester bond. The bond by which nucleotides are linked in both DNA and RNA; formed by esterification of the phosphoric acid residue, which is already esterified to the 5′-position of the sugar of one nucleotide, to the 3′-position of the sugar of an adjacent nucleotide. *Aka* 3′,5′-phosphodiester link.

phosphoenolpyruvate carboxylase. The enzyme that catalyzes the anaplerotic reaction whereby phosphoenolpyruvic acid is carboxylated to oxaloacetic acid.

phosphoenolpyruvic acid. A high-energy compound, the dephosphorylation of which to pyruvic acid leads to the synthesis of ATP from ADP in the second stage of glycolysis. *Abbr* PEP.

phosphogluconate oxidative pathway. HEXOSE MONOPHOSPHATE SHUNT.

phosphogluconate pathway. HEXOSE MONO-PHOSPHATE SHUNT.

phosphoglyceric acid. A phosphate ester of glyceric acid, various forms of which are intermediates in glycolysis. *Abbr* PGA. *See also* 1,3-diphosphoglyceric acid; *sn*-glycero-3-phosphoric acid.

phosphoglyceride. A phospholipid derived from phosphoglyceric acid and containing at least one *O*-acyl or related group.

phosphoguanidine. A high-energy compound, such as phosphocreatine or phosphoarginine, that contains a phosphorylated guanido group.

phosphoinositide. 1. A lipid that contains groups derived from inositol and phosphoric acid. 2. A phosphoglyceride that contains inositol.

phospholipase. An enzyme that catalyzes the hydrolysis of fatty acids or other groups from phosphoglycerides. Phospholipase *A* catalyzes the hydrolysis of the fatty acid from position 2; phospholipase *B* catalyzes the hydrolysis of the fatty acid from position 1; phospholipase *C* catalyzes the cleavage of the bond between the phosphate group and glycerol; and phospholipase *D* catalyzes the hydrolysis of phosphoglycerides to yield phosphatidic acid.

phospholipid. A lipid that contains one or more phosphate groups, particularly a lipid derived from either glycerol or sphingosine. Phospholipids are polar lipids that are of great importance for the structure and functioning of biological membranes. *Abbr* PL.

phospholipoprotein. A conjugated protein that contains phospholipid and that is soluble in aqueous solutions.

phosphomonoesterase. An enzyme that catalyzes the hydrolysis of a once-esterified phosphoric acid (phosphomonoester). The enzyme is called specific if it catalyzes the hydrolysis of a small number of phosphomonoesters, and it is called nonspecific if it catalyzes the hydrolysis of a large number of phosphomonoesters at similar rates.

phosphonolipid. A lipid that contains a carbon-to-phosphorus bond; a lipid containing the phosphoryl group attached to a carbon atom

4′-phosphopantetheine. The prosthetic group of the acyl carrier protein.

phosphoprotein. A conjugated protein in which the nonprotein portion is a residue of phosphoric acid.

phosphor. FLUOR.

phosphorescence. The emission of radiation by an excited molecule in which the excited molecule first undergoes an electronic transition to a long-lived excited state and then slowly returns from that state to the ground state, dissipating the excitation energy by the emission of radiation at the same time. The emitted radiation is of a different wavelength than that of the exciting radiation, and the time interval between excitation and emission is usually several seconds or longer. *See also* fluorescence.

α-5-phosphoribosyl-1-pyrophosphate. A compound that is formed by the transfer of a pyrophosphate group from ATP to ribose-5-phosphate and that serves as a key intermediate in the biosynthesis of both purine and pyrimidine nucleotides. *Abbr* PRPP.

phosphorimetry. The measurement of phosphorescence.

phosphoroclastic reaction. A cleavage reaction by means of inorganic phosphate, as the cleavage of pyruvate by inorganic phosphate to acetyl phosphate, carbon dioxide, and hydrogen.

phosphorolysis. The cleavage of a covalent bond of an acid derivative by reaction with phosphoric acid H_3PO_4, so that one of the products combines with the H of the phosphoric acid and the other product combines with the H_2PO_4 group of the phosphoric acid.

phosphorus. An element that is essential to all plants and animals. Symbol P; atomic number 15; atomic weight 30.9738; oxidation states -3, $+3$, $+5$; most abundant isotope P^{31}; a radioactive isotope P^{32}, half-life 14.3 days, radiation emitted—beta particles.

phosphorylase. An enzyme that catalyzes a phosphorolysis reaction.

phosphorylase a. The active form of glycogen phosphorylase of skeletal muscle; consists of four subunits, each of which contains a phosphoserine residue that is essential for activity.

phosphorylase b. The less active form of glycogen phosphorylase of skeletal muscle; consists of two subunits and represents one half of the phosphorylase *a* molecule, minus two phosphate groups.

phosphorylase kinase. The enzyme that catalyzes the phosphorylation of phosphorylase *b* by ATP and thereby leads to its conversion to phosphorylase *a*.

phosphorylase phosphatase. The enzyme that catalyzes the hydrolysis of a phosphate group from phosphorylase *a*, and thereby leads to its conversion to phosphorylase *b*.

phosphorylation. The introduction of a phosphate group into a compound through the formation of an ester bond between the compound and phosphoric acid; more precisely referred to as the introduction of a phosphoryl group.

phosphoryl group. The radical $-P{=}O$ with OH groups, which, under physiological conditions, is dissociated to give the corresponding anionic forms.

phosphoryl-group carrier. The ADP-ATP system that serves as an intermediate for phosphoryl-group transfer between high- and low-energy phosphate compounds.

phosphoryl-group transfer. A reaction in which a phosphoryl group is transferred from one compound to another, as in the reactions involving high- and low-energy phosphate compounds. *Aka* phosphate-group transfer.

phosphoryl transfer potential. The group transfer potential for the phosphoryl group.

phosphosphingolipid. A phospholipid derived from sphingosine.

phosvitin. A phosphoglycoprotein in hen's egg yolk. *Aka* phosphovitin.

phot. A unit of illumination that is equal to 1 lm/cm^2 of surface.

photo-. Combining form meaning light.

photoaffinity labeling. Affinity labeling in which a chemical labeling reagent $R{-}P$ is used such that R can bind specifically and reversibly to an active site, and P is a chemical group that is unreactive in the dark. Upon photolysis, R—P is converted to a highly reactive intermediate R—P^*, which can then form a covalent bond with a group at the active site before R—P^* dissociates from the site. In true photoaffinity labeling, the rate of formation of this covalent bond is much greater than the rate at which R—P^* dissociates from the site; the reverse is the case for pseudophotoaffinity labeling. The latter is essentially identical to ordinary affinity labeling, with the exception that the labeling agent R—P^* is produced by photolysis.

photoautotroph. A phototrophic autotroph.

photocell. A device, the electrical properties of which, such as voltage or resistance, are altered in response to changes in the intensity of light that impinges upon it.

photochemical. Of, or pertaining to, photochemistry.

photochemical action spectrum. A plot of photochemical efficiency, such as photosynthetic efficiency, as a function of the wavelength of the incident light.

photochemical effect. The initiation of a chemical reaction by the absorption of light.

photochemical reaction. A chemical reaction initiated by the absorption of light.

photochemical sensitizer. PHOTOSENSITIZER.

photochemistry. The area of chemistry that deals with the interaction of radiant energy and chemical processes. *See also* first law of photochemistry; second law of photochemistry.

photochromism. The reversible change of color by a compound upon excitation by either ultraviolet or infrared radiation.

photocoupler. A photoreceptor that, upon stimulation by light, initiates a reaction that is driven by energy derived from the absorbed light energy. *See also* photosensor.

photodisintegration. A nuclear reaction in which an atomic nucleus absorbs a high energy photon and ejects a neutron, a proton, or an alpha particle.

photodynamic action. The oxidation of biologically important molecules in the presence of molecular oxygen, a photodynamic substance, and visible light. *See also* photodynamic substance; photosensitization.

photodynamic dye. A pigment that can serve as a photodynamic substance.

photodynamic inactivation. The inactivation of a molecule, a virus, or a cell by photodynamic action.

photodynamic substance. A substance, frequently a pigment, that sensitizes a biologically important molecule toward oxidation and achieves this by absorbing light energy and transferring this energy by means of various mechanisms to the target molecule. *See also* photosensitization.

photoelectric cell. PHOTOCELL.

photoelectric effect. The ejection of an orbital electron from an atom as a result of the impingement on the atom of a photon of sufficient energy; all of the energy of the photon is used to eject the electron and to impart kinetic energy to it.

photoelectron. The electron ejected from an atom in a photoelectric effect.

photographic rotation technique. A technique for determining the symmetry of a structure from its photograph; used for the determination of the symmetry of a virus from its electron micrograph. The micrograph is printed n times and the printing paper is rotated by $360°/n$ between successive exposures. A structure that has an n-fold radial symmetry will yield a photograph in which the details have been reinforced, while this is not the case for a structure that has an $n - 1$ or $n + 1$ fold radial symmetry.

photoheterotroph. A phototrophic heterotroph.

photoinhibition. The inhibition of photosynthesis by light.

photoisomerization. An isomerization brought about by the absorption of light.

photolithotroph. An organism or a cell that utilizes primarily light as its source of energy, inorganic compounds as electron donors, and carbon dioxide as its source of carbon atoms.

photolysis. The fragmentation of a molecule into smaller parts by irradiation with light; the dissociation of water in photosynthesis is an example.

photolytic. Of, or pertaining to, photolysis.

photometer. 1. An instrument for the measurement of light scattered at different angles by means of a photomultiplier tube that can be rotated around the sample cell. 2. An instrument for the direct measurement of light intensities; two basic types are the filter photometer and the spectrophotometer.

photometry. The measurement of light intensity by means of a photometer; applicable to ultraviolet, visible, and infrared radiations.

photomicrograph. A photograph taken through a light microscope.

photomicrography. The methodology for obtaining photomicrographs.

photomultiplier tube. An electronic tube that amplifies the beam of electrons released by the incident radiation and that is used in high quality spectrophotometers.

photon. A corpuscular unit of light; a quantum of light energy that is equal to $h\nu$, where h is Planck's constant (6.625×10^{-27} erg-sec) and ν is the frequency of the light in cycles per second.

photoneutron. A neutron ejected from an atomic nucleus by photodisintegration.

photon fluence. The number of photons that cross a unit area; the photon fluence rate refers to the number of photons that cross a unit area per unit time.

photonuclear reaction. PHOTODISINTEGRATION.

photoorganotroph. An organism or a cell that utilizes primarily light as its source of energy, organic compounds as electron donors, and organic compounds as well as carbon dioxide as its source of carbon atoms.

photooxidation. An oxidation reaction that is caused by light.

photoperiodism. The periodicity in the response of an organism that results from changes in either light intensity or the length of days.

photophobia. Lack of tolerance for light; unusual sensitiveness to light that is characteristic of conditions which result from vitamin A deficiency.

photophosphorylation. The synthesis of ATP that is coupled to the operation of an electron transport system in photosynthesis.

photopic vision. Vision in bright light in which the cones of the retina function as light receptors.

photoprotection. The protection of cells against the damaging effects of ultraviolet irradiation by prior exposure of the cells to light in the wavelength range of 310 to 370 nm; the effect may be due to an inhibition of cellular growth which allows more time for the repair of damaged DNA. *Abbr* PP.

photoreaction. PHOTOCHEMICAL REACTION.

photoreactivating enzyme. The enzyme that catalyzes the photoreactivation reaction whereby thymine dimers are excised from the DNA in which they were formed as a result of irradiation with ultraviolet light.

photoreactivation. The recovery of cells from the damage caused by irradiation with ultraviolet light that is brought about by a photoreactivating enzyme when the damaged cells are exposed to visible light. *Abbr* PHR.

photoreceptor. A receptor that is stimulated by light; a photoreceptor may be either a photosensor or a photocoupler.

photorecovery. PHOTOREACTIVATION.

photoreduction. A reduction reaction that is caused by light.

photorespiration. The respiration of plant cells that occurs in the presence of light and while the cells are concurrently carrying on photosynthesis. Photorespiration utilizes reducing power generated by photosynthesis for the reduction of molecular oxygen; it does not involve mitochondria and does not yield ATP.

photosensitive. Sensitive to light; capable of being stimulated by light.

photosensitization. The process whereby a substance, frequently a dye, sensitizes a biologically important molecule toward oxidation and achieves this by absorbing light energy and transferring this energy by means of various mechanisms to the target molecule. The oxidation of the target molecule may then occur either in the presence of molecular oxygen, resulting in a photodynamic action, or in the absence of oxygen, but in the presence of appropriate electron and/or hydrogen acceptors, resulting in a dye-sensitized photooxidation.

photosensitizer. 1. A substance which, when added to a biological system, will increase the damage to the system when it is exposed to a subsequent dose of radiation. 2. A substance that brings about photosensitization. 3. PHOTODYNAMIC SUBSTANCE.

photosensor. A photoreceptor which, upon stimulation by light, initiates a reaction that is driven by energy derived from sources other than the absorbed light energy. *See also* photocoupler.

photosynthate. A product obtained as a result of photosynthesis.

photosynthesis. The reaction whereby solar energy is captured by an organism and converted to chemical energy and which, in its most general form, can be written as: $H_2D + A$ $\xrightarrow{\text{light}} H_2A + D$, where D is an electron donor and A is an electron acceptor. Photosynthesis is carried out by a large number of organisms, both procaryotic and eucaryotic, including plants, algae, and bacteria, and involves a variety of electron donors and electron acceptors. In green plants, photosynthesis takes place in chloroplasts and leads to the synthesis of carbohydrates from water and carbon dioxide and to the evolution of oxygen; it occurs in the presence of light and several chlorophyll pigments that are assembled in two photosystems (I and II) which are connected with electron transport systems.

photosynthetic. 1. Of, or pertaining to, photosynthesis. 2. PHOTOTROPHIC.

photosynthetic electron transport. The flow of electrons through a chain of electron carriers that is induced by the light reaction of photosynthesis.

photosynthetic organism. *See* photolithotroph; photoorganotroph.

photosynthetic phosphorylation. PHOTOPHOSPHORYLATION.

photosynthetic quotient. A measure of the photosynthetic activity of a system that is equal to the number of moles of oxygen evolved divided by the number of moles of carbon dioxide taken up.

photosynthetic unit. The number of chlorophyll molecules that are required for the fixation of one molecule of carbon dioxide in photosynthesis.

photosystem. A photosynthetic pigment or group of pigments that absorb light energy and that participate in the light reaction of photosynthesis. The photosynthetic reactions of chloroplasts are carried out by two such systems, designated I and II.

photosystem I. The photosystem of chloroplasts that is based on the P_{700} pigment, requires light of longer wavelength and is associated with the reduction of $NADP^+$ and with photophosphorylation.

photosystem II. The photosystem of chloroplasts that is based on chlorophyll *a*, a second chlorophyll, and accessory pigments; it requires light of shorter wavelength and is associated with the dissociation of water and the evolution of oxygen.

phototaxis. A taxis in which the stimulus is light.

phototroph. A cell or an organism that uses light as a source of energy.

phototropism. A tropism in which light is the stimulus.

photovoltaic cell. An electrical cell that operates on the principle that when light strikes certain metals or semiconductors, the flow of electrons produced is proportional to the intensity of the light; used in photometers and in some spectrophotometers.

pH paper. Paper that is impregnated with indicator dyes and that is used for the approximate measurement of the pH values of solutions.

PHR. Photoreactivation.

pH scale. The range of pH values from 0 to 14 in which the value of 7 is that for pure water and the values of 0 and 14 represent approximately the hydrogen ion concentrations of 1.0 and 1.0×10^{-14} eq/l, respectively. *See also* pH.

pH-stat. An instrument for maintaining a constant pH during the course of a chemical reaction. The acidic or basic groups released as the reaction proceeds signal the addition of titrant (base or acid) to the reaction mixture. The amount of titrant added as a function of time provides an assay of the kinetics of the reaction.

pH unit. A change of 1.0 between two pH values on the pH scale.

phycobilin. A linear tetrapyrrole derivative that occurs in conjugation with proteins and that functions in the form of a phycobiliprotein as an accessory pigment of photosynthesis in algal chloroplasts.

phycobiliprotein. A conjugated protein that functions as an accessory pigment of photosynthesis in algal chloroplasts. Phycobiliproteins occur in phycobilisomes and include the phycocyanins, phycoerythrins, and allophycocyanins.

phycobilisome. A granule in algal cells that serves to harvest light energy and to transfer it to chlorophyll. Phycobilisomes contain the phycobiliprotein pigments: phycocyanin, allophycocyanin, and frequently phycoerythrin.

phycocyanin. A blue accessory pigment of algal chloroplasts that consists of a protein conjugated to a phycobilin.

phycoerythrin. A red accessory pigment of algal chloroplasts that consists of a protein conjugated to a phycobilin.

phylloquinone. Vitamin K_1.

phylogenesis. PHYLOGENY.

phylogenetic. Of, or pertaining to, phylogeny.

phylogenetic tree. A diagrammatic representation of the development of species which indicates their interrelationships and the times of their evolutionary divergence. Phylogenetic trees have been constructed on the basis of amino acid sequences in selected proteins, such as cytochrome *c*.

phylogeny. The evolutionary development of a group of organisms, such as a species. *Aka* phylogenesis.

physical adsorption. Adsorption that is brought about by physical forces, such as Van der Waals forces.

physical biochemistry. A branch of biochemistry that deals with the transformations of physical and chemical energies in biological systems, particularly as they relate to macromolecules.

physical half life. *See* half life (1).

physicochemical. Physical-chemical.

physics. The science that deals with matter and energy, their interactions, and their changes.

physiological. Of, or pertaining to, physiology.

physiological chemistry. Biochemistry, particularly that of higher animal organisms.

physiological conditions. The normal conditions pertaining to an organism such as, in the case of man, a temperature of 37° and a pH of about 7.0.

physiological saline. A 0.9% (w/v) solution of sodium chloride that is approximately isotonic to the blood and lymph of mammals and that is used for the temporary maintenance of living cells and tissues.

physiology. The science that deals with the processes and the activities of living organisms or parts of organisms.

phyto-. Combining form meaning plant.

phytoagglutinin. A lectin that is derived from plants and that agglutinates cells.

phytochemistry. The science that deals with the chemistry of plant materials.

phytochrome. A chromophore-protein complex that occurs in plants and that is associated with photoperiodism.

phytoestrogen. A compound, such as a specific flavonoid, that is derived from plants and that has estrogenic activity even though it is not a steroid.

phytohemagglutinin. A lectin that is derived from plants and that agglutinates red blood cells.

phytohormone. PLANT HORMONE.

phytol. A long chain alcohol that occurs in chlorophyll.

phytology. Botany.

phytopathology. The science that deals with the origin, the nature, and the course of plant diseases.

phytosterol. A sterol that occurs in plants.

phytotoxin. 1. PLANT TOXIN. 2. A proteinaceous plant toxin; a toxalbumin.

phytylmenaquinone. Vitamin K_1. *Aka* phylloquinone; phytonadione; phytylmenadione.

P_i. Inorganic phosphate.

pi. 1. A mathematical constant that is equal to 3.1416 . . . and that expresses the ratio of the circumference of a circle to its diameter. 2. Osmotic pressure. *Sym* π.

pI. Isoelectric pH.

pi bond. A chemical bond formed by electrons that are in pi orbitals.

pico-. Combining form meaning 10^{-12} and used with metric units of measurement. *Sym* p.

picornavirus. A small, naked, icosahedral animal virus that contains single-stranded RNA; the poliovirus belongs to this group.

4-pi counter. A geometrical arrangement for the standardization of beta radiation sources; produced by placing two windowless Geiger-Mueller detectors face-to-face and suspending the radiation source between them.

pi electron delocalization. RESONANCE.

piericidin A. An antibiotic that resembles coenzyme Q in its structure and inhibits the electron transport system between NADH dehydrogenase and cytochrome *b*.

PIF. *See* prolactin regulatory hormone.

pigment. 1. A naturally-occurring coloring matter in an animal, a plant, or a microorganism; a biochrome. 2. A synthetic coloring matter; a dye.

PIH. *See* prolactin regulatory hormone.

pilus (*pl* pili). A small filamentous projection attached to the surface of a bacterium.

pinealectomy. The surgical removal of the pineal gland.

pineal gland. A small endocrine gland in the brain.

ping pong mechanism. The mechanism of an enzymatic reaction in which two or more substrates and two or more products participate, and the enzyme shuttles back and forth between its original and a modified form. According to this mechanism, after the binding of the first substrate by the enzyme, a product is released and the enzyme is converted to a modified form. The second substrate then binds to the modified form of the enzyme, and this is followed by the release of a second product and the regeneration of the original form of the enzyme.

pinocytosis. The taking up of droplets of liquid by a cell.

pinosome. A cytoplasmic vesicle that is formed in endocytosis by invagination of the cell membrane and that contains liquids taken up by the cell through pinocytosis.

pi orbital. A molecular orbital that is a delocalized bond orbital, spread over two or more atoms, or over an entire molecule.

PIPES. Piperazine-N,N'-bis(2-ethanesulfonic acid); used for the preparation of buffers in the pH range of 6.1 to 7.5.

pipet. A graduated open tube, usually made of glass, and used for measuring and transferring small and definite volumes of liquids. *Var sp* pipette.

pi pi-star transition. The excitation of an electron from a pi orbital to a pi-star orbital; such transitions are responsible for the most intense absorption bands of molecular spectra.

Pirani gauge. A thermal conductivity vacuum gauge.

piscine. Of, or pertaining to, fish.

pi-star orbital. *See* antibonding orbital.

pitch. The distance between two identical points along the axis of a helix; equal to the number of residues per turn of the helix, multiplied by the distance per residue along the axis of the helix.

pitocin. OXYTOCIN.

Pitressin. Trade name for vasopressin.

pituitary basophilism. A tumor of the pituitary gland that is sometimes associated with Cushing's disease.

pituitary dwarfism. Dwarfism that is caused by a deficiency in the secretion of growth hormone by the pituitary gland.

pituitary gland. An endocrine gland, located below the brain, that regulates a large portion of the endocrine activity of vertebrates. The gland consists of an anterior lobe called adenohypophysis and a posterior lobe called neurohypophysis. *See also* adenohypophysis; neurohypophysis.

Pituitrin. Trade name for hypophysin.

pK. The negative logarithm, to the base 10, of an equilibrium constant based on activities.

pK′. The negative logarithm, to the base 10, of

an apparent equilibrium constant based on molar concentrations.

pK$_a$. The negative logarithm, to the base 10, of an acid dissociation constant.

pkat. Picokatal.

pK$_b$. The negative logarithm, to the base 10, of a base dissociation constant.

P-K reaction. Prausnitz-Kuestner reaction.

PKU. Phenylketonuria.

pK$_w$. The negative logarithm, to the base 10, of the ion product of water.

PL. 1. Phospholipid. 2. Pyridoxal. 3. Placental lactogen.

placebo. 1. An inactive substance that is identical in appearance to a biologically active one and that is given to a number of individuals out of a group while the remainder of the individuals receive the biologically active substance. The individuals receiving the placebo thus serve as controls which permit an evaluation of the effectiveness of the biologically active substance given to the other individuals. 2. An inert medication given to an individual for its suggestive and psychological effect.

placenta. The structure by which the fetus is attached to the uterus and through which it exchanges materials with the maternal circulation, receiving nutrients and excreting waste products.

placental barrier. A semipermeable membrane that restricts the type and quantity of material exchange between the fetus and the mother and that represents a partial block to the passage of antibodies from the mother to the fetus.

plain dispersion. SIMPLE DISPERSION.

planchet. A thin disk, commonly of metal, used for the deposition and counting of radioactively labeled material.

planchet counter. A radiation counter for radioactive samples deposited in planchets.

Planck's constant. A universal constant that relates the energy of a photon to its frequency; equal to 6.625×10^{-27} erg-sec. *Sym* h.

Planck's law. The law that the energy of a photon is equal to the frequency of the radiation multiplied by Planck's constant.

plane of symmetry. An imaginary plane that divides a symmetrical body into two mirror image halves.

plane-polarized light. Light in which the electric field vectors oscillate in a plane that passes through the axis along which the light is being propagated.

planetesimals. Small bodies of matter, believed to have been formed from primordial dust and gas, and to have consolidated subsequently to form the terrestrial planets and the asteroids.

planimeter. A device for measuring the area under a curve; a graphical integrator that is used in chromatography and electrophoresis for estimating the relative amounts of separated components.

plant agglutinin. A lectin extracted from plants. *See also* lectin.

plant bile pigment. PHYCOBILIN.

plant hormone. *See* hormone.

plant lectin. *See* lectin.

plant pigment. A pigment of plant origin. *See also* carotenoid; chlorophyll; flavonoid.

plant sulfolipid. A sulfonic acid derivative of a glycosyldiacylglycerol that occurs in plants.

plant toxin. A toxin of plant origin.

plant-type ferredoxin. IRON-SULFUR PROTEIN (b).

plant virus. A virus that infects plants and multiplies in them.

plaque. 1. A clear region in a culture plate that represents an area of cell lysis, devoid of intact cells. *See also* plaque assay; plaque technique. 2. An atheromatous deposit.

plaque assay. An assay for counting the number of infectious bacterial or animal viruses. The host cells are mixed with the virus in a gel, and the virus particles diffuse through the gel, infect, and lyse the host cells. The viral progeny thus produced in turn infect and lyse adjacent host cells, resulting in the formation of a plaque or clear region that is devoid of intact cells at each site of infection by a viral particle.

plaque-forming cell. *See* plaque assay; plaque technique.

plaque technique. A technique, devised by Jerne, for counting antibody-producing lymphocytes. Lymphocytes from animals that have been immunized with red blood cells are mixed with some of the same red blood cells in a gel. The red blood cells become bound to the lymphocytes by means of the antibodies synthesized by the lymphocytes; the addition of complement then causes the complex to lyse and leads to the formation of a plaque around each antibody-producing cell.

plaque-type mutant. A phage mutant that gives rise to a plaque of changed morphology.

plasma. The fluid obtained from blood by removal of the formed elements by means of centrifugation; serum plus fibrinogen.

plasma cell. A lymphocyte that is capable of synthesizing antibodies.

plasma clearance. CLEARANCE.

plasmacyte. PLASMA CELL.

plasmagene. The smallest heritable unit in the plasmon.

plasmalemma. CELL MEMBRANE.

plasmalogen. A phosphoglyceride that contains a phosphatidal group, such as phosphatidal choline, phosphatidal serine, or phosphatidal ethanolamine; plasmalogens are abundant in the membranes of muscle and nerve cells. *See also* phosphatidal group.

plasma membrane. CELL MEMBRANE.

plasmapheresis. A technique for decreasing the concentration of the plasma proteins of an animal; achieved by bleeding the animal repeatedly, collecting the blood cells, and reinjecting the blood cells, suspended in saline, into the animal.

plasma protein. One of a group of proteins present in blood plasma.

plasma-specific enzyme. An enzyme that is present in blood plasma and that has a specific function in plasma; an enzyme that functions in the reactions of blood clotting is an example.

plasma thromboplastic factor. ANTIHEMO-PHILIC FACTOR.

plasma thromboplastic factor B. CHRISTMAS FACTOR.

plasma thromboplastin antecedent. The factor that is activated by the Hageman factor in the intrinsic system of blood clotting. *Abbr* PTA.

plasma thromboplastin component. CHRISTMAS FACTOR.

plasmid. Any extrachromosomal self-replicating genetic element.

plasmid curing. The release of an episome, such as the fertility factor, from the bacterial chromosome by treatment with intercalating acridine dyes. *See also* cure.

plasmin. The proteolytic enzyme that catalyzes the hydrolysis of fibrin and thereby leads to the dissolution of intravascular blood clots.

plasminogen. The inactive precursor of plasmin.

plasmogeny. The artifical production of microscopic structures, the properties of which bear some resemblance to those of living cells.

plasmolysis. The shrinking of cellular protoplasm that occurs when a cell is placed in a hypertonic solution so that water moves out of the cell.

plasmon. A collective term for the total extrachromosomal hereditary complement of a cell.

plasmoptysis. The swelling and rupturing of a cell and the escape of its protoplasm that occur when a cell is placed in a hypotonic solution so that water moves into the cell.

plasmosome. NUCLEOLUS.

plastic chlorophyll. A chlorophyll molecule that undergoes somewhat different photochemical reactions from the normal ones because it differs from ordinary chlorophyll in its conformation, its packing within the chloroplast, or its environment.

plastid. A DNA-containing self-replicating subcellular organelle; used particularly in reference to those organelles present in higher plants, some of which function in photosynthesis while others serve as storage vessels. Plastids that contain pigments are known as chromoplasts and those devoid of pigments are known as leucoplasts.

plastocyanin. A copper-containing protein that serves as an electron carrier in chloroplast photosynthesis.

plastogene. The plasmagene of a plastid.

plastoquinone. A compound that is closely related to coenzyme Q and that functions as a hydrogen donor and acceptor in the photosynthetic electron transport system. *Abbr* PQ.

plate. 1. PETRI PLATE. 2. The tail plate of a T-even phage.

plateau. 1. A region in a solution in which the concentration remains uniform but changes with time. This region is below the boundary in sedimentation, above the boundary in flotation, and between boundaries in moving boundary electrophoresis. A plateau is formed in these cases if the initial concentration was uniform throughout the solution. 2. That portion of the characteristic curve of a radiation detector in which the count rate is almost independent of the applied voltage.

plateaued rat. A rat that has a slow rate of growth and that is used in the assay of growth hormone.

plate count. A viable count of bacteria that is based on the number of colonies that develop on a solid nutrient medium when appropriately diluted aliquots of the original culture are plated out.

platelet. A small, irregularly shaped disk that is present in the blood and that functions in blood clotting by releasing thromboplastin.

platelet cofactor I. ANTIHEMOPHILIC FAC-TOR.

platelet cofactor II. CHRISTMAS FACTOR.

plate theory. The application of the theoretical plate concept, derived from countercurrent distribution, to chromatography and particularly to gas chromatography. *See also* theoretical plate.

plating. The cultivation of microorganisms on a solid nutrient medium in a petri dish.

plating efficiency. *See* absolute plating efficiency; relative plating efficiency.

pleated sheet. A configuration of protein molecules in which the polypeptide chains are partially extended. The chains are held together by means of interchain hydrogen bonds between the *CO* and *NH* groups of all the peptide bonds. The pleated sheet structure occurs predominantly in fibrous proteins; the pleated sheet is referred to as being parallel or antiparallel, depending on whether the polypeptide chains are parallel or antiparallel.

plectonemic coiling. The coiling of two threads in the same direction so that they cannot be separated except by uncoiling the threads; applies to the two strands in double helical DNA. *See also* paranemic coiling.

pleiotropism. The production of multiple, and apparently unrelated, phenotypic effects by a single gene. *Aka* pleiotropy.

pleiotypic control. The regulation of the reactions of the pleiotypic program; believed to be based on a coordinated influence by cyclic AMP. Malignant or transformed cells are viewed as being defective in their pleiotypic control.

pleiotypic program. The mechanistically unrelated parameters of membrane transport and the overall rates of protein synthesis, RNA synthesis, and protein degradation which, in normal cells, fluctuate coordinately with changes in the rate of cell growth.

pleomorphism. The occurrence of two or more forms, such as the different forms of an organism during its life cycle. *See also* doctrine of pleomorphism.

pleromer. A component that can replace another component in a polymer with respect to the overall "balance" of components in that polymer. Thus in a DNA molecule in which the mole percent of guanine equals that of cytosine plus 5-methylcytosine, the components cytosine and 5-methylcytosine are considered to be pleromers.

plexiglass. Polymethylmethacrylate; a plastic.

PLI. Pulsed-laser interferometry.

PLK. Polylysine-kieselguhr.

-ploid. Combining form indicating the multiple of the chromosome set in the nucleus, as in diploid and polyploid.

ploidy. The degree of chromosome multiplicity; the chromosome state in which each chromosome is represented once, twice, etc. *See also* aneuploid state; euploid state; heteroploid state; polyploid state.

PLP. Pyridoxal phosphate.

plus strand. 1. The nucleic acid strand of a phage that contains single-stranded nucleic acid. 2. CRICK STRAND.

PM. 1. Pyridoxamine. 2. Puromycin.

PMF. Proton motive force.

PMP. Pyridoxamine phosphate.

PM-particle. A ribosomal subparticle isolated from bacterial cells in which protein synthesis was inhibited by puromycin.

PMR. Proton magnetic resonance.

PMS. 1. Phenazine methosulfate; a reducible dye that can serve as an electron acceptor. 2. Pregnant mare's serum; *see* pregnant mare's serum gonadotropin.

PMSG. Pregnant mare's serum gonadotropin.

PN. Pyridoxine.

PNA. Pentose nucleic acid.

pneometer. SPIROMETER.

PNP. 1. *p*-Nitrophenol. 2. Pyridoxine phosphate.

PNPP. *p*-Nitrophenylphosphate.

PNS. Peripheral nervous system.

P_{O_2}. 1. Oxygen tension. 2. Partial pressure of oxygen.

pocket ionization chamber. A small dosimeter designed to be worn by an individual and used for monitoring the amount of radiation to which the individual has been exposed. *Aka* pocket dosimeter.

pock method. A method for counting the number of infectious viral particles by counting the lesions produced in the choriallantoic membrane of chick embryos following the infection of the membrane with the viral particles.

-poiesis. Combining form meaning formation or production.

poikilocyte. A red blood cell of irregular shape.

poikilothermic. Descriptive of an organism whose temperature varies with the temperature of its environment. *Aka* cold-blooded.

point group. A symmetry class to which objects may belong by virtue of possessing elements of symmetry that pass through, or are arranged about, a single point which serves as the center of symmetry.

point mutation. A mutation in which there is a change in only one nucleotide of a nucleic acid. *See also* transition (1); transversion.

poise. 1. *n* A unit of dynamic viscosity; one hundredth of this unit is called the centipoise. 2. *v* To buffer an electrode potential. *See also* poising.

Poiseuille's law. An equation for the volume rate of flow dV/dt of a liquid through a capillary; specifically, $dV/dt = \pi a^4 P/8\eta l$, where π is a constant equal to $3.1416\ldots$, a is the radius of the capillary, P is the pressure, η is the viscosity of the liquid, and l is the length of the capillary.

poising. The resistance to change in electrode potential that is shown by an oxidation-reduction couple at and near the midpoint potential of the couple. At the midpoint potential the concentrations of the oxidant and the reductant are equal, and an oxidation-reduction couple acts as a potential buffer at and near this point, much as a weak acid and its conjugate base act as a pH buffer at and near the pK value where their two concentrations are equal.

poison. A substance that alters the normal metabolism of an organism, is injurious to health, and may be lethal when a small amount of it is either taken into, or comes in contact with, the organism.

Poisson distribution. A probability distribution

in which the variance equals the mean. If the total number of objects observed under certain conditions varies according to a Poisson distribution with a mean m, then the probability P_x of obtaining x objects is given by $P_x = e^{-m}m^x/x!$, where x is a whole number between zero and infinity, and e is the base of natural logarithms. *Aka* Poisson's law.

poky mutant. A slow-growing mutant of the mold *Neurospora crassa.*

polar. Possessing polarity; having a permanent dipole moment.

polar amino acid. An amino acid that has a polar side chain.

polar bond. A covalent bond in which the electron pair or pairs of the bond are held with unequal strength by the two bonded atoms.

polarimeter. An instrument for measuring optical rotation. The instrument contains two nicol prisms (polarizer and analyzer) and is generally operated using light from a sodium lamp; the light passes in succession through the polarizer, the solution of the compound being studied, and the analyzer. *Aka* polariscope.

polarity. 1. The property of having two poles, specifically in reference to a molecule in which the center of the positive charges does not coincide with the center of the negative charges; the degree to which a molecule is polar and possesses a permanent dipole moment. 2. The phenomenon in which a mutant gene leads to a decrease in the synthesis of proteins that are specified by genes which belong to the same operon as the mutant gene but are farther removed from the operator. 3. The sense in which a polynucleotide, or other biopolymer, is transversed, synthesized, or functioning; thus the two strands in double helical DNA are said to be antiparallel or to have opposite polarity. 4. The existence of two mating types in a unicellular organism due to either the presence or the absence of a fertility factor.

polarity gradient. The variation with distance along the operon of the effect of a polarity mutation in one gene on the expression of the remaining genes in the operon.

polarity mutation. POLAR MUTATION.

polarity ratio. The ratio of polar to nonpolar amino acid residues in a protein.

polarizability. A measure of the tendency of a substance to have dipoles induced in it when placed in an electric field; equal to the magnitude of the induced dipole moment per unit strength of the applied electric field.

polarization. 1. The state of charge separation, as that across a biological membrane, that results from the orientation and the distribution of ions and molecules. 2. The state of light, as that produced by passing light through certain substances, in which the electric and magnetic

field vectors of the light oscillate only in specific directions.

polarization curve. POLAROGRAPHIC WAVE.

polarization fluorescence. *See* fluorescence polarization.

polarization microscope. A microscope used for studying the anisotropic properties of objects and for visualizing objects by virtue of their anisotropic properties. *Aka* polarizing microscope.

polarized electrode. An electrode, the potential of which varies with the current passing through it.

polarized light. Light in which the electric and magnetic field vectors oscillate only in specific directions.

polarizer. The nicol prism in a polarimeter that is used for producing plane-polarized light. *See also* analyzer.

polar lipid. An amphipathic lipid.

polar mutation. A mutation in a gene that reduces the rate of synthesis of proteins which are coded by genes that belong to the same operon as the mutant gene but are farther removed from the operator.

polar-nonpolar. AMPHIPATHIC.

polarogram. The record of a polarographic wave, either in the form of a direct visual display or in the form of a plot.

polarograph. An instrument for conducting polarographic measurements.

polarographic wave. The variation of electrode current as a function of potential that is produced in polarography.

polarography. A method for electroanalytical studies of chemical substances, including the reduction of anions and cations, and for qualitative and quantitative microanalysis; based on measurements of the current produced at a microelectrode, such as at a dropping mercury electrode, as a function of the changing potential applied to an electric cell.

polar requirement. The slope of the line that is obtained by plotting the logarithm of the R_m value of an amino acid in a series of pyridine solvents as a function of the logarithm of the mole percent of water in the solvent. *See also* R_m value.

polar solvent. A solvent that contains charged groups and/or dipoles.

Polenske number. A measure of the volatile fatty acids in a fat that is equal to the number of milliliters of 0.1 N alkali which are required to neutralize the volatile water-insoluble fatty acids in 5 grams of fat. *Aka* Polenske value.

poliomyelitis. The disease caused by the poliovirus; infantile paralysis.

poliovirus. A virus that causes poliomyelitis and

that belongs to the enterovirus subgroup of picornaviruses.

poly-. 1. Combining form meaning many. 2. Combining form meaning excessive.

poly A. Polyadenylic acid.

polyacrylamide gel. A cross-linked acrylamide gel prepared from the monomer acrylamide and the cross-linking compound, N,N'-methylenebisacrylamide, in the presence of a polymerizing agent such as ultraviolet light.

polyacrylamide gel electrophoresis. A zone electrophoretic technique of high resolution in which a polyacrylamide gel is used as the supporting medium. *Abbr* PAGE.

polyaffinity theory. THREE POINT ATTACHMENT.

polyamine. A long chain aliphatic compound that contains multiple amino and/or imino groups. Polyamines are widely distributed in nature and include compounds such as spermine, spermidine, cadaverine, and putrescine. Polyamines affect ribosomes, DNA, RNA, and other biological components, and their action is frequently attributed to an electrostatic interaction between the polyamine cation and a negatively charged molecule.

polyampholyte. A polyelectrolyte that can function as either a proton donor or a proton acceptor.

polyanion. A molecule that possesses a large number of negative charges.

poly C. Polycytidylic acid.

polycarbonate. Polybisphenol-A-carbonate; a plastic.

polycation. A molecule that possesses a large number of positive charges.

polychlorinated biphenyl. One of a group of industrial chemicals that are used as lubricants, heat-exchange fluids, insulators, and as plasticizers in paints, synthetic resins, and plastics. Polychlorinated biphenyls are of ecological interest, since residues of these compounds are found in a wide variety of tissues in fish, wildlife, and man. Polychlorinated biphenyls have been shown to induce steroid hydroxylases, drug-metabolizing enzymes, and several cytochromes. *Abbr* PCB.

polycistronic messenger RNA. A messenger RNA molecule that serves as a template for the translation of two or more polypeptide chains which are specified by adjacent cistrons in the DNA. *Aka* polycistronic message.

polycythemia. A condition characterized by the presence of abnormally large numbers of circulating red blood cells.

polydeoxyribonucleotide. A linear polymer of more than 10 deoxyribonucleotides that are linked by means of $3',5'$-phosphodiester bonds; a polynucleotide.

polydipsia. Excessive thirst.

polydisperse. Consisting of macromolecules that fall into a large number of classes with respect to their size.

polyelectrolyte. A linear polymer in which each monomer carries one or more ionic groups so that the polymer is a polyvalent ion with the charges distributed all along the chain.

polyene antibiotic. An antifungal antibiotic that is characterized by its high content of alternating single and double bonds and that functions by damaging the cell membrane.

polyenoic fatty acid. A polyunsaturated fatty acid.

polyestrous. Having more than one estrous cycle per year.

polyetiological theory. A theory of cancer according to which cancer can be caused by a variety of chemical, physical, and biological events.

polyfunctional. *See* bifunctional.

poly G. Polyguanylic acid.

polygene. One of a group of genes that control a quantitative trait such as size, weight, or pigmentation, and that are believed to function together and to have a cumulative effect.

polyglucosan. GLUCAN.

polygon. A plane and closed figure that is bounded by many straight lines.

polyhead. A long, hollow cylinder, produced by some phage mutants, that has the diameter of a normal phage head but differs from it in its properties.

polyhedron. A solid that is bounded by many plane faces.

polyhydroxy. Containing two or more hydroxyl groups.

polyisoprene. A polymer of isoprene; the *cis*-isomer is natural rubber and the *trans*-isomer is gutta percha.

polykaryocyte. A multinucleated cell.

polyketide. ACETOGENIN.

polylysine-kieselguhr. An adsorbent that consists of polylysine bound to diatomaceous earth and that is used for column chromatographic fractionation of nucleic acids. *Abbr* PLK.

polymer. A high-molecular weight compound consisting of long chains that may be open, closed, linear, branched, or cross-linked. The chains are composed of repeating units, called monomers, which may be either identical or different.

polymerase. *See* DNA-dependent DNA polymerase; DNA-dependent RNA polymerase; RNA-dependent DNA polymerase; RNA-dependent RNA polymerase.

polymerization. The repetitive reactions whereby the repeating units of a polymer are linked together to form long chains; the formation of a polymer.

polymetaphosphate. LINEAR POLYPHOS-PHATE.

polymetaphosphate ethyl ester. A polyphosphate compound, the hydrolysis of which has been used to drive the polymerization of amino acids in studies on the origin of life.

polymorphic gene. A gene that exists in the form of several prevalent alleles.

polymorphism. 1. The occurrence of two or more forms, such as the different forms of a protein in individuals of the same species. 2. The occurrence of two or more genetically different individuals in the same breeding population.

polymyxin. A cyclic peptide antibiotic, produced by *Bacillus polymyxa,* that is surface-active and that damages the bacterial cell membrane.

polyneuritis. A disease of birds caused by a deficiency of thiamine.

polynuclear complex. A metal ion-ligand complex of the type —*M*—*L*—*M*—*L*—*M*— where the metal ions *M* are held together in chains by means of ligands *L*, each of which binds to two metal ions.

polynucleotidase. A polynucleotide phosphatase.

polynucleotide. A linear polymer of more than 10 nucleotides that are linked by means of 3′,5′-phosphodiester bonds. *See also* polydeoxyribonucleotide; polyribonucleotide.

polynucleotide ligase. DNA LIGASE.

polynucleotide phosphorylase. The enzyme that catalyzes the random polymerization of ribonucleoside diphosphates to polyribonucleotides, a reaction that is useful for the synthesis of synthetic messenger RNA molecules.

polyol. A polyhydroxy alcohol.

polyoma virus. A small, naked, icosahedral, oncogenic virus that contains double-stranded DNA and belongs to the group of papovaviruses; produces tumors in rodents.

polypeptide. A linear polymer of more than 10 amino acids that are linked by means of peptide bonds.

polyploid state. The chromosome state in which each type of chromosome is represented more than twice. *Aka* polyploidy.

polyribonucleotide. A linear polymer of more than 10 ribonucleotides that are linked by means of 3′,5′-phosphodiester bonds; a polynucleotide.

polyribosome. POLYSOME.

polysaccharide. A linear or branched polymer of more than 10 monosaccharides that are linked by means of glycosidic bonds.

polysaccharide phosphorylase. *See* glycogen phosphorylase; starch phosphorylase.

polysheath. A long phage sheath that is produced by some phage mutants in the absence of a phage tube.

polysome. A strand of messenger RNA with two or more ribosomes attached to it.

polysome profile. The tracing that shows the types and the relative amounts of different polysomes in a sample, and that is obtained by monitoring the sample material after it has been fractionated by density gradient centrifugation.

poly T. Polythymidylic acid.

polytailtube. A long fiber, produced by some phage mutants, that has the diameter of a normal phage tube.

polytene chromosome. An exceptionally large chromosome that has a band-like appearance and that contains numerous strands of DNA attached side-by-side in the form of a giant cable.

poly U. Polyuridylic acid.

polyunsaturated. Highly unsaturated; containing many double and/or triple bonds between carbon atoms.

polyuria. The excretion of excessive amounts of urine.

polyvalent. 1. Having a high valence. 2. Having more than one valence.

polyvalent allosteric inhibition. *See* multivalent allosteric inhibition.

polyvalent antiserum. An antiserum that contains antibodies against many different kinds of antigens.

polyvalent vaccine. A vaccine that contains antigens derived from two or more different types of bacteria or viruses.

polywater. A liquid that has a density of about one and a half times that of ordinary water and is prepared by condensation of water in fine capillaries; originally described as a new, stable form of water, but now known to be an ordinary aqueous solution containing substances dissolved from the capillaries by the condensing vapor.

Pompe's disease. A genetically inherited metabolic defect in man that is due to a deficiency of the enzyme amylo-α-1,4-glucosidase; glycogen storage disease type II.

pontal atom. BRIDGING ATOM.

pool. The total amount of a substance, or a group of similar substances in equilibrium with each other, that is not covalently bound and that is available for, and participates in, the anabolic and the catabolic reactions of the steady state; may refer to substances in a cell, an organ, a tissue, or an organism.

POPOP. 1,4-*bis*-2-(5-Phenyloxazolyl)-benzene; a secondary fluor.

population. A collection of organisms, cells, or molecules that have some quality or characteristic in common; generally refers to a large collection contained within a particular space.

por. The bare 16-membered ring of the porphyrin ring system.

P:O ratio. A measure of oxidative phosphorylation in a system that is equal to the number of moles of ATP formed per gram-atom of oxygen taken up.

porcine. Of, or pertaining to, swine.

pore. A minute opening through a solid.

porosity. The porous quality of a solid.

porous. Having a large number of pores.

porous disk method. A method for measuring the apparent translational diffusion coefficient of a macromolecule by an application of Fick's first law. The apparatus consists of two chambers connected by a porous disk across which diffusion takes place. The diffusion coefficient is calculated from the mass transfer across the disk. The method may be used for impure preparations provided an assay for the macromolecule of interest is available. *Aka* porous diaphragm method; porous plate method.

porphin. The parent compound of the porphyrins.

porphobilinogen. The monopyrrole that is formed by the condensation of two molecules of δ-aminolevulinic acid and that serves as an intermediate in the biosynthesis of the porphyrins. *Abbr* PBG.

porphyria. One of a number of pathological conditions that are due to abnormalities in the metabolism of heme and porphyrins and that are characterized by the presence of excessive amounts of porphyrin in the urine.

porphyrin. The heterocyclic compound, present in hemoglobin, cytochromes, and other hemoproteins, that has a tetrapyrrole ring structure in which iron is chelated.

porphyrinuria. The presence of excessive amounts of porphyrins, particularly of coproporphyrin, in the urine.

porphyropsin. A visual pigment, present in fresh water fish, that consists of rod opsin and retinal$_2$ and that has an absorption maximum at 522 nm.

Porter diagram. The representation of an immunoglobulin molecule by means of straight and parallel lines for the light and heavy chains.

position isomer. One of two or more isomers that differ from each other in the position of either substituents or functional groups on a chain or on a ring.

positive catalysis. Catalysis that leads to an increase in the rate of a chemical reaction, as in the case where the binding of a substrate by an enzyme leads to a decrease of the activation energy for the conversion of the substrate to products.

positive control. The initiation of a biological activity by the presence of a particular molecule; the initiation of gene expression in an induction-repression system by the presence of a particular regulatory protein is an example.

positive cooperativity. Cooperative binding in which the binding of one ligand to one site on the molecule increases the affinity for the binding of subsequent ligands to other sites on the same molecule.

positive electron. POSITRON.

positive feedback. A feedback mechanism, as in an autocatalytic reaction, in which there is a direct relationship between the magnitude of the input into, and the output of, a system; a small input leads to a small increase in the subsequent output, while a large input leads to a large increase in the subsequent output.

positive hydration. Hydration in which the water molecules in the primary hydration shell of an ion have lesser mobility than those in pure water.

positive staining. A staining technique, used in electron microscopy, in which components of the sample are visualized through their binding of an electron-dense material; the staining of nucleic acids with uranyl acetate and the staining of antigens with ferritin-labeled antibodies are two examples.

positron. A positively charged electron.

postabsorptive state. The state of a person or an animal after a fast that was of sufficient duration so that all of the last nutrients have been absorbed through the intestinal wall.

posterior. 1. Behind, or in the back part of, a structure. 2. After, in relation to time.

postmortem. Of, or pertaining to, the period after death.

postnatal. Of, or pertaining to, the period after birth.

postoperative. Of, of pertaining to, the period following a surgical operation.

postpartum. Of, or pertaining to, the period after childbirth.

postprandial. Of, or pertaining to, the period following a meal.

postreplication repair. The repair of damaged DNA that occurs after the DNA has been replicated. The replicated molecule contains gaps in one of the strands corresponding to the damaged sections of a complementary parental strand, and these gaps are then filled in by means of a repair mechanism. *See also* repair replication.

potassium. An element that is essential to all plants and animals. Symbol K; atomic number 19; atomic weight 39.102; oxidation state +1; most abundant isotope K^{39}; a radioactive isotope K^{42}, half-life 12.4 hours, radiation emitted—beta particles and gamma rays.

potency. The degree of effectiveness of a drug in terms of the quantities required to produce certain effects.

potential. A measure of the electrical energy of a half cell in comparison to an arbitrary standard; the difference in electrical energy between an indicator and a reference electrode.

potential difference. The difference in electrical potential between two points in an electrical circuit.

potential-drop method. HIGH-RESISTANCE-LEAK METHOD.

potential energy barrier. ENERGY BARRIER (1).

potential energy diagram. A graphical representation of the potential energy barrier of a molecule in which the potential energy of the molecule is plotted as a function of the internuclear distance of its atoms.

potential energy well. The ground state and the low energy levels of a molecule as represented by a potential energy diagram.

potential gradient. The rate of change of potential with distance in a specified direction.

potential mediator. An electromotively active system that is added during potentiometric titrations to accelerate the establishment of equilibrium.

potential well. See potential energy well.

potentiation. 1. The increase in the effectiveness of a drug, a hormone, or a carcinogen that is produced by either prior or simultaneous treatment of the organism with another agent. 2. The increase in the reaginic antibody response of an animal that is produced by injecting it with certain parasites.

potentiometer. An instrument for measuring electrical potentials.

potentiometric titration. A titration in which the electrical potential is measured as a function of titrant added.

potentiometry. The measurement of either electromotive force or electrical potential and the application of these measurements to the study of oxidation-reduction systems.

Potter-Elvehjem homogenizer. A homogenizer that consists of a glass tube in which a tightly fitting pestle is rotated; the shear forces that develop between the pestle and the wall of the tube lead to the homogenization of materials introduced into the tube.

pour plate. A solid medium that contains bacteria and that is prepared by adding a bacterial inoculum to melted nutrient agar, pouring the mixture into a petri dish, and allowing it to solidify.

powder method. A method of x-ray diffraction in which a sample is used that is in powdered form.

powder pattern. The x-ray diffraction pattern that is obtained from a sample in powdered form and that is equivalent to the aggregate pattern that would have been obtained from the

same sample if it were present in the form of a large number of small, randomly oriented crystals. *Aka* powder diagram.

poxvirus. The largest and most complex of the animal viruses that is brick-shaped and contains a double-stranded DNA core surrounded by membranes. Poxviruses infect man, mammals, and birds, and some are oncogenic.

PP. 1. Inorganic pyrophosphate. 2. Protoporphyrin. 3. Photoprotection.

P particle. *See* kappa (1).

ppb. Parts per billion; a measure of concentration equal to the number of parts of a component per billion parts of the total sample, such as parts of solute per billion parts of solution.

PPD. 1. Purified protein derivative. 2. *p*-Phenylenediamine.

PP factor. Pellagra-preventative factor.

PP$_i$. Inorganic pyrophosphate.

ppm. Parts per million; a measure of concentration equal to the number of parts of a component per million parts of the total sample, such as parts of solute per million parts of solution.

PPO. 2,5-Diphenyloxazole; a primary fluor.

ppt. Precipitate.

PQ. Plastoquinone.

PRA. 5-Phosphoribosyl-1-amine; an intermediate in the biosynthesis of purines.

practical. Denoting a chemical that has not been rigorously purified.

Prausnitz-Kuestner reaction. A skin test for the detection of human reagins in serum; performed by injecting the test serum intradermally into a healthy person and then eliciting a wheal and erythema response by injecting a dose of allergen into the same site 24 hours later.

prealbumin. A plasma albumin that moves ahead of the major albumin fraction when subjected to electrophoresis under alkaline conditions.

prebiotic. Pertaining to the period prior to the occurence of life on the earth.

Precambrian era. The geologic time period that extended over about 1.6 billion years and that ended about 600 million years ago; it is divided into the Proterozoic and Archeozoic eras.

precancerous. Descriptive of a cell or a tissue that is presently benign but from which a malignant tumor is expected to develop with a high degree of probability.

precession diagram. An x-ray diffraction pattern obtained by means of a precession camera.

precipitant. A substance that, when added to a solution, causes the formation of a precipitate.

precipitate. The deposit of insoluble material that is obtained from a solution by an alteration of the conditions or by the addition of specific substances.

precipitating agent. PRECIPITANT.

precipitating antibody. PRECIPITIN.

precipitation membrane. An artificial membrane that can act as a selective ion barrier and that is produced by the diffusion-controlled precipitation of two ions of opposite charge.

precipitin. 1. An antibody that forms a precipitate with an antigen in a precipitin reaction. 2. The precipitate formed in a precipitin reaction.

precipitin curve. A plot of the amount of antibody precipitated as a function of increasing amounts of antigen added to the solution.

precipitin reaction. The formation of an insoluble precipitate by a reaction between antigens and antibodies.

precision. A measure of the reproducibility of a measurement; the degree of agreement between two or more measurements made in an identical fashion.

precursor. 1. A compound that precedes another compound in a metabolic pathway. 2. A simple low-molecular weight molecule in the environment, such as carbon dioxide or nitrogen, that is used by living organisms for the synthesis of biomolecules.

precursor of serum prothrombin conversion accelerator. PROCONVERTIN.

preearly RNA. A virus-specific RNA that is synthesized by RNA polymerase very soon after the infection of the host cell by the virus (within 1 min in the case of phage infection). *Aka* immediate early RNA; prereplicative RNA; very early RNA.

prefolic A. SERUM FOLATE.

preformation. The concept that an organism develops through the appearance and growth of structures and functions that are already present in the egg. *See also* epigenesis.

preformed gradient isodensity centrifugation. DENSITY GRADIENT SEDIMENTATION VELOCITY.

pregnane. The parent ring system of the progestogens, mineralocorticoids, and glucocorticoids.

preganediol. A major metabolite of progesterone.

pregnant mare's serum gonadotropin. A gonadotropic hormone, present in the serum of pregnant mares, that is produced by the endometrium and that has similar biological effects to those of follicle-stimulating hormone. *Var sp* pregnant mare's serum gonadotrophin. *Abbr* PMSG.

pregnenolone. A precursor of the steroid hormones. *See also* desmolase.

preincubation. The incubation of a reaction mixture prior to the test incubation, the effect of which is being measured; preincubation may be for such purposes as the depletion of an endogenous component or the establishment of a binding equilibrium.

preinductive phase. The time period that precedes the administration of an antigen to an animal.

prelumirhodopsin. A structurally altered form of rhodopsin that is formed after exposure of rhodopsin to light and prior to its dissociation into opsin and retinal$_1$.

premessenger RNA. The nuclear DNA-like RNA that is a precursor of messenger RNA and that consists of both messenger and nonmessenger RNA sequences. The nonmessenger RNA sequences, called pseudomessenger RNA, are destroyed inside the cell nucleus without reaching the cytoplasm.

premix burner. A burner used in atomic absorption spectrophotometry and designed so that the gases are mixed and the sample is atomized before entering the flame.

prenol. A long-chain isoprenoid alcohol.

preparative method. A method, such as ultracentrifugation, electrophoresis, or chromatography, that requires relatively large amounts of sample and that is used primarily for the isolation and purification of specific substances.

preparative ultracentrifuge. An ultracentrifuge, equipped with rotors of varying capacities, that is used for the preparative fractionation of macromolecules.

prereplicative RNA. PREEARLY RNA.

pressor agent. HYPERTENSIVE AGENT.

pressor amine. An amine, such as vasopressin, that functions as a hypertensive agent.

pressor effect. An increase in blood pressure.

pressor principle. VASOPRESSIN.

pressure dialysis. Dialysis in which there is either an application of pressure to the dialysis bag, or an application of a vacuum to the space surrounding the dialysis bag.

pressure-jump method. A relaxation technique in which pressure is the variable that disturbs the equilibrium of a system. *See also* relaxation technique.

presteady-state kinetics. The kinetics of an enzymatic reaction proceeding under conditions that precede the establishment of a steady-state; generally investigated by means of rapid flow or relaxation techniques that permit a study of both the initial and the intermediate steps of the reaction.

pretransfer RNA. A precursor of transfer RNA that occurs in eucaryotic cells and that contains 20 to 30 nucleotides more than transfer RNA.

previtamin. A precursor of a vitamin that is formed either during the in vivo conversion of a provitamin to the vitamin or during the in vitro conversion of a synthetic compound to the vitamin. *See also* provitamin.

PRF. *See* prolactin regulatory hormone.

PRH. *See* prolactin regulatory hormone.

PRIF. *See* prolactin regulatory hormone.

PRIH. *See* prolactin regulatory hormone.

primaquine. An antimalarial drug.

primaquine sensitivity. A genetically inherited metabolic defect in man that is characterized by the tendency of erythrocytes to hemolyze upon the administration of a variety of compounds, including primaquine, to an individual; due in most cases to a deficiency of the enzyme glucose-6-phosphate dehydrogenase. *See also* glucose-6-phosphate dehydrogenase deficiency.

primary acidosis. A deviation from the normal acid-base balance in the body that is due to overproduction, ingestion, or retention of acid. In the absence of compensatory mechanisms, such conditions lead to a lowering of the blood pH.

primary alkali deficit. PRIMARY ACIDOSIS.

primary alkali excess. PRIMARY ALKALOSIS.

primary alkalosis. A deviation from the normal acid-base balance in the body that is due to excessive loss of acid or an overdose of sodium bicarbonate. In the absence of compensatory mechanisms, such conditions lead to an increase in the blood pH.

primary carbon dioxide deficit. RESPIRATORY ALKALOSIS.

primary carbon dioxide excess. RESPIRATORY ACIDOSIS.

primary charge effect. The charge effect in a solution containing charged macromolecules that results from the differential movement of the charged macromolecules and their oppositely charged counterions. The primary charge effect leads to a decrease in the sedimentation rate and to an increase in the diffusion rate of a charged macromolecule.

primary culture. A culture that is started from cells, tissues, or organs that are derived directly from an organism.

primary deficiency. DIETARY DEFICIENCY.

primary derived protein. *See* derived protein.

primary filament. MYOSIN FILAMENT.

primary fluor. A fluor that is excited by the radiation from a radioactive sample and that produces a flash of light during scintillation counting.

primary food producer. A photosynthetic organism; a photolithotroph or a photoorganotroph.

primary hydration shell. The layer of water molecules that are closest to an ion and which, in the case of a metal ion, are frequently considered to be molecules acting as ligands.

primary immune response. *See* primary response.

primary ionization. The ionization of matter that is produced by the ionizing radiation that impinges upon it or passes through it.

primary isotope effect. An isotope effect in which the isotope itself is involved in bond cleavage, such that the bond to the isotope is broken in the transition state of the reaction.

primary pigment. The major photosynthetic pigment of an organism; the primary pigment of plants is chlorophyll and that of bacteria is bacteriochlorophyll. *See also* accessory pigment.

primary plot. A direct plot of experimental enzyme kinetics data, such as a Lineweaver-Burk plot. *Aka* primary kinetic plot.

primary prostaglandin. A prostaglandin of either type PGE or type PGF; so called, since prostaglandins of types PGA and PGB can be derived from them.

primary protein derivative. *See* primary derived protein.

primary response. The immune response of an animal that is produced when the animal is first exposed to, or injected with, an antigen.

primary solvent. The solvent used in scintillation counting for the transfer of energy from the radioactive sample to the fluor.

primary standard. 1. A purified chemical that can be weighed out and used for the preparation of solutions of known concentrations. 2. A reference, such as a source of radiation, that is used for purposes of calibration.

primary stimulus. The immunogen that stimulates an animal to produce a primary immune response.

primary structure. The basic structure of a polypeptide chain or a polynucleotide strand that is described by the type, the number, and the sequence of either the amino acids in the polypeptide chain or the nucleotides in the polynucleotide strand. The primary structure of proteins excludes the spatial arrangement of the atoms except for the configuration about the alpha carbon atoms of the amino acids; it likewise excludes disulfide bonds and hence is not equivalent to the covalent structure of the molecule.

primary tissue culture. A short-term tissue culture.

primary tumor. The original tumor as contrasted with a secondary tumor, formed through metastasis.

primate. A mammal of the order Primates, which includes man, apes, and monkeys.

primed cell. A lymphocyte that has "recognized" an antigen; a Y-cell according to the X-Y-Z cell theory.

primer. A macromolecule that stimulates the

synthesis of another macromolecule by participating in the initiation of polymerization and that is linked covalently to the product of the reaction.

prime strain. A viral strain that is not well neutralized by antibodies which are specific for the wild-type strain.

primeval. PRIMITIVE.

priming. 1. The administration of antigens to an animal organism in such a fashion that the responsive immunocytes are activated. 2. The conversion of X-cells to Y-cells according to the X-Y-Z cell theory.

priming dose. The first dose of an antigen administered to an animal organism to produce an immune response.

primitive. Of, or pertaining to, the early stages in the development of the earth and the evolution of life.

primitive atmosphere. The atmosphere that surrounded the primitive earth and that is considered by many to have been a reducing atmosphere.

primitive earth. The earth at an early stage of development at which time the first organic compounds are believed to have been formed.

primordial. PRIMITIVE.

primordial soup. The primitive oceans and other primitive bodies of water that contained a variety of organic compounds and that are thought to have been the site for the reactions in the primitive earth which led to the synthesis of the first macromolecules and to the assembly of the first living cell.

principle of evolutionary continuity. The concept that the formation of biomolecules, subcellular structures, and ultimately living cells required a large number of small, but reasonably probable, steps.

private antigen. A rare blood group antigen that occurs only in one or in a few individuals.

privileged site. A region in an organism that lacks normal lymphatic drainage and that constitutes a location where a tissue transplant may persist for extended periods of time without inducing transplantation immunity.

PRL. Prolactin.

pro-. Prefix indicating an inactive precursor, as that of an enzyme, a vitamin, or a hormone.

Pro. 1. Proline. 2. Prolyl.

proaccelerin. The factor in blood clotting that is activated to accelerin by the action of the Stuart factor.

probability. The relative frequency of occurrence of a specific type of event out of the total number of occurrences of this and other types of events, all equally likely to take place. *Sym* p.

probability curve. NORMAL DISTRIBUTION.

probability distribution. A frequency distribution in which a variable is divided into classes and the probabilities for these classes are indicated.

probability paper. Graph paper in which one scale has been changed so that a plot of a normal distribution curve will yield a straight line.

probability value. The probability expressed as a decimal fraction.

probable error. An error, equal to 0.6745 times the standard deviation, such that there is a 50–50 chance that any other error will be larger than it.

probe. A group of atoms or a molecule that is attached to other molecules or cellular structures and that is used as an aid in studying the properties of these molecules and structures. *See also* reporter group; spin labeling.

probiogenesis. Primordial biosynthetic reactions; primordial biogenesis.

procarboxypeptidase. The inactive precursor of carboxypeptidase.

procaryon. The nuclear region of a procaryotic cell.

procaryote. A simple, unicellular organism, such as a bacterium, that lacks a discrete nucleus surrounded by a nuclear membrane, and that contains its genetic material within a single chromosome. *See also* eucaryote.

procaryotic. Of, or pertaining to, procaryotes.

process. An outgrowth or an extension, such as an axon or a dendrite of a neuron.

processive exonuclease. An exonuclease that, once bound to a polynucleotide strand, completely degrades the strand before it is released.

proconvertin. The factor that is activated by thromboplastin in the extrinsic system of blood clotting.

proctodone. An insect hormone that terminates diapause.

product. An atom, an ion, or a molecule that is produced in a chemical reaction. *Sym* P.

product inhibition. The inhibition of an enzyme by a product of the reaction that the enzyme catalyzes.

productive cell. A cell that produces viral progeny when it is infected with a virus.

productive complex. An enzyme-substrate complex in which the substrate is bound to the enzyme in such a fashion that catalysis is possible and that products can be formed.

productive infection. A viral infection that leads to the formation of infectious viral progeny.

productive phase. The stage in an immune response that follows the first appearance of antibodies in the serum and that corresponds to the time during which antibodies are synthesized vigorously.

proelastase. The inactive precursor of elastase.

proenzyme. ZYMOGEN.

profibrinolysin. PLASMINOGEN.

profile. *See* elution profile; polysome profile; thermal denaturation profile.

proflavin. A mutagenic acridine dye that causes frameshift mutations.

progeny. 1. The offspring of an organism or of a cell. 2. The newly formed DNA molecules or viral particles.

progestational. Preceding gestation.

progesterone. The major female sex hormone required for the maintenance of pregnancy; a 21-carbon steroid that is secreted largely by the corpus luteum.

progestin. Any 21-carbon steroid sex hormone required for the maintenance of pregnancy; progesterone is the major progestin.

progestogen. A substance that induces progestational changes in the uterus, such as a progestin or a related synthetic compound.

progress curve. A plot of the concentration of either a reactant or a product of a chemical reaction as a function of the time that the reaction has been allowed to proceed.

progression. *See* tumor progression.

prohormone. The inactive precursor of a hormone.

proinsulin. The inactive, cyclic precursor of insulin that is converted to active insulin by hydrolytic removal of a peptide of 33 amino acids.

projection formula. A two-dimensional representation of a molecule in which bonds projecting forward with respect to the plane of the page are indicated by horizontal lines, and bonds projecting backward are indicated by vertical lines.

prokaryon. Variant spelling of procaryon.

prokaryote. Variant spelling of procaryote.

prolactin. A protein hormone, secreted by the anterior lobe of the pituitary gland, that is essential for the initiation of lactation in mammals. Prolactin also has a gonadotropic effect and stimulates progesterone secretion by the corpus luteum. *Abbr* PRL.

prolactin regulatory hormone. One of two hypothalamic hormones (or factors) that, respectively, stimulate or inhibit the release of prolactin from the pituitary gland. The prolactin releasing hormone (or factor) is abbreviated PRH (PRF); the prolactin release-inhibiting hormone (or factor) is abbreviated as PIH (PIF) or PRIH (PRIF).

prolactin release-inhibiting hormone. *See* prolactin regulatory hormone.

prolactin releasing hormone. *See* prolactin regulatory hormone.

prolamin. A simple, globular protein of plant origin that is insoluble in water but is soluble in 50 to 90% ethanol solutions.

prolate ellipsoid of revolution. An ellipsoid of revolution formed by rotation of an ellipse about its major axis.

prolidase. An exopeptidase that is specific for N-terminal proline or hydroxyproline.

proline. A heterocyclic, nonpolar alpha imino acid. *Abbr* Pro; P.

prolonged test. A toxicity test that is performed on laboratory animals and that requires the administration of a chemical at least once daily for periods of about 1 to 3 months.

promitochondrion. An abnormal mitochondrion that is present in yeast grown under anaerobic conditions and that is characterized by having a modified inner membrane and by lacking certain cytochromes.

promoter. 1. The site on the DNA molecule to which RNA polymerase attaches and at which transcription is initiated; the promoter separates the operator from the first structural gene of the operon. 2. A carcinogenic agent that brings about the second stage (promotion stage) in a two-stage or multistage mechanism of carcinogenesis; a cocarcinogen.

promoter gene. PROMOTER (1).

promotion. The second state in a two-stage or multistage mechanism of carcinogenesis, and the stage during which a precancerous cell is converted to a dependent cancer cell, possibly through the action of a promoter.

pronase. A nonspecific proteolytic enzyme isolated from *Streptomyces griseus*.

prooxidant. A substance that accelerates an autoxidation reaction.

propagation. ELONGATION.

properdin. A normal serum globulin that has bactericidal properties against gram-negative bacteria in the presence of complement and magnesium ions.

prophage. The stable, inherited, noninfectious, provirus form of a temperate phage in which the phage DNA has become incorporated into, and replicates with, the host bacterial DNA.

prophage map. The genetic map of a phage as determined from recombination studies between prophages.

prophage-mediated conversion. *See* conversion.

prophase. The first stage in mitosis during which the nuclear membrane breaks down.

propinquity effect. PROXIMITY EFFECT.

propionate rule. A rule that accounts for the number of methyl side chains in the aglycone portion of macrolide antibiotics which are produced by bacteria of the genus *Streptomyces*. The rule states that propionate may replace acetate in the building of the carbon skeleton of the antibiotic, and that one methyl side chain occurs every time that a propionate unit is incorporated.

propionibacteria. *See* propionic fermentation.

propionic fermentation. The fermentation of glucose, and generally also of lactic acid, that yields propionic acid and other products and that is characteristic of propionic acid bacteria (propionibacteria).

proportional counter. A radiation counter designed for operation in the proportional region.

proportional region. That portion of the characteristic curve of an ionization chamber in which, during gas amplification, the chamber yields a charge that is proportional to the initial charge produced by the radiation.

ProSPCA. Precursor of serum prothrombin conversion accelerator.

prostaglandin. One of a group of biologically active lipids that are derived from a 20-carbon unsaturated fatty acid containing a five-membered ring. Prostaglandins were first found in the prostate gland but are now known to occur in most, if not all, mammalian tissues. Prostaglandin effects appear to be hormonal in nature and include lowering of blood pressure, stimulating smooth muscle contraction, and opposing the action of other hormones. *Abbr* PG. *See also* PGA; PGB; PGE; PGF.

prostanoic acid. A 20-carbon carboxylic acid that is the parent compound of the prostaglandins.

prosthetic group. The cofactor of an enzyme or the nonprotein portion of a conjugated protein that is bound so tightly to either the enzyme or the protein that it cannot be removed by dialysis.

protamine. A small, simple, and globular protein that is virtually devoid of sulfur but contains large amounts of arginine. Protamines are basic proteins that have a molecular weight of about 5000 and that are found in association with nucleic acids, primarily in sperm cells of fish.

protamine zinc insulin. The zinc salt of a protamine-insulin complex which is less soluble than insulin. *Abbr* PZI. *See also* NPH insulin.

protean. An insoluble, primary derived protein that is obtained by treatment of a protein with heat, acid, enzymes, or other agents.

protease. PROTEOLYTIC ENZYME.

protecting group. A chemical group that is reacted with, and bound to, a functional group in a molecule to prevent the functional group from participating in subsequent reactions of the molecule.

protective antigen. An antigen that is derived from a pathogenic microorganism and that, when injected into an animal, will produce an immune response that will provide protection for the animal against infection by that microorganism.

protective colloid. A colloid that is added to a food to prevent the separation of components in that food.

proteid. PROTEIN.

protein. A high molecular weight polypeptide of L-amino acids that is synthesized by living cells. Proteins are biopolymers with a wide range of molecular weights, structural complexity, and functional properties. Proteins are variously classified on the basis of their (a) solubility (albumins, globulins, scleroproteins, etc.); (b) function (transport proteins, storage proteins, contractile proteins, enzymes, hormones, antibodies, etc.); (c) shape (globular proteins and fibrous proteins); (d) composition (simple proteins, conjugated proteins, and derived proteins).

proteinaceous. Consisting in part, or entirely, of protein.

proteinase. PROTEOLYTIC ENZYME.

protein biosynthesis. *See* protein synthesis.

protein-bound iodine. The iodine in the blood that is conjugated to protein and that is a measure of the concentration of circulating thyroid hormone. *Abbr* PBI.

protein coat. The protein shell that surrounds the nucleic acid of a virus. *See also* capsid.

protein conformation. *See* chain conformation; primary structure; secondary structure; tertiary structure; quaternary structure.

protein efficiency ratio method. A method for determining the nutritive value of a protein by measuring the gain in weight of young rats that are fed a diet containing 10% of the particular protein. *Abbr* P.E.R. method.

protein error. The change in the relative amounts of the undissociated and dissociated forms of an indicator that is brought about by the binding of one of these forms to a protein. The change in the relative amounts of indicator forms leads to a change in color; such a color change forms the basis of the albustix test.

protein evolution. The molecular evolution of proteins. *See also* chemical evolution.

protein factor. The factor 6.25 that, when multiplied by the weight of nitrogen (in grams) derived from a sample containing protein, gives the approximate weight (in grams) of the protein in the sample.

protein fractionation. The separation of a mixture of different proteins for the purpose of isolating one particular type of protein; requires the use of one or more physical-chemical techniques such as precipitation, centrifugation, or electrophoresis.

protein-free filtrate. A liquid, such as plasma or serum, from which protein has been removed by precipitation.

protein index. The polarographic wave height, expressed in terms of current density, for a plasma filtrate divided by that for a plasma digest. The plasma filtrate is obtained by treating plasma with sulfosalicylic acid; the plasma digest is obtained by treating plasma with potassium hydroxide.

proteinoid. A protein-like polymer formed by thermal polymerization of amino acids in the dry state.

protein overloading. IMMUNOLOGICAL UNRESPONSIVENESS (2).

proteinpolysaccharide. MUCOPROTEIN.

protein release factor. *See* release factor.

protein-sparing action. The decrease in protein catabolism that is produced by the intake of dietary carbohydrates or lipids.

protein structure. *See* chain conformation; primary structure; secondary structure; tertiary structure; quaternary structure.

protein synthesis. The process whereby proteins are synthesized on ribosomes according to the genetic information contained within messenger RNA. It is synonymous with protein biosynthesis and includes amino acid activation and the three stages of translation: chain initiation, chain elongation, and chain termination. The amino acids are brought to the ribosome in the form of aminoacyl transfer RNA molecules and are then polymerized on the ribosome. The ribosome moves along the messenger RNA molecule and the amino acids are polymerized in the order dictated by the codons in the messenger RNA. *See also* translation; amino acid activation; initiation; elongation; termination.

protein-synthesizing system. CELL-FREE AMINO ACID INCORPORATING SYSTEM.

proteinuria. The presence of protein in the urine.

protein value. BIOLOGICAL VALUE.

proteoglycan. MUCOPROTEIN.

proteolipid. A conjugated protein that contains phospholipid and that is soluble in some nonpolar solvents but is insoluble in aqueous solutions. *See also* lipoprotein.

proteolysis. The hydrolysis of proteins, particularly that due to the action of proteolytic enzymes.

proteolytic. Of, or pertaining to, proteolysis.

proteolytic coefficient. A measure of peptidase activity that is equal to the unimolecular rate constant of the reaction catalyzed by a peptidase, divided by the peptidase concentration.

proteolytic enzyme. An enzyme that catalyzes the hydrolysis of peptide bonds.

proteolytic quotient. The ratio of two proteolytic coefficients that are determined with one enzyme and two different substrates.

proteose. A partially hydrolyzed protein that is water soluble and precipitable with ammonium sulfate; an intermediate form between a protein and a peptone.

Proterozoic era. The more recent of the two subdivisions of the Precambrian era that extended over about one billion years and ended about 600 million years ago; an era during which primitive invertebrates and algae evolved.

prothrombin. The inactive precursor of thrombin that is converted to thrombin by the action of accelerin.

prothrombin derivatives theory. A theory of blood clotting according to which different factors, similar to those of the cascade mechanism, participate in the clotting reactions but are not present as such in the blood; the factors are newly made molecules, derived from prothrombin during the process of clot formation.

prothrombin factor. Vitamin K_1.

protic solvent. 1. PROTOPHILIC SOLVENT. 2. PROTOGENIC SOLVENT.

protist. A unicellular or multicellular organism that lacks the tissue differentiation and the elaborate organization that is characteristic of plants and animals; some protists are plant-like, some are animal-like, and some have properties common to both kingdoms.

protium. The ordinary isotope of hydrogen that contains one proton and no neutrons in the nucleus. *Sym* H.

protobiochemistry. The developments in the science of biochemistry that preceded the foundation of modern chemistry by Lavoisier and Dalton.

protobiont. A primitive living cell.

protocell. A primitive forerunner of a living cell.

protoenzyme. A primitive forerunner of an enzyme.

protofibril. MYOFILAMENT.

protofilament. A filament that occurs in cilia and flagella.

protogen. LIPOIC ACID.

protogenic solvent. A nonaqueous, acidic solvent that has the capacity of donating protons to a solute.

protoheme. HEME (2).

protolysis. A reaction in which there is a transfer of a proton from an acid to a base; an acid-base reaction; the Bronsted concept of neutralization.

protolyte. 1. ACID. 2. BASE.

protolytic. Of, or pertaining to, protolysis.

protomer. 1. The individual polypeptide chain in an oligomeric protein, such as the alpha or beta chain in hemoglobin. 2. One of the identical subunits, or monomers, of a regulatory enzyme, each of which has a catalytic site. *See also* subunit; monomer.

proton. An elementary particle of the atomic

nucleus that has a charge of plus one and a mass of 1.0073 amu; identical to the nucleus of the hydrogen atom. *Sym* p.

proton abstraction. The removal of a proton from a compound.

protonate. To add protons to a group of atoms or to a compound, as in the conversion of an $-NH_2$ group to an $-NH_3^+$ group.

protonic acid. BRONSTED ACID.

proton magnetic resonance. *See* nuclear magnetic resonance.

proton motive force. The force arising from an energy-rich gradient of protons and potential. *See also* chemiosmotic hypothesis.

proton translocation. TRANSPROTONATION.

protophilic solvent. A nonaqueous, basic solvent that has the capacity of accepting protons from a solute.

protoplanet. A body of matter considered to have been a forerunner of the planets.

protoplasm. The living matter that forms the basis of animal, plant, and microbial cells; the substance of the cell that is surrounded by the cell membrane; the cytoplasm and the nucleus or the nuclear zone.

protoplast. A bacterial cell that has been freed entirely of its cell wall. Protoplasts are prepared artificially by lysozyme digestion of gram-positive bacteria; they can survive only in hypertonic media and generally cannot multiply.

protoporphyrin. The biochemically most important porphyrin derivative that occurs in hemoglobin in the form of protoporphyrin IX. *Abbr* PP.

prototroph. 1. An organism or a cell that is capable of synthesizing all of its metabolites from inorganic compounds. 2. A microorganism that has no nutritional requirements over and above those of the wild-type strain from which it is derived.

prototropy. Tautomerism in which the two isomers differ from one another in the position of a hydrogen atom and of a double bond. *Aka* prototropism.

protovirus theory. A theory of cancer according to which the RNA of oncogenic RNA viruses has become part of the genome of animals in the form of highly mutable segments that have the potential of becoming mutated to cancer-causing genes. The incorporated RNA (protovirus; virogene) directs the formation of a messenger RNA copy and its packaging with a reverse transcriptase in a form that can infect the nucleus of a neighboring cell. The reverse transcriptase then catalyzes the synthesis of a DNA molecule in the infected cell and its incorporation into the DNA of the infected cell. Cancer may arise from either mutations or mistakes related to the transcription and/or the integration of the protovirus. *See also* oncogene theory.

protozoan (*pl* protozoa). A unicellular animal organism; a nonphotosynthetic, eucaryotic protist.

provirus. A virus that is integrated into the chromosome of a host cell and is transmitted in that form from one host cell generation to another without leading to the lysis of the host cells. *See also* prophage.

provitamin. A naturally occurring precursor of a vitamin that is transformed to the vitamin in the animal body; beta carotene of plants, for example, is a provitamin of vitamin A. *See also* previtamin.

provitamin A. BETA CAROTENE.

provitamin A carotenoid. A generic descriptor for all carotenoids that exhibit qualitatively the biological activity of beta carotene.

provitamin D. A sterol such as ergosterol that is converted to vitamin D upon irradiation with ultraviolet light.

Prower factor. STUART FACTOR.

proximal. Close to a particular location or to a point of attachment.

proximity effect. The contribution to the catalytic activity of an enzyme that results from the reactants being brought closer together on the surface of the enzyme. *See also* orientation effect.

prozone. 1. The concentration range in some agglutination reactions in which an undiluted, or slightly diluted, cell suspension fails to lead to agglutination, while a more dilute cell suspension produces a normal agglutination reaction. 2. AUTOINTERFERENCE.

PRPP. 5-Phosphoribosyl-1-pyrophosphate.

PS. 1. Peptide synthetase. 2. Phosphatidylserine.

pseudoalleles. Closely linked genes that behave in the complementation test as if they were alleles but that can be separated by crossing over.

pseudocholinesterase. CHOLINESTERASE.

pseudocyclic photophosphorylation. A process that is similar to cyclic photophosphorylation but is dependent on the presence of oxygen; it leads to the reduction of an added hydrogen acceptor and the reoxidation of this acceptor by oxygen.

pseudofeedback inhibition. Feedback inhibition that is caused by the analogue of a metabolite which is produced by, or participates in, a biosynthetic reaction. The feedback inhibition produced by a nucleotide that is formed by the incorporation of a base analogue is an example.

pseudoglobulin. A globulin that is sparingly soluble in water.

pseudohemoglobin. An artificially-prepared hemoglobin-like molecule in which the globin molecules are combined with iron porphyrins other than protoporphyrin IX. *See also* hybridization (2).

pseudomessenger RNA. *See* premessenger RNA.

pseudonucleoprotein. Obsolete designation for a phosphorus-containing protein that is not a nucleoprotein.

pseudophotoaffinity labeling. *See* photoaffinity labeling.

pseudoreversion. The restoration of biological activity, lost as a result of a mutation, by means of a second mutation of the mutated base which leads to the formation of a codon that differs from the normal codon of the wild-type organism but constitutes an acceptable missense codon.

pseudouridine. An unusual nucleoside, occurring in transfer RNA, in which the normal base uracil is linked to ribose through carbon atom 5 of the uracil. *Sym* ψ; ψU; ψrd.

pseudouridylic acid. The nucleotide formed from pseudouridine.

pseudoverification. The deacylation of a correctly acylated, but noncognate, transfer RNA molecule by an aminoacyl-tRNA synthetase; can be brought about under special conditions in a mixed solvent.

psilocin. A naturally occurring hallucinogenic drug that is an indolealkylamine.

psilocybin. A naturally occurring hallucinogenic drug that is an indolealkylamine.

P site. PEPTIDYL SITE.

PSP test. Phenolsulfonphthalein test.

P^{32} suicide. *See* suicide.

psychedelic drug. HALLUCINOGENIC DRUG.

psycholytic drug. HALLUCINOGENIC DRUG.

psychosine. A glycoside of sphinganine; a glycosphingolipid without the fatty acid.

psychotomimetic drug. HALLUCINOGENIC DRUG.

psychrophile. An organism that grows at low temperatures in the range of 0 to 25°C, and that has an optimum growth temperature in the range of 20 to 25°C.

psychrophilic. Of, or pertaining to, psychrophiles; preferring low temperatures.

PTA. Plasma thromboplastin antecedent.

PTC. 1. Phenylthiocarbamyl group. 2. Plasma thromboplastin component.

PTC-amino acid. Phenylthiocarbamyl amino acid.

pteridine. A nitrogen-containing compound that consists of two fused, six-membered rings and that is a structural component of biopterin, folic acid, and riboflavin.

pterin. One of a group of widely-distributed derivatives of the parent compound, 2-amino-4-hydroxypteridine.

pteroic acid. A structural component of folic acid that consists of a pterin attached to *p*-aminobenzoic acid.

pteroylglutamic acid. FOLIC ACID.

PTH. 1. Parathyroid hormone. 2. Phenylthiohydantoin.

PTH-amino acid. Phenylthiohydantoin amino acid.

ptomaine. A group of toxic substances, specifically amines, that are formed by microbial decomposition of proteins.

ptyalin. Salivary alpha amylase.

PU. Pregnancy urine; the urine of pregnant individuals that is used as a source of hormones.

public antigen. A blood group antigen that occurs in a great number of individuals.

puff. *See* chromosome puff.

pulmonary. Of, or pertaining to, the lungs.

pulsating ribosome. A model for the ribosome during protein synthesis according to which the two ribosomal subunits move apart and come together repeatedly as the amino acids are being polymerized.

pulse. 1. A brief exposure to a radioactive isotope, as that used to label messenger RNA in bacterial cultures. 2. The amount of radioactive isotope used for the labeling of a substance by means of a brief exposure to the isotope. 3. The energy of the discharge in an ionization chamber. 4. The electric current produced by a discharge in an ionization chamber or by scintillations in a scintillation counter.

pulse-chase experiment. *See* pulse; chase.

pulse discriminator. *See* discriminator.

pulsed laser interferometry. An optical technique that can be used in conjunction with the analytical ultracentrifuge, and in which intermittent laser illumination is used to produce interference patterns. *Abbr* PLI.

pulse height. The intensity of the electric current produced in a scintillation counter.

pulse-height analyzer. A device, consisting of two discriminators, that accepts pulses that have intensities that lie between the setting of the two discriminators. *See also* differential counting.

pulse-height shift method. CHANNELS RATIO METHOD.

pulse-labeled RNA. An RNA that is labeled by means of a brief exposure to a radioactive isotope and that is considered to represent largely, if not entirely, messenger RNA that has a short half-life.

pump. The structure and/or the mechanism that mediates the active transport of a given substance across a biological membrane.

punctuation. Those elements of the genetic code that serve as initiation and termination signals of the messages, and that do not code for amino acids.

Puo. Purine nucleoside.

Pur. Purine.

pure. Containing no contaminating material. *See also* purification; purity.

pure culture. 1. A culture containing only microorganisms from one species. 2. A culture derived from a single cell.

pure line. A strain of organisms that is homozygous as a result of continued inbreeding.

purging. The replacement of one type of a gaseous environment by another; the flushing out of one atmosphere by another.

purification. 1. The process whereby a preparation is freed of certain types of molecules to yield a sample that is enriched in, or consists solely of, molecules of a single type; the process whereby a specific enzyme, a nucleic acid, etc. is being isolated. 2. The ratio of the specific activity at a given step in the isolation of a substance divided by the specific activity at a reference step; applicable to the isolation of enzymes and other macromolecules, the activity of which can be measured. *See also* purity.

purified diet. SYNTHETIC DIET.

purified parathyroid extract. PARATHORMONE.

purified protein derivative. A protein fraction obtained by ammonium sulfate precipitation of a culture of the tubercle bacillus, *Mycobacterium tuberculosis*, that had been grown in a synthetic medium. *Abbr* PPD.

purine. 1. A basic, heterocyclic, nitrogen-containing compound that occurs in nucleic acids; common purines are adenine and guanine. *Abbr* Pur. *Aka* base; nitrogenous base. 2. The parent compound of adenine, guanine, and related compounds.

purine nucleotide cycle. The group of reactions whereby AMP is deaminated to IMP and the latter is reaminated to form AMP.

purine salvage. *See* salvage metabolic pathway.

purity. 1. The state of a preparation in which all the molecules are those of a single type, as in a preparation containing only palmitic acid molecules. 2. The state of a preparation in which all the macromolecules are those of a single type, as in a preparation containing buffer ions and small molecules, but containing ribonuclease molecules as the sole macromolecules. 3. The degree to which a preparation consists of macromolecules of a single type. *See also* purification.

puromycin. An antibiotic, produced by *Streptomyces alboniger*, the structure of which resembles the terminal grouping of the amino acid joined to adenosine in an aminoacyl transfer RNA molecule. Because of this structural similarity, puromycin acts as an analogue of aminoacyl transfer RNA and inhibits protein synthesis by binding to the growing polypeptide chain and causing its premature release from the ribosome in the form of peptidyl puromycin. *Abbr* PM.

purple membrane. A section of the cell membrane of the halophile *Halobacterium halobium* that contains bacteriorhodopsin as its only protein.

putrefaction. The formation of foul-smelling products by microbial decomposition of high-protein materials such as meat and eggs.

putrescine. A four-carbon polyamine that contains two amino groups.

p value. Probability value.

PVC. Polyvinylcarbonate; a plastic.

pycnometer. A small glass vessel that has a definite volume and that is used for the weighing of different liquids to determine specific gravities and densities. *Var sp* pyknometer.

pycnosis. The shrinkage and condensation of the cell nucleus into a compact, densely-staining structure that occurs when the cell dies.

Pyd. Pyrimidine nucleoside.

pyknosis. Variant spelling of pycnosis.

Pyr. Pyrimidine.

pyran. A heterocyclic compound, the structure of which resembles the ring structure of the pyranoses.

pyranose. A monosaccharide having a six-membered ring structure.

pyranoside. A glycoside of a pyranose.

pyrenoid. A proteinaceous granule around which starch may accumulate in chromatophores.

pyrenoid starch. Starch that accumulates around pyrenoids in chromatophores.

pyridine-linked dehydrogenase. A dehydrogenase that requires a pyridine nucleotide, NAD^+ or $NADP^+$, as a coenzyme.

pyridine nucleotide. Any one of the oxidized or reduced forms of nicotinamide adenine dinucleotide or nicotinamide adenine dinucleotide phosphate; $NAD^+(NADH)$ or $NADP^+(NADPH)$.

pyridoxal. The aldehyde form of pyridoxine; a form of vitamin B_6. *Abbr* PL.

pyridoxal phosphate. The coenzyme form of vitamin B_6 that functions in the metabolism of amino acids, as in the transamination reaction. *Abbr* PLP; PAL; PALP.

pyridoxamine. The amine form of pyridoxine; a form of vitamin B_6. *Abbr* PM.

pyridoxine. The alcohol form of vitamin B_6. *Abbr* PN.

pyridoxol. PYRIDOXINE.

pyrimidine. 1. A basic, heterocyclic, nitrogen-containing compound that occurs in nucleic acids; common pyrimidines are cytosine, thymine, and uracil. *Abbr* Pyr. *Aka* base; nitrogenous base. 2. The parent compound of cytosine, uracil, and related compounds.

pyrimidine dimer. The dimer formed by the linking of two adjacent pyrimidines, such as two thymines, in a nucleic acid strand as a result of ultraviolet irradiation.

pyrimidineless mutant. *See* -less mutant.

pyrocondensation. The condensation of molecules that is brought about by heat.

pyrolysis. The transformation of one substance into another that is brought about by heat alone; includes such processes as thermal decomposition, isomerization, and synthesis.

pyronin Y. A basic dye used in cytochemistry for the staining of RNA.

pyrophosphatase. The enzyme that catalyzes the hydrolysis of inorganic pyrophosphate to two molecules of orthophosphate.

pyrophosphate. *See* inorganic pyrophosphate.

pyrophosphate cleavage. The hydrolytic removal of a pyrophosphate group from either a nucleoside diphosphate or a nucleoside triphosphate.

pyrophosphorolysis. The reaction catalyzed by pyrophosphorylase.

pyrophosphorylase. 1. The enzyme that catalyzes the formation of a nucleoside-5′-diphosphate sugar and pyrophosphate from a sugar-1-phosphate and a nucleoside-5′-triphosphate. 2. A nucleotidyl transferase.

pyrrole. A five-membered, heterocyclic, nitrogen-containing building block of the porphyrins.

pyruvate carboxylase. The enzyme that catalyzes the anaplerotic reaction whereby pyruvic acid is carboxylated to oxaloacetic acid.

pyruvate dehydrogenase system. A multienzyme system, consisting of a large number of three different enzymes and five different coenzymes, that catalyzes the conversion of pyruvic acid to acetyl coenzyme A. *Aka* pyruvate dehydrogenase complex.

pyruvate kinase. The enzyme that catalyzes the phosphorylation of ADP to ATP by phosphoenolpyruvate.

pyruvate oxidation factor. LIPOIC ACID.

pyruvic acid. The three-carbon keto acid that is the end product of glycolysis under aerobic conditions.

PZI. Protamine zinc insulin.

Q

Q. 1. Ubiquinone. 2. Glutamine. 3. Q-value.

Q_{10}. The ratio of the velocity of a reaction at a particular temperature to the velocity at a temperature that is lower by 10 deg; the Q_{10} is approximately 2 for chemical reactions.

$Q\beta$. A phage that infects *Escherichia coli* and that contains single-stranded RNA.

Q_{CO_2}. *See* metabolic quotient.

Q notation. A method used in the past to denote enzyme activity, especially that of respiratory enzymes. The Q_s value of an enzyme was taken to be the number of microliters, at standard temperature and pressure, of the substrate used up per hour per milligram of enzyme. For nongaseous substrates, 1 μmole of substrate was considered to be equivalent to 22.4 μl.

Q_{O_2}. *See* metabolic quotient.

quadri-. Combining form meaning four.

Quaking mutation. An autosomal, recessive mutation in mice that produces a myelin-deficient animal.

quantasome. An elementary photosynthetic particle that is believed to be present in the thylakoid disks of chloroplasts and to contain both photosystems I and II as well as the electron transport system.

quantify. 1. QUANTITATE. 2. To transform a relationship from a qualitative into a quantitative form.

quantitate. 1. To measure the quantity of an item. 2. To express a relationship in quantitative terms.

quantized. In the form of discrete units, or quanta; used in reference to energy.

quantum (*pl* quanta). The unit amount of energy that is released during the emission of radiation and that is taken up during the absorption of radiation; equal to $h\nu$, where h is Planck's constant, and ν is the frequency of the radiation in cycles per second

quantum efficiency. QUANTUM YIELD.

quantum mechanics. The description of atomic and molecular phenomena in terms of energy quanta and quantized energy states rather than in terms of classical Newtonian mechanics; achieved by Heisenberg by using matrices and linear operators, and achieved by Schroedinger by considering the wave aspects associated with subatomic particles. The Schroedinger approach was originally called wave mechanics, but the term is now used as a synonymous expression for quantum mechanics.

quantum requirement. The reciprocal of the quantum yield.

quantum theory. The theory that energy can be radiated and absorbed only in discrete packets, called quanta, the energy of which is proportional to the frequency of the radiation.

quantum yield. The number of molecules that react chemically in a photochemical reaction, divided by the number of photons absorbed; the number of moles that react chemically in a photochemical reaction, divided by the number of einsteins absorbed.

quartet. A quadruple peak, such as a nuclear magnetic resonance peak that has split into four peaks.

quartz. A glassy silicon dioxide that is used for the production of cuvettes that are utilized in absorbance measurements of ultraviolet light.

quaternary nitrogen. A positively charged nitrogen atom that is linked to other atoms by means of four covalent bonds.

quaternary structure. The structure of a protein that results from the interaction between individual polypeptide chains to yield larger aggregates; the arrangement in space of the subunits of a protein and the intersubunit contacts and interactions without regard to the internal structure of the subunits.

Quellung reaction. The precipitin reaction that occurs between polysaccharides of bacterial capsules and antibodies to these polysaccharides.

quench correction curve. A plot of counting efficiency versus the ratio of counts in two channels; used to correct the observed counts in liquid scintillation for quenching.

quenching. 1. The process whereby secondary and subsequent ionizations in an ionization detector are stopped so that the detector becomes again sensitive to new, incoming ionizing radiation. 2. A decrease in the counting efficiency in liquid scintillation. 3. The decrease in fluorescence that results from an absorption of some or all of the emission energy. *See also* fluorescence quenching.

quinine. An alkaloid drug used in the treatment of malaria.

quinone. 1. *p*-Dioxybenzene. 2. A derivative of *p*-dioxybenzene.

Q-value. The total energy per atom that is released in a nuclear reaction in which a nuclide is transformed into another, and ground-state, nuclide.

R

r. 1. Ribo-, as in ribothymidine monophosphate (rTMP). 2. Ribosomal, as in ribosomal RNA (rRNA). 3. Roentgen.

R. 1. R-group. 2. Gas constant. 3. R-configuration. 4. Purine nucleoside. 5. Arginine. 6. Bacterial colony of rough morphology. 7. Resistance of a bacterial strain to an inhibitor or a phage (used as a superscript). 8. Conformational form of a regulatory enzyme. 9. Roentgen.

Ra. Radium.

racemase. An enzyme that catalyzes the interconversion between two optical isomers, each of which has more than one asymmetric center. *See also* epimerase.

racemate. RACEMIC MIXTURE.

racemic mixture. An equimolar and optically inactive mixture of the two enantiomers of an optically active compound.

racemization. The conversion of an optically active compound to a racemic mixture.

rachitis. RICKETS.

rachitic. Of, or pertaining to, rickets.

Racker band. An absorption band that is produced by the binding of NAD^+ to a dehydrogenase and that is thought to be due to the formation of a charge transfer complex.

rad. Radiation absorbed dose.

radial chromatography. CIRCULAR CHROMATOGRAPHY.

radial dilution. The dilution of sedimenting components that is produced in the analytical ultracentrifuge due to the sectorial shape of the centrifuge cell and to the variation of the centrifugal force with distance from the center of rotation. *Aka* square dilution law.

radian. The angle subtended by an arc that is equal in length to the radius of the circle.

radiation. 1. The emission and propagation of waves of electromagnetic energy such as visible light, x-rays, or gamma rays. 2. The emission and propagation of corpuscles such as alpha particles, beta particles, or electrons.

radiation absorbed dose. The quantity of ionizing radiation that results in the absorption of 100 erg/g of irradiated material. *Abbr* rad. *See also* exposure dose.

radiation biochemistry. An area of biochemistry that deals with the effects of radiation on biochemical compounds and biochemical systems.

radiation chimera. A chimera produced experimentally by first irradiating an organism so as to destroy its antibody-producing cells, and by then injecing it with antibody-producing cells from a different organism.

radiation curing. *See* cure.

radiation dose. The amount of radiation to which a specified tissue area or an entire organism is exposed.

radiation sickness. A pathological condition that results from exposure to x-rays or other ionizing radiations; characterized in its mild form by nausea, vomiting, and weakness, and in its severe form by damage to blood-forming tissues and by loss of red and white blood cells.

radical. A univalent group of atoms that acts as a unit in chemical reactions. *See also* carbon radical; free radical.

radical amino acid replacement. RADICAL SUBSTITUTION.

radical anion. An anion that is also a free radical.

radical cation. A cation that is also a free radical.

radical ion. An ion that is also a free radical.

radical scavenger. A chemical compound that reacts readily with free radicals and that, when added to a biological system, provides protection against the indirect effects of radiation.

radical substitution. The replacement in a protein of one amino acid by another, chemically different, amino acid such as the replacement of a polar amino acid by a nonpolar one or vice versa. A radical substitution is generally expected to lead to significant changes in the properties of the protein. *See also* conservative substitution.

radio-. 1. Combining form meaning radiation. 2. Combining form meaning radioactive radiation.

radioactivation analysis. ACTIVATION ANALYSIS.

radioactive contamination. The deposition of radioactive material in preparations and/or in places where it was not intended to be deposited.

radioactive decay. The changes occurring in the nucleus of a radioactive atom that lead to transformation of the nucleus into a different one and to the emission of ionizing radiation.

radioactive disintegration. RADIOACTIVE DECAY.

radioactive half-life. *See* half-life (1).

radioactive isotope. An unstable isotope that undergoes radioactive decay.

radioactive radiation. The electromagnetic or the corpuscular radiation emitted by radioactive isotopes.

radioactive series. A succession of radioactive nuclides, each one decaying to the next by radioactive disintegration until a stable nuclide is formed.

radioactive suicide. *See* suicide.

radioactive tracer. *See* tracer (1).

radioactivity. The spontaneous disintegration of certain unstable nuclides that is caused by changes in the atomic nucleus and that results in a transformation of the nucleus into a different one and in the emission of one or more types of ionizing radiation, such as alpha particles, beta particles, or gamma rays.

radioassay. An assay in which radioactive isotopes are employed.

radioautograph. The photographic record obtained in radioautography.

radioautographic efficiency. The number of activated silver grains that are produced in a photographic emulsion that covers a radioactively labeled tissue section for every 100 radioactive disintegrations which occurred in that tissue section during the exposure interval.

radioautography. A technique for studying the location of radioactive isotopes in macromolecules and in larger structures. The material to be studied is labeled with a radioactive isotope and is placed in contact, in the dark, with a photographic film or a photographic emulsion; the latent image produced by the radioactive radiation is subsequently developed.

radiobiology. A branch of biology that deals with the effects of radiation on biological systems.

radiocarbon dating. A method for establishing the age of archeological, geological, or biological remains by determining the relative amounts of C^{12} and C^{14} in the specimen and by calculating its age from the known natural abundance and the known half-life of C^{14}.

radiochemical. 1. *n* A chemical that is labeled with a radioactive isotope. 2. *adj* Of, or pertaining to, radiochemistry.

radiochemical purity. The degree of contamination of a radioactively labeled compound with other radioactive substances.

radiochemistry. A branch of chemistry that deals with the chemistry of radioactive isotopes and their compounds, and with the application of radioactive isotopes in other areas of chemistry.

radiochromatogram. A chromatogram that contains substances labeled with radioactive isotopes.

radiochromatography. Any chromatographic technique in which substances, labeled with radioactive isotopes, are separated.

radiocolloid. An aggregate formed in solution by the clumping of molecules that contain radioactive isotopes.

radiogenic. Produced by radioactivity, such as an element that is formed from another element by radioactive decay.

radiogram. RADIOGRAPH.

radiograph. The photographic record obtained in radiography.

radiography. A photographic technique in which radiation other than light is passed through an object and a photograph is obtained which reflects the selective absorption of the radiation by various parts of the object.

radioimmunoassay. The measurement of either antigen or antibody concentration that is based on the competitive inhibition of labeled antigens on the binding of unlabeled antigens, or vice versa, to specific antibodies. *Abbr* RIA.

radioimmunochemistry. The use of immunochemical techniques in which one or more of the components are radioactively labeled.

radioimmunoelectrophoresis. Immunoelectrophoresis in which either the antigens or the antibodies used are radioactively labeled.

radioisotope. RADIOACTIVE ISOTOPE.

radioisotopic enzyme assay. An enzyme assay based on the measurement of radioactivity in a product of the enzyme-catalyzed reaction when one of the reactants is radioactively labeled.

radiolysis. A chemical decomposition that is caused by radiation; the self-decomposition of aged tritium or of C^{14}-labeled compounds are examples.

radiometer. An instrument for measuring the intensity of radiation.

radiometric analysis. The determination of an unknown compound by either reacting the unlabeled, unknown compound with a labeled reagent, or reacting the labeled, unknown compound with an unlabeled reagent; in either case, a radioactively labeled product formed in the reaction is then isolated and determined.

radiomimetic drug. A chemical immunosuppressive agent, such as an alkylating agent, the effect of which on nucleic acids resembles that of ionizing radiation.

radionuclide. A radioactive nuclide; a radioactive isotope.

radiopaque. Describing material that does not transmit radioactive radiation.

radiophosphorus decay. The radioactive disintegration of P^{32}, particularly that in P^{32}-labeled phage nucleic acid. The disintegrations lead to breaks in the sugar-phosphate backbone of the nucleic acid. Single-stranded nucleic acid is inactivated when a chain break occurs, but double-stranded nucleic acid is inactivated only when a break occurs in both strands.

radioresistance. The relative resistance of cells,

tissues, organs, or organisms to the harmful effects of radiation.

radiorespirometry. The measurement of the kinetics of oxygen uptake and/or carbon dioxide evolution in a tissue or in an organism by means of radioactive isotopes.

radiosensitivity. The relative sensitivity of cells, tissues, organs, or organisms to the harmful effects of radiation.

radiotherapy. Therapy by means of x-rays or other radioactive radiations.

radiotoxemia. RADIATION SICKNESS.

radiotracer. RADIOACTIVE TRACER.

radius of gyration. A measure of the spatial extension of a polymer that is related to the distribution of mass in the polymer and to the shape of the polymer. For a molecule that consists of an assembly of mass elements $\dot{m}_i$, each located at a distance r_i from the center of mass, the radius of gyration R is given by $R^2 = \Sigma m_i r_i^2 / \Sigma\ m_i$. *See also* average radius of gyration.

Ramachandran plot. A plot of the degrees of rotation about the bond between the alpha carbon and the carbonyl carbon in the peptide bond versus the degrees of rotation about the bond between the alpha carbon and the nitrogen atom; constructed on the basis of Van der Waals contact distances and the bond angles of the peptide bond, and used for indicating allowed and forbidden conformations of proteins.

Raman effect. The light scattering that is produced when incident light leads to rotational and vibrational transitions of molecules; the scattered light has different frequencies from those of the incident light. *Aka* Raman scattering.

Raman spectrum. The spectrum of the light that is emitted in the Raman effect.

Ramon method. A method for determining the equivalence zone of a precipitin reaction by mixing a constant amount of antigen with varying dilutions of antibodies, and taking the tube in which precipitation occurs most rapidly to be indicative of the equivalence zone. *See also* Dean and Webb method; method of optimal proportions.

rancidity. The development of unpleasant odors and tastes from fats and oils by the oxidation of the unsaturated fatty acid components and/or the hydrolysis of the triglycerides to diglycerides, monoglycerides, glycerol, and free fatty acids.

random coil. A linear polymer in a relatively compact, irregular conformation in which there is little interaction between the side chains of the polymer. A random coil exhibits little resistance to rotation about single bonds and is continually contorted by impact of the solvent

molecules. A random coil has no unique three-dimensional structure, only average dimensions, and its time-average shape is spherical.

randomization. The process in which a compound, labeled at a given position, gives rise to a product in which half of the molecules are labeled in one position, while the other half are labeled in another, symmetrical position.

randomly labeled. GENERALLY LABELED.

random mechanism. The mechanism of an enzymatic reaction in which two or more substrates participate, such that each substrate can readily undergo an association-dissociation reaction with the enzyme to form a binary complex prior to the formation of the ternary complex in which both substrates are associated with the enzyme.

random sample. A sample of items that are selected from a population in such a fashion that all the items in the population have an equal chance of being included in the sample.

random walk. The path in space taken by a molecule in which each step is uncorrelated with the preceding one. The path traced by a molecule in solution due to Brownian motion is an example; the conformation of an ideal, freely jointed random coil can likewise be treated as if the segments represented the random walk of a molecule.

Raney nickel. A preparation of finely-divided nickel used as a catalyst for hydrogenation reactions.

range. 1. The thickness of absorber required to absorb all of the radiation of a particular type. 2. The highest and lowest values for a set of results.

Raoult's law. The law that the lowering of the vapor pressure of the solvent by the solute is proportional to the mole fraction of the solute in the solution.

rapid flow kinetics. The kinetics of a chemical reaction which are determined by means of rapid flow techniques.

rapid flow technique. A technique for studying fast chemical reactions in which the reactants are forced out of two syringes into a mixing chamber and the mixture is then allowed to flow through a tube for spectroscopic, or other, analysis. The distance along the tube is proportional to the reaction time. *See also* stopped flow technique.

rapidly labeled RNA. PULSE-LABELED RNA.

rapidly reannealing DNA. REPETITIOUS DNA.

rapidly reassociating DNA. REPETITIOUS DNA.

rapid lysis mutant. A phage mutant that does not show lysis inhibition.

rapid mixing technique. RAPID FLOW TECHNIQUE.

rapid reaction. *See* fast reaction.

rare amino acid. An amino acid that occurs in only a few proteins, such as hydroxylysine or hydroxyproline, which are found in collagen and gelatin.

rare base. MINOR BASE.

rare earth. An element belonging to a group of 15 metals (atomic numbers 57 to 71) that have very similar properties.

RAST. Acronym for "radioallergosorbent test"; an isotopic technique for the demonstration of reagins directed against specific allergens.

rat antidiuresis assay. A bioassay for the activity of neurohypophyseal hormones in which the reduction of urine formation is measured in hydrated rats.

rat antispectacle eye factor. INOSITOL.

rate. *See* reaction rate.

rate constant. A proportionality constant between the velocity of a chemical reaction and the concentrations of the reacting species; denoted k_{+n} for the forward, and k_{-n} for the reverse, reaction at the nth step of a reaction sequence. *Aka* rate coefficient.

rate-determining step. The slowest step in a sequence of reactions; the step with the smallest rate constant.

rate equation. A mathematical expression for the rate of a chemical reaction in terms of the rate constants of the various steps and the concentrations of the reactants and the products.

rate-limiting step. RATE-DETERMINING STEP.

ratemeter. A radiation detector that indicates the rate of emission of radioactive radiation.

rate-of-change method. A method of amplifying the ion current produced in an ionization chamber when high sensitivity is required.

rate of shear. The variation in the velocity of flow of a liquid flowing through a tube with the radial distance from the center of the tube.

rate zonal centrifugation. DENSITY GRADIENT CENTRIFUGATION.

ratio. The supply of a nutrient to a tissue divided by the requirement of the tissue for the nutrient.

Rauscher leukemia virus. A mouse leukemia virus that belongs to the leukovirus group.

Rayleigh fringe. An interference fringe obtained with a Rayleigh interferometer.

Rayleigh interferometer. *See* interferometric optical system.

Rayleigh quotient. RAYLEIGH RATIO.

Rayleigh ratio. A measure of the intensities of incident and scattered light in Rayleigh scattering; specifically, $R = (i_\theta/I_0)r^2$, where R is the Rayleigh ratio, i_θ is the scattered light intensity

at angle θ, I_0 is the incident light intensity, and r is the distance from the observer to the source of the scattered light.

Rayleigh scattering. The light scattering that is produced by solutes in dilute solutions when the solute particles can be considered to be independent scatterers, when they are small compared to the wavelength of the incident light, and when the scattering is due to elastic collisions between photons and orbital electrons. The greatest dimension of the scattering particles must be less than about 0.05 times the wavelength of the incident light, and the scattered light is of the same frequency as that of the incident light.

Rb. Ribosome.

RBC. Red blood cell.

RBE. Relative biological effectiveness.

RCF. Relative centrifugal force.

R-configuration. The relative configuration of an asymmetric center in a molecule in which the substituents lie in a clockwise (right-handed, rectus) array of decreasing atomic number.

RDE. Receptor destroying enzyme.

rDNA. The DNA that codes for ribosomal RNA.

R-DNA polymerase. DNA POLYMERASE III.

R1,5 DP. Ribulose-1,5-diphosphate.

reactant. An atom, an ion, or a molecule that enters into a chemical reaction.

reaction coordinate. The abscissa that represents the progress of a reaction as measured by some quantity indicated on the ordinate.

reaction kinetics. The rate behavior of a reaction.

reaction mixture. The mixture composed of sample material and reagents that is allowed to undergo a reaction under controlled conditions.

reaction of identity. The complete fusion of two precipitin bands in either two-dimensional double immunodiffusion or immunoelectrophoresis; obtained when two indistinguishable antigens react with an antibody in an adjacent field.

reaction of nonidentity. The complete crossing of two precipitin bands in either two-dimensional double immunodiffusion or immunoelectrophoresis; obtained when two unrelated antigens react with an antibody in an adjacent field.

reaction of partial identity. The partial fusion of, and spur formation by, two precipitin bands in either two-dimensional double immunodiffusion or immunoelectrophoresis; obtained when two cross-reacting antigens react with an antibody in an adjacent field.

reaction order. The sum of the powers of the reactant concentrations to which the reaction rate is proportional. *See also* reaction rate; order with respect to concentration; order with respect to time.

reaction paper chromatography. A chromatographic technique for determining the number of various functional groups in a molecule, chiefly in an aromatic compound, on the basis of the chromatographic behavior of the unreacted molecule and of the molecule after it has reacted with appropriate reagents.

reaction rate. The rate at which either a product is formed or a reactant is used up in a chemical reaction; for the first order reaction $A \rightarrow B$, the rate is given by $v = -d[A]/dt = d[B]/dt = k[A]$, where brackets indicate molar concentrations, and k is the rate constant. For the second order reaction $A + B \rightarrow C$, the rate is given by $v = k[A][B]$. *Aka* reaction velocity.

reactivation. 1. The restoration of activity to cells or viruses that have suffered photochemical damage; photoreactivation, thermal reactivation, and multiplicity reactivation are examples. 2. The restoration of activity to an inactivated poxvirus, the protein coat of which has been denatured but the DNA of which has not been damaged; achieved by the presence of an infectious poxvirus, the enzymes of which lead to an uncoating of the inactivated virus. 3. The restoration of activity to an inhibited enzyme by removal of the inhibitor through a chemical reaction. The process has sometimes been termed reversible inhibition. *See also* irreversible inhibitor; reversible inhibitor.

reactive residue. An amino acid residue in a protein that is accessible to and can undergo a reaction with a specific reagent.

reading. The process whereby the sequence information in one polymer is used to produce a defined sequence in another polymer; replication, transcription, and translation are examples of processes that entail reading.

reading frame shift. The shift in reading produced by a frame-shift mutation.

reading mistake. MISTRANSLATION.

readout. The act of reading out. *See also* reading.

read-through protein. A protein that is produced as a result of a failure in the termination of translation of a polycistronic messenger RNA; such a protein consists of the regular amino acid sequence specified by its cistron plus a sequence of amino acids that corresponds to a translated intercistronic region.

reagent. A substance that participates in a chemical reaction.

reagin. A homocytotropic antibody of the IgE immunoglobulin class that is formed in response to an allergen and that, upon combination with the allergen, causes the release of histamine and other vasoactive agents of immediate-type hypersensitivity.

reaginic antibody. REAGIN.

rearrangement reaction. A chemical reaction in which there is an alteration in the distribution of the atoms in a molecule.

recapitulation theory. The theory that an organism during its development passes through and recapitulates the stages that have occurred in the development of the species. *Aka* ontogeny recapitulates phylogeny.

receptor. 1. A target site at the molecular level to which a substance becomes bound as a result of a specific interaction. As an example, the site may be on the cell wall, on the cell membrane, or on an intracellular enzyme, and the substance bound may be a virus, an antigen, a hormone, or a drug. 2. A site in an organism that responds to specific stimuli such as a chemoreceptor, an osmoreceptor, or a photoreceptor.

receptor destroying enzyme. NEURAMINIDASE.

receptor gradient. An arrangement of viruses in a series based on their reaction with, and their destruction of, receptor sites on red blood cells; any virus in the series will react with its own receptor sites and with those specific for viruses that precede it in the gradient, but will not react with receptor sites for viruses that follow it in the gradient.

recessive. 1. RECESSIVE GENE. 2. The trait produced by a recessive gene in the homozygous state.

recessive gene. A gene the expression of which is either partially or entirely suppressed when the dominant allelic gene is present.

reciprocal ion-atmosphere radius. The term kappa κ of the Debye-Hueckel theory that is equal to the reciprocal of the distance from the surface of the central ion to the maximum in the charge distribution of the ion atmosphere.

reciprocal lattice. The three-dimensional crystal lattice deduced from a two-dimensional x-ray diffraction pattern; used to obtain the dimensions of the unit cell in the real crystal lattice and so called because the positions of the spots in the x-ray diffraction pattern are an inverse measure of the spacings in the real crystal. *Aka* reciprocal space.

reciprocating shaker. A shaker that provides a forward and backward motion, as that of a piston.

reciprocity. The condition that exists when the product of dose rate, specifically that of radiation, and time of exposure is constant; thus (dose rate)$_1$ × time$_1$ = (dose rate)$_2$ × time$_2$. *Aka* Bunsen-Roscoe law.

rec⁻ mutant. Recombination-deficient mutant.

recognition. A specific binding interaction occurring between macromolecules, as that between a transfer RNA molecule and an aminoacyl-tRNA synthetase, or that between an immunocyte and an antigen.

recognition site. 1. tRNA SYNTHETASE RECOGNITION SITE. 2. AMINOACYL SITE.

recombinant. One of the progeny formed as a result of genetic recombination.

recombinase. An enzyme that participates in the process of genetic recombination.

recombination. The production of progeny that derives some of its genes from one parent and some from another, genetically different, parent; as a result, the combination of genes in the progeny is different from that of either of the parents. In higher organisms, recombination occurs by way of independent assortment or crossing-over; in lower organisms, it occurs by way of transformation, conjugation, or transduction.

recombination-deficient mutant. A mutant that is unable to produce recombinants. *Abbr* rec⁻ mutant.

recombination frequency. *See* frequency of recombination.

recombinationless mutant. RECOMBINATION-DEFICIENT MUTANT.

recombination value. FREQUENCY OF RECOMBINATION.

recommended dietary allowance. A recommended quantity for the daily intake of calories, a food, or a vitamin that has been established by the Food and Nutrition Board of the National Research Council; recommended for a normal individual engaged in average activity and living in a temperate climate.

recon. The unit of genetic recombination; the smallest section of a chromosome, which may be as small as a single nucleotide, that is capable of recombination and that cannot be divided by recombination.

reconstitute. To reassemble a particle from its fragments or to reassemble a system from its fractions, as in the assembly of viruses, ribosomes, and protein-synthesizing systems.

reconstituted ghost. An erythrocyte ghost that has been loaded with specific substances, and the membrane of which has been allowed to shrink back to its normal size and to return to its normal state of permeability.

recording spectrophotometer. A spectrophotometer with an attached recorder for graphical representation of the data obtained.

recovery heat. The heat produced by a muscle that relaxes after a single contraction.

recovery time. COINCIDENCE TIME.

recycling chromatography. A column chromatographic technique in which resolution is improved by passing the column effluent back onto the same column; fractions may be collected and fresh solvent may be added during this process.

red cell agglutination. HEMAGGLUTINATION.

red cell lysis. HEMOLYSIS.

red drop. The decrease in photosynthetic efficiency (the quantum yield) of chloroplasts that occurs at wavelengths longer than 680 nm.

red muscle. A dark skeletal muscle that has a relatively high content of myoglobin and cytochromes.

redox. Oxidation-reduction.

redox carrier. ELECTRON CARRIER.

redox couple. The electron donor and the electron acceptor species of a given half-reaction.

redox lipid. A lipid, such as ubiquinone or tocopherol, that undergoes oxidation-reduction reactions and that contains polyisoprenoid chains.

redox pair. REDOX COUPLE.

red plaque. A plaque that, in the plaque assay, is stained excessively with neutral red due to the increased binding of the dye by the lysosomes present in the virus-infected cells.

red shift. BATHOCHROMIC SHIFT.

reduced mean residue rotation. The mean residue rotation corrected to that in a medium of unit refractive index; specifically, $[m'] = 3[m]/(n^2 + 2)$, where $[m']$ is the reduced mean residue rotation, $[m]$ is the mean residue rotation, and n is the refractive index of the medium.

reduced osmotic pressure. The osmotic pressure of a solution divided by the concentration of the solution.

reduced scattered light intensity. RAYLEIGH RATIO.

reduced viscosity. The ratio of the specific viscosity of a solution to either the solute concentration or the volume fraction of the solute. *Aka* reduced specific viscosity.

reducing agent. REDUCTANT.

reducing atmosphere. An atmosphere that is rich in gases that are readily oxidized; a reducing atmosphere consisting of water, hydrogen, ammonia, nitrogen, methane, and hydrogen sulfide is believed by some to have been the primitive atmosphere of the earth about 4.5 billion years ago.

reducing end. The end of an oligo- or a polysaccharide that carries the hemiacetal or the hemiketal grouping.

reducing equivalent. A measure of reducing power equal to either one electron or one hydrogen atom.

reducing power. The capacity of a substance to function as a reducing agent.

reducing sugar. A sugar that will reduce certain inorganic ions in solution, such as the cupric ions of Fehling's or Benedict's reagent; the reducing property of the sugar is due to its aldehyde or potential aldehyde group.

reductant. The electron donor species of a given half-reaction; the species that undergoes oxidation in an oxidation-reduction reaction.

reductase. A dehydrogenase for which the transfer of hydrogen from the donor molecule is not readily demonstrated.

reduction. The change in an atom, a group of atoms, or a molecule that involves one or more of the following: (a) loss of oxygen; (b) gain of hydrogen; (c) gain of electrons.

reduction division. MEIOSIS.

reduction potential. The electrode potential that is used in biochemistry and that measures the tendency of an oxidation-reduction half-reaction to occur by way of a gain of electrons. *Sym* E. *See also* electrode potential; standard electrode potential.

redundant cistron. A cistron that is present in multiple copies on the same chromosome.

reference. BLANK.

reference electrode. An electrode against which the potential of another electrode is being measured. *Aka* reference half-cell.

reflection. The partial or complete return of light waves or other types of radiation from a surface.

reflection symmetry. The symmetry of a body that exists if an identical structure of the body is produced when it is rotated about an axis and reflected through a plane perpendicular to that axis; the order in which these two processes are carried out is not significant. The rotation-reflection axis is denoted S_n, indicating that identical structures are produced by a rotation of $360°/n$.

refolding. RENATURATION.

refractile. Capable of refracting light.

refraction. The change in the velocity and in the direction of light waves which pass obliquely from one medium into another.

refractive increment. REFRACTIVE INDEX INCREMENT.

refractive index (*pl* refractive indices). A measure of the light-retarding property of a medium, equal to the ratio of the velocity of light in a vacuum to that in the medium; also equal to the ratio of the sine of the angle of incidence to the sine of the angle of refraction for light passing obliquely from a vacuum into the medium. *Sym* n. *Aka* index of refraction.

refractive index increment. The rate of change of the refractive index of a solution with the concentration of the solution. *See also* specific refractive index increment.

refractometer. An instrument for measuring refractive indices.

refractory. Resistant to a given treatment or cure.

refractory period. The period after the passage of an action potential during which a nerve axon or a muscle fiber is resistant to stimulation.

regeneration. 1. The repair and replacement of damaged or lost tissue, as in the formation of liver tissue following partial hepatectomy. 2. The restoration of an ion-exchange resin to its original ionic form.

regression. A decrease in the size of a tumor or in the manifestations of a disease.

regression coefficient. The rate of change of a dependent variable with respect to an independent variable as indicated by a regression line.

regression curve. *See* regression line.

regression line. A plot of the average of a dependent variable Y as a function of an independent variable X; a plot of $\bar{Y}_x$ versus X. The line defines the amount of change of one variable per unit change in the other; if the plot does not yield a straight line, it is referred to as a regression curve.

regulator-constitutive mutant. A mutant that results from a constitutive mutation in which the regulator gene has been mutated in such a way as to prevent the formation of a repressor or to produce a defective repressor; as a result, a previously inducible enzyme becomes a constitutive one.

regulator gene. A gene that is responsible for the synthesis of a repressor that, in turn, controls an operator. The regulator gene need not be adjacent to the operator. *Abbr* R gene. *Aka* regulatory gene. *See also* enzyme induction; enzyme repression.

regulatory enzyme. An enzyme that has a regulatory function in metabolism and that has the capacity of having its catalytic activity modified through the binding of one or more metabolites; it is generally an oligomeric protein that readily undergoes conformational changes. A regulatory enzyme is frequently the first enzyme in a reaction sequence or the enzyme at a branch point of metabolic pathways. It is frequently an allosteric enzyme, having two or more topologically distinct sites, either interacting catalytic sites or interacting catalytic and regulatory sites, and, as a result of such interactions, the enzyme frequently exhibits sigmoid, rather than hyperbolic, kinetics. The metabolites that bind to the regulatory enzyme may be activating or inhibiting and are called effectors. *Aka* allosteric enzyme. *See also* sequential model; symmetry model.

regulatory protein. ALLOSTERIC PROTEIN.

regulatory site. A site on a regulatory enzyme to which an effector binds, as distinct from a catalytic site to which the substrate binds. *Aka* allosteric site.

regulatory subunit. The subunit of the regulatory enzyme aspartate transcarbamylase that has no enzymatic activity but binds the negative effector CTP. *See also* catalytic subunit.

regulon. A group of genes that are not associated as an operon but yet are responsible for the coordinate induction of a number of enzymes. Although these genes are scattered along the chromosome, they are apparently under the control of a single regulator gene.

Reichert-Meissl number. A measure of the volatile fatty acids in a fat; equal to the number of milliliters of 0.1 *N* alkali required to neutralize the volatile, water-soluble fatty acids in 5 grams of fat. *Aka* Reichert-Meissl value.

reiterated DNA sequences. REPETITIOUS DNA.

rejection. *See* immunological rejection.

relative biological effectiveness. The ratio of the biological effect produced by one ionizing radiation to that produced by an identical dose of a different ionizing radiation; also equal to the ratio of the doses of two different ionizing radiations that produce the same biological effect. For such calculations, the biological effect produced by x-rays, gamma rays, or beta particles is generally assigned a value of unity. *Abbr* RBE.

relative centrifugal force. The magnitude of the centrifugal force compared to the gravitational force; expressed in terms of multiples of *g*, as in $100,000 \times g$, where *g* is the gravitational acceleration. The relative centrifugal force (in multiples of *g*) is equal to $(1.12 \times 10^{-5})r(\text{rpm})^2$, where *r* is the distance from the center of rotation in cm, and rpm is the speed of the rotor in revolutions per minute. *Abbr* RCF.

relative configuration. 1. The comparative spatial arrangement of the atoms about two or more asymmetric carbon atoms in one molecule. 2. The arrangement of the atoms in one molecule compared to that of the atoms in a different molecule.

relative counting. The counting of radiation such that only a fraction of the actual radioactive disintegrations which occur in the sample are detected; consequently, the results are expressed as counts per minute rather than as disintegrations per minute.

relative deviation. A deviation expressed in relative terms such as a percentage average deviation.

relative error. The number of standard deviations in the error.

relative infectivity. The fraction of the initial infectivity of a virus preparation that remains when a virus is neutralized with antiviral antibodies and the reaction mixture is sampled as a function of time.

relative migration distance. *See* retardation coefficient.

relative plating efficiency. The percentage of cells that give rise to colonies when plated on a nutrient medium, compared to a control for which the absolute plating efficiency is arbitrarily taken as 100%.

relative retention. The retention volume of a component that is separated by gas chromatography relative to the retention volume of a standard.

relative specific activity. The ratio of the specific activity of the sample to that of a reference substance.

relative standard deviation. A deviation of measurements that is equal to $\sqrt{n} \times 100/n$, where *n* is either the mean of a number of measurements or the magnitude of a single measurement.

relative viscosity. The ratio of the viscosity of a solution to that of the solvent. *Sym* η_r.

relaxation. 1. The transition of a system from a suddenly disturbed equilibrium position to a new equilibrium position. 2. The return of a muscle from its contracted to its resting state.

relaxation effect. The retardation of the electrophoretic mobility of a charged particle that results from the electric field set up by the differential movement of the charged particle and its ion atmosphere.

relaxation kinetics. The kinetics of a system that undergoes relaxation.

relaxation technique. A technique for studying either a rapid reaction or the intermediate steps in a complex reaction by means of relaxation; performed by allowing the system to come to equilibrium and then disturbing the system suddenly by means of a rapid change in one variable, following which the system is allowed to come to a new equilibrium position. Depending on the variable that is being altered, the technique is referred to as temperature jump, pressure jump, etc.

relaxation time. A measure of relaxation that is equal to the time interval between the disturbance of the original equilibrium of the system and the achievement of the new equilibrium position.

relaxed control. The continued synthesis of RNA that occurs in some bacterial mutants after removal of an essential amino acid from the medium. *See also* stringent control.

relaxed helix. A nontwisted helix; a helix that is not a superhelix.

relaxed muscle. A muscle that has returned to its resting state following a contraction.

relaxed strain. A bacterial strain that shows relaxed control.

relaxin. A polypeptide hormone, produced by the corpus luteum, that causes relaxation of the symphyseal ligaments in mice and guinea pigs.

relaxing factor. The calcium pump of the sarcoplasmic reticulum of muscle.

relaxing protein. The complex formed between troponin and tropomyosin B.

release factor. 1. One of a group of protein factors that function in the release of the polypeptide chain from the ribosome at the termination stage of translation. 2. RELEASING HORMONE.

releasing hormone. A hormone that causes the release of another hormone. *Aka* releasing factor. *See also* hypothalamic hormone.

reliability. The degree to which experimental data, or methods leading to such data, reflect both accuracy and precision.

relic model. A model for the evolution of the genetic code according to which the early development of the code resulted from mechanistic processes while the later development resulted from stochastic processes.

REM. 1. Roentgen equivalent man. 2. Roentgen equivalent mammal.

remission. A temporary decrease in the size of a tumor or in the manifestations of a disease.

renal. Of, or pertaining to, the kidneys.

renal clearance. *See* clearance.

renal compensation. One of a number of mechanisms whereby the kidneys counteract the effects of either acidosis or alkalosis.

renal diabetes. RENAL GLUCOSURIA.

renal glucosuria. A pathological condition that is characterized by the recurrent excretion of glucose in the urine while the plasma concentration of glucose is either normal or slightly elevated; due to impaired reabsorption of glucose by the renal tubules.

renal threshold. *See* threshold.

renaturation. The reformation of all, or part of, the native conformation of either a protein or a nucleic acid molecule after the molecule has undergone denaturation; a reversal of denaturation.

renatured. Having undergone renaturation.

renin. A proteolytic enzyme produced by the kidney that has hormone-like properties and catalyzes the conversion of angiotensinogen to angiotensin I. *Aka* rennet.

rennin. A proteolytic enzyme from the fourth stomach of the calf that has properties similar to those of pepsin.

renotropic. Exerting an effect on the kidneys.

reovirus. A naked, icosahedral animal virus that contains double-stranded RNA.

REP. Roentgen equivalent physical.

repair enzyme. A DNA-dependent DNA polymerase that catalyzes the replacement of damaged and excised segments of single strands in double-stranded DNA. The enzyme uses the undamaged strand as a template and the repair is completed by a ligase that catalyzes the joining of the newly synthesized segments to the existing strand.

repair polymerase. DNA POLYMERASE I.

repair replication. The synthesis, by means of a repair enzyme, of single stranded DNA segments to replace damaged segments that have been excised from double-stranded DNA; in this process, the undamaged strand serves as a template. *See also* repair enzyme; cut and patch repair; patch and cut repair.

repair synthesis. REPAIR REPLICATION.

reparase. REPAIR ENZYME.

repeating polymer. A polymer that consists of identical repeating units.

repeating unit. The structural unit of a polymer, a large number of which are linked together to form the polymer; repeating units may be either identical or similar.

repeat pipet. AUTOMATIC PIPET.

repeat unit. A major periodicity in molecular structure as deduced from x-ray diffraction patterns.

repetitious DNA. A DNA that constitutes a significant fraction of the total DNA of eucaryotic cells and that is characterized by its content of an extremely large number of copies of different nucleotide sequences; the number of copies of a given nucleotide sequence may be as high as one million. *Aka* repetitive DNA.

replacement. *See* amino acid replacement; conservative substitution; radical substitution.

replica plating. A method for producing a large number of identical patterns of bacterial colonies by pressing a cylinder, covered with sterile velvet, first against the bacterial colonies in a petri dish and then against a number of plain agar surfaces in other petri dishes.

replicase. An RNA-dependent RNA polymerase that catalyzes the synthesis of RNA from the ribonucleoside-5′-triphosphates, using RNA as a template.

replicating fork. The Y-shaped region of a replicating DNA molecule in which strand separation and synthesis of new strands takes place.

replicating form. *See* replicative form.

replication. 1. The process whereby a new daughter DNA molecule is synthesized from a parent DNA molecule which serves as a template for the synthesis; one or two daughter molecules will be synthesized depending on whether the parental DNA molecule was single-

or double-stranded. 2. The process whereby a new daughter molecule is synthesized from either a parent DNA or a parent RNA molecule, with the parent molecule serving as a template for the synthesis. 3. A technique used in electron microscopy in which either a plastic or a carbon film is spread over the surface of the specimen, after which the specimen is removed and the remaining surface is subjected to shadow-casting.

replication fork. *See* replicating fork.

replication order. The number of replicating Y-forks per DNA molecule that is undergoing replication.

replication polymerase. DNA POLYMERASE II.

replicative form. A double-stranded intermediate that is formed during the replication of single-stranded DNA or RNA viruses; consists of the original viral nucleic acid strand which is hydrogen-bonded to a complementary strand. *Abbr* RF.

replicative intermediate. An intermediate in the replication of the single-stranded DNA of phage ϕX 174 that is larger than the replicative form. *Abbr* RI. *See also* rolling circle model.

replicator. A chromosome locus that forms part of the replicon and that, when acted upon by an initiator, initiates the replication of the DNA attached to it.

replicon. A section of DNA that controls the replication of the DNA; consists of a structural gene, the product of which is the initiator of replication, and of a site at which replication is initiated.

replot. SECONDARY PLOT.

reporter group. A group of atoms or a molecule that can be introduced into a protein and that has a characteristic property—such as pK value, ultraviolet absorbance, or fluorescence—which is sensitive to changes in the polarity of the medium. The changes that occur in this characteristic property when the reporter group is attached to the protein are used to explore the nature of the immediate environment of the reporter group in the protein molecule.

repressible enzyme. An enzyme, the synthesis of which is decreased when the intracellular concentration of specific metabolites reaches a certain level. *Aka* repressed enzyme. *See also* enzyme repression.

repressible system. The regulatory system consisting of the components which function in enzyme repression. *See also* enzyme repression.

repressing metabolite. COREPRESSOR.

repressor. 1. A protein molecule produced by a regulatory gene that either by itself, or in conjunction with a corepressor, prevents the synthesis of an enzyme by inhibiting the operator of

the enzyme. *Aka* aporepressor. *See also* enzyme repression. 2. IMMUNITY SUBSTANCE.

reproducibility. The degree to which an experimental measurement or result may be obtained repeatedly.

reproductive death. The death of a cell that results from a failure of a DNA molecule to replicate.

reproductive mycelium. AERIAL MYCELIUM.

repulsion. 1. The repelling electrostatic force between two like charges. 2. The tendency of linked genes to be inherited separately on different chromosomes.

RER. Rough-surfaced endoplasmic reticulum.

RES. Reticuloendothelial system.

resealed ghost. RECONSTITUTED GHOST.

residual air. The volume of air in the lungs that cannot be expelled voluntarily.

residual relative infectivity. The fraction of the initial infectivity of a virus preparation that remains after the neutralization reaction of the virus with antiviral antibodies has reached its equilibrium value.

residue. That portion of a monomer that is present in a polymer; the monomer minus the atoms removed from it in the process of polymerization. *See also* amino acid residue.

resin. 1. A polymerized support used in chromatography. *See also* ion-exchange resin; electron-exchange resin. 2. A water-insoluble, heterogeneous plant toxin.

resinoid. 1. *n* A resin-like substance. 2. *adj* Resin-like.

resinous. Of, or pertaining to, a resin.

resistance. 1. The capacity of bacteria to resist infection by phage particles; results from the inability of the phage particles to adsorb to, and to inject their DNA into, the bacterial cells. 2. DRUG RESISTANCE.

resistance factor. A bacterial episome that endows a recipient bacterium with resistance to an antibiotic; consists of a resistance-transfer factor (RTF) together with the genes for drug resistance (R genes). *Abbr* R factor; RF.

resistance-transfer factor. *See* resistance factor.

resolution. 1. The separation of enantiomers from a racemic mixture. 2. The minimum distance between two points in a microscopic specimen or in an x-ray diffraction pattern at which the points are seen as two distinct spots; resolution in which the minimum distance is small is referred to as high resolution, and resolution in which the minimum distance is large is referred to as low resolution. 3. The degree of separation between two components in a mixture as achieved by chromatographic, electrophoretic, or other separation techniques.

resolve. To achieve resolution.

resolving power. 1. The capacity of a magnifying system to reveal detail. *See also* resolution (2). 2. The capacity of a fractionating system to separate components. *See also* resolution (3).

resolving time. COINCIDENCE TIME.

resolving time loss. COINCIDENCE LOSS.

resonance. The phenomenon that a compound which can be represented by two or more equivalent, or nearly equivalent, electronic formulas has in reality a structure that is a composite of all the possible electronic formulas and that is more stable than any of the separate structures.

resonance hybrid. The true structure of a compound for which resonance structures may be formulated; the composite of all the possible resonance structures of a compound.

resonance stabilization. The stabilization of a compound as a result of resonance; due to the fact that in resonance the pi electrons of the compound are less localized and hence have a lower potential energy level than in the absence of resonance.

resorcinol test. SELIWANOFF TEST.

respiration. 1. The cellular oxidative reactions of metabolism, particularly the terminal steps, by which nutrients are broken down; the reactions which require oxygen as the terminal electron acceptor, produce carbon dioxide as a waste product, and yield utilizable energy. The major pathway of respiration consists of (a) the formation of acetyl coenzyme A from carbohydrate, fatty acid, and amino acid metabolism, (b) the citric acid cycle, and (c) the electron transport system. 2. The physical and chemical processes by which an organism transports oxygen to the tissues and removes carbon dioxide from them. 3. The act of breathing; inhaling and exhaling; inspiration and expiration.

respiratory acidosis. A primary acidosis that is caused by a decrease in respiration which leads to an increase in the carbon dioxide and carbonic acid concentrations of the plasma.

respiratory alkalosis. A primary alkalosis that is caused by an increase in respiration which leads to a decrease in the carbon dioxide and carbonic acid concentrations of the plasma.

respiratory assembly. A self-contained unit in the inner mitochondrial membrane that consists of fixed numbers of molecules of the various electron carriers; the unit has been fractionated into four respiratory complexes designated complex I-IV. *See also* complex I-IV.

respiratory chain. ELECTRON TRANSPORT SYSTEM.

respiratory chain phosphorylation. OXIDATIVE PHOSPHORYLATION.

respiratory complex. *See* complex I-IV.

respiratory control. ACCEPTOR CONTROL.

respiratory-control index. ACCEPTOR-CONTROL RATIO.

respiratory enteric orphan virus. REOVIRUS.

respiratory enzyme. 1. An enzyme that functions in cellular respiration. 2. CYTOCHROME OXIDASE.

respiratory pigment. A protein pigment that functions in the reactions of cellular respiration; hemoglobin, myoglobin, and hemerythrin are examples.

respiratory protein. RESPIRATORY PIGMENT.

respiratory quotient. The number of moles of carbon dioxide produced by a tissue or an organism divided by the number of moles of oxygen consumed during the same time. *Abbr* R.Q.

respiratory reduction. The process in plants whereby sulfate and nitrate are reduced in the course of respiration and the resultant sulfide and ammonia are excreted. *See also* assimilatory reduction.

respiratory repression. The regulation of the synthesis of an enzyme by an exogenous electron acceptor that is independent of either specific or catabolite repression.

respiratory virus. A virus that infects the respiratory system.

respirometer. An instrument for measuring and/or recording respiratory movements.

respirometry. The measurement of the kinetics of oxygen uptake and carbon dioxide evolution during respiration.

resting cell. A metabolically active cell that is not in the process of dividing.

resting heat. The heat produced by a resting muscle.

resting muscle. A muscle that is not in the process of contracting or relaxing.

resting nucleus. A metabolically active nucleus that is not in the process of dividing.

resting potential. The membrane potential of an unstimulated membrane.

restitutive protection. The protection of biomolecules against damage from an ionizing radiation by chemical substances which aid in the restoration of primary lesions to their original condition, but do not alter the number of these lesions. *See also* competitive protection.

restraining autacoid. CHALONE (2).

restricted diffusion. Diffusion through a porous medium in contrast to free diffusion which occurs in the gas phase or in solution.

restricted diffusion chromatography. GEL FILTRATION CHROMATOGRAPHY.

restricted DNA. A DNA from one cell that is prevented from replicating in a related, but not identical, cell because it is degraded by endogenous endonucleases of the related cell.

restricted rotation. The limited rotation about a bond that can be attained by an atom or a group of atoms in a molecule; the rotation about a double bond and the rotation about the bond in a ring structure are examples.

restricted transduction. SPECIALIZED TRANSDUCTION.

restricted virus. A virus, the host range of which is limited as a result of a host-induced modification.

restriction. The ability of a bacterial strain to degrade the DNA from a related, but not identical, strain. *See also* restricted DNA.

restriction allele. *See* restriction gene.

restriction enzyme. An endonuclease of a bacterial strain that catalyzes the degradation of DNA from a related, but not identical, strain.

restriction gene. A gene, the product of which is a restriction enzyme.

restriction mutant. A mutant that carries a restriction gene.

restrictive cell. A cell in which a conditional lethal mutant cannot grow.

restrictive conditions. Conditions that do not permit the growth of a conditional lethal mutant.

resultant spin. The algebraic sum of the spin quantum numbers of all the electrons of the atom or the molecule; the resultant spin is zero if all of the electrons are paired.

retardation coefficient. The slope of the line that is obtained when the relative migration distance of a protein in starch gel electrophoresis is plotted as a function of the reciprocal of the starch concentration. The relative migration distance is the ratio of the migration distance at a given starch concentration to that at a standard starch concentration.

retention index. *See* Kovats retention index system.

retention of configuration. The maintenance of a given enantiomeric configuration about an asymmetric carbon atom in the course of a chemical reaction.

retention time. The time between the injection of a sample into a gas chromatographic column and the appearance of the peak maximum.

retention volume. The volume of gas in gas chromatography that is required to elute the compound of interest. *See also* hold-up volume.

reticulate evolution. An evolutionary pattern that, when diagrammed, resembles a net; such patterns are characteristic of the evolution of plant species.

reticulocyte. An immature red blood cell that actively synthesizes hemoglobin and that possesses functioning pathways of glycolysis, the hexose monophosphate shunt, and the citric acid cycle.

reticuloendothelial system. Collectively, the cells in spleen, liver, bone marrow, and other tissues that are involved primarily in phagocytosis and in the metabolism of hemoglobin. *Abbr* RES.

reticuloendothelium. RETICULOENDOTHELIAL SYSTEM.

retinal. The aldehyde form of vitamin A; retinal$_1$ is the aldehyde form of vitamin A$_1$ and retinal$_2$ is the aldehyde form of vitamin A$_2$. *See also* 11-*cis* retinal; all-*trans* retinal.

11-cis **retinal.** An isomeric form of retinal that is converted by the action of light to the all-*trans* isomer.

retinaldehyde. RETINAL.

retinene. RETINAL.

retinol. *See* vitamin A.

retrogradation. The formation of microcrystals and precipitates that occurs in starch gels and in starch solutions upon standing; caused by the separation of amylose molecules that become aligned and hydrogen bonded.

retroinhibition. FEEDBACK INHIBITION.

retrosteroid. A synthetic steroid that has a structure at carbon atoms 9 and 10 which is opposite to that in progesterone. Retrosteroids are made from lumisterols and are extremely active as progestogens.

reversal. The change of cancer cells to normal cells or to benign tumor cells.

reversed phase partition chromatography. Partition chromatography in which the mobile phase consists of a polar solvent and the stationary phase consists of a nonpolar solvent. *Aka* reversed phase chromatography.

reverse electron transport. The reduction of NAD$^+$ to NADH by mitochondria in the presence of ATP. *Aka* reverse electron flow.

reverse isotopic trapping. A technique for determining whether more than one metabolic pathway leads to the synthesis of compound A; performed by continuously administering a labeled precursor X, isolating compound A, and determining the specific activity of compound A. By comparing the specific activity of compound A with that of the administered precursor X, it is possible to decide whether A is synthesized solely from X, or also from other precursors by way of different pathways.

reverse mutation. REVERSION (1).

reverse osmosis. The movement of water or another solvent from a more concentrated to a more dilute solution across a semipermeable membrane; the desalination process in which seawater is forced across a semipermeable membrane is an example.

reverse passive anaphylaxis. The anaphylactic reaction produced in an animal by injecting it first with an antigen which is itself an immunoglobulin and, following a latent period, by injecting it with an antiserum that is specific for the immunoglobulin used as an antigen.

reverse transcriptase. An RNA-dependent DNA polymerase that catalyzes the synthesis of DNA from deoxyribonucleoside-5′-triphosphates, using RNA as a template. The enzyme has been found in RNA-containing tumor viruses and is thought to be related to the tumorigenic properties of these viruses. The reaction catalyzed by the enzyme is in contradiction to the flow of genetic information described by the central dogma $DNA \rightarrow RNA \rightarrow protein$.

reverse transcription. The reaction catalyzed by the enzyme reverse transcriptase.

reversible boundary spreading test. A test applied in moving boundary electrophoresis to assess the homogeneity of the sample. A boundary is first allowed to migrate for some distance and the electric field is then reversed; any sharpening of the boundary that occurs upon reversal of the electric field reflects heterogeneity in the sample and cannot be attributed to diffusion.

reversible inhibitor. An inhibitor that binds to an enzyme in an equilibrium reaction so that the inhibition can be reversed by removal of the inhibitor from the enzyme by such processes as dialysis or ultrafiltration. See also irreversible inhibitor; reactivation.

reversible metabolic pathway. A metabolic pathway in which the equilibrium constants of the individual reactions are such that products can be converted to the reactants in significant amounts. The pathway is called directly reversible if the reaction sequence can be traversed in both directions $A \rightleftharpoons B \rightleftharpoons C \rightleftharpoons D$ and is called indirectly reversible if it is traversed principally in one direction $A \rightleftharpoons B \rightleftharpoons C \rightleftharpoons D$.

$$\underset{E}{\swarrow \downarrow}$$

reversible process. A process in equilibrium thermodynamics in which the system goes from an initial to a final equilibrium state through a succession of equilibrium states. Thus the system is always close to equilibrium and the direction of the process can be reversed by an infinitesimal change. The net entropy change for the system plus its surroundings is zero for such a process.

reversible reaction. A chemical reaction that (a) establishes an equilibrium and can be made to proceed in either direction by a change in the conditions, or (b) has an equilibrium constant of the order of 1.0, or (c) proceeds equally well in both directions because of other factors.

reversion. 1. The change of a mutant gene to its state prior to mutation; a reverse mutation. Also referred to as true reversion as distinct from restoration of the genetic function by a suppressor mutation. 2. DEADAPTATION.

reversion index. The mutation index for a reverse mutation.

reversion spectroscope. A spectroscope that allows the formation of two spectra alongside each other, with the wavelength changing in one direction for one spectrum and in the opposite direction for the other spectrum.

revertant. 1. A cell, a virus, or an organism that carries a gene that has undergone reversion. 2. A gene that has undergone reversion.

Reynold's number. A quantity that characterizes the flow of a liquid through a cylindrical tube; equal to $2\rho ua/\eta$, where u is the average velocity, a is the radius of the tube, ρ is the density, and η is the viscosity.

RF. 1. Replicative form. 2. Resistance factor.

R factor. Resistance factor.

R_f value. The ratio of the distance traveled by a compound in flat-bed chromatography to that traveled by the solvent.

R_G. Average radius of gyration.

R gene. 1. Regulator gene. 2. Drug resistance gene; see resistance factor.

R group. 1. That portion of an organic molecule that does not contain the functional group of interest; an organic radical. 2. The side chain of an amino acid.

rhabdovirus. An enveloped, helical animal virus that contains single-stranded RNA.

rhamnose. A deoxysugar that occurs in some bacterial cell walls.

R_h blood group system. A human blood group system in which the R_h factor is the antigen and individuals are either R_h positive or R_h negative. An R_h positive baby of an R_h negative mother may be born with a hemolytic disease called erythroblastosis fetalis.

rheology. The science that deals with the deformation and flow of matter.

rheostat. A variable resistor.

Rhesus factor. R_h FACTOR.

R_h factor. The antigen of the R_h blood group system.

rhinovirus. A virus that belongs to a subgroup of picornaviruses and that causes respiratory infections.

rhizopterin. SLR FACTOR.

rhodopsin. A visual pigment, consisting of rod opsin plus retinal$_1$, that occurs in mammals and other vertebrates and that has an absorption maximum at 500 nm.

rhodopsin cycle. VISUAL CYCLE.

Rhodospirillum rubrum. A photosynthetic bacterium used for studies of photosynthesis.

RHO factor. A protein that is required for the correct termination of transcription. $Sym\,\rho$.

RHP cytochrome. A cytochrome that contains two hemes per molecule and that is considered to be a variant of cytochrome c; the heme groups are considered to be bound through thioether linkages as in cytochrome c, but to lack hemochrome linkages to extraplanar ligands of the protein at physiological pH.

rH scale. A scale for evaluating oxidation-reduction reactions that is based on rH values rather than on electrode potentials. The rH value is the negative logarithm of the hydrogen pressure in atmospheres, and runs from zero for a hydrogen pressure of 1 atm to 41 for an oxygen pressure of 1 atm. The rH value of a reaction is obtained by relating the reaction to the hydrogen half-reaction; rH values for most biochemical systems fall in the range of 0 to 25.

RI. Replicative intermediate.

RIA. Radioimmunoassay.

Rib. Ribose.

α-ribazole. The compound, 1-α-D-ribofuranosyl-5,6-dimethylbenzimidazole, that forms part of the structure of vitamin B_{12} and that has a nucleoside structure in which an uncommon nitrogenous base is linked to the sugar by means of an α-glycosidic bond.

ribitol. A five-carbon sugar alcohol that forms part of the structure of riboflavin, FMN, and FAD.

riboflavin. Vitamin B_2; a vitamin that is widely distributed in nature and the coenzyme forms of which are FMN and FAD.

riboflavin mononucleotide. FLAVIN MONONUCLEOTIDE.

riboflavin phosphate. FLAVIN MONONUCLEOTIDE.

ribonuclease. An endonuclease that catalyzes the hydrolysis of RNA. Ribonuclease T_1 leads to the production of mono- and oligonucleotides consisting of, or terminating in, a 3'-guanine nucleotide; ribonuclease T_2 leads to the production of mono- and oligonucleotides consisting of, or terminating in, a 3'-adenine nucleotide. *Abbr* RNase; RNAase. *See also* pancreatic ribonuclease.

ribonucleic acid. The nucleic acid (*abbr* RNA) that occurs in three major forms as ribosomal, transfer, and messenger ribonucleic acid, all of which function in the biosynthesis of proteins. RNA is a polynucleotide that is characterized by its content of D-ribose and the pyrimidines uracil and cytosine.

ribonucleoprotein. A conjugated protein that contains RNA as the nonprotein portion. *Abbr* RNP.

ribonucleoside. A nucleoside of D-ribose.

ribonucleotide. A nucleotide of D-ribose.

ribose. The five-carbon aldose that is the carbohydrate component of ribonucleic acid. *Abbr* Rib.

ribose nucleic acid. *See* ribonucleic acid.

riboside. A glycoside of ribose.

ribosomal. Of, or pertaining to, ribosomes.

ribosomal DNA. The DNA that codes for ribosomal RNA.

ribosomal particle. A ribosome or any of its complete subunits.

ribosomal precursor RNA. A high-molecular weight RNA, having a sedimentation coefficient of 45S, that is synthesized in the nucleus of eucaryotic cells and that serves as the precursor of both the 18 and the 28S ribosomal RNA.

ribosomal protein. A protein that forms part of the ribosome. Ribosomal proteins are linked noncovalently to the ribosomal RNA and, together with the ribosomal RNA, form the two subunits of the ribosome. In bacterial ribosomes, there are some 50 different ribosomal proteins per ribosome. *Abbr* r-protein.

ribosomal RNA. The RNA that is linked noncovalently to the ribosomal proteins in the two ribosomal subunits and that constitutes about 80% of the total cellular RNA. Three types of ribosomal RNA have been identified, having sedimentation coefficients of about 5S, 16 to 19S, and 23 to 29S, depending on the source of the ribosomes. *Abbr* rRNA; RNA_r.

ribosomal subparticle. 1. A ribosomal subunit, such as the 30S or the 50S particle of bacterial ribosomes. 2. A particle that either is a precursor of a ribosome, or is prepared from a ribosome by the removal of ribosomal proteins. *See also* intersome.

ribosomal subunit. One of the two ribonucleoprotein particles that make up the complete ribosome; the 30S or the 50S particle in bacteria, the 40S or the 60S particle in plant and animal cells.

ribosomal subunit exchange. *See* subunit exchange.

ribosome. 1. One of a large number of subcellular, nucleoprotein particles that are composed of approximately equal amounts of RNA and protein and that are the sites of protein synthesis in the cell. Each ribosome is roughly spherical in shape, has a diameter of about 200 Å, and consists of two unequal subunits linked together noncovalently by means of magnesium ions and other bonds. Four classes of ribosomes have been identified (bacterial, plant, animal, and mitochondrial) and they can be characterized by the sedimentation coefficients of the monomers, of the subunits, and of the ribosomal RNA. The bacterial ribosome contains about 50 different protein molecules and 3 different RNA molecules; the smaller subunit contains about 20 protein molecules and 1 RNA molecule; the larger one contains about 30 protein molecules and 2 RNA molecules. The ribosome has two binding sites for transfer RNA (A site, P site) and can attach to messenger RNA as well. *Abbr* Rb. 2. RIBOSOMAL SUBUNIT.

ribosome binding technique. *See* binding assay (2).

ribosome cycle. The set of reactions whereby ribosomal subunits combine to form the intact ribosome during the initiation of translation, travel along the messenger RNA as intact ribosomes, and dissociate back to the subunits during the termination of translation. *See also* subunit exchange.

ribosome dimer. An aggregate of two ribosomes, consisting of two small and two large ribosomal subunits.

ribosome dissociating factor. A factor that promotes the dissociation of the ribosome monomers into the two subunits during the termination of translation.

ribosome exchange. SUBUNIT EXCHANGE.

ribosome monomer. A complete ribosome that is composed of one small and one large subunit.

ribosome run-off. The loss from polysomes of ribosomes that have not completed the synthesis of the polypeptide chain.

ribosubstitution. The replacement of some of the deoxyribonucleotides in a DNA by ribonucleotides; achieved by the in vitro synthesis of DNA under conditions that allow the incorporation of ribonucleotides from a mixture of ribo- and deoxyribonucleoside-5′-triphosphates. Ribosubstituted DNA can be used as an aid in the determination of the base sequence of the DNA.

ribosylthymine. RIBOTHYMIDINE.

5-ribosyluracil. PSEUDOURIDINE.

ribothymidine. The ribonucleoside of thymine; an unusual nucleoside that does not, as a rule, occur in RNA.

ribothymidylic acid. The ribonucleotide of thymine.

ribulose. A five-carbon ketose that is an intermediate in the hexose monophosphate shunt. *Abbr* Rul.

ribulose-1,5-diphosphate. A five-carbon ketose that is the acceptor of carbon dioxide in the Calvin cycle. *Abbr* R1,5DP.

ribulose diphosphate carboxydismutase. DIPHOSPHORIBULOSE CARBOXYLASE.

Rice test. INDIRECT COMPLEMENT FIXATION TEST.

ricin. A plant protein in the seeds of *Ricinus communis* that is toxic to animals and man, inhibits protein synthesis, and has antitumor activity.

rickets. A disease of children that is characterized by a softening and bending of the bones and that is caused by a deficiency of vitamin D.

rifampicin. A semisynthetic antibiotic that inhibits the transcription of DNA to RNA by DNA-dependent RNA polymerase.

rifamycin. An antibiotic that inhibits the transcription of DNA to RNA by DNA-dependent RNA polymerase.

rigor mortis. The irreversible contraction of muscles upon the death of an animal.

ring. A chain of atoms in which the ends of the chain are linked together covalently.

Ringer's solution. A salt solution that is approximately isotonic to the blood and to the lymph of mammals; used for the temporary maintenance of living cells. As first proposed, it consisted of sodium chloride, potassium chloride, calcium chloride, and sodium bicarbonate; various modifications of this solution are now in use.

ring test. A rapid and simple precipitin test in which a solution of antigen is layered over a solution of antibodies in either a test tube or in a capillary, and the presence or absence of a precipitate at the interface is determined.

RISA. Acronym for "radioimmunosorbent assay"; an isotopic technique for the demonstration of minute amounts of immunoglobulins of the IgE type.

rise curve. The increase in color intensity of a sample as a function of time; used in reference to determinations with an autoanalyzer.

rise period. The time interval during which the extracellular titer of a phage multiplication cycle increases to a maximum.

R locus. The position on a chromosome of a regulator gene.

rII locus. A segment of the chromosome of T4 phage to which fine structure genetic mapping was first applied.

R-meter. A radiation meter that is calibrated to indicate roentgens.

R mutant. 1. A bacterial mutant that gives rise to a rough colony. 2. RAPID LYSIS MUTANT.

R_m value. A chromatographic term that is related to the R_f value of a compound; $R_m = (1 - R_f)/R_f$.

Rn. Radon.

RNA. Ribonucleic acid.

RNAase. Ribonuclease.

RNA coding triplet. CODON.

RNA-dependent DNA polymerase. REVERSE TRANSCRIPTASE.

RNA-dependent RNA polymerase. REPLICASE.

RNA_m. Messenger RNA.

RNA phage. An RNA-containing phage.

RNA polymerase. An enzyme that functions in the transcription of DNA and that catalyzes the synthesis of RNA from the ribonucleoside-5′-triphosphates using DNA as a template; referred to as DNA-dependent RNA polymerase to distinguish it from RNA-dependent RNA polymerase.

RNA_r. Ribosomal RNA.

RNA replicase. *See* replicase.

RNase. Ribonuclease.

RNA synthetase. RNA-DEPENDENT RNA POLYMERASE.

RNA$_t$. Transfer RNA.

RNA virus. An RNA-containing virus. *See also* oncornavirus.

RNP. Ribonucleoprotein.

Robison ester. Glucose-6-phosphate.

Rochelle salt. Potassium sodium tartrate.

rod. A light receptor in the retina of vertebrates that functions in night vision.

rodenticide. A chemical that kills rodents.

rod threshold. The lowest light intensity that the fully dark-adapted eye can detect.

roentgen. A quantity of x- or gamma radiation such that the associated corpuscular emission per cubic centimeter of dry air at 0°C and 760 mm Hg produces, in air, ions that carry one electrostatic unit of electricity of either sign. *Sym* r; R.

roentgen equivalent man. The product of the radiation absorbed dose (rad) and the relative biological effectiveness (RBE); the quantity of radiation that, when absorbed by man, produces an effect equivalent to the absorption of 1 R of x- or gamma radiation. *Abbr* REM. *Aka* roentgen equivalent mammal.

roentgen equivalent physical. The amount of ionizing radiation capable of producing 1.615×10^{12} ion pairs per gram of tissue; the amount of ionizing radiation that will result in the absorption by tissue of 93 erg/gram of tissue. *Abbr* REP.

roentgen rays. X-RAYS.

rohferment. A crude enzyme preparation from almonds that is rich in glycosidases, particularly in β-D-glucosidase and β-D-galactosidase.

Rohrschneider constant. A constant for relating the gas chromatographic retention behavior of a compound to the polarity of the liquid stationary phase.

rolling circle model. A model for the replication of duplex circular DNA. According to this model, a nuclease nick in one strand is followed by the addition of nucleotides to the 3′-end of the nicked strand, a reaction catalyzed by DNA polymerase. At the same time, the 5′-end of the strand is rolled out as a free tail of increasing length, resulting in intermediates larger than the original duplex. Small fragments are then synthesized complementary to the free tail and are eventually joined together through the action of a ligase.

Roman square. LATIN SQUARE.

root-mean-square end-to-end distance. A measure of the spatial extension of a polymer; equal to the square root of the average of the squares of the distances between the ends of the polymer, carried out over all possible conformations of the polymer.

rosette technique. IMMUNOCYTE ADHERENCE.

rotamer. A rotational isomer; a conformational isomer resulting from a rotation about single bonds.

rotary. *See* rotational.

rotary evaporator. FLASH EVAPORATOR.

rotating crystal method. A method for the analysis of a single crystal by means of x-ray diffraction. The crystal is mounted and rotated about an axis, thereby producing a large number of diffraction spots; rotation of the crystal about all of its axes produces the maximum number of diffraction spots.

rotation. The turning about an axis.

rotational base substitution. The process whereby a base in one DNA strand exchanges position with its complementary base in the second strand. This may occur if the glycosidic bonds between the bases and their sugar molecules are broken by irradiation, and the hydrogen-bonded base pair is then rotated prior to being reinserted into the strands.

rotational diffusion. The rotation of molecules about their axes that results in their achieving a random orientation. *See also* rotational relaxation.

rotational diffusion coefficient. A measure of rotational diffusion that depends on the size and shape of the diffusing particle; specifically, $\Theta = RT/N\zeta$, where Θ is the rotational diffusion coefficient, ζ is the rotational frictional coefficient, N is Avogadro's number, R is the gas constant, and T is the absolute temperature.

rotational frictional coefficient. A measure of the frictional resistance experienced by a particle in solution that is equal to the frictional force divided by the angular velocity of the particle. *See also* rotational diffusion coefficient.

rotational relaxation. The relaxation that takes place when a field which led to the orientation of molecules, otherwise randomly oriented due to Brownian motion, is suddenly turned off. The return of the molecules to their random orientation is characterized by the rotational diffusion coefficient.

rotational relaxation time. The time required, during rotational relaxation, for the average value of $cos\ \phi$ for all the solute molecules to fall to $1/e$ of its original value; e is the base of natural logarithms and ϕ is the angle through which the molecule has rotated away from its original direction of orientation.

rotational strength. A quantity that is used in calculations of circular dichroism and that is related to the integrated value, over an absorption band, of the difference between the extinction coefficients for left and right circularly polarized light.

rotational symmetry. The symmetry of a body that exists when identical structures are produced as the body is rotated about an axis. *See also* axis of rotational symmetry.

rotational transition. The transition of a molecule in which it rotates about an axis; rotational transitions require relatively small amounts of energy and are induced by infrared radiations of long wavelength.

rotation angle. TORSION ANGLE.

rotation axis. AXIS OF ROTATIONAL SYMMETRY.

rotation diagram. An x-ray diffraction pattern obtained by the rotating crystal method. *Aka* rotation photograph.

rotation-reflection axis. *See* reflection symmetry.

rotation technique. *See* photographic rotation technique.

rotatory. Of, or pertaining to, optical rotation.

rotatory dispersion. *See* optical rotatory dispersion.

rotatory power. SPECIFIC ROTATION.

rotenone. An insecticide that inhibits the electron transport system between the flavoproteins and coenzyme Q.

Rothera's test. A qualitative test for ketone bodies in urine that is based on the production of a blue-purple color upon treatment of urine with sodium nitroprusside and ammonium hydroxide.

rotometer. A device for measuring the rate of gas flow; used in gas chromatography.

rotor. The container that is rotated in a centrifuge and that holds the tubes filled with the solution which is subjected to centrifugation. *Aka* head.

rough microsomes. ROUGH-SURFACED ENDOPLASMIC RETICULUM.

rough-surfaced endoplasmic reticulum. That portion of the endoplasmic reticulum to which a large number of ribosomes is attached. *Abbr* RER.

rounds of mating. The number of matings in the line of ancestry of an average phage particle in the phage population.

Rous sarcoma. A virus-induced malignant tumor in chickens.

Rous sarcoma virus. A virus that contains single-stranded RNA and causes cancer in chickens; belongs to the group of leukoviruses. *Abbr* RSV.

royal jelly. A liquid nutrient produced by worker bees which, when fed to female larvae, leads to the production of queens.

R5P. Ribose-5-phosphate.

rpm. Revolutions per minute.

r-protein. Ribosomal protein.

R.Q. Respiratory quotient.

rRNA. Ribosomal RNA.

r-strand. CRICK STRAND.

RSV. Rous sarcoma virus.

RTF. Resistance-transfer factor.

rubber. A natural high-molecular weight polymer of isoprene units; a polyterpene.

rubella virus. The virus that causes German measles and that belongs to the paramyxovirus group.

Rubner's law. The principle that the heat produced by the metabolism of animals is proportional to the surface area of the animals; the principle has been shown to be approximately correct.

rubredoxin. IRON-SULFUR PROTEIN (a).

Ruff degradation. A degradative technique for aldonic acids whereby a carbon atom is removed by treatment with hydrogen peroxide in the presence of ferrous ions, and the sugar is converted to the next lower aldose.

Rul. Ribulose.

run-off ribosome. *See* ribosome run-off.

runt disease. GRAFT-VERSUS-HOST REACTION.

runting syndrome. GRAFT-VERSUS-HOST REACTION.

rutherford. The amount of radioactive substance that undergoes 10^6 disintegrations per second.

Rutherford scattering. The scattering of radiation that results from elastic collisions of alpha or beta particles with atomic nuclei.

R value. The fraction of the solute, in partition chromatography, that is present in the mobile phase.

S

s. 1. Sedimentation coefficient. 2. Second. 3. Standard deviation.

$s_{20,w}^0$. Standard sedimentation coefficient.

S. 1. Substrate. 2. Svedberg unit. 3. Sulfur. 4. Entropy. 5. Bacterial colony of smooth morphology. 6. Period of DNA synthesis in the cell cycle. 7. S-configuration. 8. Sensitivity of a bacterial strain to a phage or an inhibitor (used as a superscript). 9. Serine.

S^{35}. A radioactive isotope of sulfur that emits a weak beta particle and has a half-life of 87.2 days.

S.A. Specific activity.

Sabin vaccine. A poliomyelitis vaccine that is given orally and that consists of live, attenuated, poliovirus preparations.

saccharase. SUCRASE.

saccharic acid. ALDARIC ACID.

saccharide. CARBOHYDRATE.

saccharimeter. An instrument, such as a polarimeter, or a device, such as a fermentation tube, for determining the amount of sugar in a solution. *Var sp* saccharometer.

saccharin. *o*-Sulfobenzimide; an artificial sweetener.

saccharogenic amylase. BETA AMYLASE.

saccharogenic method. An assay of the enzyme amylase that is based on a determination of the amount of product formed.

Saccharomyces cerevisiae. A species of yeast that includes the strains of baker's and brewer's yeast.

saccharose. SUCROSE.

sacculus. The sack-like peptidoglycan structure of the bacterial cell wall.

sacrifice. Euphemism for "to kill"; used in reference to experiments with animals.

Sagavac. Trademark for a group of agarose gels used in gel filtration chromatography.

Sakaguchi reaction. A colorimetric reaction for arginine that is based on the production of a red color upon treatment of the sample with α-naphthol and sodium hypochlorite.

salimeter. A hydrometer that is calibrated for the determination of either the specific gravity of a saline solution or the concentration of sodium chloride in the solution. *Aka* salinometer.

saline. 1. *n* An aqueous solution of sodium chloride. 2. PHYSIOLOGICAL SALINE. 3. *adj* Of, or pertaining to, salt.

salivary juice. The digestive juice, consisting of the saliva, that is secreted by the salivary glands into the mouth and that contains the enzyme ptyalin. *Aka* salivary fluid.

Salkowski reaction. A modification of the Liebermann-Burchard reaction for cholesterol in which acetic anhydride is omitted.

Salk vaccine. A poliomyelitis vaccine that is given by injection and that consists of formalin-killed poliovirus preparations.

salmine. A protamine isolated from salmon sperm.

salt fractionation. The fractional precipitation of proteins by means of inorganic salt solutions, frequently those of ammonium sulfate.

salting in. The increase in the solubility of a protein that is produced in solutions of low ionic strength by an increase of the concentrations of neutral salts; due to a stabilization of the charged groups on the protein as a result of a decrease in the activity coefficients of these groups.

salting out. The decrease in the solubility of a protein that is produced in solutions of high ionic strength by an increase of the concentrations of neutral salts; believed to be due to a partial dehydration of the protein as a result of the competition between the protein and the salt ions for solvating water molecules.

salting-out chromatography. A chromatographic technique in which water-soluble organic compounds are separated by ion-exchange chromatography using an aqueous salt solution for elution.

salting-out constant. A constant characteristic of the solubility behavior of a protein; specifically, $log\ S = log\ S_0 - K(\Gamma/2)$, where S is the actual solubility of the protein, S_0 is the hypothetical solubility at zero ionic strength, $\Gamma/2$ is the ionic strength, and K is the salting-out constant.

salt link. IONIC BOND.

salt precipitation. SALT FRACTIONATION.

salt respiration. ANION RESPIRATION.

salvage metabolic pathway. A pathway that utilizes compounds formed in catabolism for biosynthetic purposes, even though these compounds are not true intermediates of the corresponding normal biosynthetic pathway. Thus free purines may be salvaged from the hydrolysis of nucleotides and then used for the biosynthesis of nucleotides; likewise, free choline may be salvaged from the degradation of phosphatidyl choline and then used for the biosynthesis of phosphoglycerides.

SAM. *S*-Adenosyl-L-methionine.

sample ampholyte. An ampholyte of the mixture that is fractionated by isoelectric focusing, as distinct from a carrier ampholyte that is used to form the pH gradient.

sampling error. An error due to the method of obtaining samples, to the inadequacy of the number of samples, or to the inadequacy of the size of the sample.

sandwich technique. INDIRECT FLUORESCENT ANTIBODY TECHNIQUE.

Sanger reaction. The reaction of the Sanger reagent, 1-fluoro-2,4-dinitrobenzene, with the free alpha amino group of amino acids, peptides, or proteins; the reaction yields a dinitrophenyl derivative that is useful for the chromatographic detection and quantitative estimation of amino acids, peptides, and proteins, as well as for endgroup analysis of N-terminal amino acids in peptides and proteins.

Sanger reagent. *See* Sanger reaction.

S antigen. Soluble antigen; an incomplete and noninfectious virus form that appears early in the course of certain viral infections.

sapogenin. A steroid that occurs in plants in the form of glycosides known as saponins.

saponifiable fraction. The fraction of total lipid that, after saponification, is soluble in water and insoluble in ether.

saponifiable lipid. A lipid that can be hydrolyzed with alkali to yield soap as one of the products; glycerolipids and sphingolipids are two major saponifiable lipids.

saponification. The alkaline hydrolysis of a lipid, particularly a glyceride, that yields soap as one of the products.

saponification equivalent. The number of grams of fat saponified by one mole of potassium hydroxide. *Abbr* SE.

saponification number. A measure of the average chain length of the fatty acids in a fat; equal to the number of milligrams of potassium hydroxide required to saponify 1 gram of fat. *Aka* saponification value.

saponin. A water-soluble surface-active plant substance that forms soapy solutions even at high dilutions; saponins are powerful hemolytic agents that either have a triterpenoid structure or are steroid glycosides.

sarcolemma. The membrane surrounding the fiber of a striated muscle.

sarcoma. A malignant tumor that arises from connective tissue.

sarcomere. The longitudinal repeat unit of a myofibril; the segment that extends from one Z line to the next.

sarcoplasm. The intracellular fluid that bathes the myofibrils of a muscle cell.

sarcoplasmic reticulum. The endoplasmic reticulum of a muscle cell.

sarcosine. *N*-Methylglycine; an amino acid that is an intermediate in the metabolism of one-carbon fragments and that forms part of the structure of the antibiotic actinomycin D.

sarcosome. A mitochondrion from a striated muscle.

sarcotubule. A transverse tubule of the T-system of muscle.

sardinine. A protamine isolated from sardines.

satellite DNA. A DNA fraction that has properties different from the bulk DNA and that can be separated from it by density gradient sedimentation equilibrium in a cesium chloride solution.

satellite DNA of mouse. *See* mouse satellite DNA.

satellite phenomenon. CROSS-FEEDING.

satellite RNA. An RNA that has a sedimentation coefficient of 5 to 8S, small amounts of which are found in association with plant 26S ribosomal RNA.

satellite virus. A small virus that occurs in association with another virus, such as the tobacco necrosis satellite virus or the adenovirus associated virus.

satellitism. CROSS-FEEDING.

saturated fatty acid. A fatty acid that contains a saturated alkyl chain.

saturated solution. *See* saturation (3).

saturation. 1. The state of an organic compound in which it contains only single bonds between the carbon atoms. 2. The conversion of an unsaturated organic compound to a saturated one. 3. The state of a solution in which it contains the maximum amount of solute that can be dissolved permanently in that volume of solvent under specified conditions. 4. The state of a macromolecule in which it has bound the maximum number of ligands of a given type, as in the saturation of hemoglobin or myoglobin with oxygen, or in the saturation of an enzyme with its substrate.

saturation backscattering. The maximum increase in counting rate that is observed when increasing thicknesses of backing material are placed under a radioactive sample.

saturation-backscattering thickness. The thickness of backing material required to achieve saturation backscattering.

saturation current. The current produced in an ionization chamber when the potential is of sufficient magnitude to permit the collection of all the ions; the saturation current is independent of the applied voltage.

saturation kinetics. The kinetics of a reaction in which the rate of the reaction levels off with increasing concentrations of a component, as is the case for a simple enzymatic reaction and for mediated transport.

saturation thickness. The thickness of a radioactive sample such that any additional increase in its thickness will not increase the observed counts any further.

sawhorse projection. A representation of the arrangement of the atoms in a molecule that provides a three-dimensional view of the molecule and resembles a sawhorse in outline.

SBP. Serum blocking power.

sc. 1. *adj* Subcutaneous. 2. *adv* Subcutaneously.

scalar. 1. *n* A nondirectional quantity. 2. *adj* Of, or pertaining to, a nondirectional quantity.

scalar reaction. A reaction that is nondirectional in which the components either are free to move at random or are fixed in a random order; an overall reaction in solution, but not the individual molecular event, is generally a scalar reaction.

scale method. LAMM SCALE DISPLACEMENT METHOD.

scaler. An electronic recording device that produces one output pulse for a given number of input pulses.

scanner. 1. An instrument for measuring the distribution of either color intensity or radioactivity on a chromatogram or on an electropherogram. 2. A photoelectric scanning attachment for the analytical ultracentrifuge that provides a plot of absorbance versus radial distance and that is used in conjunction with the ultraviolet absorption optical system.

scanning. Any measurement performed systematically across an experimental pattern, such as the measurement of either color intensity or radioactivity as a function of distance across a radiochromatogram.

scanning electron microscope. An electron microscope in which a three-dimensional view of the specimen is produced by the deflection of primary and secondary electrons and by the use of an electron beam to scan the specimen. *Abbr* SEM.

Scatchard plot. A graphical representation of binding data for the determination of intrinsic association constants and of the number of noninteracting binding sites per molecule; based on the Hill equation for the case of $n_H = 1$, and consists of a plot of $r/[S]$ versus r. *See also* Hill equation.

scatter diagram. A plot of data as points in a plane rectangular coordinate system that is made to determine whether there is a correlation between the two variables indicated on the axes.

scattering. The dispersion of radiation by matter in directions other than that of the incident beam as a result of collisions and/or interactions.

Schardinger dextrin. A group of oligosaccharides that are formed by the action on starch of amylase from *Bacillus macerans*; in-

cludes α-dextrins, which contain six glucose residues per molecule, and β-dextrins, which contain seven glucose residues per molecule.

Schardinger enzyme. XANTHINE OXIDASE.

Schardinger reaction. A test for oxidase activity in milk that is based on the incubation of milk with formaldehyde and methylene blue; the blue color of the solution disappears as the methylene blue is reduced in the process of serving as a hydrogen acceptor for the oxidation of formaldehyde.

Scheraga-Mandelkern equation. An expression for a function, denoted β, that is based on various physical parameters and that is used as an aid in determining the shape of a macromolecule in solution. For calculated values of β greater than 2.12×10^6, the molecule is best approximated by a prolate elliposid of revolution; for calculated values of β approximately equal to 2.12×10^6, the molecule is best approximated by an oblate ellipsoid of revolution.

Schiff's base. The condensation product formed between a primary amine and either an aldehyde or a ketone.

Schiff's reagent. A reagent for the detection of aldehydes that consists of fuchsin, bleached with sulfurous acid; produces a red color upon reaction with an aldehyde.

Schiff's test. 1. A test for uric acid that is based on the reduction of silver ions to metallic silver. 2. A test for urea that is based on the formation of a purple color upon treatment of the sample with furfural and hydrochloric acid. 3. A test for aldehydes that is based on the reaction of aldehydes with Schiff's reagent.

Schilling test. A test for measuring the rate of absorption of vitamin B_{12} from the intestine by administering vitamin B_{12}, labeled with radioactive cobalt, to an individual.

schlieren optical system. An optical system that focuses ultraviolet light passing through a solution in such a fashion that a photograph is obtained which represents a plot of refractive index gradient as a function of distance in the solution. A boundary in the solution appears as a peak on the photographic plate and measurements are made on this plate with a microcomparator. The optical system is used in the analytical ultracentrifuge and in the Tiselius electrophoresis apparatus.

Schmidt-Thannhauser procedure. A procedure for the extraction of nucleic acids from tissue by digestion of the tissue with dilute alkali; DNA and RNA are obtained as separate fractions by this procedure, since the alkali hydrolyzes RNA but not DNA.

Schneider procedure. A procedure for the extraction of nucleic acids from tissue by treatment of the tissue with either trichloroacetic acid or perchloric acid; DNA and RNA are not obtained as separate fractions by this procedure.

Schuetz-Borrisow rule. An empirical rule that states that the velocity of an enzymatic reaction is proportional to the square root of the enzyme concentration; the rule was developed for pepsin and applies, under limited conditions, to pepsin and other proteolytic enzymes.

Schultz-Dale reaction. An anaphylactic response that is produced in a sensitized animal by the introduction of antigens into its uterus.

Schwann cell. The cell surrounding a myelinated nerve axon.

scintillation. The emission of flashes of light by fluorescent substances subsequent to their excitation by means of radioactive or other radiation.

scintillation autography. A modification of radioautography in which the labeled material is placed in a solution containing a scintillator, and the scintillations produced by the radioactive disintegrations expose a photographic film.

scintillation counter. A radiation counter in which incident ionizing particles or incident photons are counted by means of the scintillations that they induce in a fluor.

scintillation detector. A solid or liquid fluor together with the electronic circuitry that converts the light flashes of the fluor to electrical pulses. *See also* external-sample scintillation counter; internal-sample scintillation counter.

scintillation spectrometer. A scintillation counter designed to permit the measurement of the energy distribution of radiation.

scintillator. FLUOR.

scission. The introduction of a break into a biopolymer, particularly a nucleic acid, which may be enzymatic or nonenzymatic.

scleroprotein. A simple and generally fibrous protein that is insoluble in aqueous solvents and that serves as a structural component of tissues. Collagen and keratin, which occur in cartilage, connective tissue, hair, nails, and the like, are two examples.

sclerosis. The pathological hardening of tissues.

S-configuration. The relative configuration of an asymmetric center in a molecule in which the substituents lie in a counterclockwise (left-handed; sinister) array of decreasing atomic number.

scorbutic. Of, or pertaining to, scurvy.

scotophobin. A memory-directing peptide isolated from brain.

scotopic vision. Vision in dim light in which the rods of the retina serve as light receptors.

scotopsin. RHODOPSIN.

Scrapie virus. A virus that infects sheep and that has great thermal stability.

screw axis of symmetry. An axis of symmetry such that a rotation about it, plus a translation parallel to the rotation axis, yields a structure that is identical to that before rotation. The screw axis is called an *n*-fold axis if the rotation involves $1/n$ of a turn, or $360°/n$.

scurvy. The disease caused by a deficiency of vitamin C.

SD. Standard deviation.

S.D.A. Specific dynamic action.

SDS. Sodium dodecyl sulfate.

SE. 1. Standard error. 2. Saponification equivalent.

sebaceous gland. A cutaneous gland that secretes sebum into a hair follicle.

sebum. The oily secretion of a sebaceous gland that softens and lubricates the hair and the skin.

sec. 1. *n* Second. 2. *adj* Secondary.

secondary acidosis. COMPENSATED ACIDOSIS.

secondary alkalosis. COMPENSATED ALKALOSIS.

secondary bond. WEAK INTERACTION.

secondary charge effect. The charge effect in a solution of charged macromolecules that is due to the differential movement of the positive and the negative ions of the supporting electrolyte. The secondary charge effect may lead either to an increase or to a decrease in the sedimentation rate and in the diffusion rate of the charged macromolecule.

secondary culture. A culture started from a primary culture.

secondary deficiency. The disorder that is caused by the inadequate intake of an essential nutrient in the diet, even though the dietary level of the nutrient is known to be adequate under normal conditions. Such exceptional conditions arise when there is an extra demand for the nutrient as may be the case during a disease or during pregnancy. *See also* conditioned vitamin deficiency.

secondary derived protein. *See* derived protein.

secondary electron. An electron ejected from an atom or a molecule by collision with a charged particle or with a photon.

secondary filament. ACTIN FILAMENT.

secondary fluor. A fluor, such as POPOP, that is used in a scintillation counter to shift the wavelength of the light emitted by the primary fluor to longer wavelengths, to which the photomultiplier tube is more sensitive.

secondary hydration shell. The layer of water molecules around an ion that is beyond the primary hydration shell, and that is not as firmly held as the primary hydration shell.

secondary immune response. *See* secondary response.

secondary ionization. The ionization of matter

that is produced by the fragments that have resulted from a primary ionization.

secondary isotope effect. An isotope effect in which the isotope itself does not participate in bond cleavage; the bond to the isotope, though not broken, has a different energy during the transition state than it does in the ground state.

secondary malnutrition. Malnutrition that results from diseases such as food allergies, ulcers, diabetes, or pernicious anemia; not a direct result of inadequate food intake.

secondary plot. A plot of derived enzyme kinetics data that are obtained from a primary plot; a plot of intercepts (intercept replot) and a plot of slopes (slope replot) that are obtained from primary plots are examples. *Aka* secondary kinetic plot.

secondary protein derivative. SECONDARY DERIVED PROTEIN.

secondary response. The enhanced immune response of an animal organism to a second administration of the same antigen.

secondary solvent. The solvent used in scintillation counting for solubilizing a sample that is insoluble in the primary solvent.

secondary standard. 1. A solution, the concentration of which is determined by means of titration against either a known weight of a primary standard, or a known volume of a standard solution of a primary standard. 2. A standard, such as a source of radiation, that has been calibrated against a primary standard.

secondary structure. The regular folding of a polypeptide chain or of a polynucleotide strand along one axis of the molecule, as in a helix, that is due to the formation of intramolecular hydrogen bonds along the length of the chain or strand; the local spatial arrangement of segments of the polypeptide chain or of the polynucleotide strand without regard to the conformation of side chains or to the relationship of one segment to other segments.

secondary tumor. A tumor that is formed through metastasis of a primary tumor.

secondary valence bond. VAN DER WAALS INTERACTION.

second critical concentration. The concentration above the critical micelle concentration at which spherical micelles begin to undergo the structural changes that lead to the formation of liquid crystals.

second law of cancer biochemistry. The principle that hormonal regulation of glycolysis at the initial hexokinase stage constitutes a major control mechanism of metabolism and of growth in both normal and malignant mammalian cells.

second law of photochemistry. The law that absorbed light does not necessarily result in a photochemical reaction, and when it does, then only one photon is required for each molecule affected. *Aka* Stark-Einstein Law.

second law of thermodynamics. The principle relating to the direction in which a process proceeds: all processes tend to proceed in such a direction that the entropy of the system plus its surroundings increases until an equilibrium is reached at which the entropy is at a maximum.

second messenger. An intracellular substance, such as cyclic AMP, that mediates the effects of extracellular stimuli, such as hormones and nerve impulses.

second moment. The position in a sedimentation boundary that must be used for precise calculations; in the case of a symmetrical boundary, the second moment corresponds to a position that is farther away from the center of rotation than the position of the peak.

second-order reaction. A chemical reaction in which the velocity of the reaction is proportional either to the product of the concentrations of two reactants, or to the square of the concentration of one reactant.

second-set rejection. The accelerated sequence of events that leads to the rejection of an allograft by an individual who has already rejected a previous allograft at the same site. *Aka* second-set reaction.

second-site reversion. INTRAGENIC SUPPRESSION.

secosteroid. A modified steroid that has a broken ring.

secretagogue. One of a group of large polypeptides that stimulate the secretion of gastric and pancreatic juices.

secretin. A polypeptide hormone of 27 amino acids that is secreted by the duodenum and that stimulates the release of pancreatic juice.

secretor. An individual who secretes water-soluble forms of the blood group substances into his body fluids.

secretor gene. A gene that controls the secretion of water-soluble forms of blood group substances into body fluids.

secretory piece. A protein that is bound to IgA immunoglobulins that occur in secretions and that is believed to function in protecting the immunoglobulins against proteolytic enzymes. *Abbr* SP.

sector cell. The standard cell used in the analytical ultracentrifuge that is designed in the shape of a sector to eliminate convection in the solution. Commonly used cells contain either a single or a double sector; the double sector cell permits the simultaneous sedimentation of two samples or of a sample and a reference solution.

secular equilibrium. The equilibrium that is established for a radioactive decay reaction if the half-life of the parent isotope is much greater than that of the daughter isotope.

sedation. The calming and quieting of the nerves.

sedative. 1. *n* An agent that causes calming and quieting of the nerves. 2. *adj* Of, or pertaining to, sedation.

sediment. 1. *n* The material removed from a solution by sedimentation. 2. *v* To subject a solution to sedimentation; to cause solute molecules of the solution to move by sedimentation.

sedimentation. 1. The movement of molecules in solution under the influence of a centrifugal field and away from the center of rotation. 2. The settling out of molecules from a solution under the influence of a gravitational field.

sedimentation coefficient. A measure of the rate of sedimentation of a macromolecule that is equal to the velocity per unit centrifugal field; specifically, $s = (dx/dt)/w^2x$, where s is the sedimentation coefficient, dx/dt is the velocity, w is the angular velocity, and x is the distance from the center of rotation. *Sym* s. *Aka* sedimentation constant. *See also* standard sedimentation coefficient.

sedimentation equilibrium. Sedimentation, generally performed in an analytical ultracentrifuge, in which the centrifuge rotor is driven at relatively low speeds and for relatively long times so that an equilibrium is established in the solution between sedimentation and diffusion. No boundaries are formed in the solution during the sedimentation, and the photographic plate does not show peaks but only a curvature of the gradient curve. Sedimentation equilibrium is used for calculations of molecular weights. *See also* approach to sedimentation equilibrium; density gradient sedimentation equilibrium; meniscus depletion sedimentation equilibrium; short column sedimentation equilibrium.

sedimentation equilibrium in a density gradient. DENSITY GRADIENT SEDIMENTATION EQUILIBRIUM.

sedimentation potential. The electrical potential produced by the sedimentation of particles through a liquid.

sedimentation velocity. Sedimentation, generally performed in an analytical ultracentrifuge, in which the centrifuge rotor is driven at high speeds and for relatively short times so that sedimentation exceeds diffusion and the macromolecules sediment through the cell. Boundaries are formed in the solution during the sedimentation and the photographic plate shows peaks. Sedimentation coefficients can be calculated from the movement of these peaks as a function of time. Sedimentation velocity is useful for studies of purity, homogeneity,

association-dissociation equilibria, reaction kinetics, and other properties of macromolecules.

sedoheptulose. A seven-carbon ketose that is an intermediate in the hexose monophosphate shunt.

segment long spacing. An artificially prepared assembly of collagen molecules in which the segments have the same length as the collagen molecule (2900 Å) and show a characteristic pattern of more than 40 cross-striations when examined with the electron microscope; produced from acidic solutions of collagen in the presence of polyanions such as ATP or chondroitin sulfate. *Abbr* SLS. *See also* fibrils long spacing.

segregation. The separation of the two members of a pair of alleles during meiosis so that each gamete receives one of the alleles of the pair.

selection. The fourth stage in a multistage mechanism of carcinogenesis in which drug- and hormone-resistant cells are selected from a population of autonomous cancer cells.

selection rules. A set of quantum mechanical rules and formulas that are used to determine which transitions between energy states of a molecule are allowed and which are forbidden.

selective marker. A marker, such as drug resistance or nutritional independence, that permits the selection of recombinants over the parental types.

selective medium. A medium designed to allow the preferential growth of some organisms over that of others.

selective plating. A method for isolating recombinants by plating two auxotrophic mutants on a minimal medium; only those recombinants that receive the normal allele of each mutant can grow under such conditions.

selective theory. 1. A theory of antibody formation according to which the information for antibody synthesis is genetically determined; the antigen is considered to react selectively with certain cells, receptors, or other biosynthetic units and to stimulate them to synthesize antibodies which they already were synthesizing at low levels or were potentially capable of synthesizing prior to immunization. *See also* instructive theory. 2. A theory according to which the evolution of the genetic code is due to natural selection, such that a large number of codes is theoretically possible, but one code has been selected over others because of its value for the survival of the organism; the code may, for example, be such as to minimize the effects of mutation or minimize the errors during translation. *See also* mechanistic theory.

selective toxicity. The variable degree of harmfulness of a chemical such that it is toxic, at a given concentration, to one organism but not to another. A chemotherapeutic agent must show selective toxicity with respect to the host and the invading microorganism.

selective variant. A mutation that enables the organism possessing it to survive under conditions which kill all other organisms not possessing the mutation.

selenium. An element that is essential to a wide variety of species in one class of organisms. Symbol Se; atomic number 34; atomic weight 78.96; oxidation states -2, $+4$, $+6$; most abundant isotope Se^{80}; a radioactive isotope Se^{75}, half-life 120.4 days, radiation emitted—gamma rays.

self-. An immunological term used in reference to the antigens and antibodies involved in autoimmunity.

self-absorption. The absorption of radiation, particularly of radioactive radiation, by the material emitting the radiation.

self-assembly. The spontaneous formation of a supramolecular structure from its component molecules or subunits, as in the assembly of ribosomes, viruses, membranes, or multienzyme systems.

self-marker theory. A theory proposed by Burnet and Fenner in 1949 to explain an organism's capacity to distinguish between "self" and "nonself." According to this theory, all the antigens in an organism carry characteristic self-markers that can be recognized by the immunocompetent cells of that organism so that formation of antibodies to these antigens is prevented.

self-quenching. INTERNAL QUENCHING.

self-recognition. AUTOIMMUNITY.

self-scattering. The scattering of radiation, particularly of radioactive radiation, by the material emitting the radiation.

Seliwanoff's test. A colorimetric test for ketohexoses that is based on the production of a red color upon treatment of the sample with resorcinol and hydrochloric acid.

SEM. 1. Scanning electron microscope. 2. Standard error of the mean.

semialdehyde. A compound formed by the conversion of one of two carboxyl groups in a molecule to an aldehyde; the aldehydes formed from aspartic and glutamic acid are examples.

semiconservative replication. A mode of replication for double-stranded DNA in which the parental strands separate and each daughter molecule consists of one parental strand and one newly synthesized strand.

semiconstitutive mutant. A mutant that synthesizes an inducible enzyme at a greater rate than does the uninduced wild-type organism, and that can be induced to synthesize the enzyme at a level that is characteristic of the fully constitutive strain.

semilethal mutation. A mutation that results in the death of more than 50 but less than 100% of the mutants.

semilogarithmic paper. Graph paper in which one scale has been distorted so that a plot of an exponential function will yield a straight line.

semipermeable membrane. A membrane that allows the passage of only certain solutes but that is freely permeable to water.

semipolar bond. COORDINATE COVALENT BOND.

semisynthetic. Descriptive of a compound of which part of the structure has been isolated from natural sources and part of it has been synthesized. Semisynthetic penicillins and cephalosporins are examples of such synthetically modified antibiotics.

Semliki forest virus. A virus that contains single-stranded RNA and that has a spherical nucleocapsid; it belongs to the group of arboviruses.

Sendai virus. A virus that contains single-stranded RNA and that belongs to the paramyxovirus group of myxoviruses; used for cell fusion studies, since it modifies the surface of cells in such a manner that they tend to fuse.

senescence. 1. The state of growing old; aging. 2. The phase of plant growth that extends from full maturity to death.

sense codon. A normal, amino acid-specifying codon.

sense strand. WATSON STRAND.

sensitive strain. A bacterial strain that can be lysed by a temperate phage or by a lysogenic culture produced with that phage.

sensitive volume. 1. That volume of a biological specimen in which an ionization must occur to produce a particular effect. 2. That volume of an ionization chamber through which the radiation must pass in order to be detected.

sensitivity spectrum. The types of antimicrobial drugs that are effective against a given microorganism. *See also* antimicrobial spectrum.

sensitization. The conditioning of an animal by administration of an allergen so that a second administration of the allergen will trigger an anaphylactic response. The sensitization may be active or passive depending on the type of anaphylaxis induced. *See also* active anaphylaxis; passive anaphylaxis.

sensitized erythrocyte. An antibody-coated erythrocyte that is used in the detection of complement.

sensitized fluorescence. The fluorescence that occurs when a photon excites a molecule, which then excites a second molecule by means of an energy transfer, and the second molecule dissipates the excitation energy by fluorescence.

sensitized phosphorescence. The phosphorescence that occurs when a photon excites a molecule, which then excites a second molecule by means of an energy transfer, and the second molecule dissipates the excitation energy by phosphorescence.

sensitizer. 1. PHOTOSENSITIZER. 2. ALLERGEN.

sensitizing agent. SENSITIZER.

sensitizing injection. The initial and harmless injection of an allergen into an animal which, if followed by a second injection, will trigger an anaphylactic response.

sentinel antibody. An antibody-like receptor site on an antibody-producing cell by which the antigen is believed to stimulate the cell to produce antibodies.

separated plasma. Plasma, obtained from whole blood, that is equilibrated with carbon dioxide at a given partial pressure.

separate package hypothesis. The hypothesis according to which photosystems I and II of chloroplasts are two separate systems such that energy transfer is possible only between the pigments within each system, but not between the pigments from one system to those of the other. *See also* spillover hypothesis.

separation cell. An analytical ultracentrifuge cell that allows for the separation and recovery from a mixture of the component having the smallest sedimentation coefficient.

separation factor. The ratio of the retention times, or the ratio of the retention volumes, of two compounds that are separated by gas chromatography.

Sephadex. Trademark for a group of cross-linked dextran gels used in gel filtration.

Sepharose. Trademark for a group of agarose gels used in gel filtration.

sepsis. The presence of pathogenic microorganisms or their toxins in the blood or in the tissues.

septanose. A monosaccharide that has a seven-membered ring structure.

septanoside. A glycoside of a septanose.

septic. Of, or pertaining to, sepsis.

septicemia. The presence of pathogenic microorganisms in the blood.

septum (*pl* septa; septums). A wall or a membrane that divides a cavity.

sequenator. An instrument for the automatic determination of amino acid sequences in a polypeptide chain; operation of the instrument is based on the repetitive application of the Edman degradation. *Aka* sequencer.

sequence. The linear order in which monomers occur in a polymer; the order of amino acids in a polypeptide chain, and the order of nucleotides in a polynucleotide strand are examples.

sequence gap. A segment, consisting of one or more amino acids, that appears to be missing from a polypeptide chain when this chain is compared with others of the same protein but isolated from different sources, and when the chains are matched up so as to provide a maximum of sequence homology.

sequence homology. The identity in sequence of either the amino acids in segments of two or more proteins, or the nucleotides in segments of two or more nucleic acids.

sequence hypothesis. The hypothesis that the sequence of nucleotides in a nucleic acid specifies the sequence of amino acids in a protein.

sequence isomer. One of two or more polymeric isomers that differ from each other in the sequence of the monomers in the chain.

sequential induction. Enzyme induction in which a single inducer brings about the synthesis of a number of inducible enzymes; the first enzyme induced acts on the inducer, thereby transforming it into an inducer for the second enzyme, which in turn acts on the second inducer, and so on. *See also* coordinate induction.

sequential mechanism. The mechanism of an enzymatic reaction in which two or more substrates participate in such a fashion that all the substrates must become bound to the enzyme before any products can be released. The mechanism is ordered if the substrates add to, and the products leave, the enzyme in an obligatory sequence; the mechanism is random if the substrates add to, and the products leave, the enzyme in a nonobligatory sequence.

sequential model. A model for regulatory enzymes—proposed by Koshland, Nemethy, and Filmer—according to which the enzyme undergoes a series of conformational changes as the various ligands become bound to the enzyme. Different types of site interactions may occur of which symmetry preservation, as in the symmetry model, may be a special case. In general, however, the symmetry of the enzyme molecule is not conserved, since a subunit changes its conformation as a ligand becomes bound to it. The capacity of the enzyme to bind substrate, positive effectors, and negative effectors is altered by the conformational changes which the subunits undergo.

sequential reactions. CONSECUTIVE REACTIONS.

sequester. To form a chelate.

sequestering agent. CHELATING AGENT.

sequestration. CHELATION.

Ser. 1. Serine. 2. Seryl.

SER. Smooth-surfaced endoplasmic reticulum.

serendipity. The gift for discovering valuable or useful things not specifically sought but recognized in the process of dealing with other things.

serial dilution. The systematic and progressive dilution that is frequently used in immunology, serology, and microbiology. A fixed volume of diluent is placed into a number of tubes and a given volume of sample, say 1.0 ml, is then added to the first tube. After mixing, 1.0 ml of solution is transferred to the second tube and, after mixing, 1.0 ml from this tube is transferred to the third tube, and so on.

seriatim. In a series; serially.

serine. An aliphatic, polar alpha amino acid that contains an alcoholic hydroxyl group and that frequently occurs at or near the active site of an enzyme. *Abbr* Ser; S.

serine convention. A method for assigning configurations to compounds by a comparison with the configuration of serine; thus (+) tartaric acid is designated as L_G when compared to glyceraldehyde, and as D_S when compared to serine.

serologic. Of, or pertaining to, serology. *Aka* serological.

serologic adhesion. IMMUNE ADHERENCE.

serological pipet. A measuring pipet in which the graduation marks extend to the tip of the pipet.

serology. The science that deals with serums, particularly immune serums, and with the reactions and properties of antigens, haptens, antibodies, and complement.

serotonin. 5-Hydroxytryptamine; a pharmacologically active mediator of immediate-type hypersensitivity. Serotonin is formed from tryptophan and is released from mast cells during the allergic response; it has hormone-like properties and causes vasodilation, increased capillary permeability, and contraction of smooth muscle.

serotonin hypothesis. The hypothesis that schizophrenia is caused by an abnormality in the metabolism of serotonin in the brain, and that most hallucinogens act by either antagonizing or mimicking the functions of serotonin.

serotype. A subdivision of a species or of a subspecies that is identifiable by serologic methods and that is distinguished by its antigenic character.

serous. Of, or pertaining to, serum.

serous fluid. A fluid resembling serum.

serous membrane. One of various thin membranes that secrete a serous fluid, line cavities, or enclose the organs within them, and form the inner layer of a blood vessel.

serum (*pl* serums; sera). The fluid obtained from blood after it has been allowed to clot; plasma without fibrinogen.

serum Ac globulin. ACCELERIN.

serum albumin. The major protein in serum, the main function of which is the regulation of osmotic pressure.

serum blocking power. The capacity of an immunoadsorbent to adsorb antibodies from a serum and to decrease the antibody titer of the serum. *Abbr* SBP.

serum converting enzyme. The peptidase that catalyzes the conversion of angiotensin I to angiotensin II.

serum factor. *See* cell factor.

serum folate. N^5-Methyltetrahydrofolic acid; a stable derivative of folic acid that appears to be the major storage form of folate coenzymes in higher organisms.

serum hepatitis. Hepatitis that is caused by a viral infection brought about by the injection of infected blood or blood products.

serum L. caseii factor. SERUM FOLATE.

serum proteins. The proteins present in blood serum.

serum prothrombin conversion accelerator. CONVERTIN.

serum prothrombin converting factor. PROCONVERTIN.

serum sickness. A disease, characterized by fever and by local swelling at the injection site, that in man may follow the injection of an immune serum prepared in an animal, and in animals may be initiated by the injection of large amounts of foreign protein. *Aka* serum disease.

serum sulfation factor. SOMATOMEDIN.

servomechanism. An automatic device for controlling the operation of a mechanism by having the output of the mechanism compared with its input, so that the error between the two quantitites can be controlled in a prescribed manner.

Sevag method. A procedure for the deproteinization of nucleoprotein in which the protein is denatured by the shaking of a solution of the nucleoprotein with chloroform and isoamyl alcohol.

sex chromatin. BARR BODY.

sex chromosome. A chromosome that is specifically connected with the determination of sex. *See also* X-chromosome; Y-chromosome.

sexduction. The process whereby a segment of genetic material is transferred from one bacterium to another by attachment to the fertility factor; the formation of F′-strains.

sex factor. FERTILITY FACTOR.

sex hormone. One of a group of hormones that are responsible for the development of secondary sex characteristics and that are capable of stimulating the development of accessory reproductive organs. Sex hormones are secreted principally by the gonads and consist of androgens, estrogens, and gestagens.

sex linkage. The linkage of genes located on a sex chromosome.

sex-linked gene. A gene located on a sex chromosome.

sexual conjugation. CONJUGATION (3).

s_f. Flotation coefficient.

s_f^0. Standard flotation coefficient.

SF$_1$ fragment. One of two fragments of the head, or globular part, of the myosin molecule; produced when myosin is treated with the enzyme papain.

S-100 fraction. A cell-free preparation obtained from a suspension of broken cells by first removing intact cells and cell debris by low-speed centrifugation, and then removing ribosomes by centrifugation at 100,000 $\times g$. The fraction contains transfer RNA and amino acyl-tRNA synthetases, and is used in studies of cell-free amino acid incorporation. *See also* pH 5 fraction.

shadowcasting. A technique for preparing specimens for electron microscopy in which the specimen is covered by metal atoms deposited onto it at a fixed angle. Areas around particles on the far side of the metal source remain free of deposited metal atoms and form shadows that provide information about the size and the shape of the particle. *Aka* shadowing.

shaker. A laboratory device for the mechanical shaking of samples for purposes of mixing and aeration.

shallow groove. MINOR GROOVE.

shape factor. *See* Oncley equation.

shear. The deformation experienced by a liquid as a result of the variations in the velocity of flow of different layers. Shear is associated with the flow of a liquid through a capillary, with the forcing of a liquid through a pipet, and with the homogenization of a suspension. *See also* rate of shear.

shear dichroism. FLOW DICHROISM.

Shemin cycle. A complex set of reactions whereby the tetrapyrroles are synthesized from glycine and succinyl coenzyme A.

SH group. Sulfhydryl group.

shield. A solid barrier for the protection of individuals from radiation or from potentially explosive laboratory set-ups.

shift. A chromosomal aberration in which a chromosome segment is removed from its normal place and is inserted elsewhere in the chromosome, with the original nucleotide sequence in the segment either maintained or reversed.

shift down. A shift experiment in which the change in the medium leads to a decrease in the rate of growth of the cells.

shift experiment. An interference with the balanced growth of cells in a culture by a precisely timed and defined change in the medium. *See also* shift down; shift up.

shift up. A shift experiment in which the change

in the medium leads to an increase in the rate of growth of the cells.

shikimic acid. A hydroxylated, unsaturated, acid derivative of cyclohexane that serves as a key intermediate in the biosynthesis of the aromatic amino acids.

shikimic acid pathway. A pathway for the synthesis of shikimic acid in bacteria and for its conversion to tyrosine, tryptophan, and phenylalanine.

shock. A circulatory failure due to loss of blood from the vascular compartment by either hemorrhage or increased capillary permeability.

shocking dose. The second injection of allergen that triggers the anaphylactic response in an animal.

Shope papilloma virus. A virus that produces papillomas in rabbits.

short column cell. An analytical ultracentrifuge cell in which the sample solution constitutes a very short liquid column (about 1 to 3 mm); used for molecular weight determinations by sedimentation equilibrium.

short column sedimentation equilibrium. A variation of the sedimentation equilibrium method for the determination of molecular weights in which short column cells are used so that the time required to reach equilibrium is greatly decreased; the method is especially useful for the simultaneous analysis of a large number of different samples.

short-range hydration. The hydration by water molecules that are located in the primary hydration shell of an ion.

short-range interactions. The attractive and repulsive forces between atoms and molecules that decrease rapidly with distance. *See also* weak interactions.

short trough technique. An immunoelectrophoretic technique for the identification of a specific antigen in a mixture.

shunt. *See* hexose monophosphate shunt.

shuttle. A mechanism whereby reducing equivalents are removed from cytoplasmic NADH and passed to the electron transport system in the mitochondrial membrane by way of intermediate compounds, since NADH itself cannot pass through the membrane.

Shwartzman reaction. The production of an inflammatory lesion in an animal by the subcutaneous injection of an endotoxin from a gram-negative bacterium, followed by a second injection, about 24 hours later, of the same endotoxin or of some other substance.

Si. Silicon.

SI. Système International d'Unités; an extension and a refinement of the metric system. The system is based on seven basic units from which other units are derived. The basic units, their symbols, and the quantities which they measure are: meter m—length, kilogram kg—mass; second s—time; ampere A—electric current; Kelvin K—thermodynamic temperature; candela cd—luminous intensity; and mole M—amount of substance. Some derived units are the newton N for force, the joule J for energy, and the liter l for volume.

sialic acid. *N*-Acylneuraminic acid or any of its esters or other derivatives of its alcoholic hydroxyl groups.

sialidase. NEURAMINIDASE.

sickle cell. An erythrocyte that has an abnormal crescent-like shape; such a cell is more fragile than the normal cell and tends to hemolyze in the blood capillaries.

sickle cell anemia. A genetically inherited metabolic defect in man that is characterized by the formation of an abnormal hemoglobin (sickle cell hemoglobin) which leads to sickling and hemolysis of erythrocytes. *See also* sickle cell disease; sickle cell trait.

sickle cell disease. A condition in which an individual is homozygous for sickle cell anemia.

sickle cell hemoglobin. The abnormal hemoglobin responsible for sickle cell anemia. It differs from normal adult hemoglobin in having valine in place of glutamic acid in the sixth position of the beta chain. *Sym* HbS.

sickle cell trait. A condition in which an individual is heterozygous for sickle cell anemia.

sicklemia. SICKLE CELL ANEMIA.

sickling. The process whereby erythrocytes take on abnormal, crescent-shaped forms, as in sickle cell anemia.

side chain. 1. A smaller chain attached laterally to a longer chain. 2. A chain attached to a ring. 3. AMINO ACID SIDE CHAIN.

side chain cleavage. The removal of some or all of the carbon atoms from the aliphatic side chain attached to the steroid nucleus.

side chain theory. EHRLICH'S RECEPTOR THEORY.

sidedness. The vectorial character of either a molecule or a system.

side reaction. A secondary reaction that occurs simultaneously with the major reaction.

siderophilin. TRANSFERRIN.

sidescattering. The scattering of radiation in directions other than those of forward or backward scattering.

sievorptive chromatography. A chromatographic technique that combines gel filtration and adsorption chromatography; ion filtration chromatography and intervent dilution chromatography are two examples.

SIF. *See* growth hormone regulatory hormone.

sigma. The symbol Σ that indicates the summation of all the quantities that follow it.

sigma bond. A chemical bond formed by the electrons in sigma orbitals.

sigma cycle. The set of reactions composed of (a) the attachment of the sigma factor to the core enzyme of RNA polymerase at the initiation site of transcription; (b) the release of the sigma factor from the core enzyme during chain elongation; and (c) the subsequent binding of the sigma factor to the same, or to another, core enzyme at an initiation site.

sigma factor. A protein subunit of the enzyme RNA polymerase that functions in the recognition of the promoter and the initiation of transcription. *Sym* σ. *See also* sigma cycle.

sigma orbital. A molecular orbital that is a localized bond orbital, spread over two bonding atoms.

sigma star orbital. *See* antibonding orbital.

sigmoid kinetics. The kinetics of an enzymatic reaction that yield a sigmoid curve for a plot of reaction velocity versus substrate concentration; the sigmoid, or S-shaped, curve is characteristic of many regulatory enzymes as well as of other systems showing cooperative-type interactions.

signal-to-noise ratio. The ratio of the electrical response of an instrument during the measurement of a sample to the response from random background electrical fluctuations.

significance of results. A measure of the deviation of results from the mean. If the probability value of a result is equal to, or less than, 0.05 but greater than 0.01, the result is considered to be significantly different from the mean; if the probability value is less than, or equal to, 0.01 but greater than 0.001, the result is considered to be highly significant; and if the probability value is equal to, or less than, 0.001, the result is considered to be very highly significant. A probability value of 0.05 means that there is a 5% chance that the result will differ from the mean by ± 2 standard deviations.

significant figures. The digits in a number, the values of which are known with certainty plus the first digit, the value of which is uncertain; the position of the decimal point is immaterial. Thus the number 12.40 is considered to have four significant figures, and its true value lies between 12.395 and 12.405.

sign mutation. FRAME SHIFT MUTATION.

SIH. *See* growth hormone regulatory hormone.

silanizing. The treatment of a chromatographic support, such as a diatomaceous earth, with trichloromethylsilane or a similar reagent that converts active silanol groups—$SiOH$ to less polar silyl ethers—$SiOR$, thereby changing the adsorptive properties of the support. *Aka* silanization.

silent allele. *See* silent gene.

silent gene. A gene that has no detectable product.

silent mutation. A mutation that does not result in a detectable phenotypic effect. A silent mutation may be due to a transition or a transversion that leads to either a synonym codon or a codon which codes for an amino acid closely related to that coded for by the original codon. As a result, the polypeptide chain will be synthesized, either without a change in its amino acid sequence, or with replacement of an amino acid by a closely related one. Such a polypeptide, while altered, may still be fully functional so that the mutation produces no detectable effect.

silica gel. An adsorbent used in column and thin-layer chromatography for the separation of nonionic organic compounds. *Aka* silica; silicic acid.

silicon. An element that is essential to several classes of animals and plants. Symbol Si; atomic number 14; atomic weight 28.086; oxidation states -4, $+2$, $+4$; most abundant isotope Si^{28}.

silk fibroin. A fibrous protein of silk that has an antiparallel pleated sheet structure.

silver nitrate chromatography. ARGENTATION CHROMATOGRAPHY.

silylation. The introduction of a trimethylsilyl group $-Si(CH_3)_3$ into an organic compound; used for making volatile derivatives for gas chromatography.

Simha equation. One of two equations that relate the viscosity increment to the axial ratio of either a prolate or an oblate ellipsoid of revolution.

simian. Of, or pertaining to, monkeys.

simian virus 40. A small, naked, icosahedral, and oncogenic virus that contains double-stranded DNA and belongs to the group of papovaviruses. *Abbr* SV-40.

7S immunoglobulin. A basic structural unit of the immunoglobulins. The IgG and IgD immunoglobulins as well as the monomeric units of the IgA and IgM immunoglobulins all have sedimentation coefficients of 7S; the IgE immunoglobulins, however, have a sedimentation coefficient of 8S.

19S immunoglobulin. IgM.

simple anhydride. *See* acid anhydride.

simple competitive inhibition. PURE COMPETITIVE INHIBITION.

simple diffusion. 1. The movement of a solute across a biological membrane that does not require the participation of transport agents; unmediated transport. 2. FREE DIFFUSION. 3. SINGLE DIFFUSION.

simple dispersion. *See* simple optical rotatory dispersion.

simple enzyme. 1. An enzyme that consists only of protein. 2. An enzyme that contains a nonprotein component which does not participate in the catalytic process.

simple goiter. A thyroid enlargement that is unaccompanied by either hyper- or hypothyroidism.

simple hapten. A low-molecular weight hapten that constitutes a separate part of a complete antigen but that does not give a visible precipitin reaction with the appropriate antibody.

simple lipid. NEUTRAL LIPID.

simple microscope. A microscope having one lens.

simple noncompetitive inhibition. PURE NONCOMPETITIVE INHIBITION.

simple optical rotatory dispersion. An optical rotatory dispersion that can be described by a one-term Drude equation.

simple protein. A protein that is composed only of amino acids.

simple sugar. MONOSACCHARIDE.

simple triglyceride. A triglyceride containing only one type of fatty acid.

simultaneous reactions. A mechanism in which one reactant can give rise to either of two different products by either of two different reactions.

single-burst experiment. A technique for studying viral multiplication by isolating single infected cells.

single carbon unit. ONE-CARBON FRAGMENT.

single diffusion. Immunodiffusion in which either the antigen diffuses through a gel containing antibody, or the antibody diffuses through a gel containing antigen.

single-displacement mechanism. SEQUENTIAL MECHANISM.

single-hit survival curve. A survival curve in which the absorption of one photon leads to loss of viability of the active unit; such a survival curve reflects simple exponential inactivation kinetics.

single-ion monitoring. MASS FRAGMENTOGRAPHY.

single reciprocal plot. A plot of X/Y versus X or a plot of Y/X versus Y, where X and Y are two variables; the Scatchard plot and the Eadie plot are two examples.

single-stranded. Descriptive of a nucleic acid molecule that consists of only one polynucleotide chain. *Abbr* ss.

singlet. A single peak, as that obtained in nuclear magnetic resonance.

singlet state. The electronic state of an atom or a molecule in which two single electrons in two orbitals have their spin in opposite directions and are spin-paired so that $S = 0$ and $2S + 1 = 1$, where S is the resultant spin and $2S + 1$ is the spin multiplicity. Such an atom or molecule has no net magnetic moment.

sintered. Descriptive of porous material that is formed by the partial fusion of a substance by heat.

siphon. An inverted U-shaped tube with legs of unequal length that is used for the delivery of liquid from one container to a second one at a lower level.

Sips distribution. A frequency distribution that closely resembles a Gaussian distribution and that is used for the treatment of data from equilibrium dialysis measurements of the antigen-antibody reaction.

site. 1. A specific region of a macromolecule at which a binding interaction with another molecule, or with an atom or an ion, takes place; the active site of an enzyme and the antibody combining site of an antibody are two examples. 2. The position of a mutation in a gene.

site heterogeneity. The heterogeneity of antibodies that results from the antigen molecule possessing several antigenic determinants; consequently, one antigen leads to the production of several types of antibodies.

skeletal band. An infrared absorption band that is characteristic of the entire molecule rather than of a specific group in the molecule. *See also* group frequency band.

skeletal muscle. A striated muscle that is attached to, or that moves parts of, the skeleton.

skin-sensitizing antibody. REAGIN.

slant culture. A bacterial culture grown in a tube that contains a solid nutrient medium which was solidified while the tube was kept in a slanted position.

slice technique. *See* tissue slice.

sliding filament model. A model proposed by Huxley and Hanson to explain the changes in the length of a striated muscle upon contraction and stretching. According to this model, thick and thin muscle filaments slide alongside each other, thereby leading to varying degrees of interpenetration without changes in the lengths of the filaments themselves.

slime. A layer of material that is external to the bacterial cell wall and that has no definite borders.

slope replot. *See* secondary plot.

slow hemoglobin. A hemoglobin which, after electrophoresis, is located closer toward the cathode than normal adult hemoglobin.

slow reacting substance-anaphylaxis. A protein-lipid complex that is a pharmacologically active mediator of the allergic response. *Abbr* SRS; SRS-A.

SLR factor. N^{10}-Formylpteroic acid; a factor that stimulates the growth of *Streptococcus lactus* R; rhizopterin.

SLS. Segment long spacing.

small-angle x-ray diffraction. A method of x-ray diffraction in which the scattering of the x-rays is measured at small angles; used for the analysis of large molecular spacings, as those between monomeric groups in a polymer.

small calorie. CALORIE.

smallpox virus. VARIOLA VIRUS.

smectic. Describing a liquid crystal that has an arrangement of parallel layers which can slide over one another; the molecules in each layer can move from side to side, as well as forward and backward, but cannot move up or down.

Smith degradation. The degradation of a polysaccharide by means of periodate oxidation, followed by reduction with sodium borohydride and acid hydrolysis; the method yields fragments that indicate the mode of linkage of the monosaccharides in the original polysaccharide.

Smithie's theory. A somatic recombination theory of antibody formation.

smooth microsomes. SMOOTH-SURFACED ENDOPLASMIC RETICULUM.

smooth muscle. An involuntary muscle such as a muscle of an internal organ or a muscle of a blood vessel.

smooth-surfaced endoplasmic reticulum. That portion of the endoplasmic reticulum to which few or no ribosomes are attached. *Abbr* SER.

S mutant. A bacterial mutant that gives rise to a smooth colony.

sn. Stereospecific numbering.

SN. Steroid number.

snapback. The rapid renaturation of heat-denatured double-stranded DNA or RNA that occurs when the nucleic acid contains a short segment in which the nucleotide pairs are intact, with the base pairing in perfect register. The mimimum size of this segment varies with the composition of the nucleic acid; the segment generally consists of between 12 and 20 nucleotide pairs. *See also* minimal stable length.

S_N1 mechanism. A unimolecular, nucleophilic substitution mechanism that can be formulated as: $RX \rightarrow R^+ + X^-$; $N^- + R^+ \rightarrow RN$; $v = k[RX]$; v is the velocity of the reaction, and k is the rate constant. Also designated SN_1 mechanism.

S_N2 mechanism. A bimolecular, nucleophilic substitution mechanism that proceeds with an inversion of configuration and that can be formulated as; $N^- + RX \rightarrow NR + X^-$; $v = k[N][RX]$; v is the velocity of the reaction, and k is the rate constant. Also designated SN_2 mechanism.

soap. The salt of a long chain fatty acid; commonly refers to either the sodium or the potassium salt.

soap bubble meter. A device for measuring low

rates of gas flow that is used in gas chromatography.

sodium. An element that is essential to several classes of animals and plants. Symbol Na; atomic number 11; atomic weight 22.9898; oxidation state +1; most abundant isotope Na^{23}; a radioactive isotope Na^{22}, half-life 2.60 years, radiation emitted—positrons and gamma rays.

sodium dodecyl sulfate. A detergent used in the solubilization of membrane fractions. *Abbr* SDS; *Aka* sodium lauryl sulfate.

sodium dodecyl sulfate polyacrylamide gel electrophoresis. An electrophoretic technique used for estimating the molecular weight of polypeptide chains from oligomeric and monomeric proteins. *Abbr* SDS-PAGE.

sodium error. The apparent decrease in the pH of a solution that is measured with most glass electrodes when they are used at high pH values; the effect is due to the fact that most glasses which are responsive to protons are also somewhat responsive to sodium ions.

sodium lauryl sulfate. SODIUM DODECYL SULFATE.

sodium pump. The structure and/or the mechanism that mediates the active transport of sodium and potassium ions across a biological membrane in higher animals; the operation of the pump requires cellular ATP and the enzyme Na,K-ATPase.

soft clot. The blood clot formed by the aggregation of fibrin molecules in the absence of calcium ions and fibrinase.

soft ice. Water that is oriented, bound, and frequently compressed by polar interactions, as distinct from water that is in the form of icebergs, in which the water is stabilized by nonpolar interactions.

soft soap. A potassium salt of a long chain fatty acid; soft soaps are more water-soluble than hard soaps.

software. The procedural specifications required for the operation of computers; includes programs, routines, translators, assemblers, etc.

soft water. Water that either contains low concentrations of calcium, magnesium, and iron ions, or is devoid of these ions; as a result, the surface-active action of ordinary soap molecules is not appreciably interfered with.

soft x-rays. Low-frequency x-rays that have long wavelengths and small penetrating power.

sol. A liquid colloidal dispersion.

solation. The transition from a gel to a sol.

solid medium. A gel that contains nutrients.

solid phase technique. An automated technique for the synthesis of polypeptides in which the chain grows while it is attached to a solid support, and excess reagents as well as by-products are washed away. The chain grows from the C-terminal to the N-terminal, and the process is initiated by attaching the C-terminal amino acid through its carboxyl group to an insoluble resin. This amino acid is then reacted with the next amino acid, the amino group of which has been blocked. After peptide bond formation, the amino-blocking group is removed and the dipeptide is then reacted with the next amino acid, and so on.

solid scintillation counter. An external-sample scintillation counter in which a solid fluor is used as a detector.

solid scintillation fluorography. A method for visualizing labeled compounds in a specimen by covering the specimen with solid fluors and allowing the photons, which are emitted by the scintillating fluors, to expose a photographic emulsion; used to locate labeled compounds in a chromatogram.

solubility. The amount of a substance that will dissolve in a given volume of solvent under specified conditions.

solubility curve. A plot of the amount of solute in solution as a function of the amount of solute added.

solubility product. The product of the concentrations of the ions of a sparingly soluble electrolyte in a saturated solution of the electrolyte; the concentration of each ion is raised to a power equal to the number of ions of that type per molecule of electrolyte. When the solubility product is exceeded, some of the electrolyte is precipitated out.

solubilization chromatography. A chromatographic technique in which compounds of insufficient solubility in water are separated by ion-exchange chromatography by using an aqueous solution of an organic solvent for elution.

solubilize. To disperse in a solution; used to describe both the dispersion of membranes by treatment with detergents, and the release of an enzyme from particulate matter.

soluble. Capable of going into solution; capable of dissolving.

soluble antigen. An antigen that is not present on a cell, a virus, or some other particle in contrast to a particulate antigen.

soluble enzyme. An enzyme that exists in the free state in the cytoplasm in contrast to a particulate enzyme, is readily extracted from cells, and can be purified by the general methods of protein fractionation.

soluble fibrin. SOFT CLOT.

soluble fraction. A subcellular fraction that contains material of the intracellular fluids but that is devoid of cellular or intracellular particulate structures.

soluble RNA. TRANSFER RNA.

soluble starch. A partially hydrolyzed starch.

solute. That component of a solution, consisting of two components, that is present in the smaller amount; for ordinary solutions of solids in liquids, it is the solid.

solution. A homogeneous mixture of two or more substances; ordinary solutions are mixtures of one or more solutes and a solvent.

solvation. The process whereby solvent molecules surround, and bind to, solute ions and molecules. Solvation may enhance or diminish the asymmetry of the solute particles in solution.

solvent. That component of a solution, consisting of two components, that is present in the greater amount; for ordinary solutions of solids in liquids, it is the liquid.

solvent demixing. The separation of a solvent system into its constituent components.

solvent drag. The increased rate of movement of a solute over that attributable to simple diffusion, and caused by the flow of the solvent; the transfer of solutes across membranes by water which moves in response to an osmotic gradient is an example.

solvent extraction. The removal of a wanted or an unwanted substance from a liquid by mixing the liquid with an immiscible, or a partially miscible, solvent, and by separating the phases.

solvent front. The line of advancing solvent in chromatography.

solvent perturbation. A technique for studying the stereochemistry and the location of chromophoric groups in a protein by measuring the changes in a property of the protein, such as absorbance or fluorescence, upon the addition of perturbing solutes to a solution of the protein. A chromophore on the surface of the protein molecule is expected to be sensitive to added perturbing solutes, while a chromophore in the interior of the molecule is expected to be insensitive to them; a chromophore in a crevice is expected to be sensitive to perturbing solutes of low-molecular weight but not to those of high-molecular weight.

solvolysis. A generalized concept for the reaction between a solvent and a solute by which new substances are formed. In most cases the solvent donates a proton to, and/or accepts a proton from, the solute. When the solvent is water, the reaction is referred to as hydrolysis.

soma. The totality of the somatic cells of an organism.

somatic. Of, or pertaining to, soma.

somatic cell. A body cell of an organism, as distinct from a germ cell.

somatic crossing over. MITOTIC RECOMBINATION.

somatic mutation. A mutation occurring in a cell that is not destined to become a germ cell.

somatic mutation theory. 1. A selective theory of antibody formation according to which antibody formation is based either on the hypermutation of specific genes or on the preferential expansion of clones in which advantageous mutations have occurred. *See also* somatic theory. 2. A theory of carcinogenesis according to which a tumor results from either a spontaneous or an induced mutation, or mutations, of the somatic cells of an organism.

somatic recombination theory. A selective theory of antibody formation according to which antibody formation is based on somatic recombinations that occur between genes that are responsible for the synthesis of antibodies. *See also* somatic theory.

somatic theory. A theory of the origin of the genes that code for the variable region of antibody molecules and hence allow for the great diversity of antibodies. According to this theory, these genes have arisen through modifications of a smaller number of germline genes. Such modifications involve mutations as in the somatic mutation theory, or recombinations as in the somatic recombination theory.

somatomedin. A low-molecular weight polypeptide that mediates the action of growth hormone on skeletal tissue and produces insulin-like effects in various target tissues. Somatomedin is present in the serum and competes with insulin for receptor sites; it also stimulates sulfate uptake by cartilage and thymidine incorporation into DNA. Somatomedin is required for the growth of tumor cells in tissue culture.

somatostatin. GROWTH HORMONE RELEASE-INHIBITING HORMONE.

somatotrophin. Variant spelling of somatotropin.

somatotropin. GROWTH HORMONE.

somatotropin regulatory factor. SOMATOTROPIN REGULATORY HORMONE.

somatotropin regulatory hormone. GROWTH HORMONE REGULATORY HORMONE.

somatotropin release-inhibiting hormone. GROWTH HORMONE RELEASE-INHIBITING HORMONE.

somatotropin releasing hormone. GROWTH HORMONE RELEASING HORMONE.

Somogyi-Nelson method. An analytical procedure for the determination of glucose in blood in which proteins are precipitated with zinc sulfate and barium hydroxide, and a blue color is produced by treatment of the protein-free filtrate with copper sulfate and an arsenomolybdate reagent.

Somogyi unit. The quantity of amylase that liberates sugars with a reducing value equivalent to 1 mg of glucose during the course of a 30-min reaction at 40°C and at a pH of 6.9 to 7.0.

sonication. ULTRASONICATION.

sonicator. An instrument for the generation of ultrasonic sound waves and for the rupture of cells by means of these sound waves.

sonic oscillation. ULTRASONICATION.
sonic oscillator. SONICATOR.
sonification. ULTRASONICATION.
sonifier. SONICATOR.
sorbate. 1. Absorbate. 2. Adsorbate. 3. The anion of sorbic acid.
sorbent. 1. Absorbent (1). 2. Adsorbent.
sorbitol. A sugar alcohol derived from glucose.
Sørensen buffer. 1. A phosphate buffer prepared by mixing x ml of M/15 KH_2PO_4 and $(100 - x)$ ml of M/15 Na_2HPO_4; pH range 5.0 to 8.2. 2. A glycine buffer prepared by mixing x ml of 0.1 M glycine in 0.1 M $NaCl$ and $(100 - x)$ ml of either 0.1 N HCl or 0.1 N $NaOH$; pH range 1.2 to 3.6 or 8.4 to 13.0. 3. A citrate buffer prepared by mixing x ml of 0.1 M disodium citrate and $(100 - x)$ ml of either 0.1 N HCl or 0.1 N $NaOH$; pH range 2.2 to 4.8 or 5.0 to 6.8.
Soret band. A characteristic absorption band of the cytochromes around 400 nm.
sorption. 1. ABSORPTION (1). 2. ADSORPTION (1).
source. The material that emits the radiation.
Sowden-Fischer synthesis. A reaction whereby a one-carbon fragment is added to a carbonyl group by means of nitromethane, as in the extension of a monosaccharide from a pentose to a hexose.
Soxhlet extractor. A device that is interposed between a flask of boiling solvent and a condenser and that allows for the continuous extraction of a specimen with the solvent.
soybean trypsin inhibitor. A protein, isolated from soybeans, that is an inhibitor of the enzyme trypsin; the Kunitz and the Bowman-Birk inhibitors are two such proteins.
SP. 1. Split protein. 2. Secretory piece.
space-filling model. A compact molecular model that shows the full bulk of each atom and the effective shape of the molecule. In this model, the bond angles are correct and the distances between the atoms are to scale based on their Van der Waals radii.
space group. A symmetry class to which objects may belong by virtue of possessing elements of symmetry over and above those of a point group; the additional elements of symmetry may be symmetry operations such as rotation and translation.
spacer DNA. A segment of DNA that does not code for any identifiable RNA species; spacer DNA, rich in adenine and thymine, occurs between transcribable segments of the DNA that codes for ribosomal RNA.
sparing effect. The decrease in dietary requirement of one substance that is occasioned by the presence of a metabolically-related substance in the diet; tyrosine, for example, exerts a sparing effect on phenylalanine.
sparking. The phenomenon that a catalytic

amount of a certain di- or tricarboxylic acid, such as oxaloacetic acid, must be present for initiation of the beta oxidation of fatty acids. The acid is oxidized, thereby yielding ATP which is required for fatty acid activation; the acid also combines with the acetyl coenzyme A formed during beta oxidation.
sparsomycin. An antibiotic, produced by *Streptomyces sparsogenes,* that inhibits protein synthesis by preventing peptide bond formation.
SPCA. Serum prothrombin conversion accelerator.
specialized transduction. Transduction in which only certain bacterial genes of the donor bacterium may become transduced.
special structure. The structural elements of the bacterial cell wall that are additional to the peptidoglycan framework and that include polysaccharides, teichoic acids, polypeptides, proteins, lipopolysaccharides, and lipoproteins. The nature of the special structure varies greatly from organism to organism and contributes to the antigenic properties of the bacterial cells and to their acceptor specificity for viruses and bacteriocins.
species (*pl* species). A taxon that forms a division of a genus and that consists of a group of individuals of common ancestry that closely resemble each other structurally and physiologically and that, in the case of sexual forms, are capable of interbreeding.
specific. 1. Of, or pertaining to, species. 2. Designating a physical or a chemical property per unit amount of substance. 3. Of, or pertaining to, specificity. 4. Constituting or falling into a category.
specific acid-base catalysis. The catalysis in solution in which the catalysts are free protons H^+, H_3O^+ and/or free hydroxyl ions; it is not affected by other acidic or basic species present. *See also* general acid-base catalysis.
specific activity. The number of activity units per unit of mass; the number of enzyme units per milligram of protein, the number of katals per kilogram of protein, and the number of microcuries per micromole are examples. *Abbr* S.A.
specific adsorption. The adsorption of specific cations to a surface that results from the formation of coordination complexes between groups of atoms on the surface and the cations. *Aka* ion pairing.
specifically labeled. Designating a compound in which one or more known atoms contain all of the label, not necessarily in an even distribution; the positions of the labeled atoms are included in the name of the compound.
specific dynamic action. The extra heat, over and above that due to the basal metabolism, that is produced by an animal upon the ingestion of a

food. The heat represents the extra energy released as a result of the metabolism of the food, and amounts to approximately 30% for proteins, 6% for carbohydrates, and 4% for lipids. *Abbr* S.D.A.

specific extinction coefficient. The extinction coefficient that attains when the concentration of the solution is expressed in terms of grams per milliliter.

specific gravity. The ratio of the weight of a given volume of a substance to the weight of an equal volume of water.

specific growth rate. The rate of growth of a bacterial population, either per cell or per unit mass of cells; equal to $(1/x)(dx/dt)$, where t is the time and x represents either the number or the mass of the cells.

specific heat. The quantity of heat required to raise the temperature of 1 gram of a substance by 1°C. *Aka* specific heat capacity.

specific immunity. The immunity that is due to the formation of antibodies in response to the recognition of specific antigens, as contrasted with nonspecific immunity which is due to nonimmunological mechanisms.

specific ionization. The number of ion pairs formed per unit distance along the path of an ionizing particle.

specificity. 1. The degree of selectivity shown by an enzyme with respect to the number and types of substrates with which the enzyme combines, as well as with respect to the rates and the extents of these reactions. 2. The degree of selectivity shown by an antibody with respect to the number and types of antigens with which the antibody combines, as well as with respect to the rates and the extents of these reactions. 3. The degree of selectivity shown by a membrane, or a membrane component, with respect to the type and the degree of permeability to substances transported across the membrane in mediated transport.

specific radioactivity. The specific activity of radioactive material, frequently expressed in terms of the number of microcuries per micromole.

specific rate constant. RATE CONSTANT.

specific reaction rate. RATE CONSTANT.

specific refractive index increment. The contribution to the refractive index per gram of solute; equal to $(n - n_0)/c$, where n is the refractive index of the solution, n_0 is the refractive index of the solvent, and c is the concentration of the solute in grams per cubic centimeter. *Aka* specific refractive increment.

specific retention volume. The volume of liquid, per gram of adsorbent, that passes through a column in displacement chromatography before a particular substance is eluted from the column.

specific rotation. A measure of the optical rotation at a particular wavelength per unit amount of a substance; specifically, $[\alpha]_D^{t°} = \alpha/dc$, where $[\alpha]_D^{t°}$ is the specific rotation at a temperature of $t°C$ for the sodium D line, α is the observed rotation in degrees, d is the optical path length in decimeters, and c is the concentration in grams per milliliter.

specific rotatory power. SPECIFIC ROTATION.

specific viscosity. A measure of the fractional change in viscosity that is produced by the solute; equal to the relative viscosity minus 1. *Sym* η_{sp}.

specific volume. The volume occupied by 1 gram of material; the reciprocal of the density.

spectral. Of, or pertaining to, a spectrum.

spectral band pass. *See* band pass.

spectral shift. The change of an absorption spectrum or an absorption band to either longer or shorter wavelengths.

spectrofluorometer. A fluorometer in which the desired excitation and emission wavelengths are selected by means of a monochromator. Spectrofluorometers may be of a corrected or an uncorrected type depending on whether the intensity of the exciting light is controlled as a function of the wavelength, and whether the response of the detector is controlled as a function of the emission wavelength. *Var sp* spectrofluorimeter.

spectrogram. The photographic record of a spectrum.

spectrograph. An instrument for separating light into its component wavelengths and for obtaining a photographic record of the spectrum thus produced.

spectrometer. 1. An instrument for measuring either the wavelengths or the frequencies of a spectrum. 2. A liquid scintillation assembly that includes a detector, scaler, sample changer, print out, and electronic circuitry.

spectrophotometer. A photometer in which a monochromator, composed of prisms or diffraction gratings, is used for the isolation of narrow band widths. Spectrophotometers allow the measurement of the selective absorption of light; are used for both qualitative and quantitative analysis of chemical substances; and cover the ultraviolet, visible, and infrared ranges of the electromagnetic spectrum. *See also* double beam in space spectrophotometer; double beam in time spectrophotometer; recording spectrophotometer.

spectropolarimeter. An instrument that is a combined spectroscope and polarimeter and that is used for measurements of optical rotation as a function of wavelength.

spectroscope. An instrument for separating light into its component wavelengths and for examining the spectrum thus produced.

spectroscopy. The production and study of spectra.

spectroscopic splitting factor. g-VALUE.

spectrum (*pl* spectra; spectrums). 1. The variation of radiation intensity as a function of either wavelength or frequency that is generally represented in graphical form. 2. A range of either wavelengths or frequencies of a radiation. 3. ELECTROMAGNETIC SPECTRUM. 4. A specific type of radiation wavelengths or frequencies, such as an absorption or an emission spectrum.

S-peptide. A short fragment of the enzyme ribonuclease that is obtained by a subtilisin-catalyzed cleavage of the molecule between amino acid residues 20 and 21.

spermatocyte. A cell that develops into a mature sperm upon meiosis.

spermatogenesis. The formation and development of spermatozoa.

spermatozoon (*pl* spermatozoa). The male reproductive cell of animals; the male gamete.

spermidine. A polyamine that contains two amino groups and one imino group.

spermine. A polyamine that contains two amino groups and two imino group.

spherical phage. *See* minute phage.

spherocyte. A spherical red blood cell that has a smaller diameter than the normal erythrocyte.

spheroplast. A bacterial cell that is largely, but not entirely, freed of its cell wall. Spheroplasts are prepared artificially from gram-negative bacteria by either lysozyme digestion of the cells or by growing them in the presence of penicillin. Spheroplasts are osmotically sensitive but may, at times, convert to an L-form. *See also* protoplast.

spherule. LIPOSOME.

sphinganine. Dihydrosphingosine.

4-sphingenine. SPHINGOSINE.

sphingolipid. A lipid that contains either sphinganine, its homologue, its isomer, or its derivative, and that is especially predominant in brain and nervous tissue.

sphingomyelin. A phosphosphingolipid that consists of sphingosine, a fatty acid, a phosphate group, and choline, and that is predominant in the myelin sheath of nerves.

sphingomyelinosis. NIEMAN-PICK DISEASE.

sphingophospholipid. *See* phosphosphingolipid.

sphingosine. A long chain, unsaturated amino alcohol that is the parent compound of the sphingolipids. *Aka* 4-sphingenine.

spike. A characteristic protrusion on viral envelopes.

spike potential. ACTION POTENTIAL.

spiking. INTERNAL STANDARDIZATION.

spillover hypothesis. The hypothesis according to which photosystems I and II of chloroplasts have specific points of contact between them so that energy transfer is possible not only between the pigments within each photosystem, but also between the pigments of one photosystem and those of the other. *See also* separate package hypothesis.

spin. The rotation about an axis, as the rotation of an electron or an atomic nucleus.

spinach ferredoxin. A ferredoxin that contains two iron atoms per molecule, has a molecular weight of about 12,000, and serves as an early acceptor for electrons from the excited P_{700} pigment in photosystem I of chloroplasts.

spin coupling. *See* spin-spin coupling.

spin decoupling. A technique used in nuclear magnetic resonance in which the effect of spin coupling with nucleus A on the spectrum of nucleus B is eliminated by irradiation with the resonance frequency of nucleus A.

spin labeling. The introduction into a protein of a substituent, the electron paramagnetic resonance of which is sensitive to changes in the environment of the substituent. Measurements of the electron paramagnetic resonance of the substituent in the protein can then be used to explore the environment of the substituent in the protein molecule. *See also* reporter group.

spin lattice relaxation. A radiationless process whereby the energy of a nucleus in a high spin state is dissipated to the lattice of surrounding nuclei by means of oscillating magnetic fields.

spin multiplicity. The term $2S + 1$, where S is the resultant spin.

spin quantum number. The value of either $+\frac{1}{2}$ or $-\frac{1}{2}$ that is arbitrarily assigned to one of the two directions of spin of an orbital electron.

spin-spin coupling. The interaction between the magnetic moment of a proton, or some other nuclear dipole, and those of neighboring dipoles in nuclear magnetic resonance; this interaction leads to the splitting of a single peak into multiple peaks. *Aka* spin-spin interaction.

spin-spin splitting. The production of multiple peaks from a single peak in nuclear magnetic resonance as a result of spin-spin coupling.

spirillum (*pl* spirilla). A bacterium having a helically shaped cell; spirilla represent one of the three major forms of eubacteria.

spirometer. An instrument for measuring the volume of air inhaled and exhaled; used in measurements of the basal metabolic rate.

spleen. A large organ in the abdominal cavity that functions in the destruction of erythrocytes and in the production of antibodies.

splenectomy. The surgical removal of the spleen.

split protein. A protein that has been removed from ribosomes, such as one removed in the disassembly of ribosomes by concentrated salt solutions. *Abbr* SP.

splitter. A device for decreasing the amount of sample that is introduced into a gas chromatographic column from the inlet.

splitting. The separation, in terms of energy, between previously degenerate energy levels in an atom or a molecule.

spontaneous generation. The doctrine that living things can come from nonliving matter; abiogenesis.

spontaneous hypersensitivity. Hypersensitivity that is produced in an animal in the absence of any known contact of the animal with an antigen.

spontaneous induction. The induction of a prophage that is caused by the random interaction between the immunity substance and the prophage.

spontaneous mutation. A naturally occurring mutation for which there is no observable cause; a mutation that results from a chance exposure of an organism to a mutagen in the environment.

spontaneous process. A process that is accompanied by an increase in entropy.

spontaneous reaction. A chemical reaction that is accompanied by a negative free energy change; an exergonic reaction.

spontaneous tumor. A tumor that arises without known exposure of the organism to a carcinogen, such as one formed in certain inbred strains of mice that are genetically prone to tumor development.

sporangium (*pl* sporangia). A special structure housing spores.

spore. A dormant cellular form, derived from a bacterial or a fungal cell, that is devoid of metabolic activity and that can give rise to a vegetative cell upon germination; it is dehydrated and can survive for prolonged periods of time under drastic environmental conditions.

spore coat. One of two outer layers that surround the spore cortex and that consist largely or entirely of cross-linked proteins having a high cystine content.

spore core. The central body of a spore; the cytoplasm of a spore.

spore cortex. A thick layer that contains glycopeptides and that is located between the spore coats and the spore membrane.

spore membrane. The membrane that surrounds the spore core.

spore peptide. The glycopeptide material released from germinating spores.

spore wall. SPORE MEMBRANE.

sporogenesis. 1. Reproduction by means of spores. 2. The production of spores.

sporulation. The process of forming spores; the division into spores.

spot. 1. *n* The location of a separated and visualized component in either chromatography or electrophoresis. 2. *v* To apply material in small amounts to either a chromatographic or an electrophoretic support.

spray-freezing. A modification of the freeze-fracturing technique in which the cooling rate is increased by spraying the sample into a liquid at very low temperatures; the frozen droplets are then collected, "glued" together with butyl benzene, and subjected to freeze-fracturing.

spreading factor. HYALURONIDASE.

spreading reaction. The change in permeability and other properties of the cell membrane that is believed to occur in a phage-infected cell in the vicinity of each bound phage particle.

S-protein. A long fragment of the enzyme ribonuclease that is obtained by a subtilisin-catalyzed cleavage of the molecule between amino acid residues 20 and 21.

S-100 protein. An acidic brain-specific protein that has been implicated in neurophysiological functions and in the process of learning; the protein has a molecular weight of 20,000 and is rich in aspartic and glutamic acids.

SP-sephadex. Sulfopropyl sephadex; a cation exchanger.

spur. The overlapping portion of two precipitin arcs in immunodiffusion; one such spur is formed in a reaction of partial identity, and two such spurs are formed in a reaction of nonidentity.

spurious counts. Counts of radioactivity that are not caused by the sample but that result from outside factors, such as malfunctioning of the apparatus.

squalene. A terpenoid that serves as the immediate precursor of sterols in their biosynthesis; it gives rise to lanosterol which is then converted to cholesterol.

square dilution law. RADIAL DILUTION.

square wave polarography. A polarographic technique in which a square-wave alternating potential is superimposed on the normal d-c applied potential and the alternating component of the current is measured.

squiggle. The symbol $\sim$ used to designate a high energy bond, specifically one involving the phosphate group.

Sr^{90}. Strontium 90.

S-region. EXTRA ARM.

SRF. *See* growth hormone regulatory hormone.

SRH. *See* growth hormone regulatory hormone.

SRIF. *See* growth hormone regulatory hormone.

SRIH. *See* growth hormone regulatory hormone.

sRNA. Soluble RNA.

4S RNA. TRANSFER RNA.

5S RNA. An RNA molecule that has a sedimentation coefficient of 5S and that forms part of the ribosomal RNA; it is similar to transfer

RNA in its base composition and sedimentation coefficient, but it does not contain minor bases.

SRS-A. Slow reacting substance-anaphylaxis.

SRS technique. Separation-reaction-separation technique; a separation technique for steroids in which the steroids are first separated in one dimension by thin-layer chromatography, are then exposed to radiation or are reacted with chemical reagents, and are then separated in the second dimension by thin-layer chromatography with the solvent used for the first dimension.

ss. Single-stranded.

SSA test. Sulfosalicylic acid test.

SSC. Standard saline citrate.

ssDNA. Single-stranded DNA.

ssRNA. Single-stranded RNA.

SSS. Steady-state stacking.

stab culture. A bacterial culture made by plunging an inoculating needle into a solid medium.

stability constant. FORMATION CONSTANT.

stable factor. PROCONVERTIN.

stable isotope. An isotope that is not radioactive.

stacking. 1. BASE STACKING. 2. ISOTACHOPHORESIS.

stacking interactions. The hydrophobic interactions between the base pairs which are arranged in parallel planes in the interior of a double-helical nucleic acid structure.

staggered conformation. The conformation of a molecule in which, in a Newman projection, few atoms are either partially or fully concealed from view by other atoms. In such a conformation, interatomic distances are relatively great and interatomic interactions are minimized; as a result, staggered conformations are more stable than eclipsed ones.

staircase reaction. A stepwise chemical reaction, such as the hydrolysis of starch, which proceeds through the formation of intermediate dextrins and oligosaccharides to the formation of glucose.

staircase response. The gradual increase in steroid excretion upon the administration of adrenocorticotropic hormone that occurs in individuals who suffer from adrenal insufficiency as a result of either pituitary or hypothalamic dysfunction.

stalk. *See* supermolecule.

standard. *See* standard solution.

standard conditions. *See* standard temperature and pressure.

standard curve. A plot of a physical property, commonly absorbance, of a compound or of its derivative as a function of the concentration of the compound; used for the determination of unknown concentrations of the compound from measurements of the physical property.

standard deviation. A measure of the scatter of values about the mean of a set of measurements; for a normal distribution curve, it is the range about the mean within which 68.27% of all the measurements will fall. The standard deviation is equal to $\sqrt{\sum(X - \bar{X})^2/(N - 1)}$, where N is the number of measurements, and $\sum(X - \bar{X})^2$ is the sum of the squared deviations from the mean $\bar{X}$. *Abbr* SD; s; σ.

standard deviation of the mean. STANDARD DEVIATION.

standard diffusion coefficient. The value of a translational diffusion coefficient that is calculated from data which have been extrapolated to zero concentration, and that is corrected to a diffusion coefficient under standard conditions which are taken as the diffusion in a medium of water at 20°C. *Sym* $D^0_{20,w}$.

standard electrode potential. The electrode potential for a half reaction at 25°C and 1 atm of pressure in which all reactants and products are present in their standard states; denoted E^0. It is usually approximated as the midpoint potential for a system in which all reactants and products are present at 1 M concentrations and is then denoted E_0. The biochemical standard electrode potential is a reduction potential for a half reaction at 25°C and 1 atm of pressure at which all reactants and products are present at 1 M concentrations except protons which, unless otherwise specified, are present at a concentration of 10^{-7} M (pH 7.0); this potential is measured as a midpoint potential and is denoted E_0'.

standard error. A measure of the reliability of the mean, expressed as $SE = \sigma/\sqrt{N}$, where SE is the standard error, σ is the standard deviation, and N is the number of individual results. *Abbr* SE.

standard error of the mean. STANDARD ERROR.

standard flotation coefficient. The value of a flotation coefficient that is calculated from data which have been extrapolated to zero concentration, and that is corrected to a flotation coefficient under standard conditions which, for many lipoproteins, are taken as the flotation in a sodium chloride solution having a density of 1.063 g/ml at 26°C. *Sym* s_f^0.

standard free energy change. The free energy change of a reaction at 25° and 1 atm of pressure in which all reactants and products are present in their standard states; denoted ΔG^0. The biochemical standard free energy change is the free energy change of a reaction at 25°C and 1 atm of pressure in which all reactants and products are present in their standard states except protons which, unless otherwise specified, are present at a concentration of 10^{-7} M (pH 7.0); this free energy change is denoted as $\Delta G^{0'}$.

standardization. The determination of the

concentration of a solution by titration against either a primary or a secondary standard.

standard oxidation potential. The standard electrode potential for an oxidation half-reaction.

standard potential. *See* standard electrode potential.

standard reduction potential. The standard electrode potential for a reduction half-reaction.

standard saline citrate. A buffer composed of 0.15 *M* sodium chloride and 0.015 *M* trisodium citrate, pH 7.0; used in studies of DNA in solution. *Abbr* SSC.

standard sedimentation coefficient. The value of a sedimentation coefficient that is calculated from data which have been extrapolated to zero concentration, and that is corrected to a sedimentation coefficient under standard conditions which are taken as the sedimentation in a medium of water at 20°C. *Sym* $s^0_{20,w}$.

standard set. The group of amino acids that commonly occur in proteins.

standard solution. A solution of known concentration.

standard state. A thermodynamic reference state in which the temperature is 25°C, the pressure is 1 atm, and the composition is such that all components are present in their defined reference states. The standard state of pure substances is the state of the pure liquid or solid at 25°C and 1 atm of pressure. The standard state for components in solution is usually taken as that in which the activities of both solutes and solvent are equal to unity; for solutes, an activity of one is approximated by a 1 *M* concentration. In biochemical systems, the standard state of hydrogen ions is taken to be the hydrogen ion concentration which corresponds to pH 7.0.

standard temperature and pressure. A temperature of 0°C and a pressure of 1 atm (760 mm Hg). *Abbr* STP.

staphylokinase. The enzyme that is present in filtrates of *Staphylococcus* cultures and that promotes the lysis of human blood clots by catalyzing the activation of plasminogen to plasmin.

star. *See* star gazing.

starch. The major form of storage carbohydrates in plants. It is a homopolysaccharide, composed of D-glucose units, that occurs in two forms: amylose, which consists of straight chains, and in which the glucose residues are linked by means of $\alpha(1 \rightarrow 4)$ glycosidic bonds; and amylopectin, which consists of branched chains, and in which the glucose residues are linked by means of both $\alpha(1 \rightarrow 4)$ and $\alpha(1 \rightarrow 6)$ glycosidic bonds.

starch gel electrophoresis. A zone electrophoretic technique of high resolving power in which a gel of partially hydrolyzed starch is used as the supporting medium.

starch granule. A storage particle of starch that

occurs in the cytoplasm of plant cells and that also contains proteins and enzymes that function in the synthesis and breakdown of starch. The granules vary in size and shape, and may be used to classify the plants from which they are derived.

starch indicator. A starch solution used as an indicator in iodometric titrations.

starch phosphorylase. The enzyme that catalyzes the successive hydrolytic removal of glucose residues in the form of glucose-1-phosphate from the nonreducing end of starch; this reaction is the first step for the utilization of starch in glycolysis.

star gazing. A technique for studying the radioactive decay of phosphorus in nucleic acids by labeling DNA or a virus with radioactive phosphorus (P^{32}) and then embedding it in a sensitive photographic emulsion. Star-shaped patterns of beta particle tracks result from the decay of the radioactive phosphorus atoms and a count of these tracks can be used to calculate the molecular weight of the nucleic acid.

Stark effect. The change in the absorption spectrum of a pigment which is placed in a strong electric field.

Stark-Einstein law. SECOND LAW OF PHOTOCHEMISTRY.

Starling's hypothesis. The hypothesis that the rate of exchange of fluid between the plasma in the blood capillaries and the interstitial fluid is governed by a balance of hydrostatic pressure and osmotic pressure, primarily the hydrostatic pressure due to the blood and the osmotic pressure due to the plasma proteins.

start codon. INITIATION CODON.

starter. 1. PRIMER. 2. INOCULUM.

starting potential. That portion of the characteristic curve of a Geiger-Mueller counter that precedes the plateau and at which there is a sharp rise in the curve.

starvation. 1. An extreme case of undernutrition due to severe and prolonged inadequate intake of most of the required nutrients. 2. Undernutrition with respect to one or more nutrients, as in the withholding of an amino acid from a growing bacterial culture.

starvation diabetes. A condition of temporary carbohydrate intolerance that is characterized by glucosuria; follows the ingestion of carbohydrate after prolonged starvation and is due to either the decreased output of insulin or the decreased ability to synthesize glycogen.

-stat. Combining form meaning constant. *See also* cryostat; pH stat; thermostat.

state function. Any one of the four thermodynamic parameters: internal energy, enthalpy, entropy, and free energy. The values of these parameters for a given process depend on the initial and the final state of the system and not on the path taken to proceed from the initial to the final state.

state of a system. The description of a system in terms of the thermodynamic state functions. The state before a process occurs is known as the initial state, and the state after the process has occurred is known as the final state.

static osmometer. An osmometer in which osmosis is allowed to take place so that a pressure difference develops.

stationary electrolysis. ISOELECTRIC FOCUSING.

stationary phase. 1. The phase of growth of a bacterial culture that follows the exponential phase and in which there is little or no growth. 2. The liquid, solid, or solid plus adsorbed liquid that serves as a support in chromatography.

stationary state. STEADY STATE.

stationary substrate. STATIONARY PHASE (2).

statistical. Of, or pertaining to, statistics.

statistical error. An error in the counting of radioactive materials that is due to the random nature of radioactive decay.

statistical mechanics. A description of the equilibrium properties of macroscopic systems that is based on the application of statistics to atomic and molecular energy states.

statistics. The science that deals with the collection, analysis, interpretation, and presentation of masses of numerical data.

Staudinger index. INTRINSIC VISCOSITY.

steady state. 1. The nonequilibrium state of a system in which matter flows in and out at equal rates so that all of the components remain at constant concentrations. In a chemical reaction sequence, a component is in a steady state if the rate at which the component is being synthesized is equal to the rate at which it is being degraded. 2. The maximum color intensity of a sample that is obtained as a function of time; used in reference to determinations with an autoanalyzer.

steady-state electrolysis. ISOELECTRIC FOCUSING.

steady-state kinetics. The kinetics of an enzymatic reaction proceeding under steady-state conditions; essentially the kinetics of the rate-determining step of the reaction.

steady-state stacking. ISOTACHOPHORESIS.

steapsin. The lipase present in the pancreatic juice.

stearic acid. A saturated fatty acid that contains 18 carbon atoms and that occurs in animal fat.

steatolysis. The hydrolysis and the emulsification of fats during digestion.

steatorrhea. A genetically inherited metabolic defect in man that is characterized by a failure to digest lipids and/or absorb them from the intestine.

steatosis. 1. ADIPOSIS. 2. FATTY DEGENERATION.

stem. 1. AMINO ACID ARM. 2. STALK.

stem cell. An undifferentiated cell from which specialized cells are subsequently derived.

stepwise development. A chromatographic technique, used particularly with paper and thin-layer chromatography, in which the sample is developed repeatedly with different solvents.

stepwise elution. The chromatographic elution in which two or more solutions that differ in composition are added to a chromatographic column by abruptly changing from one solution to the next.

stereochemical. Pertaining to the three-dimensional arrangement of atoms in a molecule.

stereochemical theory. A theory according to which the evolution of the genetic code is due to a stereochemical relationship between an amino acid and either its codons or its anticodons.

stereochemistry. The branch of chemistry that deals with the spatial arrangement of atoms in a molecule.

stereoelectronic. Pertaining to the spatial aspects of atomic and molecular orbitals.

stereoisomer. One of two or more isomers that have the same molecular composition and the same nucleus-to-nucleus sequence, but that differ from each other in the spatial arrangement of the atoms in the molecule; stereoisomers cannot be distinguished on the basis of their two-dimensional structures. *See also* structural isomer.

stereomer. STEREOISOMER.

stereomutation. The conversion of one stereoisomer into another; particularly the interconversion of geometrical isomers, as in the conversion of oleic acid (cis) to elaidic acid (trans).

stereoselective reaction. A chemical reaction that proceeds preferentially with one stereoisomer over another. *See also* stereospecific reaction.

stereospecificity. The selectivity of an enzyme for a particular stereoisomer.

stereospecific numbering. A system for numbering derivatives of symmetrical compounds, such as glycerol, that is based on representing the parent molecule always in one specific stereochemical configuration. *Sym* sn.

stereospecific reaction. A chemical reaction that produces one stereoisomer preferentially over another. *See also* stereoselective reaction.

steric. STEREOCHEMICAL.

steric factor. A mathematical factor, used in the collision theory of chemical kinetics, which allows for the fact that collisons of sufficient energy may take place between molecules without leading to a chemical reaction because of an improper steric orientation of the colliding molecules.

steric hindrance. The prevention of a molecule from undergoing a chemical reaction or from achieving a particular conformation or configuration that is ascribed to the stereochemical arrangement of the atoms and groups of atoms in the molecule; bulky substituents by tending to compress reactive groups and forcing them too close to their unbonded neighbors.

sterile. Free from viable microorganisms.

sterilization. The complete destruction of all viable microorganisms in a material by physical and/or chemical means.

sterilizer. AUTOCLAVE.

steroid. A cyclic compound of animal or plant origin, the basic nucleus of which consists of three 6-membered rings and one 5-membered ring, fused together to yield perhydrocyclopentanophenanthrene. The steroids represent a wide variety of compounds, including sterols, bile acids, adrenocortical hormones, and sex hormones.

steroid conjugate. A steroid breakdown product that is conjugated to either glucuronic acid or sulfuric acid and that is excreted in this form.

steroid diabetes. A condition produced by the prolonged administration of glucocorticoids and characterized by glucose production in the liver and by suppression of the action of insulin.

steroid glycoside. A carbohydrate derivative of a steroid, such as a cardiac glycoside or a saponin.

steroid hormone. A collective term for androgens, estrogens, and corticoids.

steroid number. A number assigned to a steroid on the basis of the number of carbon atoms and the types of functional groups that it contains; specifically, $SN = S + F_1 + F_2 + \ldots F_n$, where SN is the steroid number, S is the number of carbon atoms in the parent steroid, and F_1, $F_2 \ldots F_n$ are arbitrary values, characteristic of the functional groups of the steroid.

steroidogenesis. The biosynthesis of steroids.

sterol. A steroid in which an alcoholic hydroxyl group is attached to position 3, and an aliphatic side chain of eight or more carbon atoms is attached to position 17 of the steroid nucleus; a steroid alcohol.

Sterzl theory. X-Y-Z CELL THEORY.

STH. Somatotropic hormone.

Stickland reaction. The coupled decomposition of two amino acids such that one amino acid is oxidized while the other is reduced. The reaction occurs in some species of the genus *Clostridium*, which cannot degrade single amino acids but can degrade suitable pairs of amino acids.

sticky end. One of two complementary single-stranded segments at opposite ends of each of the two strands of a double-stranded nucleic acid molecule; the presence of these segments permits the joining of the ends of the molecule and its conversion to a double-stranded circular form.

sticky region. A region in a nucleic acid molecule that is rich in guanine and cytosine.

stilbestrol. 1. 4,4´-Dihydroxystilbene. 2. Diethylstilbestrol.

still. An apparatus for the purification of liquids by distillation.

stimulin. INSULIN-LIKE ACTIVITY.

stochastic process. 1. A process of problem solving that provides a solution which is close to the best; the steps in the process are based on an uncertainty which is indicated by the laws of probability and the movement at each step is random. Conjecture and speculation are used to select a possible solution which is then tested against known evidence, observation, and measurement. 2. A process of problem solving that consists of random trial and error steps. *See also* algorithm; heuristic process.

stoichiometric amount. The amount of a substance that is used in a chemical reaction either as a reactant or as a product.

stoichiometry. The quantitative relationships between the elements in a compound or between the reactants and the products in a chemical reaction.

stoke. A unit of kinematic viscosity; one hundredth of this unit is the centistoke.

Stoke's law. An expression for the frictional coefficient of a sphere; specifically, $f = 6\pi\eta r$, where f is the frictional coefficient, π is a constant equal to $3.1416\ldots$, η is the viscosity of the solvent, and r is the radius of the sphere.

Stoke's radius. The radius of a perfect anhydrous sphere that has the same rate of passage through a gel filtration column as the protein under study.

Stoke's reagent. A solution of ferrous sulfate, tartaric acid, and ammonia that is used in testing for hemoglobin.

Stoke's shift. The change in the wavelength of light in fluorescence from that of the exciting light to that of the emitted light. The emitted light is usually of longer wavelength than the exciting light, and the change in wavelength is attributable to energy lost as vibrational energy.

stone. CALCULUS.

stop codon. TERMINATION CODON.

stopped flow technique. A technique for studying fast chemical reactions in which the reactants are forced out of two syringes into a mixing chamber, and the mixture is then allowed to flow into an observation cell where the flow is halted abruptly and the mixture is analyzed spectroscopically or by other means. *See also* rapid flow technique.

STP. Standard temperature and pressure.

straggling. The variation in the range of particles of a specific radiation in which initially all of the particles had the same energy.

straight chain. OPEN CHAIN.

strain. 1. *n* A subdivision of a species that possesses distinguishing characteristics. 2. *n* A deformation in a molecule. 3. *v* To pass through a coarse filter such as cheesecloth.

strand. A polynucleotide chain.

strand exchange. BRANCH MIGRATION.

streak. 1. To apply material as a strip to either a chromatographic or an electrophoretic support, usually for preparative purposes. 2. To apply a bacterial inoculum as a strip to the surface of a solid nutrient medium.

streak plating. A method for isolating bacterial strains by drawing an inoculating needle, containing a culture, lightly over the surface of a solid nutrient medium.

stream birefringence. FLOW BIREFRINGENCE.

streamer sedimentation. DROPLET SEDIMENTATION.

streaming birefringence. FLOW BIREFRINGENCE.

streaming current. The electrical current produced by the movement of an electrolyte through a tube.

streaming potential. The electrical potential produced by the movement of a conducting liquid through a porous medium; an electrokinetic phenomenon that is the reverse of electroosmosis.

streamline flow. LAMINAR FLOW.

stream potential. STREAMING POTENTIAL.

Strecker synthesis. The synthesis of amino acids from aldehydes, ammonia, and hydrogen cyanide; proposed as a possible mechanism for the synthesis of amino acids under conditions existing on the primitive earth.

street virus. A virus obtained from a naturally infected animal. *See also* fixed virus.

streptococcal fibrinolysin. STREPTOKINASE.

streptokinase. The enzyme that is present in filtrates of *Streptococcus* cultures and that promotes the lysis of human blood clots by catalyzing the activation of plasminogen to plasmin.

streptolysin. A toxin, produced by the genus *Streptococcus,* that causes hemolysis.

Streptomyces. A genus of soil bacteria, many species of which produce antibiotics.

streptomycin. An aminoglycoside antibiotic, produced by *Streptomyces griseus,* that inhibits protein synthesis by binding to the 30S ribosomal subunit, and that causes misreading of the genetic code.

streptonigrin. An antibiotic, produced by *Streptomyces flocculus,* that leads to chromosome breaks.

stretched muscle. A muscle that has undergone stretching.

stretching vibration. A vibration in a molecule that results in a lengthening of an interatomic bond distance.

striated muscle. A muscle, such as a skeletal or a cardiac muscle, that is characterized by transverse striations.

strict aerobe. OBLIGATE AEROBE.

strict anaerobe. OBLIGATE ANAEROBE.

striction. The decrease in the volume of a solution, compared to the sum of the volumes of the separate solute and solvent, as a result of solute-solvent interactions. *See also* covolume; electrostriction.

stringent control. The decrease of RNA synthesis that occurs normally in wild-type bacteria after removal of an essential amino acid from the medium. *See also* relaxed control.

stringent strain. A bacterial strain that shows stringent control.

stripped particle. CORE PARTICLE.

stripped transfer RNA. An aminoacyl transfer RNA molecule from which the amino acid has been detached by hydrolysis.

stripping. 1. The hydrolytic removal of an amino acid from an aminoacyl transfer RNA molecule. 2. The removal of ribosomal proteins from ribosomes, resulting in the production of ribosomal subparticles.

stroma. 1. The matrix material between grana in a chloroplast. 2. The connective tissue framework of a gland or an organ as distinct from the parenchyma.

stroma starch. The starch of chromatophores that is not clustered around pyrenoids.

stromatin. A protein in the cell membrane of erythrocytes.

strong electrolyte. An electrolyte that is completely dissociated into ions in water.

strong interactions. The attractive forces between atoms and molecules that result in the formation of covalent and ionic bonds, and the repulsive forces of ion-ion interactions.

strontium. An element that is essential to a few species of organisms. Symbol Sr; atomic number 38; atomic weight 87.62; oxidation state +2; most abundant isotope Sr^{88}; a radioactive isotope Sr^{85}, half-life 64.7 days, radiation emitted—gamma rays.

strontium 90. The heavy radioactive isotope of strontium that occurs in the fallout from the explosion of nuclear weapons; it has a half-life of 25 years and is incorporated into biological systems.

strophantidin. OUABAIN.

strophantin G. OUABAIN.

structural formula. A two-dimensional representation of the structure of a molecule in which the atoms are connected by one or more lines, with each line indicating a pair of shared electrons.

structural gene. 1. A gene, the nucleotide sequence of which determines the amino acid sequence of a polypeptide chain; a cistron. 2. A gene, the nucleotide sequence of which determines the nucleotide sequence of a polynucleotide strand.

structural isomer. One of two or more isomers that have the same molecular composition but that differ from each other in their nucleus-to-nucleus sequence; they can be distinguished on the basis of their two-dimensional structures. *See also* stereoisomer.

structural protein. 1. A protein that functions primarily as a structural component of cells and tissues. 2. A noncatalytic protein of mitochondrial membranes, the existence of which is in doubt.

structural unit. The monomer of the viral capsomer that consists of one or more polypeptide chains. *See also* capsid; capsomer.

Stuart factor. The first factor that is common to both the intrinsic and the extrinsic systems of blood clotting. The Stuart factor is activated by the antihemophilic factor in the intrinsic system, and it is activated by factor VII in the extrinsic system. *Aka* Stuart-Prower factor.

student's t test. A statistical test for evaluating the difference between two sample means.

sturine. A protamine isolated from sturgeon.

sub-. 1. Prefix meaning below or under. 2. Prefix meaning fraction of or less than.

subacute test. PROLONGED TEST.

subcellular fraction. A preparation containing either one or more specific components of the cell, or a portion of the total cellular material; a preparation of an enzyme, or of mitochondria, or of cell membranes, etc.

subculture. A culture that is produced by transferring a portion of a stock culture to fresh medium.

subcutaneous injection. An injection under the skin. *Abbr* sc.

sublimation. The transition of a solid to a vapor without passage through the intermediate liquid state.

subnatant. The liquid below another liquid or below solid material.

subribosomal particle. RIBOSOMAL SUBPARTICLE.

subribosome. RIBOSOMAL SUBPARTICLE.

substituent. An atom or a group of atoms that is introduced into a molecule by replacement of another atom or group of atoms.

substituted enzyme. The modified enzyme that is produced by the reaction of the original enzyme molecule with the first substrate in a ping-pong mechanism.

substitution. *See* base substitution; conservative substitution; radical substitution; substitution reaction.

substitution reaction. A chemical reaction in which an atom or a group of atoms attached to a

carbon atom is removed and replaced by another atom or group of atoms.

substrate. The compound acted upon by an enzyme; the molecule or structure, the transformation of which is catalyzed by an enzyme. *Sym* S.

substrate constant. The equilibrium (dissociation) constant of the reaction $ES = E + S$, where E is the enzyme and S is the substrate. *Sym* K_s.

substrate elution chromatography. A column chromatographic technique in which an enzyme is adsorbed to a column and is subsequently eluted with a solution containing the substrate of the enzyme.

substrate inhibition. The inhibition of the activity of an enzyme by the substrate of the enzyme; generally most pronounced at high substrate concentrations.

substrate phosphorylation. The synthesis of ATP that is coupled to the exergonic hydrolysis of a high-energy compound and that is not linked to an electron transport system; the synthesis of ATP in glycolysis is an example. *Aka* substrate level phosphorylation.

subtilin. A peptide antibiotic produced by *Bacillus subtilis* that consists of 32 amino acids.

subtilisin. A nonspecific proteolytic enzyme derived from *Bacillus subtilis*.

subunit. 1. The smallest covalent unit of a protein; it may consist of one polypeptide chain or of several chains linked together covalently, as by means of disulfide bonds. 2. The functional unit of an oligomeric protein, such as the structure in hemoglobin that consists of one alpha and one beta chain. 3. A definite substructure of a macromolecule such as the 30S or the 50S unit of the 70S bacterial ribosome. *See also* monomer; protomer.

subunit exchange. The association of the small ribosomal subunit from one ribosome with the large subunit from a different ribosome, and vice versa, as a result of the operation of the ribosome cycle.

subunit model. SYMMETRY MODEL.

subvital mutation. A mutation that results in the death of less than 50% of the mutants.

succinate pathway. A catabolic pathway whereby methionine, isoleucine, and valine are converted to succinic acid which then enters the citric acid cycle.

succinic acid. A symmetrical dicarboxylic acid that is an intermediate in the citric acid cycle where it is formed from α-ketoglutaric acid.

sucrase. The enzyme that catalyzes the hydrolysis of sucrose to glucose and fructose.

sucrose. A disaccharide of glucose and fructose that is abundant in the plant world and that is the sugar used in the home.

sucrose density gradient. A density gradient prepared by using sucrose solutions of different concentrations, frequently covering a range of 5 to 25% (w/w).

sucrose gradient centrifugation. Density gradient sedimentation velocity in which a sucrose density gradient is used.

Sudan black B. A lysochrome.

sugar. CARBOHYDRATE.

sugar acid. An acid derivative of a monosaccharide.

sugar alcohol. An alcohol derivative of a monosaccharide.

suicide. The loss of infectivity or other biological activity by either a molecule or a particle as a result of radioactive decay of an incorporated radioisotope. *See also* radiophosphorus decay; star gazing.

sulfa drug. One of a class of antibacterial drugs that are derivatives of sulfonamide RSO_2NH_2 and that inhibit the growth of many bacteria. Sulfa drugs function by being competitive inhibitors of *p*-aminobenzoic acid, which is required by these bacteria for the synthesis of folic acid.

sulfanilamide. Benzene sulfonamide; a sulfa drug.

sulfatide. A monoglycosyl derivative of a ceramide that contains a sulfate group esterified to galactose. *Aka* sulfatidate.

sulfating agent. ACTIVE SULFATE.

sulfation factor. SOMATOMEDIN.

sulfhydryl group. The radical —SH.

sulfite oxidase deficiency. A genetically inherited metabolic defect in man that is associated with mental retardation and that is due to a deficiency of the enzyme sulfite oxidase.

sulfolipid. 1. Any sulfur-containing lipid. 2. PLANT SULFOLIPID.

sulfonamide. *See* sulfa drug.

sulfonic acid. An organic acid containing the radical —SO_3H.

sulfosalicylic acid test. A test for protein in urine and in other biological fluids that is based on the turbidity formed upon addition of sulfosalicylic acid to the sample. *Abbr* SSA test.

sulfur. An element that is essential to all plants and animals. Symbol S; atomic number 16; atomic weight 32.964; oxidation states -2, $+4$, $+6$; most abundant isotope S^{32}; a radioactive isotope S^{35}, half-life 87.2 days, radiation emitted—beta particles.

sulfur amino acid. An amino acid that contains sulfur, such as cysteine or methionine.

sulfur bacteria. A group of bacteria that are chemolithotrophs and that use carbon dioxide as their carbon source, oxidation-reduction reactions as their energy source, and inorganic sulfur-containing compounds as electron donors.

sulfur mustard. Di-(2-chloroethyl)sulfide; a

chemical mutagen and alkylating agent that is structurally related to nitrogen mustards but contains sulfur instead of nitrogen. *See also* alkylating agent.

Sulkowitch test. A test used for the semiquantitative determination of calcium in urine; based on measurements of the turbidity that is produced upon addition of oxalate, buffered with acetate, to urine.

Sullivan reaction. A colorimetric reaction for cysteine that is based on the production of a red color upon treatment of the sample with sodium 1,2-naphthoquinone-4-sulfonate and sodium sulfite.

SUN. Serum urea nitrogen; *see* blood urea nitrogen.

super-. 1. Prefix meaning above or in the upper part of. 2. Prefix meaning excessive.

superacid catalysis. Catalysis due to a positively charged ion, particularly a metal ion, that acts as a Lewis acid.

supercoil. SUPERHELIX.

supercool. To cool a liquid below its freezing point without the separation of solid matter.

supercritical fluid chromatography. A chromatographic technique in which compressed fluids, at temperatures slightly above their critical temperatures, are used as mobile phases; useful for the separation of substances that are not readily separated by gas or liquid chromatography. *Abbr* SFC.

superfusion. A technique in which blood, plasma, or some other fluid is allowed to drip onto, or to flow over, the surface of a perfused organ.

supergene. A segment of linked genes that is protected from crossing-over and that is transmitted intact from one generation to another.

superhelix. 1. A helical, circularly covalent, duplex nucleic acid molecule in which the molecule as a whole is twisted. Such a molecule has fewer turns than a linear double helix and this deficiency tends to be compensated for by a twisting of the molecule. 2. A helix composed of two or more component chains and produced by the winding of two or more helical chains, either polynucleotide or polypeptide chains, about each other.

superinducible mutant. A mutant that synthesizes an inducible enzyme at concentrations of inducer which are lower than those required by the wild-type organism.

superinfection. 1. An extensive invasion of an organism by pathogenic microorganisms, as that which may arise from its infection by drug-resistant microorganisms. 2. A repeated phage infection of a bacterial culture that is already infected with phage. Such a culture is characterized by possessing bacterial cells which contain more than one phage particle per cell.

superinfection curing. *See* cure.

superinfection exclusion. The phenomenon that superinfection of a bacterial culture with phage at times leads to failure of the DNA of the superinfecting phage to enter the host cell; a change in cell permeability (a "sealing reaction") has been invoked to explain this phenomenon.

supermolecule. The polymeric protein system that functions as the energy transducing unit and constitutes the structural unit of the inner mitochondrial membrane according to the electromechanochemical coupling hypothesis. The supermolecule consists of seven complexes, four of which are electron transfer complexes, and the other three are the ATP synthetase, the transprotonase, and the transhydrogenase complexes. The supermolecule is in the shape of a knoblike structure composed of a basepiece, a stalk, and a headpiece; this structure is formed by arranging six of the complexes around a central unit referred to as a tripartite repeating unit (TRU). The headpiece of the TRU is the site for synthesis or hydrolysis of ATP; the stalk of the TRU, which connects the headpiece and the basepiece, is a regulatory device which determines whether ATP will be synthesized or hydrolyzed; and the basepiece of the TRU is the membrane-forming element that functions as a linkage system around which are arranged the four electron transfer complexes, the transhydrogenase, and the transprotonase. *Aka* elementary particle.

supernatant. The liquid above sedimented material or above a precipitate. *Aka* supernate.

superoxide dismutase. The enzyme that catalyzes the reaction $O_2^- + O_2^- + 2H^+ = H_2O_2 + O_2$. The enzyme appears to be ubiquitous among aerobic organisms and protects the organism against action by the radical O_2^-, which is believed to be a mutagenic radical, produced in an organism by ionizing radiation.

superprecipitation. The contraction of an actomyosin gel to a small plug that occurs in the presence of ATP; the process is used for in vitro studies of muscle contraction.

superrepressed mutant. A mutant that synthesizes a repressible enzyme at a rate which is characteristic of the uninduced wild-type organism, regardless of the presence or absence of the inducer.

supersaturated solution. A solution that holds temporarily more solute than will be contained by an equal volume of a saturated solution under specified conditions.

supersonic. 1. ULTRASONIC. 2. Having a velocity exceeding that of sound.

supersonic oscillation. ULTRASONICATION.

supplemental air. The volume of air that can be forcibly expired from the lungs after the normal tidal air has been expired.

supplementary action. The increase in the biological value of two proteins when they are fed together, over the sum of their biological values when they are fed separately.

support. 1. The solid material that either is or holds the stationary phase in chromatography. 2. The solid material in which electrophoresis is performed.

suppressible mutation. A mutation such that the genetic function, the loss of which is caused by the mutation, can be restored by means of either an intragenic or an intergenic suppressor mutation.

suppression. The partial or complete restoration of a genetic function, lost as a result of a mutation, by means of a second mutation, occurring at a different site on the chromosome.

suppression test. A clinical test for differentiating between hyperadrenalism due to adrenocortical hyperplasia and that due to an adrenal carcinoma. The test is based on the administration of a cortisol analogue which, in normal individuals but not in those with an adrenal carcinoma, has a feedback effect and decreases the pituitary output of adrenocorticotropic hormone; this, in turn, leads to a suppression of the output of steroids by the adrenal glands.

suppressor. 1. SUPPRESSOR GENE. 2. The product of a suppressor gene.

suppressor gene. A gene that can partially or completely reverse the effect of a mutation in another gene.

suppressor mutation. A mutation that partially or completely restores a genetic function, lost as a result of another mutation, and that is located at a site other than that which sustained the primary mutation. The suppressor mutation may occur either in the gene that was originally mutated (intragenic suppressor mutation) or in a different gene (intergenic suppressor mutation).

suppressor sensitive mutant. A conditional lethal mutant, the lost genetic function of which can be restored by a suppressor of the mutation.

suppressor transfer RNA. An unusual transfer RNA molecule, produced by a suppressor gene, that brings about missense or nonsense suppression. The transfer RNA molecule leads to the incorporation of the correct amino acid that would have been specified by the codon prior to its being changed as a result of either a missense or a nonsense mutation.

supra-. Prefix meaning above.

supramolecular complex. A complex composed of several or many molecules, such as a multienzyme system, a ribosome, or a biological membrane.

suprarenal gland. ADRENAL GLAND.

supravital staining. INTRAVITAL STAIN-ING.

surface-active agent. A substance that alters the surface tension of a liquid, generally lowering it; detergents and soaps are typical examples.

surface activity. The strong adsorption of surface-active agents at surfaces or interfaces of liquids in the form of oriented monomolecular layers.

surface antigen. An antigen that is present on the surface of a cell, such as the H, O, or Vi antigen of bacteria.

surface balance. An instrument for measuring surface pressure.

surface enzyme. An exoenzyme that remains attached to the cell surface.

surface factor. HAGEMAN FACTOR.

surface potential. The difference between the electrical potential of the clean surface of a liquid and that of a monolayer on the surface of the same liquid. *Aka* surface film potential.

surface pressure. The lowering of the surface tension of a liquid by a monolayer that is present on the surface of that liquid.

surface tension. The tension exerted by the surface of a liquid as a result of the intermolecular attractive forces between the molecules of the liquid.

surfactant. SURFACE-ACTIVE AGENT.

surroundings. That part of the universe, in the thermodynamic sense, that is not under study; the part of the universe that is being studied is known as the system.

survival curve. A dose-response curve that indicates the surviving fraction of active units as a function of the radiation dose.

survival of the fittest. *See* natural selection.

surviving slice. TISSUE SLICE.

suspension. A colloidal dispersion in which the particles are so large that they will settle out of solution. *See also* colloidal dispersion; colloidal solution.

suspension culture. A cell culture, used in virology, in which animal cells are kept in suspension in a medium that is low in divalent ions.

SV-40. Simian virus 40.

Svedberg. A unit of the sedimentation coefficient equal to 10^{-13} sec. *Sym* S.

Svedberg equation. An equation used for calculating molecular weights from sedimentation and diffusion data; specifically, $M = RTs/D(1 - \bar{v}\rho)$, where M is the molecular weight, T is the absolute temperature, s is the sedimentation coefficient, D is the diffusion coefficient, $\bar{v}$ is the partial specific volume of the solute, and ρ is the density of the solution.

swinging bucket rotor. A centrifuge rotor used in a preparative analytical ultracentrifuge for density gradient centrifugation. During the cent-rifugation the buckets, which hold tubes filled with solution, swing out to a horizontal position that is at right angles to the axis of rotation; this eliminates the extensive convection produced by centrifugation in a fixed angle rotor.

swivel. A postulated device that is attached to a fixed support and that is instrumental in the unwinding of double-stranded DNA during replication; may be identical with the replicator.

symbiont. An organism that lives in symbiosis with another organism.

symbiosis. The living together in close association, and for their mutual benefit, of two organisms from different species.

symbiotic. Of, or pertaining to, symbiosis.

symbiotic nitrogen fixation. The conversion of atmospheric nitrogen to ammonia by the combined action of plants and bacteria; applicable primarily to leguminous plants and to bacteria of the genus *Rhizobium*.

symmetrical. Possessing symmetry. *Aka* symmetric.

symmetry. The geometrical regularity in the structure of a body such that the sizes, shapes, and relative positions of structural parts are distributed equally about dividing planes, axes, and centers in the body.

symmetry-breaking model. SEQUENTIAL MODEL.

symmetry-conserving model. SYMMETRY MODEL.

symmetry model. A model for a regulatory enzyme—proposed by Monod, Wyman, and Changeaux—according to which the enzyme exists in two different conformational forms (denoted R and T) that are in equilibrium with each other. The two forms (a relaxed and a constrained one) differ in their capacity to bind substrate, positive effectors, and negative effectors, but the overall symmetry of the molecule is maintained throughout the various binding steps so that no two identical subunits are in different conformational states at any given time. The binding of an effector or of the substrate shifts the equilibrium from one form to another. *Aka* concerted transition model.

sympathomimetic amine. A substance that acts like epinephrine and that produces effects which are similar to those brought about by a stimulation of the sympathetic nervous system.

symport. The linked transport in the same direction of two substances across a membrane.

syn. 1. Referring to a nucleoside conformation in which the base has been rotated around the sugar, using the $C-N$ glycosidic bond as a pivot, so that the sugar is placed directly below the base. This represents a sterically more hindered conformation than the anti conformation; in polynucleotides, it leads to the bulky portions of the bases being pointed toward the sugar-phosphate backbone of the chain. 2. Referring to a cis configuration for certain compounds containing double bonds, such as the oximes which contain the grouping $\diagup C{=}N{-}OH$. 3. Referring to the position occupied by two radicals of a stereoisomer in which the radicals are closer together as opposed to the anti position in which they are farther apart. *See also* anti.

synalbumin. An insulin-inhibitory polypeptide in the blood; may be identical to the B chain of insulin which binds to albumin under certain conditions and which acts as an inhibitor of insulin in that form.

synapse. The area of functional contact between two nerve cells.

synaptosome. A largely artificial structure that is produced by disruption of the nerve endings in a synapse.

syncarcinogenesis. Synergistic carcinogenesis.

synchronous growth. Growth in which all of the cells are at the same stage in cell division at any given time. *Aka* synchronized growth.

synchronous muscle. A muscle that yields a single contraction for every motor nerve impulse that it receives.

synchronous reaction. CONCERTED REACTION.

synchrotron. An accelerator designed to impart high kinetic energy to charged particles by means of a high-frequency electric field and a low-frequency magnetic field.

syndesine. An amino acid that has been isolated from cross-linked collagen chains and that represents the product of an aldol condensation between a molecule of hydroxyallysine and a molecule of allysine.

syndet. Synthetic detergent.

syndiotactic polymer. A polymer in which the *R* groups of the monomers alternate regularly on both sides of the plane that contains the main chain.

syndrome. A group of symptoms that occur at the same time and that characterize a disease.

syneresis. The shrinkage of a gel with the expulsion of liquid. *See also* clot retraction.

synergism. The phenomenon in which two or more agents work together cooperatively such that their combined effect is greater than the sum of the effects when either agent is acting alone.

synergistic. Of, or pertaining to, synergism.

synergy. 1. SYNTROPHY. 2. SYNERGISM.

syngeneic. Referring to genetically identical individuals of the same species; used in reference to tissue transplants.

syngraft. A transplant from one individual to a genetically identical individual of the same species.

synonym codon. One of several codons that code for the same amino acid, such as the codons, *UUU* and *UUC*, both of which code for the amino acid phenylalanine.

synovial fluid. The fluid present at the joints of vertebrates.

synthase. 1. LYASE. 2. An enzyme that is not a lyase but for which it is desirable to stress the synthetic aspects of the reaction.

synthesis. The process whereby a more complex substance is produced from simpler substances by a reaction or a series of reactions; the simpler substances, or portions thereof, are combined to form the more complex substance.

synthetase. A ligase that requires a nucleoside triphosphate for the reaction that it catalyzes.

synthetic. 1. Of, or pertaining to, synthesis. 2. Man-made; synthesized in vitro; prepared artificially as opposed to being isolated from natural sources.

synthetic auxin. A synthetic organic compound that has auxin activity.

synthetic boundary cell. An analytical ultracentrifuge cell in which solvent or a less concentrated solution is layered at low rotor speeds over the sample solution, thereby establishing a "synthetic boundary." The cell is useful for measurements of small sedimentation coefficients and for determinations of apparent diffusion coefficients.

synthetic diet. A diet composed only of known chemical ingredients.

synthetic medium. A medium composed only of known chemical ingredients.

synthetic messenger RNA. A synthetic polyribonucleotide that is used as messenger RNA in a cell-free amino acid incorporating system.

synthetic polyribonucleotide. An RNA molecule made in vitro in the absence of a nucleic acid template; this may be accomplished either by using the enzyme polynucleotide phosphorylase or by chemical synthesis.

synthon. A group of atoms used as a unit in the synthesis of an organic compound.

syntrophy. The nutritional and metabolic interactions between organisms that are placed in the same environment. The phenomena of cross-feeding and of the nutritional interdependence of plants, animals, and microorganisms with respect to the carbon, oxygen, and other cycles are examples. *Aka* syntrophism.

system. 1. A set of subcellular components that perform one main function, such as an amino acid incorporating system or a transport system. 2. That part of the universe, in the thermodynamic sense, which is under study; the rest of the universe is known as the surroundings. 3. A set of organs performing one main function, such as the digestive system, the central nervous system, or the endocrine system.

systematic name. A name created on the basis of definite rules of nomenclature to distinguish clearly one item from related ones; the name of a compound based on its structure or the name

of an enzyme based on the reaction that it catalyzes are examples.

systemic. 1. Relating to a system. 2. Relating to the entire organism and not just to one of its parts.

Szilard-Chalmers reaction. A reaction in which a chemical compound is altered by bombardment with neutrons in such a way as to allow chemical separation of the reaction products.

T

t. Student's t statistic; *see* student's t test.

t₁/₂. Half-life.

T. 1. Thymine. 2. Thymidine. 3. Tritium. 4. Transmittance. 5. Absolute temperature. 6. T-even phage. 7. T-odd phage. 8. Threonine. 9. Conformational form of a regulatory enzyme. 10. Tera.

T₃. Triiodothyronine.

T₄. Thyroxine.

T₁/₂. Melting out temperature.

tachometer. A device for measuring the angular velocity of a rotating shaft. *Var sp* tachymeter.

tachyphylaxis. A pharmacological phenomenon in which the first, or the first few, doses of a drug produce a response and lead to establishment of resistance so that subsequent doses of the drug fail to elicit any further response.

tactoid. A paracrystalline aggregate of molecules which resembles that in a crystal but is limited to one dimension; the resulting linear arrangements of molecules are packed parallel to each other.

tag. LABEL.

tail. 1. The long fibrous portion of the myosin molecule. 2. The 3′-hydroxyl end of an oligo- or a polynucleotide strand. 3. The elongated, cylindrical structure attached to the head of a T-even phage. 4. The passive portion of a condensing unit.

tailing. The chromatographic and electrophoretic phenomenon in which a peak appears lopsided and a band, or a spot, appears ill-defined; due to the fact that in the support the front edge of the region that contains the component is sharp while the back edge is diffuse. *See also* leading; trailing.

tail-to-tail condensation. The condensation of two molecules by way of their passive ends, as in the condensation of isoprene units to form carotenoids.

tailward growth. A polymerization mechanism in which the activated head of a monomer adds to the passive tail of a chain, thereby making its own tail the receptor for the next addition of monomer. *See also* head-to-tail condensation; headward growth.

Taka-amylase. An alpha amylase isolated from the mold *Aspergillus oryzae*.

Takadiastase. A ribonuclease preparation from fungi that contains both ribonuclease T₁ and ribonuclease T₂ activity.

tandem. One behind the other.

tangent. 1. A straight line that touches a curve at only one point. 2. The ratio of the length of the side facing an acute angle in a right triangle to the length of the side facing the other acute angle.

Tangier disease. A genetically inherited metabolic defect in man that is characterized by the almost complete lack of plasma high-density lipoproteins and by the excessive storage of cholesterol in many tissues.

tannin. One of a group of phenolic, nonnitrogenous plant toxins that are frequently glycosides with astringent properties.

Tanret's reagent. A reagent that contains potassium iodide, mercuric chloride, and acetic acid, and that forms a white precipitate when added to urine containing albumin.

T antigen. 1. An antigen occurring in the nuclei of cells that are infected with, or transformed by, certain oncogenic viruses, such as polyoma virus or SV-40; thought to be a virus-specific protein. 2. An antigen that is present on normal human red blood cells but that is unreactive unless the cells are first treated with the enzyme neuraminidase.

tape. An informational molecule that functions in replication, transcription, or translation; used specifically for messenger RNA.

tape theory. The theory according to which replication, transcription, and translation proceed by the currently accepted template-type mechanisms.

TAPS. Tris(hydroxymethyl)methylaminopropane sulfonic acid; used for the preparation of buffers in the pH range of 7.7 to 9.1.

tare. A counterweight for balancing a container during weighing.

target cell. 1. A receptor cell that binds cytotropic antibodies in anaphylaxis. 2. A receptor cell that is acted upon by a hormone.

target molecular weight. The molecular weight calculated from the dose of radiation that results in the survival of 37% of the irradiated units. *See also* thirty-seven percent survival dose.

target organ. 1. The receptor organ that is acted upon by a hormone. 2. The organ, the cells of which are damaged by the multiplication of an infecting virus.

target theory. The theory that a cell will be killed, or an enzyme molecule, virus particle, etc., will be inactivated if struck by radiation in a small, sensitive target volume in which one or more hits (ionizing events) will bring about the specific effect.

taurine. An aminosulfonic acid, derived from cysteine, that forms a bile salt when it is conjugated to a bile acid.

taurocholic acid. A compound formed by the conjugation of cholic acid and taurine; the salt of taurocholic acid is one of the bile salts.

tautomer. One of the two isomers that exhibit tautomerism.

tautomeric shift. A reversible change in the location of a hydrogen atom in a molecule that occurs in enol-keto tautomerism and that converts one tautomer into the other.

tautomerism. An equilibrium between two isomeric forms of a molecule that differ significantly in both the relative positions of the atoms and in the bonds between the atoms; the molecule may react in either of the two isomeric forms depending on the conditions.

taxis. The movement of an organism in response to a stimulus. The movement may be toward, or away from, the stimulus and is then referred to as either positive or negative taxis.

taxon (*pl* taxa). A taxonomic group of any rank or size.

taxonomic. Of, or pertaining to, taxonomy.

taxonomy. The scientific classification of plants and animals that is based on their natural relationships; includes the systematic grouping, ordering, and naming of the organisms.

Tay-Sachs disease. A genetically inherited metabolic defect in man that is characterized by an accumulation of gangliosides in the brain and that leads to a progressive and fatal disease associated with blindness and brain deterioration; caused by a deficiency of the enzyme hexosaminidase.

TBG. Thyroxine-binding globulin.

TBPA. Thyroxine-binding prealbumin.

TC$_{50}$. Median tissue culture dose.

TCA. 1. Tricarboxylic acid. 2. Trichloroacetic acid.

TCA cycle. Tricarboxylic acid cycle.

TΨC arm. The base-paired segment in the clover leaf model of transfer RNA to which the loop, carrying the base sequence $T\Psi C$, is attached.

TC detector. Thermal conductivity detector.

TCT. Thyrocalcitonin.

t$_D$. Doubling time.

TD$_{50}$. Median toxic dose.

TDP. 1. Thymidine diphosphate. 2. Thymidine-5′-diphosphate.

TEAE-cellulose. O-(Triethylaminoethyl)-cellulose; an anion exchanger.

Teflon. Trademark for polytetrafluoroethylene; a plastic.

teichoic acid. An accessory polymer in the cell wall of gram-positive bacteria that consists of long chains of either glycerol or ribitol which are linked by means of phosphodiester bonds and which carry various substituents, including both amino acids and monosaccharide residues.

teichuronic acid. A teichoic acid-type polymer that contains uronic acid substituents.

teleology. The doctrine that the occurrence of any structure or function in a living organism implies that the structure or function is valuable, has a purpose, and has conferred an advantage on the organism in the course of evolution. As originally formulated by Aristotle, the concept invoked a supernatural agent that had foreseen the final value of the structure or of the function. To circumvent this connotation, the term teleonomy has been introduced. *See also* teleonomy.

teleonomy. The doctrine that the occurrence of any structure or function in a living organism implies that the structure or function has conferred an advantage on the organism in the course of evolution; the genetic character of an organism is considered to have become adapted to its environment through the process of evolution rather than through theurgic forces. *See also* teleology.

telophase. The final stage in mitosis during which the two new nuclear membranes are formed and which is followed by cytokinesis to produce the two daughter cells.

TEM. Triethylene melamine; an aziridine mutagen that is also used as an insect chemosterilant.

TEMED. N,N,N,N-Tetramethyl ethylene diamine; an initiator of acrylamide polymerization.

temperate cycle. LYSOGENIC CYCLE.

temperate phage. A phage that is incorporated as a prophage into the host chromosome and allows the host cell to survive in the form of a lysogenic bacterium; it does not cause lysis of the bacterial cell that it has infected.

temperature. The degree of hotness measured on one of several arbitrary scales; a measure of the kinetic energy of molecules. *See also* temperature scale.

temperature coefficient. The rate of a reaction at one temperature divided by the rate at a second temperature; usually expressed as the Q_{10} value.

temperature jump method. A relaxation technique in which temperature is the variable that disturbs the equilibrium of a system. *See also* relaxation technique.

temperature programming. The controlled increase in temperature at a predetermined rate; used in the construction of thermal denaturation profiles and in the operation of gas chromatography columns.

temperature scale. *See* absolute temperature scale; centigrade temperature scale; Fahrenheit temperature scale; Kelvin temperature scale.

temperature-sensitive mutant. A conditional lethal mutant that grows at a lower (higher), permissive temperature but does not grow at a higher (lower), restrictive temperature.

template. A macromolecule that functions as a mold or pattern for the synthesis of another macromolecule. A template determines the composition of the product and directs its synthesis during the process of polymerization but does not have to be linked covalently to the product.

template RNA. MESSENGER RNA.

template theory. *See* antigen template theory.

tensiometer. An instrument for measuring surface and interfacial tensions.

tension. The concentration of a gas in a solution; expressed in terms of the partial pressure of the gas which is in equilibrium with the solution.

TEPA. An aziridine mutagen.

ter-. 1. Combining form meaning three or thrice. 2. Referring to three kinetically important substrates and/or products of an enzymatic reaction; thus a ter bi reaction is a reaction with three substrates and two products.

tera-. Combining form meaning 10^{12} and used with metric units of measurement. *Sym* T.

teratogen. An agent that causes birth defects.

teratogeny. The origin and production of congenital malformations. *Aka* teratogenesis.

teratology. The study of malformations and of the abnormal developments of organisms and parts of organisms.

teratoma. A tumor derived from embryonic tissue and containing tissue that usually does not grow in the region in which the tumor is located.

terminal. The end of a polymeric chain, such as the C- or N-terminal of a polypeptide chain, or the 3′- or 5′-terminal of a polynucleotide strand. *Aka* terminus.

3′-**terminal.** That end of an oligo- or polynucleotide strand which carries either a free or a phosphorylated hydroxyl group at the 3′-position of the terminal ribose or deoxyribose.

5′-**terminal.** That end of an oligo- or polynucleotide strand that carries either a free or a phosphorylated hydroxyl group at the 5′-position of the terminal ribose or deoxyribose.

terminal cisterna (*pl* terminal cisternae). A transverse, connecting channel of the sarcoplasmic reticulum.

terminal deletion. A chromosomal aberration in which genetic material is lost from the end of a chromosome.

terminal enzyme. An enzyme that catalyzes the addition of nucleotides to the terminal of a nucleic acid strand in the absence of a template, with nucleoside triphosphates serving as substrates.

terminal redundancy. The occurrence of identical base sequences at both ends of a double-stranded, linear DNA molecule. Treatment of such a DNA molecule with exonuclease produces sticky ends that can lead to the formation of double-stranded, circular DNA molecules under suitable conditions. *Aka* terminal repetition.

termination. 1. The final stage in protein synthesis during which the completed polypeptide chain is released from the ribosome, and the ribosome is released from the messenger RNA. 2. The third step in a chain reaction.

termination codon. A codon that codes for termination of the translation of a messenger RNA molecule and not for an amino acid; one of the three codons *UAA*, *UAG*, or *UGA*.

termination factor. RELEASE FACTOR.

terminology. NOMENCLATURE.

terminus (*pl* termini). TERMINAL.

termolecular reaction. A chemical reaction in which three molecules (or other entities) of reactants interact to form products.

ternary. Consisting of three parts.

ternary complex mechanism. SEQUENTIAL MECHANISM.

terpene. A hydrocarbon terpenoid. Terpenes are classified according to the number of isoprene units which they contain: monoterpene (2); sesquiterpene (3); diterpene (4); triterpene (6); tetraterpene (8); pentaterpene (10); and polyterpene (large number).

terpeneless oil. An essential oil from which less odorous terpenes have been removed by deterpenation.

terpenoid. A polyisoprenoid compound that may be linear or cyclic and in which the isoprene units are usually linked in a head-to-tail manner. *See also* terpene.

terramycin. *See* tetracycline.

terreactant reaction. A termolecular reaction in which three different reactants interact to form products.

tert. Tertiary.

tertiary response. The immune response to a third dose of antigen that is essentially similar in its characteristics to those of the secondary response.

tertiary structure. The irregular three-dimensional folding of the polypeptide chain upon itself, as in a globular protein, that results from the interaction of amino acid side chains which are either close or far apart along the chain; the arrangement in space of all the atoms of a protein or of a subunit without regard to the relationship of the atoms to neighboring molecules or subunits. The term may likewise be applied to the three-dimensional structure of a polynucleotide strand.

TES. N-Tris(hydroxymethyl)methyl-2-aminoethanesulfonic acid; used for the preparation of buffers in the pH range of 6.8 to 8.2.

testectomy. The surgical removal of a testis.

test meal. A food that consists of one or more selected items and that, after it has been eaten, is followed by the removal of the gastric contents for analysis.

testosterone. A steroid hormone that is the major male sex hormone in man; it is secreted by the testes and is responsible for the development of secondary male characteristics.

tetra-. Combining form meaning four.

tetracycline. One of a group of broad-spectrum antibiotics that inhibit the binding of aminoacyl transfer RNA to 70S ribosomes. Chlortetracycline (aureomycin) is produced by *Streptomyces aureofaciens*; oxytetracycline (terramycin) is produced by *Streptomyces rimosus*; and tetracycline (achromycin; tetracyn) is obtained by reductive dehalogenation of chlortetracycline.

tetracyn. *See* tetracycline.

tetrahedron. A polyhedron with four equal faces; a pyramid composed of four triangles. A tetrahedron is used for representation of the carbon atom, with the nucleus of the atom inside the pyramid, and with the four single bonds extending to the corners of the pyramid.

tetrahydrofolic acid. The reduced form of the vitamin folic acid and the parent compound of the coenzyme forms of this vitamin; the coenzymes function in the metabolism of one-carbon fragments. *Abbr* FH$_4$; THFA; THF.

tetrahydropteroylglutamic acid. TETRAHYDROFOLIC ACID.

tetramer. An oligomer that consists of four monomers.

tetramine. An aziridine mutagen.

tetranucleotide hypothesis. An early hypothesis of the structure of DNA according to which DNA has a uniform structure that is produced by the polymerization of one basic repeating unit, namely a tetranucleotide containing one residue each of adenine, cytosine, guanine, and thymine.

tetrose. A monosaccharide that has four carbon atoms.

T-even phage. A large phage that infects the bacterium *Escherichia coli* and that has a complex tadpole-like structure consisting of a head, tail, and tail fibers. The head is icosahedral and contains double-stranded DNA; to the head is attached the tail through which the DNA is ejected into the host cell. The tail terminates in

tail fibers by means of which the phage adsorbs to the bacterium. The tail core, or tube, is surrounded by a sheath and is attached to the head by means of a collar and to the tail fibers by means of a plate. Tail pins are attached to the tail plate.

text. 1. The sequence of nucleotides in a nucleic acid. 2. The sequence of amino acids in a protein.

TF. Transfer factor.

T factor. ELONGATION FACTOR I.

Tg. Generation time.

TGFA. Triglyceride fatty acids.

TH. Transhydrogenase.

thalassemia. A heritable disorder characterized by anemia and by the presence of large amounts of a minor, but normal, hemoglobin HbA_2; the homozygous condition is known as thalassemia major and the heterozygous condition is known as thalassemia minor. *Aka* Cooley's anemia.

thallus. The undifferentiated growth of a plant body as that of a fungus or a mold.

THAM. TRIS.

Thd. Thymidine.

Theorell-Chance mechanism. An ordered sequential mechanism of an enzymatic reaction in which two substrates and two products participate and in which the steady-state concentration of the ternary complexes is very low. The two ternary complexes in this mechanism are the enzyme plus the two substrates, and the enzyme plus the two products.

theoretical plate. The theoretical stage in countercurrent distribution at which perfect equilibrium is established between the two phases. The theoretical plate concept has been adapted for use in distillation and chromatographic columns where perfect equilibrium between the phases is not established, since the phases are constantly in motion relative to each other. Hence it is customary to refer to the length of column over which the separation effected is equivalent to that of a theoretical plate; this length of column is known as a "height equivalent to a theoretical plate" (HETP). The shorter the HETP, the more efficient is the column. *Aka* theoretical stage.

theory. A confirmed explanation of observed phenomena; a scientific doctrine; a hypothesis that has been tested and confirmed with facts not known when the hypothesis was first proposed.

theory of absolute reaction rates. The theory of chemical kinetics according to which the velocity of a chemical reaction is proportional to the concentration of the activated complex that is formed from the reactants; the reactants must first be activated by means of an activation energy to form the activated complex before they can be converted into products.

theory of antibody formation. *See* clonal selection theory; Ehrlich's receptor theory; germ line theory; instructive theory; selective theory; side chain theory; Smithie's theory; somatic mutation theory; somatic recombination theory.

theory of cancer. *See* Busch theory; catabolic deletion hypothesis; chromosome theory of cancer; deletion hypothesis; feedback deletion hypothesis; Greenstein hypothesis; imbalance theory; Mason's theory; membron theory of cancer; oncogene theory; polyetiological theory; protovirus theory; somatic mutation theory; virogene theory; Warburg theory.

therapeutic index. A measure of the safety of a drug that is equal to the ratio of the median lethal dose to the median effective dose.

thermal. Of, or pertaining to, either heat or temperature.

thermal chromatography. A column chromatographic technique in which elution is carried out with one eluent but at increasing temperatures.

thermal conductivity detector. A detector in which the basic component is a thermal conductivity cell of either the thermistor or the katharometer type; used in gas chromatography, primarily for detecting inorganic gases or large amounts of organic compounds. *Abbr* TC detector.

thermal death point. The lowest temperature required for the sterilization of a standard suspension of bacteria in 10 min.

thermal death time. The minimum time required for the sterilization of a standard suspension of bacteria at a given temperature.

thermal denaturation. The denaturation of macromolecules by heat, particularly the heat-induced breaking of the hydrogen bonds of the base pairs in double-helical nucleic acid segments which leads to the formation of random coils.

thermal denaturation profile. The plot of a hydrodynamic property, such as viscosity, or of an optical property, such as absorbance, for a nucleic acid solution as a function of temperature. The curve describes the degree of separation between the two strands of double-stranded molecules, or between parts of the same strand of folded single-stranded molecules, that results from the breaking of hydrogen bonds as the temperature is increased. The temperature at which one half of the maximum change in the measured property is observed is referred to as the melting out temperature and is denoted T_m.

thermal neutron. A neutron that has a kinetic

energy that is equivalent to the energy of a gas molecule at room temperature.

thermal noise. DARK CURRENT.

thermal polymer. A high-molecular weight compound produced by the heat-induced polymerization of monomers.

thermal quenching. CHEMICAL QUENCHING.

thermal reactivation. The increase in the survival of ultraviolet-irradiated bacteria that is manifested if the bacteria are incubated at an elevated temperature immediately following their irradiation and are then allowed to grow at a lower temperature; this is in comparison to those bacteria that, following their irradiation, are allowed to grow only at the lower temperature.

thermistor. A semiconductor device, the resistance of which changes with temperature; used for measuring temperatures.

thermo-. Combining form meaning heat.

thermobarometer. A control manometer used in conjunction with the Warburg apparatus to correct for changes in room temperature and barometric pressure during the experiment.

thermochromism. The reversible change of color by a compound as a function of temperature.

thermocouple. A device for measuring temperatures by the production of an electromotive force between the two junctions of two different metals in a closed circuit; the electromotive force is proportional to the temperature difference between the two junctions.

thermoduric. Heat-enduring; capable of surviving high temperatures.

thermodynamics. The science that deals with the relationships between heat and mechanical energy as they pertain to the initial and final states of a system and its surroundings. Thermodynamics is based upon fundamental laws, drawn principally from experience, and essentially statistical in nature. *See also* first law of thermodynamics; second law of thermodynamics; third law of thermodynamics; zeroth law of thermodynamics.

thermodynamics of irreversible processes. *See* irreversible thermodynamics.

thermogenic. Heat-producing.

thermolabile. Inactivated by treatment with heat; sensitive to high temperatures.

thermolysin. A proteolytic enzyme from the thermophilic bacterium *Bacillus thermoproteolyticus* Rokko.

thermolysis. PYROLYSIS.

thermometer. A device for measuring temperatures; usually implies either a mercury-in-glass or an alcohol-in-glass device.

thermonuclear reaction. A nuclear reaction that

is induced by a thermal activation of the reacting nuclei.

thermophile. An organism that grows at high temperatures in the range of 45 to 70°C (or higher temperatures) and that has an optimum growth temperature in the range of 50 to 55°C.

thermophilic. Of, or pertaining to, thermophiles; preferring high temperatures.

thermopile. An instrument for measuring the total amount of radiant energy irrespective of its wavelengths.

thermostable. Stable to treatment by heat; not inactivated by high temperatures.

thermostat. A device for maintaining a constant temperature by means of the automatic regulation of the source of heat.

thermotropic mesophase. A liquid crystal, the structure of which has been loosened by heat.

THF. Tetrahydrofolic acid.

THFA. Tetrahydrofolic acid.

thiamine. One of the B vitamins (vitamin B_1), the coenzyme form of which is thiamine pyrophosphate, and a deficiency of which causes beriberi. *Var sp* thiamin.

thiamine pyrophosphate. The coenzyme form of the vitamin thiamine that functions in the decarboxylation of α-ketoacids. *Abbr* TPP; ThPP; DPT.

thiazole ring. The ring structure occurring in thiamine and thiamine pyrophosphate.

thick filament. MYOSIN FILAMENT.

thin filament. ACTIN FILAMENT.

thin-layer chromatography. A chromatographic technique in which the stationary phase is a thin layer of solid, such as silica gel, spread on a flat glass plate; the technique allows for rapid analysis of very small amounts of sample. *Abbr* TLC.

thin-layer electrophoresis. An electrophoretic technique in which the supporting medium is a thin layer of silica gel, alumina, or other adsorbent spread on a flat glass plate; the technique allows for rapid analysis of very small amounts of sample. *Abbr* TLE.

thin-layer gel filtration. A separation technique in which gel filtration is carried out on thin layers of gel spread on a flat glass plate.

thio-. Combining form meaning sulfur.

thioalcohol. THIOL.

thioclastic reaction. A phosphoroclastic reaction in which lipoic acid participates. *Aka* thioclastic split.

thioctic acid. LIPOIC ACID.

thioester. A compound formed by the joining of a carboxyl group to a sulfhydryl group through the elimination of a molecule of water; a compound having the general formula $R—CO—S—R'$.

thioether. A compound in which two radicals are linked by means of a sulfur atom; a compound of the general formula $R—S—R'$.

thiokinase. *See* fatty acid thiokinase.

thiol. A compound containing a sulfhydryl group; a compound of the general formula $R—SH$.

thiolase. The enzyme that catalyzes the thiolytic cleavage reaction in beta oxidation.

thiolation. The introduction of a sulfhydryl group into an organic compound.

thiol ester. THIOESTER.

thiol ether. THIOETHER.

thiol group. SULFHYDRYL GROUP.

thiolysis. THIOLYTIC CLEAVAGE.

thiolytic cleavage. The last step, catalyzed by the enzyme thiolase, in the cyclic reaction sequence of beta oxidation; in this step, a fatty acyl coenzyme A derivative, in the presence of coenzyme A, is cleaved to produce acetyl coenzyme A and a fatty acyl coenzyme A of a fatty acid that contains two carbon atoms less than the fatty acid which entered the cycle.

thiophorase. The enzyme that catalyzes the conversion of an acyl coenzyme A ester and a free fatty acid to the opposite pair of acyl coenzyme A ester and free fatty acid; this reaction represents an alternative reaction for fatty acid activation to that catalyzed by fatty acid thiokinase.

thioredoxin. A protein that can exist as a dithiol, thioredoxin-$(SH)_2$, or as a disulfide, thioredoxin-S_2, and that serves to reduce ribonucleoside diphosphates to deoxyribonucleoside diphosphates.

thioredoxin reductase. The enzyme that catalyzes the reduction of thioredoxin-S_2 to thioredoxin-$(SH)_2$ with the simultaneous oxidation of NADPH to NADP$^+$; the enzyme is an FAD-containing flavoprotein.

thiouracil. A minor base occurring in transfer RNA.

third factor. NATRIURETIC HORMONE.

third law of thermodynamics. The principle relating to the magnitude of entropy: all substances have finite positive entropies and all simple, crystalline substances at a temperature of absolute zero have equal entropies which are assigned a value of zero.

third-order reaction. A chemical reaction in which the velocity is proportional to the product of three concentration terms involving one, two, or three different reactants.

thirty-seven percent survival dose. The dose of radiation that results in 37% survival of the irradiated units. The 37% survival value corresponds to an average of one lethal hit per sensitive target and can be used for calculations of target molecular weights. The number 37 is based on exponential inactivation kinetics, since $e^{-1} = 0.37$.

thixotropy. The property of some gels of undergoing a reversible, isothermal, gel-sol transformation upon agitation; the solid gel becomes a liquid sol as a result of the agitation.

ThPP. Thiamine pyrophosphate.

Thr. 1. Threonine. 2. Threonyl.

three-factor cross. A series of genetic crosses, involving three nonallelic linked genes, that is used primarily for purposes of chromosomal mapping.

three-point attachment. The concept that there must be at least three points of contact (binding sites) between an enzyme and its substrate in order to account for the different reactivity of identical groups that are either parts of a symmetrical substrate molecule, or are attached to a meso carbon of the substrate. *Aka* three-point landing; polyaffinity theory.

three-point cross. THREE-FACTOR CROSS.

threo configuration. The configuration of a compound in which two asymmetric carbon atoms have identical substituents on opposite sides, as in the configuration of threose.

threonine. An aliphatic, polar alpha amino acid that contains an alcoholic hydroxyl group. *Abbr* Thr; T.

threonine dehydrase. An inducible enzyme in the liver of higher organisms.

threshold. 1. A measure of the ability of the kidney to absorb a substance from the blood. *See also* threshold of appearance; threshold of retention. 2. A measure of the sensitivity of the eye to light. *See also* cone threshold; rod threshold; visual threshold.

threshold dose. The smallest dose of a radiation above which the radiation produces a detectable effect.

threshold limit value. The airborne concentration of a chemical to which humans may be exposed day after day in their working environment without suffering adverse effects. *Abbr* TLV.

threshold of appearance. The plasma concentration of a substance above which the substance cannot be fully absorbed by the kidney and appears in the urine. *Aka* threshold of excretion.

threshold of retention. The plasma concentration of a substance when it is identical to the concentration of the substance in the urine; at higher plasma concentration values, the urine will be more concentrated in the substance than the plasma, whereas at lower plasma concentration values, the plasma will be more concentrated in the substance than the urine.

threshold potential. That portion of the characteristic curve of a Geiger-Mueller counter that follows the starting potential and at which the curve begins to level off to a plateau.

threshold substance. A substance that appears in the urine in substantial amounts only when its concentration in plasma exceeds a certain value.

thrombin. The proteolytic enzyme that functions in blood clotting by catalyzing the hydrolytic cleavage of fibrinopeptides *A* and *B* from fibrinogen, thereby converting fibrinogen to fibrin.

thrombocyte. PLATELET.

thrombogen. PROTHROMBIN.

thrombokinase. THROMBOPLASTIN.

thromboplastin. A lipoprotein factor that is released from blood platelets in injured tissue and that initiates the sequence of reactions leading to the formation of a blood clot in the extrinsic system of blood clotting.

thromboplastinogen. ANTIHEMOPHILIC FACTOR.

thrombus (*pl* thrombi). A blood clot formed within a blood vessel or within the heart.

Thunberg technique. A technique for the estimation of dehydrogenase activity by a photometric measurement of the reduction of methylene blue.

Thunberg tube. A tube, used in the Thunberg technique, that can be evacuated and that has a side arm for the addition of reagents.

Thx. Thyroxine.

Thy. Thymine.

thylakentrin. FOLLICLE-STIMULATING HORMONE.

thylakoid disk. One of a large number of flattened vesicles that contain the photosynthetic pigments and the electron carriers of chloroplasts and that are stacked to form the grana of the chloroplasts.

thymectomy. The surgical removal of the thymus.

thymic. Of, or pertaining to, the thymus.

thymidine. The deoxyribonucleoside of thymine. Thymidine mono-, di-, and triphosphate are abbreviated, respectively, as TMP, TDP, and TTP (or as dTMP, dTDP, dTTP). The abbreviations refer to the 5′-nucleoside phosphates unless otherwise indicated. *Abbr* Thd; T.

thymidine factor. SOMATOMEDIN.

thymidine kinase. The enzyme that catalyzes the phosphorylation of thymidine to thymidine monophosphate and that is believed to play a major role in the control of DNA synthesis.

thymidylate kinase. The enzyme that catalyzes the phosphorylation of thymidine monophosphate to thymidine diphosphate and the phosphorylation of thymidine diphosphate to thymidine triphosphate.

thymidylate synthetase. The enzyme that catalyzes the methylation of deoxyuridine monophosphate to thymidine monophosphate.

thymidylic acid. The deoxyribonucleotide of thymine.

thymine. The pyrimidine, 5-methyl-2,3-dioxy-pyrimidine, that occurs in DNA. *Abbr* Thy; T.

thymine dimer. A pyrimidine dimer, produced by ultraviolet irradiation of DNA, in which two adjacent thymines in a DNA strand form a dimer and block the replication of the DNA at that point.

thymineless death. The death of bacteria that results from a lack of thymine; the absence of thymine leads to breaks in single strands of DNA and thereby blocks the synthesis of DNA. Thymineless death can be produced by depriving a thymine auxotroph of thymine or by exposing a bacterial culture to 5-fluorouracil which prevents the synthesis of deoxythymidylic acid. *Abbr* TLD.

thymineless mutant. *See* -less mutant.

thymine starvation. *See* starvation (2).

thymocyte. A lymphocyte that occurs within the thymus gland.

thymol turbidity test. A liver function test that is based on the production of turbidity when serum from individuals with one of several forms of hepatitis is treated with a barbiturate buffer that contains thymol.

thymosin. A protein hormone, secreted by the thymus gland, that stimulates the synthesis and maturation of lymphocytes.

thymus. A gland, located in the lower part of the neck, that produces lymphocytes; the thymus is large in young animals but atrophies after the animal attains sexual maturity.

thymus nucleic acid. An early designation for DNA.

thyrocalcitonin. That fraction of the hormone calcitonin that is secreted by the thyroid gland. *Abbr* TCT.

thyroglobulin. An iodinated protein that represents the form in which iodine is stored in the thyroid colloid and from which thyroxine and triiodothyronine are formed by proteolysis.

thyroid colloid. The gelatinous material in the follicles of the thyroid gland in which the iodine is stored, principally in the form of thyroglobulin.

thyroidectomy. The surgical removal of the thyroid gland.

thyroid gland. An endocrine gland, located in the neck, that produces the hormones thyroxine and triiodothyronine.

thyroid hormones. The hormones thyroxine and triiodothyronine.

thyroid hyperfunction. HYPERTHYROIDISM.

thyroid hypofunction. HYPOTHYROIDISM.

thyroid-stimulating hormone. THYRO-TROPIN.

thyroid-stimulating hormone releasing hormone. THYROTROPIN RELEASING HOR-MONE.

thyroid storm. A severe clinical state due to overactivity of, and excessive secretion by, the thyroid gland. *Aka* thyroid crisis.

thyrotoxic effect. A pathological condition produced by certain forms of excessive activity of the thyroid gland or by excessive doses of thyroid hormones.

thyrotrophic hormone. THYROTROPIN.

thyrotropic hormone. THYROTROPIN.

thyrotropic hormone releasing factor. THY-ROTROPIN RELEASING HORMONE.

thyrotropic hormone releasing hormone. THY-ROTROPIN RELEASING HORMONE.

thyrotropin. A protein hormone, secreted by the anterior lobe of the pituitary gland, that stimulates the synthesis of thyroid hormones and the release of thyroxine by the thyroid gland. *Var sp* thyrotrophin.

thyrotropin releasing hormone. A hypothalamic hormone that controls the secretion of thyrotropin from the pituitary gland. *Var sp* thyrotrophin releasing hormone. *Abbr* TRH. *Aka* thyrotropin releasing factor (TRF).

thyroxine. An iodinated aromatic amino acid that is the major hormone of the thyroid gland and that controls the rate of oxygen consumption and the overall metabolic rate. It is formed by the coupling of two molecules of 3,5-diiodotyrosine. *Abbr* Thx; T_4.

thyroxine-binding globulin. A glycoprotein that serves as the major and specific carrier of thyroxine in plasma. *Abbr* TBG.

thyroxine-binding prealbumin. An albumin that serves as a secondary carrier of thyroxine in plasma. *Abbr* TBPA.

TIBC. Total iron-binding capacity.

tidal air. The volume of air that enters and leaves the body with each normal respiratory movement.

tiered metabolic pathway. A metabolic pathway of the type $A \rightarrow B$
$$\downarrow$$
$$C \rightarrow D$$
$$\downarrow$$
$$E \rightarrow F.$$

tight coupling. The state of cellular respiration in which the mitochondria are characterized by having a high acceptor control ratio.

tilde. SQUIGGLE.

tilt angle. The angle between the perpendicular to the axis of a helical nucleic acid and the plane of an individual base.

time average. The average value of a quantity that is measured over a number of time intervals; used for quantities that vary as a function of time, such as atmospheric temperature, blood flow, and caloric intake.

time constant of a reaction. A measure of the duration of a reaction; for an exponential decay, the time constant is the time required for

63% (the fraction $1 - 1/e$) of the total change to occur.

time factor effect. The effect of exposure time on the results produced by irradiation when reciprocity does not hold, so that the product of dose rate and exposure time is not constant; under these conditions, a short-time irradiation at a high intensity is frequently more effective than a long-time irradiation at a low intensity.

time-lapse microcinematography. *See* microcinematography.

time-lapse photography. A photographic technique in which exposures are taken on a photographic film at selected time intervals and the film is then projected at a faster speed so that the actions shown are speeded up.

tin. An element that is essential for living organisms but the function of which is unknown. Symbol Sn; atomic number 50; atomic weight 118.69; oxidation states $+2$, $+4$; most abundant isotope Sn^{120}; a radioactive isotope Sn^{119}, half-life 250 days, radiation emitted—gamma rays.

tincture. An alcoholic solution of a chemical or a drug.

TIP. Translation inhibitory protein.

Tiselius apparatus. An apparatus that can be used for either diffusion or electrophoresis measurements; consists of a three-piece U-tube that allows the formation of sharp boundaries between solvent and solution. The spreading of these boundaries, or their movement under the influence of an electric field, is then analyzed by means of either a schlieren or an interferometric optical system.

tissue. An aggregate of cells and intercellular material that form a definite structure; the cells are generally of similar structure and function.

tissue culture. The maintenance of living and metabolizing cells, tissues, or organs in artificial media.

tissue factor. AUTACOID.

tissue mince. Tissue that is cut or chopped into small pieces.

tissue slice. A thin slice of tissue that can carry out metabolic reactions and that freely exchanges gases and metabolites with the suspending medium. The metabolic reactions in such a slice are frequently studied in a Warburg apparatus and monitored by manometric techniques.

tissue-specific enzyme. An enzyme that is present in appreciable concentrations in only one tissue or in one organ; the plasma concentration of such an enzyme is of clinical value in assessing the functional state of that tissue or of that organ.

tissue thromboplastin. THROMBOPLASTIN.

titer. 1. The amount of a standard solution that is required for a defined titration of a given

volume of a second solution. 2. The greatest dilution of a sample solution that gives a positive test under defined conditions; the greatest dilution of a virus sample that gives a positive hemagglutination test, and the greatest dilution of an antiserum that gives a positive precipitin test are two examples.

titrant. The solution that is added to a given volume of a second solution in the course of a titration.

titratable. Capable of being titrated. *Aka* titrable.

titratable acidity. A measure of the acidity of urine, particularly that due to $H_2PO_4^-$, expressed in terms of the number of milliliters of 0.1 N sodium hydroxide required to neutralize a 24-hour volume of urine.

titrate. To carry out a titration.

titration. A method of volumetric analysis in which a test solution is added from a buret to a known volume of a standard solution, or a standard solution is added from a buret to a known volume of a test solution.

titration curve. A plot of the amount of acid or base added during a titration as a function of the pH of the solution.

titration equivalent weight. The average molecular weight of a fatty acid in a mixture of fatty acids; equal to the number of milligrams of fatty acids in the sample, divided by the number of milliequivalents of alkali used in the titration of the sample to a phenolphthalein end point.

titrimetry. Chemical analysis by means of titration.

TLC. Thin-layer chromatography.

TLD. Thymineless death.

TLE. Thin-layer electrophoresis.

t-like RNA. 5S RNA.

TLV. Threshold limit value.

T_m. 1. Melting out temperature. 2. Transport maximum.

TMP. 1. Thymidine monophosphate (thymidylic acid). 2. Thymidine-5'-monophosphate (5'-thymidylic acid).

TMV. Tobacco mosaic virus.

TNV. Tobacco necrosis virus.

tobacco mosaic virus. A helical plant virus that contains single-stranded RNA and infects tobacco leaves. *Abbr* TMV.

tobacco necrosis virus. A spherical plant virus that contains single-stranded RNA and infects seed plants. *Abbr* TNV.

tocopherol. VITAMIN E.

T-odd phage. A phage that infects the bacterium *Escherichia coli*. A T-odd phage is smaller than a T-even phage and contains DNA that has the same base composition as that of the host cell.

Toepfer's reagent. An alcoholic solution of *p*-

dimethylaminobenzene that is used in testing for free hydrochloric acid in gastric juice.

tolerance. 1. IMMUNOLOGICAL TOLERANCE. 2. The decrease in, or the loss of, the response of an animal to a dose of a chemical to which it responded on a prior occasion. 3. The limit of error permitted in the graduation of measuring instruments, standardized products, or analytical evaluations.

tolerance test. *See* epinephrine tolerance test; galactose tolerance test; glucose tolerance test; lactose tolerance test.

tolerogenic antigen. An antigen that produces immunological tolerance.

Tollen's test. A test for reducing sugars that is based on the reduction of silver ions to metallic silver in an alkaline solution.

tomato bushy stunt virus. A plant virus that contains single-stranded RNA.

tophus (*pl* tophi). A deposit of sodium urate in cartilage that occurs in gout.

topography. The description and the mapping of the features of a surface; the delineation of the configurations and the structural relationships of a surface.

topology. A branch of mathematics that deals with those properties of shapes and figures that remain unchanged if the shape or the figure is subjected to one-to-one continuous transformations.

torr. A unit of pressure equal to $1/760$ atm; a pressure of 1 mm mercury.

torsion angle. The angle between the plane containing atoms A, B, and C and the plane containing atoms B, C, and D for a system that consists of four atoms linked in the sequence $A—B—C—D$. The torsion angle can also be described as the angle between the projection of $A—B$ and that of $C—D$ when the system of four atoms is projected onto a plane normal to the bond $B—C$.

tosyl group. A p-toluenesulfonyl group; used to block the amino group of amino acids.

total activity. The total number of activity units in a sample, such as the total number of enzyme units or the total number of microcuries.

total cell count. The total number of bacteria, both viable and nonviable, in a sample.

total consumption burner. A burner used in atomic absorption spectrophotometry and designed so that the gases and the sample are not mixed before entering the flame.

total inhibition. The inhibition of the activity of an enzyme in which a saturating concentration of inhibitor can cause the reaction velocity to become zero.

total iron-binding capacity. The concentration of iron that is equal to the sum of the actual iron concentration in plasma and the unsaturated iron-binding capacity; generally expressed in terms of micrograms per 100 ml of plasma. *Abbr* TIBC.

totally labeled. GENERALLY LABELED.

total osmotic pressure. The osmotic pressure produced when a membrane separating two solutions is impermeable to all solutes regardless of their size.

total titer. The titer of phage particles that is obtained after disruption of the infected cells and that is a measure of the maturation of phage particles in the host cells. *See also* extracellular titer; intracellular titer.

totipotence. The capacity of a cell to express all of its genetic information under appropriate conditions and to develop into a complete and fully differentiated organism. *Var sp* totipotency.

Townsend avalanche. The flood of ions produced by gas amplification in an ionization chamber.

toxalbumin. PHYTOTOXIN.

toxemia. A pathological condition characterized by the presence of toxins in the blood.

toxic. 1. Of, or pertaining to, a poison. 2. Poisonous.

toxic goiter. A thyroid enlargement accompanied by hyperthyroidism.

toxicity. The degree of harmfulness of a substance for an organism; the capacity of a substance to produce injury.

toxicity index. The ratio of the highest dilution of a germicide that kills the cells of chick heart tissue in 10 min to the highest dilution of the germicide that kills the test microbial organisms in the same time and under the same conditions.

toxicology. The science that deals with the harmful effects of chemicals on biological systems.

toxicosis. Systemic poisoning.

toxigenicity. The production of a toxin, particularly the production of a toxin by bacteria. *Aka* toxicogenicity.

toxin. A high-molecular weight compound of plant, animal, or bacterial origin that is toxic and generally antigenic in animal species.

toxohormone. A toxic substance that inhibits the activity of the enzyme catalase; apparently a polypeptide that can be extracted from cancer cells and that may also occur in normal cells.

toxoid. A toxin that has lost its toxic properties as a result of denaturation or chemical modification but that has retained its antigenic properties.

toxophore. The group of atoms in a toxin molecule that is responsible for the toxic properties of the molecule.

TP. 1. Transport piece. 2. Transprotonase.

T-phage. *See* T-even phage; T-odd phage.

TPN⁺. Triphosphopyridine nucleotide. *Aka* NADP⁺.

TPNH. Reduced triphosphopyridine nucleotide. *Aka* NADPH.

TPP. Thiamine pyrophosphate.

trace element. An element, such as cobalt, copper, manganese, or zinc, that is an essential nutrient for an organism but that is required only in minute amounts.

tracer. 1. An isotope, either radioactive or stable, that is used to label a compound. 2. A compound labeled with either a radioactive or a stable isotope.

track radioautography. The radioautography of beta particle tracks that are produced by decaying radiophosphorus atoms incorporated into nucleic acids. *See also* radiophosphorus decay; star gazing.

trailing. The chromatographic and electrophoretic phenomenon in which the material that is separated not only appears in peaks, spots, or bands, but also appears along part of, or the entire, migration path. *See also* tailing; leading.

trans. 1. Referring to the configuration of a geometrical isomer in which two groups, attached to two carbon atoms linked by a double bond, lie on opposite sides with respect to the plane of the double bond. 2. Referring to two mutations, particularly those of pseudoalleles, that lie on different chromosomes.

transacetylase. ACETYLTRANSFERASE.

transacylase. An enzyme that catalyzes the transfer of an acyl group from one compound to another.

transaldolase. An enzyme that catalyzes the transfer of a dihydroxyacetone group from a ketose phosphate to carbon number one of an aldose phosphate.

transamidinase. AMIDINOTRANSFERASE.

transamidination. The reaction in which the guanido group of arginine is transferred to glycine, thereby forming guanidinoacetic acid.

transaminase. An enzyme that catalyzes the transfer of an amino group from an amino acid to a keto acid, thereby giving rise to the opposite pair of keto acid and amino acid. Transaminases require a derivative of vitamin B_6 as coenzyme.

transaminase-type mechanism. PING-PONG MECHANISM.

transamination. The reaction catalyzed by the enzyme transaminase.

transcarboxylation. A reaction in which a carboxyl group is transferred from one compound to another; the reaction is catalyzed by a biotin-requiring enzyme.

transcellular transport. The transport across a cell or a layer of cells that moves material both into and out of a cell; the transport across the kidney tubules and the transport across the gastric mucosa are two examples. *See also* intracellular transport; homocellular transport.

transcortin. An α_1-globulin that binds and transports both cortisol and corticosterone in the blood.

transcriptase. DNA-DEPENDENT RNA POLYMERASE.

transcription. The process whereby the genetic information of DNA is copied in the form of RNA; a sequence of deoxyribonucleotides in a strand of DNA gives rise to a complementary sequence of ribonucleotides in a strand of RNA. *See also* reverse transcription.

transduce. 1. To transfer genetic material from one bacterium to another by transduction. 2. To transform one form of energy into another.

transduced element. The chromosomal fragment that is transferred from one bacterium to another during transduction.

transducer. A device that transforms one form of energy into another; a photocell that converts light energy into electrical energy is an example.

transducer cell. One of a number of neurosecretory cells in the hypothalamus that are stimulated by the central nervous system to secrete the hypothalamic releasing hormones which act on the adenohypophysis.

transducing phage. A phage that brings about transduction.

transduction. 1. The genetic recombination in bacteria in which DNA from a donor cell is transferred to a recipient cell by means of a phage. A segment of the donor DNA is first incorporated into the phage DNA and is then incorporated, by recombination, into the recipient DNA. 2. The transformation of one form of energy into another.

trans effect. The influence of one gene on the expression of another gene which is located on a different chromosome.

transeliminase. PECTATE LYASE.

transesterification. INTERESTERIFICATION.

transfection. The "transformation" of competent bacterial cells by "infection" with phage DNA.

transferase. 1. An enzyme that catalyzes a reaction in which there is a transfer of a functional group from one substrate to another. *See also* enzyme classification. 2. ELONGATION FACTOR.

transferase I. An aminoacyl transfer RNA binding factor that, together with GTP, binds aminoacyl transfer RNA to ribosomes.

transferase II. TRANSLOCASE (2).

transfer factor. 1. ELONGATION FACTOR.

2. A factor that can be extracted from lymphocytes and that can be used to transfer delayed hypersensitivity in man. *Abbr* TF.

transfer pipet. VOLUMETRIC PIPET.

transfer potential. *See* group transfer potential.

transfer rate coefficient. The permeability coefficient divided by the volume from which diffusion takes place.

transferrin. A globulin that binds two atoms of iron and that serves to transport iron in the blood.

transfer RNA. A low-molecular weight RNA molecule, containing about 70 to 80 nucleotides, that binds an amino acid and transfers it to the ribosomes for incorporation into a polypeptide chain during translation. Transfer RNA has a sedimentation coefficient of about 4S, is characterized by having a high content of minor bases, and binds to a codon in messenger RNA by way of a complementary anticodon that is present in the transfer RNA. The transfer RNA molecule is believed to exist in a clover leaf configuration, with the chain folded back upon itself and held together by means of hydrogen bonding. *Abbr* tRNA; RNA_t.

transfer RNA multiplicity. The occurrence of two or more forms of the same transfer RNA molecule, all of which can be charged with the same amino acid.

transformant. A bacterial cell that has undergone transformation.

transformation. 1. The genetic recombination in bacteria in which a DNA fragment of a purified DNA preparation is incorporated into the chromosome of a recipient bacterial cell. 2. The change of a normal cell to a malignant one, as that resulting from infection of normal cells by oncogenic viruses.

transforming principle. The purified DNA that is incorporated into the recipient bacterial cell during transformation; first described for the transformation of *Pneumococcus*. *Aka* transforming factor.

transgenosis. The overall phenomenon of the transfer of bacterial genes to plants and their subsequent expression within the plant cells; includes gene transfer, gene maintenance, transcription, translation, and metabolic function.

transhydrogenase. The mitochondrial enzyme that catalyzes the interconversion of NAD^+ and NADPH to NADH and $NADP^+$. *Abbr* TH. *See also* supermolecule.

transhydrogenation. The reaction $NAD^+ + NADPH = NADH + NADP^+$ that is catalyzed by a transhydrogenase and that occurs in mitochondria.

transient. Temporary; transitory; short-lived.

transient dipole-induced dipole interactions. DISPERSION FORCES.

transient equilibrium. The equilibrium that is established for a radioactive decay reaction if the half-life of the parent isotope is not much greater than that of the daughter isotope.

transient kinetics. PRESTEADY-STATE KINETICS.

transient phase. A temporary phase in enzyme reactions which precedes the steady-state phase. *See also* presteady-state kinetics.

transient time. The time required to reach equilibrium in a sedimentation equilibrium experiment.

transinhibition. The inhibition of transport across a membrane by an intracellular substrate of the transport system; kinetically similar to product inhibition of an enzymatic reaction.

trans isomer. *See* trans (1).

transition. 1. A point mutation in either DNA or RNA in which there is a replacement of one purine by another or a replacement of one pyrimidine by another. In double-stranded nucleic acids, a complementary base is then inserted into the second strand so that a new base pair is produced. 2. A change from one state to another, such as a change in the electronic configuration of an atom or a molecule upon excitation, or a change in the conformation of a macromolecule upon denaturation.

transitional mutant. A mutant that differs from the wild-type organism by a transition.

transition dipole moment. The dipole moment induced in a molecule as a result of the displacement of charge during the absorption of a photon by the molecule.

transition element. One of a group of metal elements, including iron and cobalt, in which filling of the outermost shell to 8 electrons within a period is interrupted to bring the penultimate shell from 8 to either 18 or 32 electrons. These elements can use outermost, as well as penultimate, shell orbitals for bonding, and they form chelates of importance in biochemistry. *Aka* transition metal.

transition moment. *See* magnetic transition moment; transition dipole moment.

transition probability. The probability that a molecule will undergo a transition from one energy state to another if it is supplied with the energy required for this transition.

transition state. ACTIVATED COMPLEX.

transition state theory. THEORY OF ABSOLUTE REACTION RATES.

transition temperature. MELTING OUT TEMPERATURE.

transketolase. An enzyme that catalyzes the transfer of a glycolaldehyde group from a 2-

ketosugar to carbon number one of various aldoses.

translation. 1. The process whereby the genetic
information of messenger RNA is used to
specify and direct the synthesis of a polypeptide
chain on a ribosome; the sequence of codons in
messenger RNA gives rise to a sequence of
amino acids in a polypeptide chain. Synthesis
of the polypeptide chain includes chain initiation, chain elongation, and chain termination.
2. A motion along a line without rotation.

translational control. The regulation of the
expression of a gene that is achieved by a control of the rate at which the messenger RNA,
specified by that gene, is being translated.

translational diffusion. The macroscopic flow of
material from a region of high concentration to
a region of lower concentration that results
from the random Brownian motion of the
molecules.

translational diffusion coefficient. A measure of
translational diffusion that depends on the size
and the shape of the diffusing particle; specifically, $D = RT/Nf$, where D is the translational
diffusion coefficient, f is the translational frictional coefficient, N is Avogadro's number, R
is the gas constant, and T is the absolute
temperature. *Aka* diffusion coefficient. *See also*
standard diffusion coefficient.

translational frictional coefficient. A measure of
the frictional resistance experienced by a
particle in solution that is equal to the frictional
force divided by the translational velocity of the
particle. *Aka* frictional coefficient. *See also*
translational diffusion coefficient.

translation error model. A model for the evolution of the genetic code according to which the
code evolved so as to minimize errors during
translation.

translation inhibitory protein. ANTIVIRAL
PROTEIN.

translocase. 1. TRANSPORT AGENT. 2. The
enzyme that catalyzes the GTP-dependent
translocation reaction in protein synthesis in
which the peptidyl transfer RNA is shifted
from the A site to the P site on the ribosome,
and in which the messenger RNA is shifted simultaneously by one codon.

translocation. 1. TRANSPORT. 2. The reaction
catalyzed by the enzyme translocase. 3. An
interchromosomal aberration in which a chromosome fragment becomes inserted into
another, nonhomologous chromosome.

translocation factor. TRANSLOCASE.

translocon. An unlinked cluster of genes that is
believed to specify a portion of the immunoglobulin molecule; three such clusters are thought
to represent the basic unit of immunoglobulin
evolution.

translucent. Descriptive of a substance that
permits only a partial passage of the light rays
striking it; partially transparent.

transmembrane potential. MEMBRANE
POTENTIAL.

transmethylation. The reaction whereby a labile
methyl group is transferred from one compound to another.

transmission. 1. TRANSMITTANCE. 2.
NERVE IMPULSE TRANSMISSION.

transmittance. The ratio of the intensity of the
transmitted light to that of the incident light.
Sym T.

transmittancy. The ratio of the transmittance of
the solution to that of the solvent.

transmitter. *See* chemical transmitter.

transmutation. The transformation of one nuclide into the nuclide of a different element, as
that which occurs during radioactive decay.

transparent. Descriptive of a substance that
permits the passage of light rays striking it and
that can be seen through.

transpeptidase. PEPTIDYL TRANSFERASE.

transpeptidation. The reaction catalyzed by
peptidyl transferase.

transphosphorylation. The transfer of a phosphoryl group, a pyrophosphoryl group, or an
adenylyl group from one molecule to another.

transpiration. The passage of water vapor
through the surface of an organism, as the
passage through a leaf or the skin.

transplant. The part of an animal that is
transferred within the same, or to a different,
animal by transplantation.

transplantation. The transfer of a part of an
animal, such as a tissue or an organ, to another
site of the same animal or to a site of a different animal.

transplantation antigen. HISTOCOMPATI-
BILITY ANTIGEN.

transplantation immunity. The reactions of the
immune response that occur in the recipient of
a transplant and that are caused by the antigens
of the transplant.

transplant rejection. The immunological reactions by which the recipient of a transplant
brings about the destruction of the cells of the
transplant.

transport. The movement of material from one
place to another, particularly in reference to the
movement of material within a biological fluid
or across a biological membrane. *See also* active transport; mediated transport; passive
transport.

transport agent. A substance, generally a protein
or an enzyme, that is instrumental in transporting material across a biological membrane or
within a biological fluid. A transport agent may
act as an actual carrier, affect a translocation,

or in some other way affect the transport process. *Aka* transporter.

transport maximum. The maximum rate at which the kidney can either absorb or secrete a particular substance.

transport-negative mutant. A mutant that has lost the ability to synthesize an enzyme, a group of enzymes, or some other protein that is required for a transport system.

transport piece. SECRETORY PIECE.

transport process. A physical property that is based on the transport of macromolecules in solution; sedimentation, diffusion, electrophoresis, and viscosity are examples.

transport system. The various components, including transport agents, that function in a given type of transport.

transposable code. An early version of the genetic code according to which the DNA strand that is not transcribed consists either of all nonsense codons, or of codons that code for the same amino acids as those coded for by the complementary codons in the transcribed strand. In this fashion each genetic locus would give rise to only one type of protein, regardless of whether one or both of the DNA strands were being transcribed.

transprotonase. An enzyme that catalyzes a transprotonation reaction. *Abbr* TP. *See also* supermolecule.

transprotonation. The transport of protons across a membrane, as that postulated by the chemiosmotic hypothesis.

transudate. The fluid exuded during transudation.

transudation. The movement of solvent plus solutes through the pores of a membrane or through the interstices of a tissue.

transverse system. T SYSTEM.

transversion. A point mutation in either DNA or RNA in which there is a replacement of a purine by a pyrimidine or vice versa. In double-stranded nucleic acids, a complementary base is then inserted into the second strand so that a new base pair is produced.

transversional mutant. A mutant that differs from the wild-type organism by a transversion.

Trappe's eluotropic series. *See* eluotropic series.

Traube's covolume. *See* covolume.

Traube's rule. The rule that, for a homologous series of surfactants, the concentration of surfactant required to produce an equal lowering of surface tension in a dilute solution decreases by about a factor of three for each additional CH_2 group in the surfactant molecule.

Trautman plot. A graphical method for the representation of data obtained by the Archibald method that is particularly useful for the comparison of a number of different experiments.

trehalose. A nonreducing disaccharide of glucose that occurs in the hemolymph of many insects.

TRF. Thyrotropin releasing factor; *see* thyrotropin releasing hormone.

TRH. Thyrotropin releasing hormone.

tri-. Combining form meaning three or thrice.

trial and error. A method for obtaining a desired result by trying various ways and/or values, noting the errors and causes of failure, and eliminating them in subsequent steps.

triangulation number. The total number of small equilateral triangles into which the surfaces of an icosadeltahedron have been divided.

tricarboxylic acid cycle. CITRIC ACID CYCLE.

tricine. *N*-Tris(hydroxymethyl)methylglycine; used for the preparation of buffers in the pH range of 7.4 to 8.8.

tridentate. Describing a ligand that is chelated to a metal ion by means of three donor atoms.

trigger reaction. A reaction that initates another reaction or a sequence of reactions.

triglyceride. A glyceride formed by the esterification of one glycerol molecule with three fatty acid molecules.

triiodothyronine. A minor hormone of the thyroid gland that has the same functions as thyroxine but is present in much smaller amounts. *Sym* T_3.

triketohydrindene hydrate. Ninhydrin. *See also* ninhydrin reaction.

trimer. An oligomer that consists of three monomers.

triose. A monosaccharide that has three carbon atoms.

tripartite repeating unit. *See* supermolecule.

triphosphopyridine nucleotide. NICOTINAMIDE ADENINE DINUCLEOTIDE PHOSPHATE.

triple bond. A covalent bond that consists of three pairs of shared electrons.

triple-chain length. A crystalline form of glycerides in which three acyl groups form a unit structure.

triple helix. A structure that is composed of three intertwined helical chains, as that of tropocollagen and certain synthetic polynucleotides.

triplet. 1. CODON. 2. A triple peak, as that obtained in nuclear magnetic resonance.

triplet code. A genetic code in which an amino acid is specified by a codon that is composed of a sequence of three adjacent nucleotides.

triplet state. The excited electronic state of an atom or a molecule in which two single electrons in two orbitals have their spin in the same direction and are not spin-paired so that $S = 1$ and $2S + 1 = 3$, where S is the resultant spin and $2S + 1$ is the spin multiplicity. A triplet state is a long-lived, metastable state with a lifetime of about 10^{-3} to 1 sec; a triplet state atom or molecule has a net magnetic moment.

triplex. A triple helix.

tris. Tris(hydroxymethyl)aminomethane; used for the preparation of buffers in the pH range of 7.2 to 9.2. *Aka* THAM.

tritiated. Labeled with tritium.

tritium. The heavy radioactive isotope of hydrogen that contains one proton and two neutrons in the nucleus; a weak beta emitter having a half-life of 12.26 years. *Sym* T.

tritium gel filtration. A method for studying the hydrogen exchange of macromolecules by gel filtration chromatography of a tritiated sample and analysis of the eluate for tritium content.

triton. 1. The tritium nucleus that consists of one proton and two neutrons. 2. Trademark for a series of organic, nonionic surface-active agents.

tritosome. A lysosome that has engulfed molecules of the detergent triton.

triturate. To grind or rub to a powder, usually with the aid of a liquid.

trityl group. A triphenylmethyl group; used to block the amino group of amino acids.

trivial name. A working or common name that is not based on rules of nomenclature, such as the name of an enzyme that is not based on the classification rules of the Enzyme Commission. *See also* systematic name.

Trizma. Trademark for a group of tris buffers.

tRNA. Transfer RNA.

tRNA^{AA}. Transfer RNA that is specific for the amino acid AA.

tRNA arm. *See* arm.

tRNA^{fmet}. Methionine transfer RNA that can be enzymatically formylated; also abbreviated tRNA$_f^{met}$.

tRNA^{met}. Methionine transfer RNA that cannot be enzymatically formylated.

tRNA stem. *See* stem.

tRNA synthetase. AMINOACYL-tRNA SYNTHETASE.

tRNA synthetase recognition site. The site on the transfer RNA molecule to which the aminoacyl-tRNA synthetase becomes bound.

-trophic. Combining form meaning related to nutrition; a common ending for the names of many hormones where it is used interchangeably with -tropic.

trophic hormone. A hormone, the main function of which is to stimulate the secretion of another hormone from an endocrine gland.

trophic value. The value of a nutrient in terms of the raw material which it furnishes for the building and the maintenance of the metabolic machinery; it is greater than that described by the caloric value of the nutrient.

-tropic. Combining form meaning a turning; a common ending for the names of many hormones where it is used interchangeably with -trophic.

tropism. An involuntary response of an organism to a stimulus. The response, such as bending, turning, or directional growth, may be either toward, or away from, the stimulus and is then referred to as positive or negative tropism.

tropocollagen. The basic unit of collagen that consists of a triple helix having a molecular weight of about 300,000. Tropocollagen units associate to form collagen fibrils and the fibrils associate to form larger fibers.

tropomyosin. A minor protein of the myofilaments of striated muscle.

tropomyosin A. A water-insoluble form of tropomyosin that is present in the catch muscles of mollusks.

tropomyosin B. A water-soluble form of tropomyosin that is present in the thin filaments of the I bands of striated muscle.

troponin. A minor protein component of the thin filaments of striated muscle.

Trp. 1. Tryptophan. 2. Tryptophanyl.

TRU. Tripartite repeating unit.

true fat. NEUTRAL FAT.

true order. ORDER WITH RESPECT TO CONCENTRATION.

true plasma. Plasma that is obtained from whole blood after the blood has been equilibrated with carbon dioxide at a specific partial pressure.

true toxin. An exotoxin against which an antitoxin can be produced.

trypsin. An endopeptidase that acts primarily on peptide bonds in which the carbonyl group is contributed by either arginine or lysine.

tryptamin. A compound derived from tryptophan by decarboxylation.

tryptic. Of, of pertaining to, trypsin.

trypticase. A tryptic digest of casein.

tryptic peptide. A peptide obtained by digestion of a protein with trypsin.

tryptone. A peptone produced by the proteolytic action of trypsin.

tryptophan. An aromatic, heterocyclic, and nonpolar alpha amino acid. *Abbr* Trp; W.

tryptophan load test. *See* load test.

tryptophan pyrrolase. An inducible enzyme in the liver of higher organisms.

tryptophan synthetase. The enzyme that catalyzes the terminal step in the biosynthesis of tryptophan; it consists of four polypeptide chains one of which, the A chain, has been used for studies of amino acid replacements.

T$_s$. *See* elongation factor.

TSH. Thyroid-stimulating hormone.

ts mutant. Temperature-sensitive mutant.

T system. A system of tubules that interconnects the sarcolemma and the myofibrils of striated muscle; thought to be instrumental in communicating the depolarization of the sarcolemma almost simultaneously to all the myofibrils of the muscle fiber.

t test. *See* student's t test.

TTP. 1. Thymidine triphosphate. 2. Thymidine-5′-triphosphate.

T$_u$. *See* elongation factor.

tube dilution method. A method for determining the sensitivity of a microorganism to an antimicrobial drug; based on measuring the concentration of drug that prevents microbial growth in one or more tubes of a series of tubes that contain identical microbial inocula but different concentrations of the drug.

tuberculin. A protein preparation obtained from the tubercle bacillus *Mycobacterium tuberculosis* that is used in the tuberculin test for delayed-type hypersensitivity. *See also* old tuberculin; purified protein derivative.

tuberculin hypersensitivity. Delayed-type hypersensitivity to inoculation with tuberculin following an infection by the tubercle bacillus *Mycobacterium tuberculosis.*

tuberculin PPD. Tuberculin purified protein derivative; *see* purified protein derivative.

tuberculin test. A test, performed in man or animals, for delayed-type hypersensitivity to tuberculin.

tubular ion exchange mechanism. A set of reactions believed to represent the major mechanism for the acidification of urine by the kidney tubules.

tumor. NEOPLASM.

tumor antigen. T ANTIGEN (1).

tumorigenesis. The formation of a tumor.

tumorigenic. Capable of causing tumor formation.

tumor progression. 1. The series of changes whereby a cancerous lesion becomes more and more malignant with time. 2. The third stage in a multistage mechanism of carcinogenesis in which a dependent cancer cell is converted to an autonomous cancer cell.

tumor virus. ONCOGENIC VIRUS.

tunneling. The transfer of a particle, such as an electron or a proton, across a potential energy barrier without the particle acquiring sufficient energy to surmount the energy barrier. The transfer occurs as a result of the probability distribution of the particle on both sides of the barrier.

turbidimetry. The quantitative determination of a substance in suspension that is based on measurements of the decrease in light transmission by the suspension due to the scattering of light by the suspended particles. *See also* nephelometry.

turbidity. 1. A measure of the scattering of light by a solution; equal to the quantity τ in $I = I_0 e^{-\tau l}$, where I is the intensity of the transmitted light, I_0 is the intensity of the incident light, e is the base of natural logarithms, and l is the length of the light path. *Aka* turbidity coefficient. 2. The cloudiness of a solution that is caused by fine suspended particles.

turbid plaque. A plaque produced in the plaque assay or in the plaque technique when a mixture of cells is present so that not all of the cells in the area of the plaque are lysed.

turbid plaque mutant. A plaque-type mutant that produces turbid plaques.

turbulent flow. The disturbed flow of a liquid along a tube; produced by obstacles and/or high rates of shear. *See also* laminar flow.

turnover. 1. The rate at which a substrate is acted upon by an enzyme and measured by the turnover number. 2. TURNOVER TIME.

turnover number. A measure of enzymatic activity, expressed as either molecular activity or as catalytic center activity. *See also* molar activity.

turnover time. The time required for the transfer of a substance into or out of a pool under steady-state conditions such that the amount of the substance transferred is equal to the amount of the substance in the pool.

Tween. Trademark for a series of nonionic detergents that consist of fatty acid esters of polyoxyethylene sorbitan.

twin ion technique. A mass spectrometric technique in which a substance is labeled with two isotopes and the substance and its metabolic transformations are recognized and identified by the characteristic twin ions in their mass spectra.

twist. A specific conformation of the furanose ring.

two-carbon fragment. 1. A group of atoms or a compound containing only two carbon atoms. 2. An early designation of acetyl coenzyme A.

two-dimensional chromatography. A flat-bed chromatographic technique in which the compounds are first separated in one direction and, after rotation of the chromatogram by 90°, are then separated in a second direction.

two-dimensional electrophoresis. A flat-bed electrophoretic technique in which the compounds are first separated in one direction and, after rotation of the electropherogram by 90°, are then separated in a second direction.

two-factor cross. A genetic recombination experiment involving two nonallelic genes.

tygon. A copolymer of vinyl chloride and vinyl acetate.

TYMV. Turnip yellow mosaic virus.

Tyndall effect. The phenomenon that the path of a beam of light through a solution containing colloidal particles becomes visible as a result of the scattering of light by the particles.

Tyndallization. FRACTIONAL STERILIZA-TION.

type A RNA virus. *See* oncornavirus.

type B RNA virus. *See* oncornavirus.

type C RNA virus. *See* oncornavirus.

type I immunoglobulin. TYPE K IMMUNO-GLOBULIN.

type II immunoglobulin. TYPE L IMMUNO-GLOBULIN.

type K immunoglobulin. An immunoglobulin containing light chains of the kappa type.

type L immunoglobulin. An immunoglobulin containing light chains of the lambda type.

type-specific antigen. An antigen that is found in only one subdivision of a family of viruses.

Tyr. 1. Tyrosine. 2. Tyrosyl.

tyrosinase. The enzyme that catalyzes the oxidation of tyrosine to dopa and the oxidation of dopa to dopa quinone which is then converted to melanin.

tyrosine. An aromatic, polar alpha amino acid that contains a phenolic hydroxyl group. *Abbr* Tyr; Y.

tyrosinemia I. TYROSINOSIS.

tyrosinemia II. A genetically inherited metabolic defect in man that is due to a deficiency of the enzyme tyrosine aminotransferase.

tyrosinosis. A genetically inherited metabolic defect in man that is characterized by an excessive excretion of p-hydroxyphenylpyruvic acid and that is apparently due to a deficiency of the enzyme p-hydroxyphenylpyruvic acid oxidase.

U

U. 1. Uracil. 2. Uridine. 3. Unit of enzyme. 4. Uranium.

Ubbelohde viscometer. A viscometer that permits the measurement of viscosity as a function of concentration, since the solution can be diluted directly in the viscometer.

ubiquinone. COENZYME Q.

UDP. 1. Uridine diphosphate. 2. Uridine-5'-diphosphate.

UDPG. Uridine diphosphate glucose; *see* UDP-glucose.

UDP-Gal. Uridine diphosphate galactose.

UDP-glucose. Uridine diphosphate glucose; a nucleoside diphosphate sugar that serves as the donor of a glucose residue in the biosynthesis of glycogen.

UFA. Unesterified fatty acids.

UIBC. Unsaturated iron-binding capacity.

ultimate carcinogen. The form of a carcinogen in which it ultimately reacts with cellular macromolecules; believed to be a form that contains an electrophilic center (such as a carbonium ion, free radical, epoxide, and metal cation) that attacks electron-rich centers in proteins and nucleic acids.

ultimate precision. *See* method of ultimate precision.

ultra-. 1. Prefix meaning excessive. 2. Prefix meaning beyond the range of.

ultracentrifugation. Centrifugation, performed in an ultracentrifuge, in which high centrifugal forces are used.

ultracentrifuge. A high-speed centrifuge capable of generating speeds of approximately 60,000 rpm and centrifugal forces of approximately $500,000 \times g$.

ultracryotomy. The preparation of thin sections from frozen tissue specimens.

ultrafiltration. The filtration of a solution through a filter or a membrane that will retain macromolecules.

ultrasonic. Of, or pertaining to, ultrasound.

ultrasonication. The exposure of material to ultrasound; used for the rupture of cells and the denaturation of proteins.

ultrasound. Sound waves that have frequencies greater than those of audible sound; refers particularly to sound waves that have frequencies of 500,000 cps or higher.

ultrastructure. The fine structure of tissues, cells, and subcellular particles beyond that revealed by the light microscope.

ultraviolet dichroism. The dichroism produced when ultraviolet light is absorbed by oriented samples such as nucleic acid preparations.

ultraviolet optics. ABSORPTION OPTICAL SYSTEM.

ultraviolet spectrum. That part of the electromagnetic spectrum that covers the wavelength range of about 1.3×10^{-6} to 4×10^{-5} cm and that includes photons which are emitted or absorbed during electronic transitions; radiation of the shorter wavelengths is known as far ultraviolet, and that of the longer wavelengths is known as near ultraviolet. *Abbr* uv spectrum.

umber codon. The codon UGA; one of the three termination codons.

umber mutation. A mutation in which a codon is mutated to the umber codon, thereby causing the premature termination of the synthesis of a polypeptide chain.

UMP. 1. Uridine monophosphate (uridylic acid). 2. Uridine-5'-monophosphate (5'-uridylic acid).

uncertainty principle. The concept that any physical measurement disturbs the system that is being measured and limits the accuracy of the measurement; particularly applicable to measurements at the atomic and subatomic levels.

uncharged polar amino acid. A polar amino acid that carries no charge in the intracellular pH range of about 6 to 7.

uncoating. The removal of the protein coat from a virus that occurs extracellularly in the case of bacterial viruses and, apparently, intracellularly in the case of animal viruses.

uncoating enzyme. An enzyme that functions in the removal of the protein coat from a virus.

uncoded amino acid. An amino acid, such as hydroxyproline or hydroxylysine, for which no codon exists. Such an amino acid is derived by enzymatic modification of the parent amino acid (e.g., proline or lysine) after the parent amino acid has become incorporated into a polypeptide chain in response to its codon.

uncompensated acidosis. An acidosis in which the blood pH falls due to insufficient compensatory mechanisms.

uncompensated alkalosis. An alkalosis in which the blood pH rises due to insufficient compensatory mechanisms.

uncompetitive inhibition. The inhibition of the activity of an enzyme that is characterized by a decrease in the maximum velocity compared to that of the uninhibited reaction and by a reciprocal plot (1/velocity versus 1/substrate concentration) which is parallel to that of the uninhibited reaction.

uncoupler. UNCOUPLING AGENT.

uncoupling. The separation of the two processes that constitute oxidative phosphorylation such that ATP synthesis is dissociated from the electron transport system at one or more of the phosphorylation sites. As a result of uncoupling, ATP synthesis is inhibited but electron transport and respiration are allowed to proceed and may even be stimulated.

uncoupling agent. An inhibitor that brings about the uncoupling of ATP synthesis from the electron transport system at one or more of the phosphorylation sites.

undecaprenyl phosphate. A compound that functions as an acceptor and as a donor of sugar groups in the process of peptidoglycan synthesis.

undernutrition. Malnutrition resulting from the consumption of inadequate amounts of food so that one or more of the essential nutrients are lacking in the diet. *Aka* undernourishment.

uneconomic species. An undesirable species; the species that is eliminated by another, desirable, species through the use of chemicals.

uni-. 1. Referring to one kinetically important substrate and/or product of a reaction; thus a uni bi reaction is a reaction with one substrate and two products. 2. Prefix meaning one.

uniformitarianism. *See* doctrine of uniformitarianism.

uniformly labeled. Designating a compound that is labeled in all of the positions in the molecule in a statistically uniform or nearly uniform manner; generally refers to C^{14}-labeled compounds. *Sym* U.

unifunctional feedback. A feedback mechanism that affords control in only one direction so that the input of a system is affected by either an increase or a decrease of the output; a system in which the pH will be adjusted if it rises above the normal value is an example.

unimolecular reaction. A chemical reaction in which one molecule (or other entity) of a single reactant is transformed into products.

uniport. The transport of a substance across a membrane that is not linked to the transport of another substance across the same membrane. *See also* antiport; symport.

unitarian hypothesis. The hypothesis that the injection of an antigen results in the formation of a single type of antibody that has multiple functions and that can react with antigens under all conditions, such as those of agglutination, precipitation, and complement fixation.

unit cell. The smallest portion of a crystal that embodies the structural characteristics of the crystal and that, when repeated indefinitely, will generate the entire structure.

unit evolutionary period. The average number of years required for the occurrence of one amino acid replacement in a given protein.

unit membrane hypothesis. The hypothesis that the structure of biological membranes can be described in terms of a unit membrane, about 90 Å thick. The unit membrane is considered to be a bilayer of polar lipids, such as phospholipids, surrounded on both sides by protein. The lipids are arranged with their nonpolar portions inward and their polar portions outward where they are coated with protein molecules that are in their beta configuration. *Aka* Danielli-Davson model; Danielli-Davson-Robertson model.

unit of complement. The amount of complement that lyses 50% of the sensitized erythrocytes.

unit of enzyme. *See* enzyme unit.

unity of biochemistry. The phenomenon that widely different microbial, plant, and animal cells show a high degree of similarity with respect to both the types of molecules that they contain and the metabolic reactions that these molecules undergo.

universal buffer. A mixture of several buffers that can be used over a relatively wide pH range.

universal code. A code in which the same codons code for the same amino acids in all types of organisms.

universal donor. An individual of the *O* type in the *ABO* blood group system, who can donate blood to any recipient.

universal recipient. An individual of the *AB* type in the *ABO* blood group system, who can receive blood from any donor.

universe. The thermodynamic system plus its surroundings.

unmasking. The conversion of an unreactive amino acid residue or an unreactive site on a protein molecule to a reactive one that is accessible to specific reagents and that can participate in specific binding or other reactions. The process may involve conformational changes in the molecule and/or proteolytic removal of blocking groups.

unmixing. 1. The sorting out of solute molecules from solvent molecules in a solution, a process that is accompanied by a decrease in entropy. 2. The formation of a single complex from separate particles, a process that is accompanied by a decrease in entropy.

unpaired electron. A single electron, instead of the usual two, in an orbital.

unresponsiveness. IMMUNOLOGICAL UNRESPONSIVENESS.

unrestricted. Descriptive of a restricted mutant

that has undergone the opposite modification. *See also* restricted virus.

unsaponifiable lipid. The fraction of lipid in a sample that remains insoluble after saponification of the sample with alkali; it consists principally of steroids and terpenes.

unsaturated fatty acid. A fatty acid that contains one or more double bonds in the alkyl chain.

unsaturated iron-binding capacity. The difference in the concentration of iron between the maximum potential concentration (transferrin fully saturated with iron) and the actual concentration (transferrin only about 25 to 30% saturated with iron); generally expressed in terms of micrograms per 100 ml of plasma. *Abbr* UIBC.

unscheduled DNA synthesis. The synthesis of DNA during some period of the cell cycle other than the *S* period. *See also* cell cycle.

unselective marker. A marker that does not affect the growth of the organism on a selective medium.

unwinding protein. A protein that stimulates DNA synthesis, presumably by aiding in the unwinding of double-stranded DNA during replication.

UP. Uroporphyrin.

UPG. Uroporphyrinogen.

uphill reaction. ENDERGONIC REACTION.

upward flow. A column chromatographic technique, frequently used in gel filtration, in which the eluent is passed upward through the column. This minimizes compression of the gel bed and allows for finer control of the flow rate.

Ura. Uracil.

uracil. The pyrimidine 2,4-dioxypyrimidine that occurs in RNA. *Abbr* U; Ura.

uracil-rich code. An early version of the genetic code according to which the codons of all the amino acids contain at least one uracil nucleotide.

Urd. Uridine.

urea. A compound that is formed in the urea cycle during amino acid catabolism and that is one of the major forms in which nitrogen is excreted in the urine; urea is also the end product of purine catabolism in most fishes and amphibia.

urea cycle. The cyclic set of reactions whereby two amino groups, derived from the catabolism of two amino acids, and a molecule of carbon dioxide are converted to urea which is then excreted.

urease. An enzyme that appears to have an absolute specificity and that catalyzes only the single reaction whereby urea is hydrolyzed to carbon dioxide and ammonia.

urenzyme. A rudimentary form of an enzyme in a primitive cell.

ureotelic organism. An organism, such as a mammal, that excretes the nitrogen from amino acid catabolism primarily in the form of urea.

urethane. Ethyl carbamate; a carcinogenic agent.

Urey equilibrium. The equilibrium established in the reaction in which the reactants are carbon dioxide and calcium silicate, and the products are calcium carbonate and silica. The reaction has been proposed by Urey to be the one by means of which the carbon dioxide pressure was maintained at low levels in the primitive atmosphere of the earth.

-uria. 1. Combining form meaning the presence of a substance in the urine. 2. Combining form meaning the presence of excessive amounts of a substance in the urine.

uric acid. A purine that does not occur in nucleic acids but that is an intermediate of purine catabolism in some organisms, and an end product of purine catabolism in man and other organisms.

uricosuria. The presence of excessive amounts of uric acid in the urine.

uricosuric drug. A drug that tends to promote the excretion of uric acid.

uricotelic organism. An organism, such as a bird, that excretes the nitrogen from amino acid catabolism primarily in the form of uric acid.

uridine. The ribonucleoside of uracil. Uridine mono-, di-, and triphosphate are abbreviated, respectively, as UMP, UDP, and UTP. The abbreviations refer to the 5′-nucleoside phosphates unless otherwise indicated. *Abbr* Urd; U.

uridine nucleotide coenzyme. *See* nucleotide coenzyme.

uridylic acid. The ribonucleotide of uracil.

urinalysis. The chemical and physical analysis of urine.

urinometer. A hydrometer designed for measuring the specific gravity of urine.

urochrome. The principal pigment of urine that is composed of a peptide of unknown structure and either urobilin or urobilinogen.

uronic acid. A monocarboxylic sugar acid of an aldose in which the primary alcohol group has been oxidized to a carboxyl group.

uroporphyrin. The urinary pigment derived from uroporphyrinogen. *Abbr* UP.

uroporphyrinogen. An intermediate in the biosynthesis of the porphyrins that is formed by the linking of four molecules of porphobilinogen. *Abbr* UPG.

U.S.P. United States Pharmacopeia; denotes a chemical that meets specifications set out in the U.S. pharmacopeia.

USPHS. United States Public Health Service.

Ussing's short circuit. An experimental setup for the measurement of membrane potentials and of concentration gradients across a membrane.

UTP. 1. Uridine triphosphate. 2. Uridine-5′-triphosphate.

uv. Ultraviolet.

uv-induced dimer. A pyrimidine dimer that is formed in DNA by exposure of the DNA to ultraviolet light. *See also* pyrimidine dimer; thymine dimer.

uv monitor. A monitor, the operation of which is based on the measurement of the absorption of ultraviolet light by a solution.

V

v. Reaction rate.

v_0. 1. Initial velocity. 2. Control velocity; initial velocity in the absence of an inhibitor.

$\bar{v}$. Partial specific volume.

V. 1. Maximum velocity. 2. Volume. 3. Volt. 4. Valine. 5. Vanadium.

vacant lattice point model. A model for the structure of water according to which the water structure is an essentially crystalline system that is closely related to an open, expanded, ice-like structure into which free and nonassociated water molecules can easily fit.

vaccination. An immunization in which a vaccine is administered to man or to an animal for the purpose of establishing resistance to an infectious disease.

vaccine. A suspension of antigens that are derived from infectious bacteria or viruses and that, upon administration to man or to an animal, will produce active immunity and will provide protection against infection by these, or by related, bacteria or viruses.

vaccinia virus. A virus of the poxvirus group that infects man and many animals.

vacuole. A space or a cavity within a cell in which substances may become concentrated.

vacutainer. An evacuated tube used in the drawing of blood.

vacuum evaporator. A vacuum chamber in which metal atoms are evaporated in the process of shadowcasting specimens for electron microscopy.

vacuum ultraviolet. The range of the ultraviolet spectrum that encompasses wavelengths less than 1.8×10^{-5} cm.

Val. 1. Valine. 2. Valyl.

valence. 1. The number of electrons of an atom or a group of atoms that participate in the formation of chemical bonds. 2. ANTIGEN VALENCE. 3. ANTIBODY VALENCE.

valence bond theory. The theory of chemical bonding that is developed by considering the atoms to have intact atomic orbitals, and then moving the atoms closer to each other with a resultant perturbation of the atomic orbitals.

valence electron. An electron that is located in the outermost energy shell of an atom and that participates in the formation of chemical bonds.

valency. Variant spelling of valence.

-valent. Combining form meaning valence; used either with mono, di, . . . poly to indicate the chemical valence of atoms and ions, or with uni, bi, . . . multi to indicate the immunological valence of antigens or antibodies.

valine. An aliphatic, nonpolar alpha amino acid. *Abbr* Val; V.

valinomycin. An ionophorous antibiotic, produced by *Streptomyces fulvissimus,* that acts as an uncoupler of oxidative phosphorylation.

vanadium. An element that is essential to a wide variety of species in one class of organisms. Symbol V; atomic number 23; atomic weight 50.942; oxidation states $+2, +3, +4, +5$; most abundant isotope V^{51}; a radioactive isotope V^{48}, half-life 16 days, radiation emitted—positrons and gamma rays.

Van Deemter plot. A plot of height equivalent to a theoretical plate as a function of average gas flow rate in a gas chromatographic column.

Van den Bergh reaction. A colorimetric reaction for bilirubin that is based on the formation of a diazo dye from bilirubin and diazotized sulfanilic acid. The direct Van den Bergh reaction measures direct-acting bilirubin, and the indirect Van den Bergh reaction measures indirect-acting bilirubin.

Van der Waals distance. The distance between two atoms at which the attractive Van der Waals force is just balanced by the repulsive force due to the orbital electrons of the two atoms; the distance is equal to the sum of the Van der Waals radii of the two atoms. *Aka* Van der Waals contact distance.

Van der Waals forces. VAN DER WAALS INTERACTIONS.

Van der Waals interactions. The attractive and repulsive forces between atoms and molecules that are described by three components: a dispersion effect, an induction effect, and an orientation effect. The three components are classified as weak interactions, since the energy of these interactions is proportional to r^{-6} where r is the distance between the interacting species. *See also* dispersion effect; induction effect; orientation effect; weak interactions.

Van der Waals radius. One half of the Van der Waals distance between two like atoms. *Aka* Van der Waals contact radius.

Van der Waals shell. The space surrounding the nucleus of an atom that is described by the Van der Waals radius.

vanilmandelic acid. 3-Methoxy-4-hydroxymandelic acid; a compound that is quantitatively the most important metabolite of the catecholamines and that is used to assess the endogenous production of catecholamines. *Abbr* VMA.

Van Slyke method. A method in which a chemical reaction is measured by either the volume or the pressure of a gas released during the reaction.

Van't Hoff equation. An equation that describes the variation of the equilibrium constant K with the absolute temperature T at constant pressure; specifically, $d\ ln\ K/dT = \Delta H^0/RT^2$, where ΔH^0 is the standard enthalpy change and R is the gas constant.

Van't Hoff factor. A measure of the deviation of a solution from ideality; equal to the ratio of the measured value of a colligative property of the solution to the value calculated on a molar basis.

Van't Hoff isobar. VAN'T HOFF EQUATION.

Van't Hoff isochore. VAN'T HOFF EQUATION.

Van't Hoff limiting law. An expression for the limiting value of the osmotic pressure of an ideal solution; specifically, $lim_{c \to 0} \Pi = RT/M$, where Π is the osmotic pressure, c is the concentration of the solute, M is the molecular weight of the solute, R is the gas constant, and T is the absolute temperature.

Van't Hoff plot. A plot of the logarithm of the equilibrium constant as a function of the reciprocal of the absolute temperature. *See also* Van't Hoff equation.

V antigen. A surface antigen of some viruses.

vapor diffusion method. A method for the slow crystallization of a substance; used in the preparation of transfer RNA crystals.

vapor liquid partition chromatography. GAS CHROMATOGRAPHY (2).

vapor phase chromatography. GAS CHROMATOGRAPHY.

vapor pressure osmometer. An osmometer, the operation of which is based on the lowering of the vapor pressure of a solvent by the addition of a solute.

variable. *See* dependent variable; independent variable.

variable arm. EXTRA ARM.

variable region. That part of the immunoglobulin molecule in which differences in amino acid sequences are found when immunoglobulins from different sources are compared. The region comprises portions of both the light and the heavy chains and includes the two antibody combining sites. *See also* constant region.

variable substrate. The substrate the concentration of which is varied while the concentrations of other substrates are maintained at fixed

values; used in kinetic studies of multisubstrate enzyme systems.

variance. 1. The square of the standard deviation. 2. DEGREES OF FREEDOM (1).

variant. 1. *n* One of several forms of a protein, occurring within one species or distributed among several species, that differ from each other either in their state of aggregation, as in the case of isozymes, or in their amino acid sequence, as in the case of abnormal hemoglobins. *See also* multiple forms of an enzyme. 2. *adj* Dissimilar; showing variation. Said of nonidentical amino acid residues that occupy similar positions in the same polypeptide chain which is isolated from different sources.

varicella. The primary form of the disease produced by varicella virus in a host that was not previously infected by the virus.

varicella virus. A virus of the herpesvirus group that leads to varicella in hosts without immunity to the disease and to herpes zoster in hosts with immunity. *Aka* varicella-zoster virus.

variola virus. The virus of the poxvirus group that causes smallpox.

vascular. Of, or pertaining to, vessels that conduct a biological fluid; such vessels include those that carry the blood and those that conduct the sap of plants.

vasoconstriction. A decrease in the diameter of the blood vessels.

vasoconstrictor. An agent that causes vasoconstriction.

vasodepressor. An agent that brings about a lowering of the blood pressure.

vasodilation. An increase in the diameter of the blood vessels.

vasodilator. An agent that causes vasodilation.

vasopressin. A cyclic peptide hormone, consisting of nine amino acids, that increases the blood pressure and increases the absorption of water by the kidneys; it is secreted by the posterior lobe of the pituitary gland and occurs in mammals. *Abbr* VP.

vasopressor principle. VASOPRESSIN.

vasotocin. A cyclic peptide hormone, consisting of nine amino acids, that is related to vasopressin in its structure and in its function; it is secreted by the posterior lobe of the pituitary gland and occurs in birds, reptiles, and some amphibians.

vector. 1. A directional quantity. 2. An organism that serves to transfer a parasite from one host to another.

vectorial. Of, or pertaining to, a vector.

vectorial enzyme. An enzyme that is directional in its action, such as one that is fixed in a membrane.

vectorial reaction. A reaction that is directional and in which the components either are not free to move at random or are fixed in a nonrandom order; a reaction across a membrane is generally a vectorial reaction.

vegetative. Pertaining to growth, particularly of plants, as opposed to reproduction.

vegetative cell. An actively growing cell, as distinct from one forming spores.

vegetative DNA. The pool of genetically competent DNA of a phage that is produced during the vegetative state of the phage but that has not yet been assembled into complete phage particles.

vegetative map. The genetic map of a phage that is deduced from its vegetative replication.

vegetative mycelium. That portion of the mycelium of a fungus that penetrates into the medium and absorbs nutrients.

vegetative nucleus. 1. The macronucleus of a ciliate. 2. The tube-nucleus of a pollen grain. 3. The nucleus of a cell during interphase.

vegetative phage. A phage in its vegetative state.

vegetative replication. The replication of vegetative DNA.

vegetative reproduction. 1. Asexual reproduction. 2. The reproduction in plants without true seeds.

vegetative state. The state of a phage during which it replicates actively and autonomously within the host cell; a noninfective state during which infective phage particles have not yet been assembled. During this state the phage controls the synthesis by the host of components that are necessary for the production of infective phage particles.

vein. A blood vessel that transports blood from the tissues to the heart.

velocity. REACTION RATE.

velocity constant. RATE CONSTANT.

venom substrate. STUART FACTOR.

venous. Of, or pertaining to, veins.

ventilation. The process whereby oxygen is supplied to the lungs and to the blood in the capillaries of the lungs; aeration of the blood in the lungs.

verification. The deacylation of a misacylated transfer RNA molecule by its aminoacyl-tRNA synthetase, thereby regenerating the uncharged transfer RNA molecule.

vernier. A small, movable, auxiliary scale that is attached to a larger scale and that has divisions of slightly different width than those of the larger scale; used for increasing the accuracy of a measurement.

Veronal. Trademark for barbital.

Versene. Trademark for ethylene dinitrolotetraacetic acid.

vertebrate. 1. *n* An animal that has a backbone; includes fishes, amphibia, reptiles, birds, and mammals. 2. *adj* Of, or pertaining to, an animal that has a backbone.

vertical transmission. The transmission of viruses from one generation of hosts to the next. *Aka* vertical infection.

very early RNA. PREEARLY RNA.

very high-density lipoprotein. An extremely high-density lipoprotein that contains approximately 60% protein, 20% phospholipid, and 20% neutral lipid; these lipoproteins have molecular weights of about 200,000 and a density greater than 1.21 g/l. *Abbr* VHDL; VHD.

very high-lipid lipoprotein. VERY LOW-DENSITY LIPOPROTEIN.

very low-density lipoprotein. An extremely low-density lipoprotein that contains approximately 10% protein, 20% phospholipid, and 70% neutral lipid; these lipoproteins have molecular weights of about 5 to 100×10^6, a flotation coefficient of 20 to 400S, and a density of about 1.01 to 1.02 g/l. *Abbr* VLDL; VLD.

very low-lipid lipoprotein. VERY HIGH-DENSITY LIPOPROTEIN.

vesicle. A small sac; a membranous cavity.

v_H. The variable region of the heavy chains of the immunoglobulins.

VHD. Very high-density lipoprotein.

VHDL. Very high-density lipoprotein.

v_i. Initial velocity in the presence of an inhibitor.

viable. Describing a cell or an organism that is alive and capable of reproduction.

viable count. The number of viable cells in a bacterial culture.

Vi antigen. A surface antigen of bacteria that is different from the *O* antigen.

vibrating reed electrometer. A sensitive electrometer that is used for measuring the small currents produced in an ionization chamber.

vibrational transition. The transition of a molecule in which, as a result of the stretching or the bending of a bond, the molecule undergoes a vibration so that the atomic nuclei are temporarily displaced, but their equilibrium positions are not changed. Vibrational transitions require energies that are intermediate between those of electronic and those of rotational transitions, and they are induced by short wavelength infrared radiation.

vic-. Combining form meaning vicinal.

vicinal. Referring to two substituents on adjacent carbon atoms. *Abbr* vic.

villikinin. A gastrointestinal hormone that controls the movement of the villi.

vinegar souring. ACETIFICATION.

vinyl. The radical $-CH=CH_2$ that is derived from ethylene.

viosterol. CALCIFEROL.

viral. Of, or pertaining to, viruses.

viral coat. CAPSID.

viral interference. The inhibition of viral multiplication that may occur upon multiple infection of cells with the same virus or upon mixed infection with two or more different viruses; due to a large extent to the action of interferon that is produced in response to the infecting viruses.

viral multiplication cycle. *See* multiplication cycle.

viral particle. VIRUS.

viral-specific. VIRUS-SPECIFIC.

viral yield. The average number of infectious viral particles obtained per productive cell.

viremia. The presence of virus particles in the blood. Viremia is said to be primary or secondary depending on whether it is due to progeny virions produced principally at the site of the initial infection, or whether it is due to those produced in other organs.

virial coefficient. *See* virial expansion.

virial expansion. A mathematical expansion of the general form $Y = A + BX + CX^2 + \ldots$, where X is a variable and $A, B, C,$ etc. are the first, second, third, etc., virial coefficients. *Aka* virial equation.

viricide. A chemical or a physical agent that inactivates viruses.

virion. A complete viral particle consisting of a nucleocapsid and any additional structural proteins and/or envelopes; a virus.

virogene theory. ONCOGENE THEORY.

viroid. A virus-like infectious particle that consists of single-stranded RNA, does not have a protein coat, and has a molecular weight of about 20,000 to 50,000. *Aka* infectious RNA; pathogene; pathogenic RNA.

virology. The science that deals with viruses and viral diseases.

virulence. 1. PATHOGENICITY. 2. The multiplication of virulent viruses.

virulent. Of, or pertaining to, virulence.

virulent virus. A virus that causes lysis of the host cell that it infects; such a virus multiplies in the host cell and forms progeny viral particles which are released into the medium when the cell bursts. In the case of bacteria, a virulent phage cannot become a prophage.

virus. An infectious agent that consists of protein and either DNA or RNA, both of which are arranged in an ordered array and are sometimes surrounded by a membrane. A virus is generally smaller than a bacterium and is an obligate intracellular parasite at the genetic level; it uses the cell machinery to produce viral products specified by the viral nucleic acid.

virus antigen. A surface antigen of some viruses.

virus-specific. Designating a protein or a nucleic acid molecule that is coded for by a viral gene rather than by a host cell gene, and that is produced in the host cell after its infection by the virus.

virus theory of cancer. *See* oncogene theory; protovirus theory.

viscid. VISCOUS.

viscoelastic. Descriptive of a substance, such as a concentrated polymer solution, that exhibits properties of both liquids and solids.

viscometer. An instrument for measuring the viscosity of a liquid.

Visconti-Delbrueck hypothesis. The hypothesis according to which the types and frequencies of phage recombinants that are obtained upon phage infection of bacterial cells are due to the multiplication and the repeated mating of the phages in the host cells. Mating occurs in pairs and at random, and each mating produces phage recombinants as a result of the exchange of genetic material by one or more crossovers between the parental phages. The total intracellular collection of vegetative phage genomes is called a mating pool.

viscose rayon. Fibers of regenerated cellulose.

viscosimeter. VISCOMETER.

viscosity. The resistance of a fluid to flow; the internal friction of a fluid. For Newtonian fluids, the viscosity is the force required to maintain unit velocity between two parallel plates of unit area each and a unit distance apart. *Sym* η. *Aka* coefficient of viscosity.

viscosity increment. A measure of the asymmetry of a solute molecule that is equal to the ratio of the intrinsic viscosity of the solution to the partial specific volume of the solute.

viscosity number. REDUCED VISCOSITY.

viscosity ratio. RELATIVE VISCOSITY.

viscous. 1. Possessing viscosity. 2. Thick; sticky; glutinous.

viscous drag. The frictional force that counteracts and balances either the electrical driving force in electrophoresis or the centrifugal force in sedimentation.

visible dichroism. The dichroism that is produced when visible polarized light is absorbed by oriented samples.

visible mutation. A mutation that results in some alteration of the morphology of an organism.

visible spectrum. That part of the electromagnetic spectrum that covers the wavelength range of about 4×10^{-5} to 7.5×10^{-5} cm and that includes photons that are emitted or absorbed during electronic transitions.

visual cycle. A cyclic set of reactions that occur in both the rods and the cones of the retina whereby (a) light leads to the isomerization of 11-*cis* retinal to the all-*trans* retinal and to its dissocation from the appropriate opsin, and (b) the all-*trans* isomer is reconverted enzymatically to the 11-*cis* isomer which recombines with the opsin.

visual pigment. One of several conjugated proteins that consist of an opsin and a form of vitamin A aldehyde and that function in the biochemical reactions that pertain to vision.

visual purple. RHODOPSIN.

visual threshold. The minimum light intensity required to produce a visual sensation.

vital capacity. The greatest volume of air that can be expired after a forced inspiration; includes the tidal, supplemental, and complemental airs.

vitalism. The doctrine that life and its phenomena are not fully explicable in terms of the laws and processes of chemistry and physics, and that they require special vital forces which are found only in living organisms. *See also* mechanistic philosophy.

vital stain. A stain that can penetrate the cell membrane of a living cell and that can stain the contents without injury to the cell.

vitamer. One of two or more forms of a vitamin; vitamins A_1 and A_2 are examples of vitamers.

vitamin. An organic compound that (a) occurs in natural food in extremely small concentrations and is distinct from carbohydrates, lipids, proteins, and nucleic acids; (b) is required by the organism (generally restricted to animals) in minute amounts for normal health and growth, and generally functions as a component of a coenzyme; (c) when absent from the diet, or improperly absorbed from the food, leads to the development of a specific deficiency disease; (d) cannot be synthesized by the organism and must, therefore, be obtained exclusively through the diet.

vitamin A. A generic descriptor of all β-ionine derivatives, other than provitamin A carotenoids, that exhibit qualitatively the biological activity of retinol. Retinol is a fat-soluble vitamin that is structurally related to the carotenes and that is required for certain aspects of metabolism, particularly the biochemistry of vision. Vitamin A_1 (retinol$_1$) predominates in higher animals and marine fish, and vitamin A_2 (retinol$_2$) predominates in fresh water fish; the two forms differ by one double bond in the molecule. A deficiency of vitamin A causes night blindness and xerophthalmia.

vitamin A_1. *See* vitamin A.

vitamin A_2. *See* vitamin A.

vitamin A aldehyde. RETINAL.

vitamin B. 1. VITAMIN B COMPLEX. 2. The original antiberiberi activity.

vitamin B_1. THIAMINE.

vitamin B_2. RIBOFLAVIN.

vitamin B_3. PANTOTHENIC ACID.

vitamin B_4. An activity that prevents muscular weakness in rats and chicks and that is replaceable by a mixture of either arginine and glycine or riboflavin and pyridoxol.

vitamin B_5. A growth-stimulating activity in pigeons that is probably identical with nicotinic acid.

vitamin B_6. A generic descriptor for all 2-methyl pyridine derivatives that exhibit qualitatively the biological activity of pyridoxine. Major forms of the vitamin are pyridoxine, pyridoxamine, and pyridoxal. Vitamin B_6 is widely distributed in nature, and its phosphorylated forms function as coenzymes in amino acid metabolism, as in the transamination reaction.

vitamin B_7. An activity that prevents digestive disturbances in the pigeon.

vitamin B_8. The nucleotide adenylic acid that is no longer classified as a vitamin.

vitamin B_9. A designation that has not been used for a B vitamin.

vitamin B_{10}. An activity that promotes growth and feathering in the chick and that is probably a mixture of folic acid and vitamin B_{12}.

vitamin B_{11}. An activity that promotes growth and feathering in the chick and that is probably a mixture of folic acid and vitamin B_{12}.

vitamin B_{12}. A generic descriptor for all corrinoids that exhibit qualitatively the biological activity of cyanocobalamine. Vitamin B_{12} is a cobalt-containing vitamin and its inadequate absorption from the intestine causes pernicious anemia. The coenzyme form of vitamin B_{12} is cobamide. *See also* cobamide; cobalamine; cyanocobalamine.

vitamin B_{13}. An unconfirmed vitamin.

vitamin B_{14}. An unconfirmed vitamin.

vitamin B_{15}. An unconfirmed vitamin.

vitamin B_c. An activity that prevents nutritional anemia in the chick and that is known to be folic acid.

vitamin B complex. A group of diverse water-soluble vitamins that are classified as a group primarily for historical reasons though, to some extent, they are found together in nature. The complex includes niacin, riboflavin, folic acid, thiamine, pyridoxine, pantothenic acid, biotin, and cobalamine. The compounds choline, lipoic acid, inositol, and *p*-aminobenzoic acid are usually also classified as B vitamins. Most, if not all, of the B vitamins function as components of coenzymes.

vitamin B group. VITAMIN B COMPLEX.

vitamin B_p. An activity that prevents perosis in the chick and that is replaceable by a mixture choline and manganese.

vitamin B_t. An activity that promotes insect growth and that contains carnitine as the active component.

vitamin C. A generic descriptor for all compounds that exhibit qualitatively the biological activity of ascorbic acid. Vitamin C is a water-soluble vitamin that occurs in fruits and vegetables; a deficiency of vitamin C causes scurvy.

vitamin C₂. An antipneumonia activity.

vitamin D. A generic descriptor for all steroids that exhibit qualitatively the biological activity of cholecalciferol; a group of fat-soluble vitamins, structurally related to the sterols, that are active in the prevention and cure of rickets. Since they can be derived from sterols by ultraviolet irradiation, vitamin D is not required in the diet if the organism has adequate access to ultraviolet light (present in sunlight).

vitamin D₁. Originally considered to be a pure vitamin but later shown to be a mixture of vitamin D₂ (calciferol) and lumisterol.

vitamin D₂. CALCIFEROL.

vitamin D₃. CHOLECALCIFEROL.

vitamin E. A generic descriptor for all tocol and tocotrienol derivatives that exhibit qualitatively the biological activity of α-tocopherol; a group of fat-soluble vitamins that are required for normal growth and fertility of animals. Tocopherols are widely distributed in nature and function primarily as antioxidants.

vitamin F. 1. Obsolete designation for the activity of the essential fatty acids as reflected in the prevention of atherosclerosis in animals. 2. Obsolete designation for thiamine.

vitamin G. Obsolete designation for riboflavin.

vitamin H. Obsolete designation for biotin.

vitamin hypothesis. The hypothesis that certain compounds having the properties of vitamins constitute an essential dietary requirement of an organism.

vitamin I. Obsolete designation for vitamin B₇.

vitamin J. Obsolete designation for vitamin C₂.

vitamin K. A generic descriptor for 2-methyl-1,4-naphthoquinone and all of its derivatives that exhibit qualitatively the biological activity of phylloquinone (phytylmenaquinone); a group of widely distributed, fat-soluble vitamins that have quinone-type structures and that are required for the formation of prothrombin. A deficiency of vitamin K causes prolonged clotting times and hemorrhagic disease.

vitamin L₁. A liver filtrate activity believed to be necessary for lactation and probably related to anthranilic acid.

vitamin L₂. A yeast filtrate activity believed to be necessary for lactation and probably related to adenosine.

vitamin M. An activity that prevents nutritional anemia and leucopenia in the monkey; the compound is an active form of folic acid.

vitamin N. Obsolete designation for a mixture obtained from either brain or stomach and believed to inhibit cancer.

vitamin P. The group of compounds known as the bioflavonoids which includes flavanones, flavones, and flavonols. These are water-soluble vitamins that are not included in the vitamin B complex and that are credited with maintaining capillary integrity and with decreasing capillary permeability and fragility.

vitamin PP. Obsolete designation for nicotinic acid.

vitamin R. An activity that promotes bacterial growth and that is apparently related to folic acid.

vitamin S. An activity that promotes the growth of chicks and that is related to the peptide streptogenin.

vitamin T. A group of activities isolated from termites, yeasts, or molds and reported to improve protein assimilation in the rat.

vitamin U. An activity from cabbage that reportedly cures ulcers and that is probably a mixture containing vitamin B₆ and folic acid activities.

vitamin V. An activity from tissue that promotes bacterial growth and that is probably due to nicotinamide adenine dinucleotide.

vitellenin. The lipid-free protein of lipovitellenin.

vitellin. The principal protein in hens' egg yolk; the lipid-free protein of lipovitellin.

vitelline membrane. A membrane that surrounds the ovum.

vitellogenesis. The formation of egg yolk.

vitellogenic hormone. ALLATUM HORMONE.

vitellogenin. An insect yolk protein.

vitreous. Glass-like.

vitreous humor. The gel-like material that fills the posterior cavity of the eye.

vivisection. The cutting of, or operating on, a living animal for purposes of experimentation.

v_L. The variable region of the light chains of the immunoglobulins.

VLD. Very low-density lipoprotein.

VLDL. Very low-density lipoprotein.

VMA. Vanilmandelic acid.

V_max. Maximum velocity.

Voges-Proskauer test. A test for organisms that carry out butylene glycol fermentation; based on the production of a pink color by the reaction of acetoin, formed during the fermentation, with creatine in an alkaline solution.

void volume. 1. The volume of solvent in column chromatography that is equal to the total bed volume of the column minus the volume occupied by the particles of the support; the volume of solvent that is external to the support particles. 2. OUTER VOLUME.

vol. Volume.

volatile buffer. A buffer that can be evaporated without leaving a residue, as one consisting of ammonium formate and ammonium hydroxide.

volt. A unit of electrical potential which is equal to the potential required to make a current of 1 A flow through a resistance of 1 Ω. *Sym* V.

voltage clamp technique. A technique for studying the effects of membrane potentials that are intermediate between the resting potential and the peak of the action potential; based on the production of a sudden displacement of the membrane potential from its resting value by means of a pair of electrodes, and on holding the potential across the membrane at this new level by means of a feedback amplifier. The current that flows through a definite area of the membrane, maintained by a space clamp, under the influence of this applied voltage is then measured with a separate pair of electrodes and a separate amplifier.

voltammetry. The electroanalytical study of chemical reactions by the measurement of the currents produced by the electrolysis of oxidizable or reducible substances as a function of the applied voltage.

volume fraction. The fraction of the solution volume that is occupied by the solute.

volume receptor. A receptor in the central nervous system that responds to changes in the volume of the blood.

volumetric analysis. A method of chemical analysis that is based on the measurement of the volume of a standard solution which is required to react completely with a sample of the substance that is being determined.

volumetric pipet. A pipet that is enlarged to a bulb in its center and that is used for the transfer of a fixed volume of liquid.

volumetric technique. A gasometric technique in which gas volumes are measured.

volutin granule. A bacterial storage granule of polymetaphosphate that stains metachromatically.

Von Gierke's disease. GLYCOGEN STORAGE DISEASE TYPE I.

vortex stirrer. A device for mixing a solution in a test tube by forming a vortex in the solution.

VP. Vasopressin.

VPC. Vapor phase chromatography.

V region. Variable region.

V system. A system composed of a regulatory enzyme and an effector, such that the effector alters the maximum velocity of the reaction but does not alter the substrate concentration at which one half of the maximum velocity is obtained.

vulcanization. The cross-linking of natural rubber chains by means of sulfur.

v/v. The concentration of a solution that is expressed in terms of volume per unit volume.

W

W. 1. Tryptophan. 2. Watt.

Walden inversion. The alteration of the configuration of an asymmetric center in a molecule as a result of a bimolecular displacement reaction.

wall effect. The spreading and curving of a zone as it migrates downward in a chromatographic column as a result of the increased flow rate at the wall of the tube compared to that at the center.

walling-off effect. AFFERENT INHIBITION.

Warburg apparatus. A manometer that is used for studying the cellular respiration of tissue slices or cells by measuring oxygen uptake and/or carbon dioxide evolution. *Aka* Warburg manometer; Warburg-Barcroft apparatus.

Warburg coefficient. METABOLIC QUOTIENT.

Warburg-Dickens-Horecker cycle. HEXOSE MONOPHOSPHATE SHUNT.

Warburg-Dickens pathway. HEXOSE MONOPHOSPHATE SHUNT.

Warburg effect. 1. The inhibition of photosynthesis by high concentrations of oxygen. 2. The overproduction of lactic acid that occurs in many tumors. *See also* Warburg theory.

Warburg method. A manometric method for studying the cellular respiration of tissue slices or cells by means of a Warburg apparatus; the tissue slices or cells are maintained at a constant temperature and the changes in gas pressure are measured with a constant volume manometer. *See also* Warburg's direct method.

Warburg's direct method. A manometric method for studying the cellular respiration of tissue slices or cells in which the only gases exchanged are oxygen and carbon dioxide, and in which respiration is measured by the absorption of carbon dioxide in alkali.

Warburg's yellow enzyme. OLD YELLOW ENZYME.

Warburg theory. A theory of cancer that is based on the universal occurrence and importance of glycolysis in the metabolism of cells. According to this theory, cancer results from an irreversible injury to respiration, specifically to the electron transport system, which is followed by a changeover in the cell to an anaerobic, glycolytic, and fermentative-type metabolism for energy production; the main difference between a tumor and a normal cell is, therefore, a shift in metabolism toward that

of an anaerobic state, in which glycolysis is emphasized.

Warburg vessel. A receptacle for the tissue slices or the cells, the cellular respiration of which is being measured in a Warburg apparatus.

Waring blender. Trademark for a blender used in the preparation of tissue homogenates.

warm antibody. An antibody that has a higher titer at elevated temperatures.

warm-blooded. HOMOIOTHERMIC.

Wasserman test. A complement fixation test that is used in the diagnosis of syphilis and that is based on the reaction of cardiolipin with the Wasserman antibody; cardiolipin serves as an antigen and is mixed with lecithin and cholesterol to form micelle-like structures that enhance its reactivity.

water balance. The reactions and factors involved in maintaining a constant internal environment in the body with respect to the distribution of water between the various fluid compartments and with respect to the establishment of an equilibrium between the intake of water and its output.

water bath. A water-filled container for maintaining immersed tubes, flasks, etc., at a constant temperature.

water compartment. FLUID COMPARTMENT.

waterfall sequence. CASCADE MECHANISM.

water hydrate model. A model for the structure of water according to which the water structure results from the formation of clathrate-type structures which consist of ordered, but highly random, labile frameworks in which the cages are occupied by unbonded water molecules.

water intoxication. An extreme case of hypotonic expansion that may lead to convulsions and death.

water of hydration. Water that surrounds and binds to solute molecules and/or forms hydrates.

water regain. The uptake of water by a dry gel expressed as grams of water taken up per gram of gel.

water-soluble B. An early designation for a fraction of water-soluble vitamins prepared from egg yolk.

water-soluble vitamin. One of a group of vitamins that are soluble in water and that include those of the vitamin B complex as well as some other vitamins such as vitamin C and vitamin P; most, if not all, of the water-soluble vitamins function by being components of coenzymes.

water structure. *See* distorted bond model; flickering cluster model; vacant lattice point model; water hydrate model.

Watson-Crick model. The model proposed by Watson and Crick in 1953 for the structure of DNA. According to this model, DNA consists of two right-handed helical polynucleotide chains coiled around the same axis to form a double helix. The chains are antiparallel, with the deoxyribose-phosphate backbone on the outside and the purines and pyrimidines stacked perpendicularly to the fiber axis on the inside. The bases are held together by hydrogen bonds and they are specifically paired: adenine and thymine by means of two hydrogen bonds, and guanine and cytosine by means of three hydrogen bonds; the two chains are, therefore, complementary.

Watson-Crick type DNA. A DNA molecule that can be described by the Watson-Crick model of DNA.

Watson strand. The DNA strand of Watson-Crick type DNA that is transcribed in vivo. *Abbr* W strand.

watt. A unit of power equal to 1 J/sec. *Sym* W.

wavelength. The distance, along the axis of propagation, between two identical points of a wave.

wave mechanics. *See* quantum mechanics.

wave number. The number of waves per unit length; the reciprocal of the wavelength.

wave shifter. SECONDARY FLUOR.

wax. A neutral lipid consisting of esters formed from fatty acids and long chain alcohols.

WBC. White blood cell.

weak electrolyte. An electrolyte that is only partially dissociated into ions in water.

weak interactions. The attractive and repulsive forces between atoms and molecules that are less strong than those of covalent bonds, ionic bonds, and ion-ion interactions. Weak interactions include the forces of hydrogen bonds, hydrophobic bonds, Van der Waals interactions, and charge fluctuation interactions.

weight average molecular weight. An average molecular weight that is weighted toward the heavier molecules in a mixture of molecules; specifically, $\bar{M}_w = \sum n_i M_i^2 / \sum n_i M_i$, where n_i is the number of moles of component i, and M_i is the molecular weight of component i. *Sym* $\bar{M}_w$. *See also* average molecular weight.

weighted mean. A mean derived from a set of measurements in which different measurements are given varying weights by being multiplied with different factors.

wetting agent. A surface-active agent that enhances the spreading of a liquid on a solid surface.

wet weight. The weight of a tissue, or of collected cells, from which water has not been removed.

Wharton's jelly. A viscous solution from the umbilical cord that is frequently used as a starting material for the preparation of hyaluronic acid.

Whatman. Trademark for a group of filter papers.

wheal and erythema response. A local, cutaneous anaphylactic reaction in man that is produced by the intracutaneous injection of an allergen.

white muscle. A pale skeletal muscle that has a relatively low content of myoglobin and cytochromes.

white plaque. A plaque that does not stain with neutral red in the plaque assay; the lack of staining is due to a destruction of lysosomes in the virus-infected cells.

White's solution. A synthetic medium for the growth of plant cells.

whole blood. Blood that has not been fractionated in any way.

whole body counter. A large, external-sample-type liquid scintillation counter that is designed for the determination of total body radioactivity in man or in animals.

whole plasma. Plasma that has not been fractionated in any way.

whole serum. Serum that has not been fractionated in any way.

Wijs iodine number. An iodine number determined by the use of a solution of iodine in glacial acetic acid, with iodine chloride serving as an accelerator of the reaction.

wild-type. 1. The typical, most frequently encountered phenotype of an organism in nature. 2. The phenotype of an organism that is used as a standard of comparison for mutants of the same organism.

wild-type allele. *See* wild-type gene.

wild-type gene. 1. The normal, most frequently encountered allele of a given gene of an organism. 2. An allele of a given gene that is arbitrarily selected as a standard of comparison for mutant alleles of the same gene.

Wilson chamber. CLOUD CHAMBER.

Wilson's disease. A genetically inherited metabolic defect in man that is due to a deficiency of ceruloplasmin, resulting in an increase in the level of copper in the brain and in the liver.

Wilzbach method. A method for the random labeling of a compound with tritium by exposing the compound to tritium gas in a sealed container for several weeks. A modification of this method entails the exposure of the compound to tritium gas in the presence of a silent electric discharge. *Aka* Wilzbach gas exposure.

window. CHANNEL.

windowless counter. A radiation counter in which the sample is not separated from the ionization detector by either a window or a membrane.

windowless gas flow counter. A radiation counter that incorporates the characteristics of both a gas flow and a windowless counter.

wobble base. The third base (the 5′-end) in the anticodon of transfer RNA; the wobble base can bind with one of several possible bases at the 3′-end of the codon.

wobble hypothesis. A hypothesis proposed by Crick to explain how one transfer RNA molecule can "recognize" more than one codon. According to this hypothesis, the first two bases (the 3′-end) of the antidocon in transfer RNA bind to the first two bases (the 5′-end) of a codon in regular base-pairing fashion. The third base, however, of the transfer RNA anticodon (the 5′-end), while hydrogen bonding, has a certain amount of play or wobble that permits it to bind to one of several possible bases at the 3′-end of the codon.

wobble pairing. The base-pairing that is allowed according to the wobble hypothesis.

Wohl-Zemplen degradation. A degradative technique for aldoses whereby the carbon of the reducing group is removed by treatment with hydroxylamine, and the sugar is converted to the next lower aldose.

Wolfson's method. An analytical procedure for determining gamma globulins in serum by precipitating them with ammonium sulfate and sodium chloride.

Wolman's disease. A genetically inherited metabolic defect in man that is due to a deficiency of the enzyme acid lipase; one of the lysosomal diseases.

word. CODON.

working hypothesis. A hypothesis, the formulation of which provides a basis for further experiments.

wormlike coil model. A model that is frequently invoked to describe the DNA molecule in solution. The model is characterized by a contour length and a persistence length; the latter increases with increasing stiffness of the molecule but is independent of the former. According to this model, the DNA molecule is best approximated as a rod having a continuum of flexible distortions; such a structure is intermediate between that of a rigid rod and that of a random coil. *See also* Gaussian chain.

W-strand. Watson strand.

wt. Weight.

w/v. The concentration of a solution that is expressed in terms of weight per unit volume.

w/w. The concentration of a solution that is expressed in terms of weight per unit weight.

X

X. 1. Xanthine. 2. Xanthosine. 3. An unidentified amino acid in an amino acid sequence.

Xan. Xanthine.

xanthine. A purine that is formed in catabolism by the deamination of guanine, and that does not occur in nucleic acids. *Abbr* Xan; X.

xanthine oxidase. An enzyme of purine catabolism that catalyzes the oxidation of xanthine to uric acid and the oxidation of hypoxanthine to xanthine; a molybdenum-containing flavoprotein that also catalyzes the oxidation of aldehydes. *Abbr* XO.

xanthine oxidase factor. Obsolete designation for inorganic molybdate.

xanthinuria. A genetically inherited metabolic defect in man that is characterized by the presence of excessive amounts of xanthine in the urine and that is due to a deficiency of the enzyme xanthine oxidase.

xanthophyll. An oxygenated carotenoid.

xanthoproteic reaction. A reaction for the qualitative determination of proteins that is based on the successive production of a white, yellow, and orange precipitate upon treatment of the sample with nitric acid, followed by heating and the addition of alkali.

xanthopterin. A pterin that functions as a pigment in insects.

xanthosine. The ribonucleoside of xanthine. Xanthosine mono-, di-, and triphosphate are abbreviated, respectively, as XMP, XDP, and XTP. The abbreviations refer to the 5′-nucleoside phosphates unless otherwise indicated. *Abbr* Xao; X.

xanthylic acid. The ribonucleotide of xanthine.

Xao. Xanthosine.

X-chromosome. A sex chromosome of which the female generally carries two and the male carries one per cell.

XDP. 1. Xanthosine diphosphate. 2. Xanthosine-5′-diphosphate.

xenobiotic. Of, or pertaining to, compounds that are foreign to living systems; the xenobiotic metabolism of DDT is an example.

xenogeneic. Referring to a transplant from one species to another.

xenograft. HETEROGRAFT.

xenotropic virus. An endogenous virus that does not, under most conditions, replicate in the species where it originates.

xerogel. A gel that has shrunk as a result of loss or removal of the solvent.

xerophthalmia. A pathological change in the eye that results from a deficiency of vitamin A.

XMDIC. Xylylene-*m*-diisocyanate; a bifunctional reagent used to label antibodies with ferritin.

XMP. 1. Xanthosine monophosphate (xanthylic acid). 2. Xanthosine-5′-monophosphate (5′-xanthylic acid).

XO. Xanthine oxidase.

x-ray. An electromagnetic radiation that is produced when high-speed electrons are suddenly stopped by a target. The impinging electrons raise target electrons to higher energy levels, and when these electrons fall back to lower energy levels the excitation energy is given off in the form of x-rays. The wavelength of x-rays extends from about 10^{-10} to 2.5×10^{-6} cm. *See also* hard x-rays; soft x-rays.

x-ray analysis. X-RAY CRYSTALLOGRAPHY.

x-ray crystallography. The study of the three-dimensional structure of molecules in a crystal by means of x-ray diffraction.

x-ray diffraction. The diffraction patterns obtained when x-rays are reflected from the atoms in a crystal; a useful method for determining the structure of macromolecules.

x-ray diffraction pattern. The pattern of spots, arcs, and rings that is produced by x-ray diffraction.

x-ray film. A photographic film that is coated with a sensitive emulsion on both sides.

x-ray structure. The structural arrangement of the atoms in a molecule or in a crystal as deduced from x-ray diffraction patterns.

XTP. 1. Xanthosine triphosphate. 2. Xanthosine-5′-triphosphate.

Xul. Xylulose.

Xyl. Xylose.

xylan. A homopolysaccharide of xylose that occurs in plants.

xylose. A five-carbon aldose. *Abbr* Xyl.

xylulose. A five-carbon ketose that is an intermediate in the hexose monophosphate shunt. *Abbr* Xul.

X-Y recorder. A recorder in which two signals are recorded simultaneously by one pen; the pen is driven in one direction, the *X*-axis, in response to one signal, and it is driven in the other direction, the Y-axis in response to the second signal.

X-Y-Z cell theory. A theory according to which immunocytes are classified into three categories: X-cells are immunologically competent cells that are not yet engaged in any specific immunological response; Y-cells are immunologically activated, or primed, cells as a result of an interaction between X-cells and antigen; and Z-cells are antibody-producing cells, formed as a result of a second stimulation of the Y-cells by antigen.

Y

Y. 1. Pyrimidine nucleoside. 2. Tyrosine. 3. Yttrium.

Y-chromosome. A sex chromosome that is generally the mate of the X-chromosome in the male.

yeast. A lower fungus that reproduces by budding and that is characterized by either short or nonexistent mycelia; refers particularly to fungi of the genus *Saccharomyces.*

yeast eluate factor. Obsolete designation for vitamin B_6.

yeast filtrate factor. Obsolete designation for pantothenic acid.

yeast nucleic acid. An early designation for RNA.

yellow enzyme. One of a group of flavoprotein dehydrogenases that contain a yellow flavin prosthetic group.

yellow protein reaction. XANTHOPROTEIC REACTION.

Y-fork. REPLICATING FORK.

yield. 1. For a general chemical reaction: the weight of product obtained divided by the theoretical yield of product. 2. For the isolation of an enzyme: the total activity at a given step in the isolation divided by the total activity at a reference step.

yield coefficient. The weight of bacteria obtained from a culture divided by the weight of a limiting material that was utilized by the bacteria during their growth. *Aka* yield constant.

yogurt. A fermented milk product, generally made by adding a culture of *Lactobacillus bulgaricus* to milk and incubating the mixture. *Var sp* yoghurt; yohourt.

Young-Helmholtz trichromatic theory. The theory according to which color vision is due to a set of at least three pigments in the cones for the perception of red, green, and violet, respectively, and the perception of other colors results from the combined stimulation of two or more of these pigments.

Yphantis method. MENISCUS DEPLETION SEDIMENTATION EQUILIBRIUM.

Z

Z. 1. Average net charge of an ion. 2. Atomic number. 3. The sum of glutamic acid and glutamine when the amide content is either unknown or unspecified. 4. Impedance.

Z-average molecular weight. An average molecular weight that is weighted toward the heavier molecules in a mixture of molecules; specifically, $\bar{M}_z = \sum n_i M_i^3 / \sum n_i M_i^2$, where n_i is the number of moles of component i, and M_i is the molecular weight of component i. *Sym* $\bar{M}_z$. *See also* average molecular weight.

Zeeman effect. The splitting of the degeneracies of the excited states of a chromophore by an external magnetic field.

zein. A seed protein of corn.

zeolite. A naturally occurring alkali metal- or alkaline earth-aluminum silicate that has a network structure and ion-exchange capacity; used as an ion exchange resin for water softening and as a molecular sieve. *See also* permutite.

zero layer line. EQUATOR.

zero meniscus concentration method. MENISCUS DEPLETION SEDIMENTATION EQUILIBRIUM.

zero mobility position. The position occupied by an uncharged substance in an electrophoresis experiment.

zero-order kinetics. The kinetics of a zero-order reaction.

zero-order reaction. A chemical reaction in which the velocity of the reaction is independent of the concentrations of the reactants.

zero point. The point on an x-ray diffraction pattern where the incident beam strikes the photographic film.

zero-point mutation. A mutation that is expressed immediately following the irradiation of cells with a mutagenic radiation.

zeroth law of thermodynamics. The law that establishes a quantitative concept of temperature so that the state of every thermodynamic system includes temperature either explicitly or implicitly. The law can be phrased as follows: if body A is in temperature equilibrium with body C, and body B is in temperature equilibrium with body C, then bodies A and B are in temperature equilibrium with each other. This principle is assumed whenever a thermometer is used to compare the temperature of two systems.

zero time control. A control used in enzyme studies in which the enzyme is inactivated prior to addition of, and incubation with, the substrate.

zeta potential. The potential difference across the plane of motion between two phases, particularly the potential across the surface of shear of a charged particle in electrophoresis.

zigzag scheme. Z-SCHEME.

Zimm plot. A double extrapolation used in the analysis of light scattering data when the scattering particles are larger than those involved in Rayleigh scattering. A plot of Kc/R_θ versus $\sin^2(\theta/2) + kc$ is extrapolated both to $c = 0$ and to $\theta = 0$, where K is an optical constant, c is the concentration, R_θ is the Rayleigh ratio, θ is the angle at which scattering is observed, and k is an arbitrary constant chosen to provide a convenient spread of the data. Both the molecular weight and the radius of gyration can be obtained from the plot. *Aka* Zimm grid.

Zimm viscometer. A Couette-type viscometer in which the inner cylinder is a self-centering float; used for viscosity studies of DNA.

zinc. An element that is essential to all plants and animals. Symbol Zn; atomic number 30; atomic weight 65.37; oxidation state +2; most abundant isotope Zn^{64}; a radioactive isotope Zn^{65}, half-life 243.7 days, radiation emitted—positrons and gamma rays.

zinc sulfate turbidity test. A liver function test that is based on the production of turbidity when serum from individuals with one of several forms of hepatitis is treated with a barbiturate buffer containing zinc sulfate.

zippering. The formation of a helical DNA or RNA duplex from the two separated and complementary strands. *Aka* zippering up.

Z line. The dark line that bisects the I band of the myofibrils of striated muscle.

Zn. Zinc.

zonal centrifugation. DENSITY GRADIENT CENTRIFUGATION.

zonal centrifuge. A specially designed centrifuge that allows large scale and continuous fractionation of material by density gradient centrifugation.

zonal diffusion. A method for determining diffusion coefficients from the diffusion profile that is produced by the diffusion of a thin layer of macromolecules which have been placed in a shallow density gradient in a zonal centrifuge rotor.

zonal electrophoresis. ZONE ELECTRO-PHORESIS.

zonal rotor. A high-capacity rotor used for zonal centrifugation in a preparative ultracentrifuge.

zone centrifugation. DENSITY GRADIENT CENTRIFUGATION.

zone convection electrofocusing. A technique for conducting horizontal isoelectric focusing in free solution rather than in a density gradient.

zone electromigration. ZONE ELECTRO-PHORESIS.

zone electrophoresis. An electrophoretic technique in which components are separated into zones or bands in a buffer that is generally stabilized by a solid, porous, supporting material such as filter paper, starch gel, agar gel, or polyacrylamide gel.

zone precipitation. The precipitation of a protein as a zone in a gel filtration column that is brought about by using a gradient of a protein precipitating agent.

zone spreading. The broadening of a zone in either chromatography or electrophoresis as a result of processes such as eddy diffusion and eddy migration.

zoopherin. Obsolete designation for vitamin B_{12}.

zootoxin. ANIMAL TOXIN.

Z scheme. The series formulation for the photosynthetic reactions of photosystems I and II of chloroplasts.

zwischenferment. GLUCOSE-6-PHOSPHATE DEHYDROGENASE.

zwitterion. DIPOLAR ION.

zygospore. A spore formed by the conjugation of two other spores.

zygote. 1. The cell produced by the union of the male and female gametes in reproduction. 2. The organism that develops from a zygotic cell.

zygotic induction. The induction of a prophage that is transferred during conjugation from a lysogenic *Hfr* cell to a nonlysogenic F^- cell.

zymase. A heat-labile enzyme fraction that is obtained from yeast and that catalyzes the reactions of alcoholic fermentation.

zymogen. The inactive precursor form of an enzyme that is generally converted to the active form by limited proteolysis.

zymogen granule. A membrane-surrounded, cytoplasmic particle that is formed by the Golgi apparatus. Zymogen granules serve to store, and subsequently to secrete, the zymogens synthesized by the ribosomes of the endoplasmic reticulum.

zymogram. The record of a zone electrophoresis experiment in which the enzymes in a sample, particularly esterases, are separated according to their charge and molecular dimensions, and in which the activity of these enzymes is indicated by specific staining reactions. A zymogram thus provides a measure of the types and the relative amounts of various enzymes in the sample; zymograms prepared from bacterial samples may be used as an aid in bacterial taxonomy.

zymology. The science that deals with fermentations.

zymosan. A polysaccharide derived from yeast cells that inactivates complement.

zymosis. FERMENTATION.

zymurgy. The application of fermentation to the manufacture of alcoholic beverages.